Locus of a Boat Designer 2

Kotaro Horiuchi

The Boat Association of Japan

Preface

This book is the English version of the story and illustrated record of development of new vehicles on water, land and in the sky, in which the author has been actively involved. Until now, the author has exerted his best efforts in each of his creations, enjoyed its success, and rejoiced when this process was repeated. He is delighted to share this pleasure with all the readers of this book. Nothing could give him more pleasure than to share it with young readers keenly interested in creative development.

After graduating from school, the author started his design career with a boat manufacturer. Soon, he became a member of the airplane and glider design team in another company and worked there for a while. He remembers that this experience proved very useful when he returned to boat design again in the first company he joined. Above all, the experience of learning the value of human life at the design stage -- where even one mistake could prove fatal -- helped him to acquire the skill of preparing layouts adhering to rational design procedures while ensuring safety and performance. This experience also tempered him to take up new challenges fearlessly in new developments and design, and helped him build many new vehicles. Thanks to the assistance and support of many people around him, he was afforded opportunities to develop many products. The author parrticularly wishes to thank Hideto Eguchi, the then president of Yamaha Motor Co., Ltd., for setting up a laboratory which bore his name and permitted him to work on a series of projects there for about ten years.

The original Japanese version of this book was published in 2002, and is a compilation of articles written by the author in the marine monthly magazine "Kazi" from January 1998 to December 2001. Fifteen years ago, the author wrote his first book "The Locus of a Boat Designer" in the same manner. This second book was named "The Locus of a Boat Designer 2". Unfortunately, the first version hasn't yet been translated into English. Despite the author's earnest desire to do so, it now seems unlikely considering that his age presently is 80.

The author requested his old friend and ex-colleague Kensuke Sakamoto for the first English translation of this book because Sakamoto worked with him for a long time at Yamaha Motor, knew almost all details of the technology and the development environment, knew the author well, and naturally understood the Japanese nuances in the book. Subsequently, the original translation was extensively revised and edited by Gururaj Rao, an ex-Naval Architect, who trained for a short while at Yamaha Motor, and is presently a professional technical translator. The book includes a fairly large number of drawings and photographs to facilitate the reader's understanding.

Two presidents of Kazi Publishing Co., Katsuyoshi Doi and Yoshio Doi, encouraged the author to launch the Japanese version of the book. Yoshio Doi, together with Buko Negishi, also arranged for this English language publication, and the author is indebted to these persons for their support and assistance. In addition to Sakamoto, the author would like to thank Kenji Ito, John Wilson, and Mike Timmons for partial proofreading and English corrections in the initial stages.

March 27, 2006

Kotaro Horiuchi
email:khoriuti@gmail.com

Table of Contents

Pubilished in Japan by The Boat Association of Japan
1-2-17 Hamamatucho Minato-ku Tokyo (105 - 0013)

From the Editor's Desk

When Mr. Kotaro Horiuchi requested me to assist in the proofreading and editing of his latest book, my joy knew no bounds. I had long been an admirer of his ideas, his designs, and his creativity, and was fortunate to have worked under him for a short time at Yamaha Motor. As a young designer eager to learn the ropes, and the only foreigner at that time undergoing training in the design of boats, I remember approaching him one day on tips to design a catamaran speedboat. His eyes sparkled with intensity as he spoke to me at length about the advantages of a catamaran. He quickly pulled out a lines plan of a successfully built catamaran from his desk, and presented it to me. Even today, nothing gives him more joy than explaining minute, technical details of a device to anyone who cares to listen.

This book is certain to be a source of valuable information, especially to boat designers, because it is full of data, plans, illustrations and photos. I wish I had this book in my hand when I just started out as a boat designer.

It has given me great pleasure to work on this book with Mr. Horiuchi. Although I had to finish the proofing work in record time, and I worked after my regular working hours to do so, I enjoyed the work. I also enjoyed the detailed telephonic conversations with Mr. Horiuchi in the course of the extensive proofing and editing of this book -- all the more since his requirements were extraordinary -- the book should not only be clear in its technical content to readers in English-speaking countries, but also easily understandable to people in countries where English is not the first language, such as Russia, China, South Korea and so on. In the short time at my disposal, I have tried to keep the text as simple as possible, making extensive revisions to the original translation in the process. Should there be awkward expressions in the text, it is entirely my responsibility for giving priority to finishing the work quickly.

I sincerely hope that Mr. Horiuchi also publishes in English his first Japanese book, "Locus of a Boat Designer" for the benefit of his admirers in other countries.

March 27, 2006

Gururaj Rao
email : honyakusha@gmail.com

1

HATERUMA

Hateruma is my dreamboat – a boat that I built to satisfy my own whims and fancies – in which I indulged myself by sailing as well as cruising at high speeds, and also used it as a second house. It took me over 10 years to finalize the layout of the boat and I designed it free from encumbrances, utilizing my design experience. The hull shape, construction, sailing system, pulpit, and fuel oil tank incorporate new ideas, of which some have never been tried out earlier. This boat initially customized for an individual, later went into production.

1) Prologue

In the early 1980s, I owned the sailboat YAMAHA30C, which I named HATERUMA, and enjoyed sailing it in Lake Hamana. HATERUMA is named after an island at the southern end of Japan. The Japanese characters in the name evoke a scene of the sun shinning through the clouds and glittering wave crests. Both my wife and I loved this name very much.

I had heard Tadahiro Matsuda, one of my junior friends in Yamaha Motor, often mention his wish to use this name on his boat when he purchased it. Before he purchased it, I called him in Sendai, a city far off from Hamamatsu, and asked his permission to take over that name and use it for my boat.

I enjoyed making various improvements to the boat. I designed an arrangement that enabled the 60 kg mast to be raised and lowered single-handedly. I developed a tiller clutch system that enabled the boat to sail on autopilot when the tiller was swung up, and sail on manual steering when the tiller was down.

A long cruise however, often proved inconvenient for me. The sturdy young fellows in the company often set off from Lake Hamana in the evening and arrived at Nishiizu or the Shima peninsula the next morning so that they could enjoy the whole day at sea. I, on the other hand, was nervous of sailing at night with only my wife, as she had low blood pressure and was weak in the early mornings. I often started on the cruise at around ten in the morning with my wife. Since the open sea turned rough around that time, we had to continue sailing near the coastline all day long looking at the same, dull scenes of the sandy beach of the Sea of Ensyu. We needed two whole days for a round trip, leaving us very little time for our leisure activities at the destination. More time was wasted when we encountered rough weather.

Before owning YAMAHA30C, I had designed and driven various other prototypes. I developed the itch again to design and build my own boat capable of cruising the Sea of Ensyu at 20 knots on power, and which could be used as a sailboat upon arriving at the destination. It had to be a home where my wife and I could relax, unlike one with many cabins for guests. It had to be spacious, and well insulated from heat with mechanical equipment such as engines, generator, and fuel oil tanks isolated from the living spaces. The boat had to be light in weight, slender, and fitted with outboard motors. The end result was the Philosopher.

During the reorganization of Yamaha Motor[1] in 1986 after a fierce battle with Honda Motor Co., Ltd to capture the motorcycle and scooter market, I resigned from the Board of Directors and took up the post of the Head of the Horiuchi Laboratory so that I could concen-

1) Yamaha Motor is an independent company separated from Nippon Gakki Co.,3 Ltd. (presently Yamaha Co.,.Ltd.) in 1955.

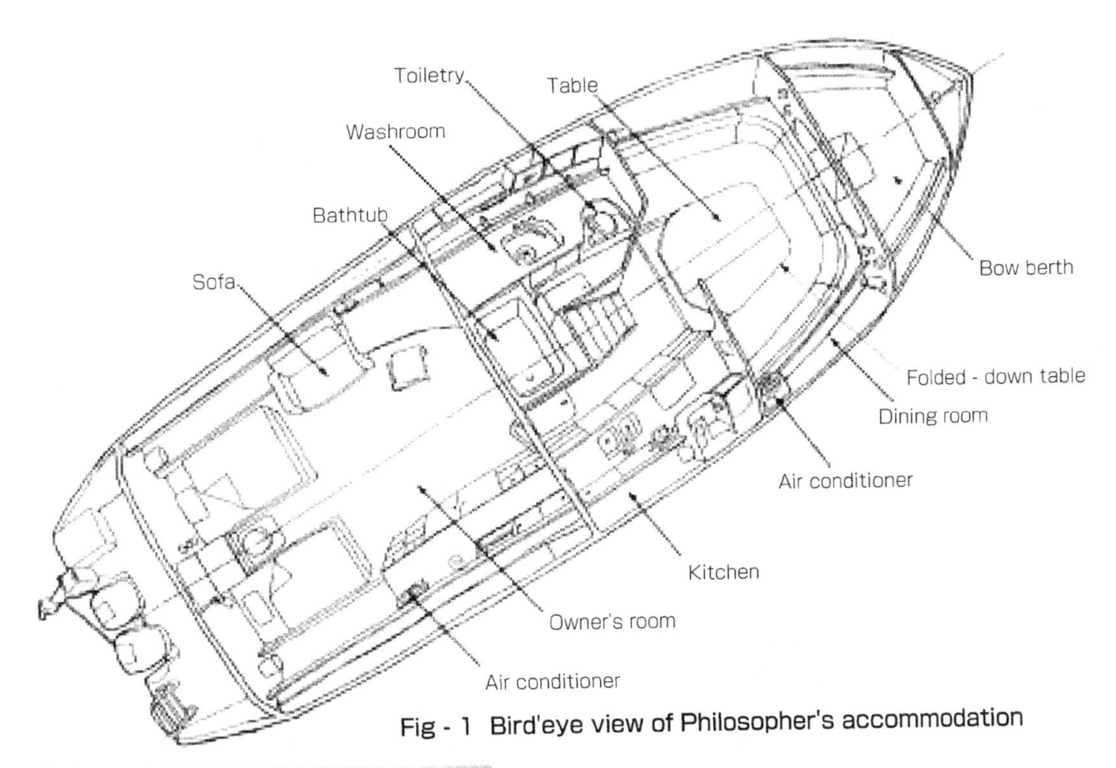

Fig - 1 Bird'eye view of Philosopher's accommodation

Fig - 2 Aft side of owner's room

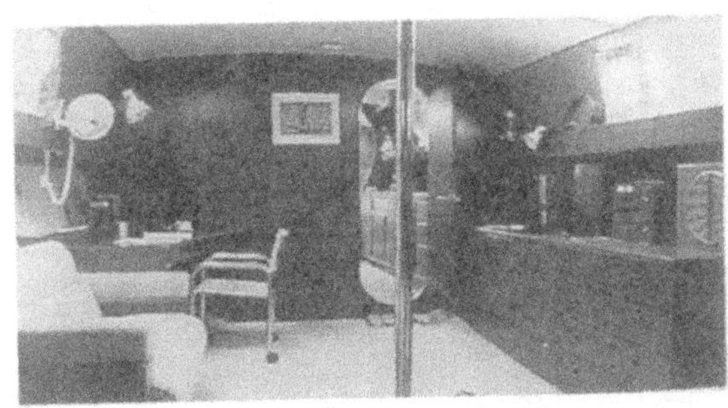

Fig - 3 Forward side of owner's room

Fig - 4 Dining room (Bow berths beyond)

trate on the design and development of boats with a team of young engineers. Two years later, I received retirement allowance. At that time, I decided to place an order on my own company to build the Philosopher. Since this was a one-off construction, a simple female mold was built at a shipyard in Matsuzaka because a special sandwich construction technique was used to build the boat. The female mold was brought to the Gamagori factory to build the boat. The boat was built only during the slack periods when the factory was idle, so it took one and a half years for the boat to reach the launching stage. The boat was christened HATERUMA during the launch, and the name Philosopher faded. During the ten or more years since then, I have spent time on the boat after work whenever possible and have invited many people on board HATERUMA. I have conducted various tests and trials of this boat, tried out ideas that have flashed into my mind, reconsidered those that did not work out well, and have explored new avenues that would lead me toward better solutions.

2) Layout

The main engines on the Philosopher were two 200 hp outboard motors This arrangement reduced the weight and cost by more than 40%, and no engine room was required (see Fig - 1). The disadvantage was that the fuel consumption would be higher compared to inboard engines, but my calculations showed that the difference was negligible.

Since there was no engine room, I could arrange a spacious and comfortable owner's room. In this room, we have a semi-double and a single bed. On the port side, we have a cozy sofa and a wide working desk. On the starboard side, we have a cabinet with a bookshelf, a shelf for TV or music system on top, drawers, and spaces with hinged doors underneath for storing tools (Fig - 2, 3). The ceiling height at the entrance is 192 cm, adequate for even a tall person. I brought in and placed two big chairs, but there's still ample space.

Passing through the door to forward area, there is a kitchen of length 2.5 m on the starboard side. A refrigerator is built into the wall on the left side of the passage. On the port side, there is a washroom of 2.5 m width with a sink large enough to use for washing clothes too. At the end of the space and near the centerline, there is bathtub of length 1.2 m. I had a gas water heater installed on the bridge to supply hot water to the bathtub.

I provided a large U-shaped seating arrangement in the dining room (Fig - 4). This arrangement probably hampers access to the bow berths a bit, but since this space is used only for sleeping, I gave more importance to creating a friendly and cozy atmosphere with a circular seating arrangement. The enclosed bow berth space also looks like a secret chamber, children enjoy this space, it is suitable for taking naps, and accommodates three adults too.

The starboard side of the dining table is a drop leaf type to enable easy access to the bow berths. The total area of the dining table is 1.2 m^2 and it can seat 8 to 9 persons. A large acrylic hatch just above the table permits adequate light to fall on the table. The seating cushions change into two 60 cm wide bunks when the backrests are removed. That means seven persons in all can stay onboard. The teak - paneled wall in the dining room matches the dark blue seat cushions, and affords a feeling of quiet sophistication.

The continuous, flat upper deck is wide and spacious. There is adequate space to sit under the awning – probably the best location of the boat in summer (Fig - 5). In addition, there is a permanent awning over the bridge, which is open and gives the feel of a cruiser's flying

Fig - 5 General arrangement

Length overall	15.350m	Fuel capacity	600 ltr
Beam	3.534m	Fuel consumption	0.5km/ ltr
Depth	2.320m	Cruising range	300km
Displacement	7.5ton	Gross tonnage	10ton
Fresh water	400 ltr	Max. speed	30.2kt

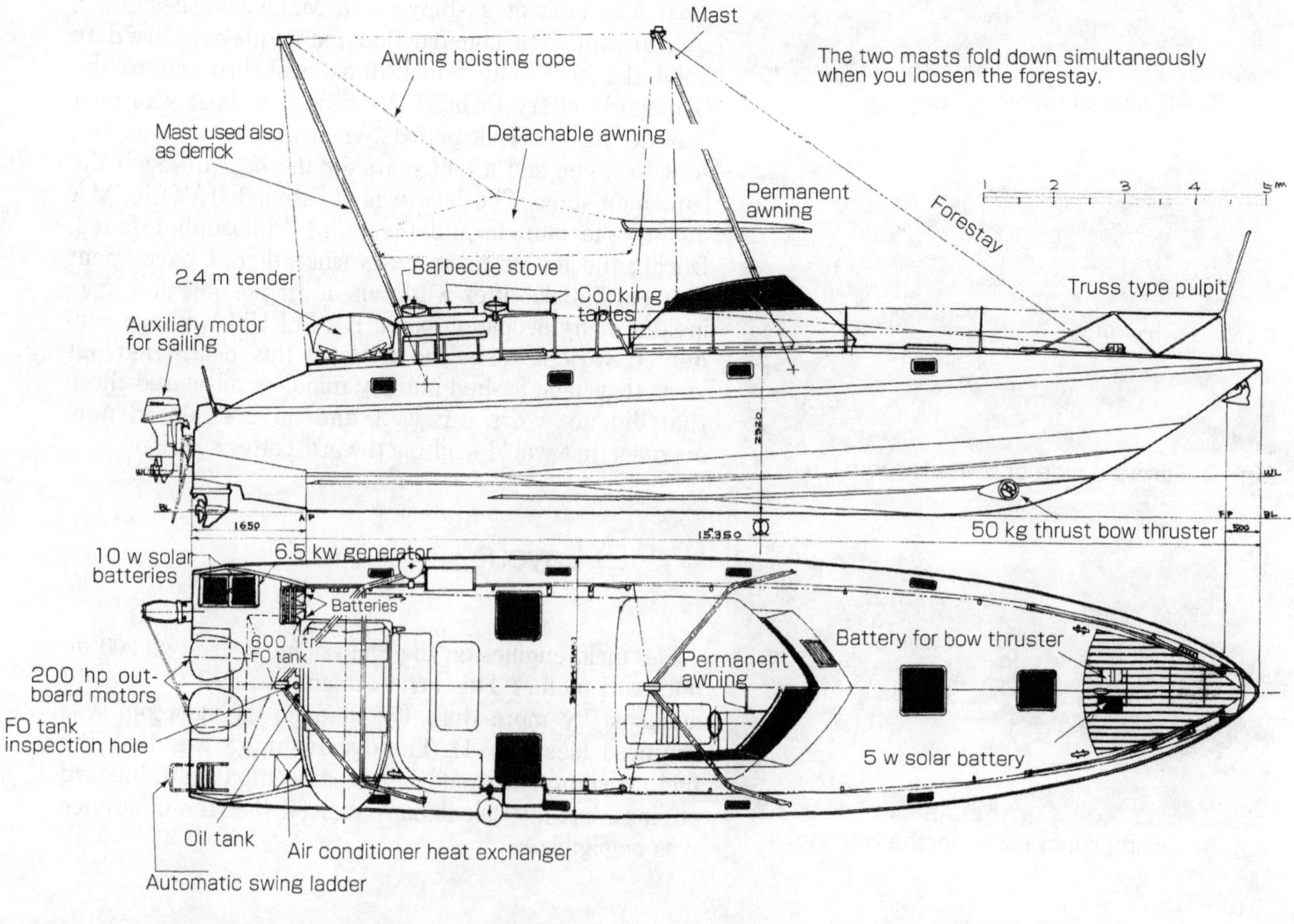

bridge. A detachable awning can be extended up to the table and the L-shaped seat behind the bridge. The total awning area thus formed becomes 15 m^2. A barbecue stove and simple cooking table are fitted to the handrails on both sides of the table. By arranging some folding chairs, a barbecue zone with seating for 7 to 8 persons becomes available.

A large working deck space is provided in front of the bridge, where leisure items such as surfboards can also be stored. The anchor windlass is installed under the deck. A clear, flat deck with practically no obstacles except a mast forestay (tension line from mast top to bow deck) is presented to a visitor who steps on board through the gated pulpit at the bow (bow guardrails). A tender (small boat), which can be lowered to the water by a mast used as a derrick, is stowed behind the barbecue space.

If you go down a ladder beside the tender, you come to the machinery area of length 1.65 m where two 200 hp outboard motors (main engines), a 10 hp auxiliary motor for sailing, a generator, batteries, solar batteries, lubricating oil tanks, and a 600 liter under-floor fuel tank are installed. A double wall separates this area and the owner's room preventing odors, oils, and noise from reaching the forward living spaces. No instruments are installed in the accommodation spaces except ones that are normally used in a house. The machinery area is small, but its deck space serves as a passageway and also as base for leisure activities. At the aft end on the starboard side is a swing ladder that reaches 60 cm below the water surface when swung down, facilitating access to water. An arrangement that swings the ladder up automatically by using a shock cord is provided such that if the boat starts moving when the ladder is still in the water, the ladder is swung up automatically. The system is an essential item in a water ski boat, and I used it on Hateruma for performing various tests; all the same, it is convenient for this boat too.

3) Hull shape

While this boat was being built, I was enthusiastically developing a new hull form named Duo - phaser and its drawing method. This hull shape has an ideal planing performance with comfortable, soft riding characteristics, but the lines cannot be drawn well by the conventional drawing method.

The newly developed drawing method consists of drawing the hull lines according to specific rules only after determining 15 to 16 parameters (specific numerical values) such as boat length, beam, bottom deadrise angle and so on. The required performance of the boat can be obtained by proper selection of these parameters. If these parameters are input into a com-

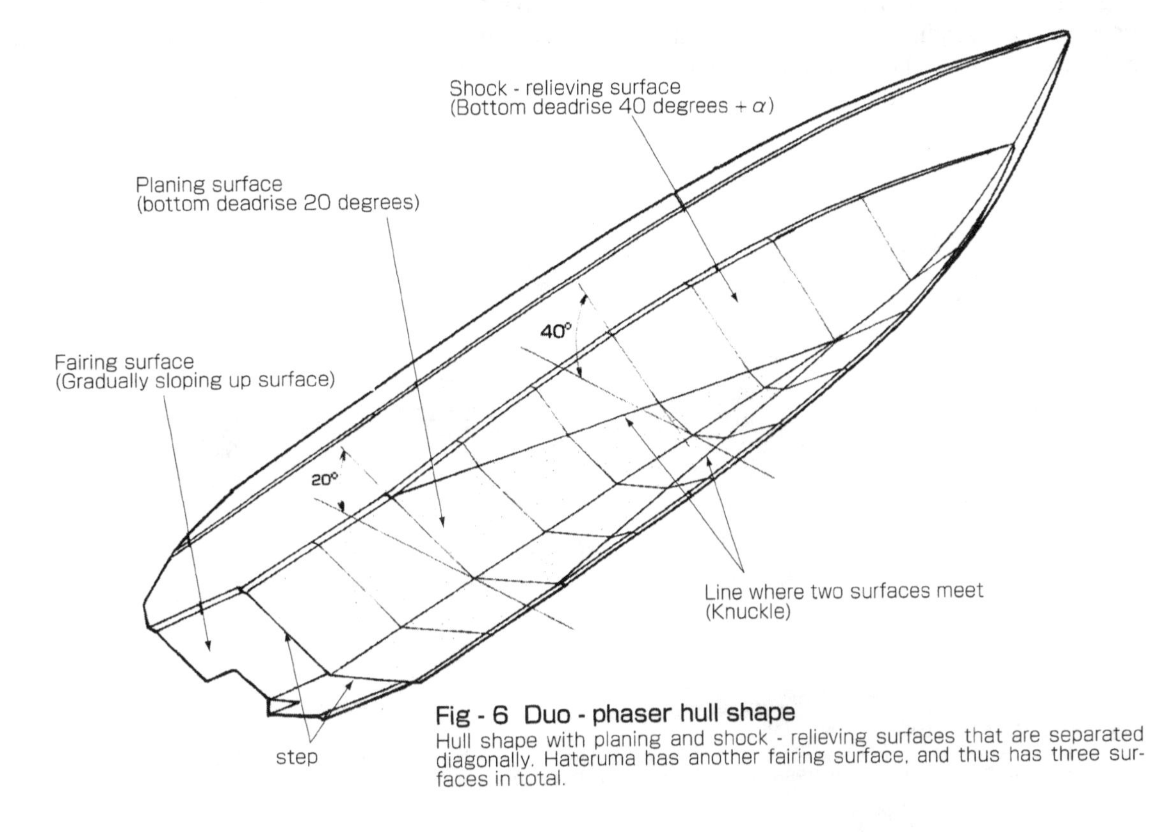

Fig - 6 Duo - phaser hull shape
Hull shape with planing and shock - relieving surfaces that are separated diagonally. Hateruma has another fairing surface, and thus has three surfaces in total.

puter, the hull lines can be drawn by the computer. Additionally, if the lines of a hull are drawn approximately similar to existing hull forms by this method, the performance and the general features of that hull shape can be understood from the selected parameters. At that time, I thought that this method was a revolutionary method.

I wanted to try the new method on the design of my own boat, and also test it out practically. I prepared the lines plan accordingly. The lines plan, as you see in Fig - 6, has a planing surface with a 20 degree dihedral angle at the bottom aft. The water flow strikes the surface, which generates lift. A planing surface with constant dihedral angle is ideal for good planing performance.

On the other hand, the forward surface at the bottom is meant for shock relief and has a 40 degree dihedral angle, which minimizes wave impact. Two surfaces meet at a diagonal line, each surface designed such that the planing surface lies below the running waterline while the shock-relieving surface remains above the running waterline when planing. There is no equivalent to this diagonal line in the conventional drawing method, so it is difficult for both planing and shock relieving surfaces to co-exist. Thanks to the hull shape, the Philosopher can cruise at a speed of 20 knots without shocks, similar to sailing boats.

Outboard motors are rarely used as main engines in large boats of length exceeding 15 m. Since the weight of the engine accounts for only about 5% of the total displacement, the center of gravity tends to move forward, which is not suitable for a planing boat running at speeds of over 30 knots. To resolve this problem, I placed a step at a distance of 1.65 m from the aftermost end of the boat. From the step to the aft, I made the bottom surface to taper up gradually (Fig - 6). With this arrangement, the water after the step separates from the bottom at the planing speed, similar to a flying boat just before the take-off. The distance from the aft end of the planing surface to the center of gravity reduces by 1.65 m, which has the same effect as moving the center of gravity toward the aft by the same distance. This sloping up surface had other advantages too. The water flow behind the boat followed this sloping up surface, and vortex did not occur at the sailing speed.

Normally, when a boat of this type is run at a speed of 5 knots the resistance is about 350 kg, but for this boat, the resistance was only 50 kg, thanks to the sloping-up surface, which contributed significantly to reducing the resistance at sailing speed. I have named this sloping up surface as the "fairing surface". Also, since the distance from the step to the propeller is almost 1.5 m, the level of the water flow behind the boat that separates at the step, and rises gradually as the boat planes. So there is no concern about air draw of the propeller even if the engine is mounted at a higher position (Fig - 7).

Thanks to the arrangement described above, the propellers and gear cases remain fully above the water when the outboard motors are tilted up while the boat is moored, and they do not become dirty. This is an advantage that will be welcome over the long term. Thus,

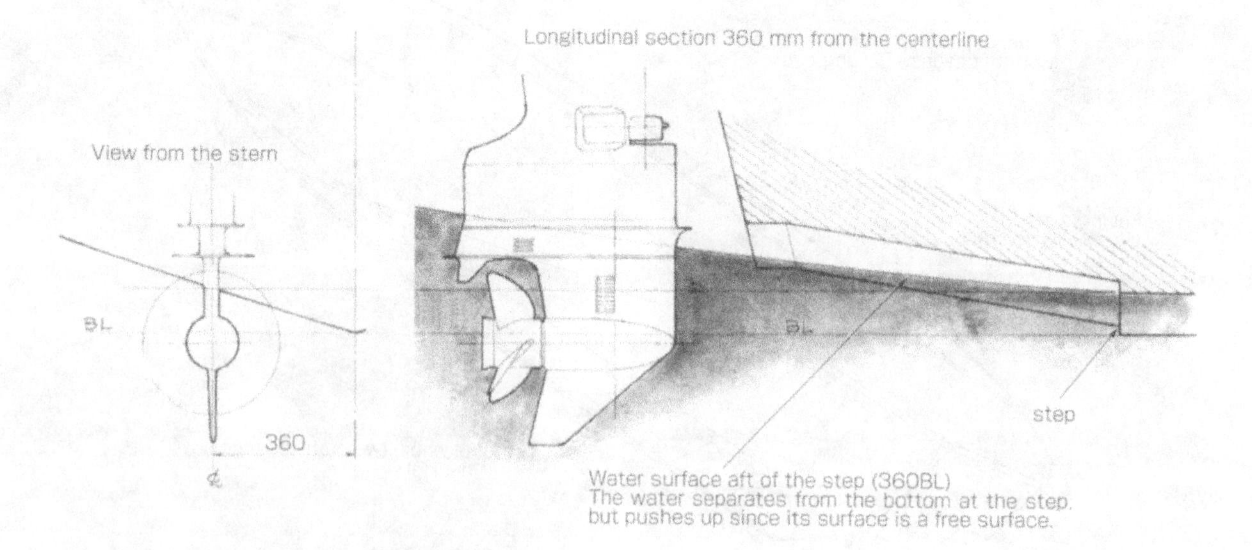

Fig - 7 Effect of step
When Hateruma planes, water flow aft of the step separates from the hull bottom. The level of water surface that has become free from the hull bottom pressure gradually rises above the propeller.

Fig - 8 Hateruma cruising at a speed of 20 knots

I derived three major advantages by fitting a step to this boat.

I calculated its transverse stability to confirm safety while sailing. The result showed that the range of stability was 110 degrees, indicating that the boat can right itself even after a knockdown, which I thought was adequate since the boat had small sails. The boat that finally took shape had both the sailing and power performance that I had anticipated. (Fig - 8).

4) Construction and habitability

I decided to go in for a sandwich construction considering the three concepts of spacious interior, good heat insulation, and preventing wave noise. (Fig - 9). With a thick outer shell, a cylindrical form, and four bulkheads, the hull structure was like that of a bamboo structure that provided adequate buoyancy and was quite sturdy as well.

The outer shell is of sandwich construction consisting of two layers of 25 mm Divinycell (PVC form) laminated with 2.5 mm internal fiberglass layer and 3 mm external fiberglass layers making up a total thickness of about 60 mm. The bending strength and stiffness of planking are both high, and no deck beams and longitudinal stiffeners are necessary. Wallpaper can be directly glued on to the inner surface to form the lining. The keel is stiffened locally, but the bottom, hull sides, and deck have no stiffeners at all, making the internal spaces roomy.

Visitors are often surprised at the spacious interior in spite of the slender outer hull, especially the internal spaces with ample headroom, a 17 m^2 owner's room, and a Japanese - style domestic bathtub. After launching the boat, I tested the habitability of the boat, by installing thermometers and hygrometers outside and inside the boat, and measuring data for a year.

Dew had often formed on the inside walls of my previous boat YAMAHA - 30, forming fungus and turning the surface black, which I hated. Therefore, I thought of keeping the interiors of this boat dry and comfortable by using desiccants or by using the Azekura system, which is a traditional system for ventilating ancient Japanese treasure storehouses. This system ventilates a space automatically when it is dry and shuts down when wet, making use of the shrinkage and expansion characteristics of wood with the changes in humidity.

The sandwich construction provided numerous advantages. When the temperature outside

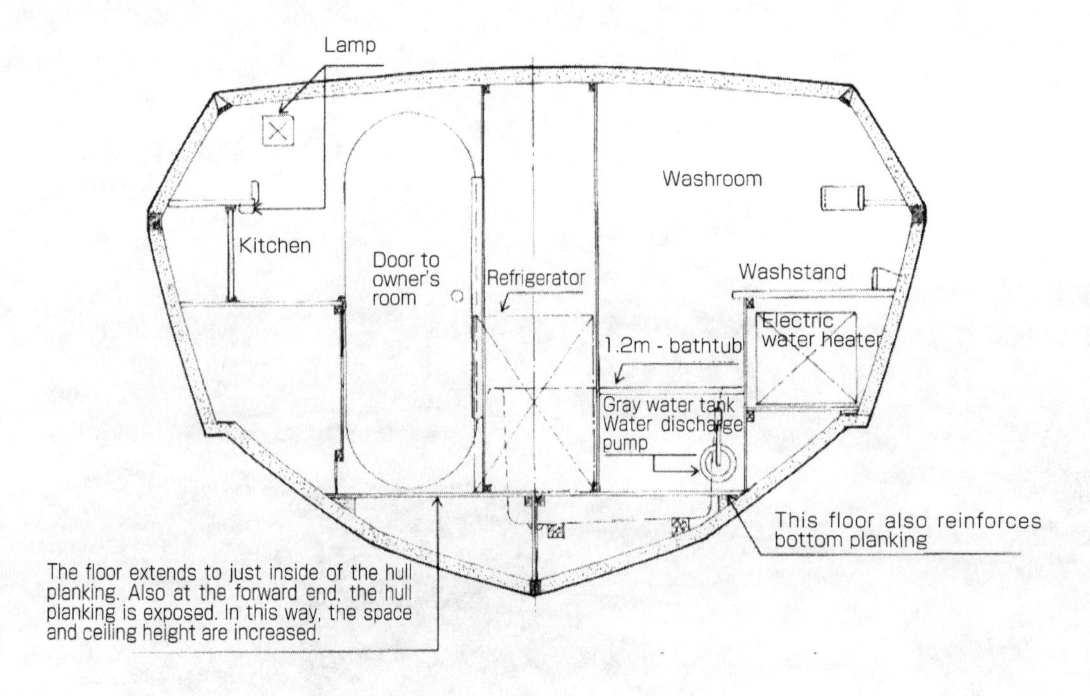

Fig - 9 Section view (in front of main bulkhead)

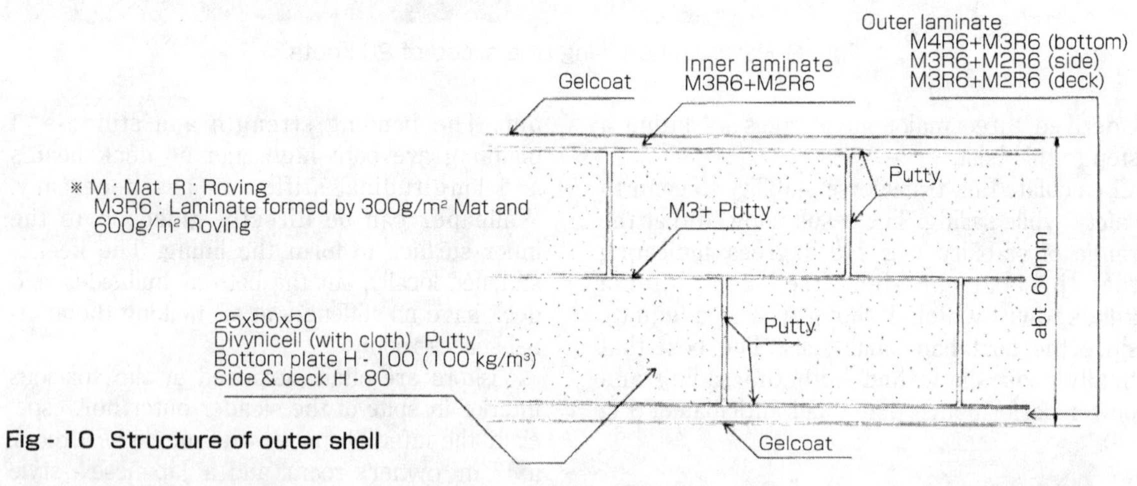

Fig - 10 Structure of outer shell

in summer was 35 degrees Centigrade, the inside room temperature was much lower and the humidity remained constant. So, the special Azekura system was not necessary. For the past thirteen years, the boat has not had any odors, and it has been kept clean. However, a 0.4 m² part of the superstructure in front of stairs not made of sandwich construction developed fungus and turned black.

Another problem that I did not expect was the sound of waves that came through the hull bottom, which was not insulated. Since the sandwich construction made use of 50 mm square Divinycell blocks as core and fiberglass laminate as the skin (Fig - 10), the sound of the waves was probably being transmitted through the solid putty between the blocks. Shutting off this noise appeared to be difficult unless the entire bottom shell was sound insulated using large sheets instead of 50 mm square blocks of Divinycell.

5) Sailing

When I drew up the initial plan of Philosopher in 1979 (Fig - 11), my emphasis on its design as a sailing boat and as a powerboat was in the ratio 50:50. My intention was to get a speed of 20 to 30 knots during powered running on a boat that was fitted with a retractable keel and swing rudder for sailing. I inserted the plan of this boat in a frame and looked at it time and again, thinking about the

Fig - 11 General arrangement of Philosopher - 1

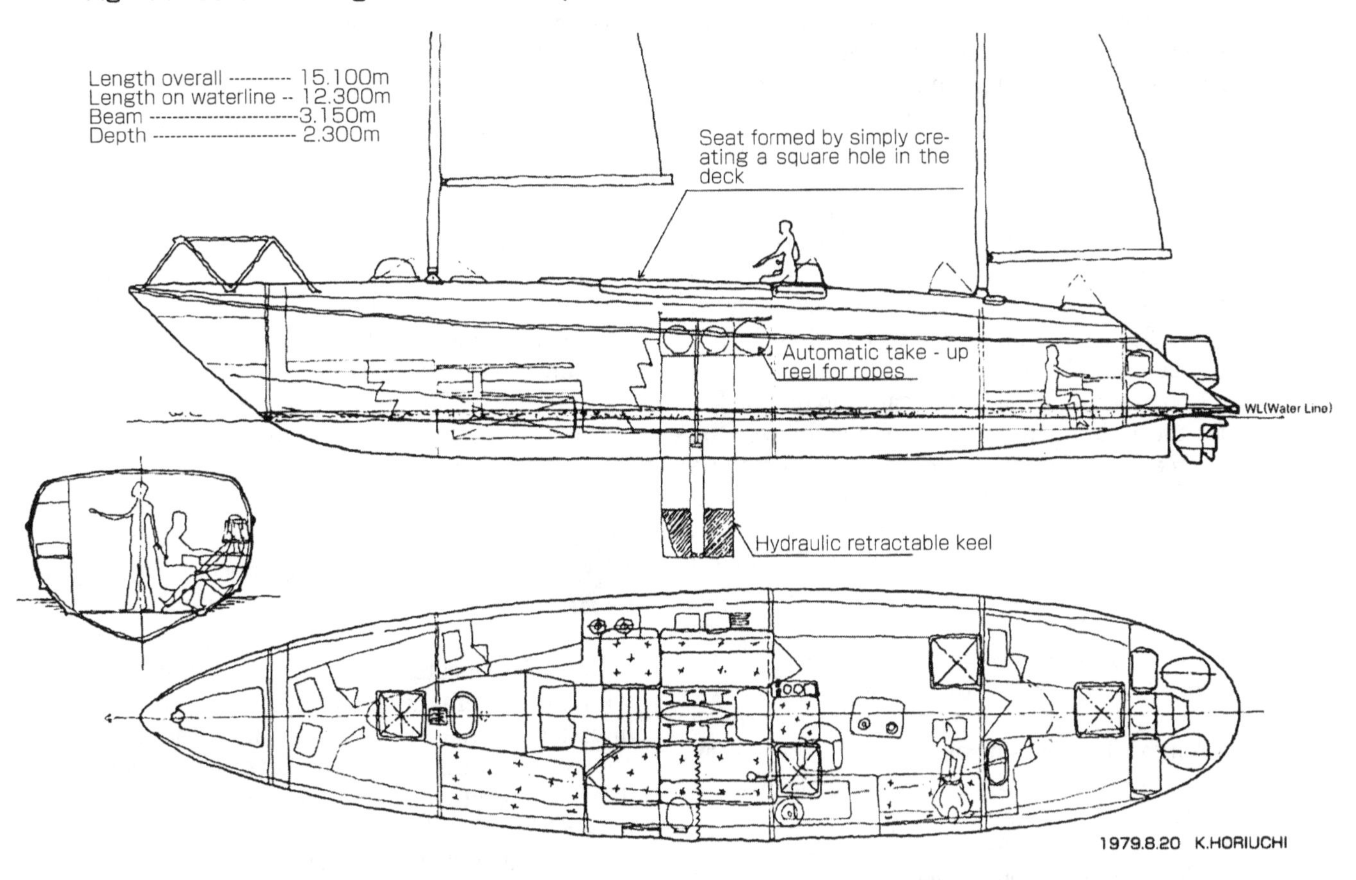

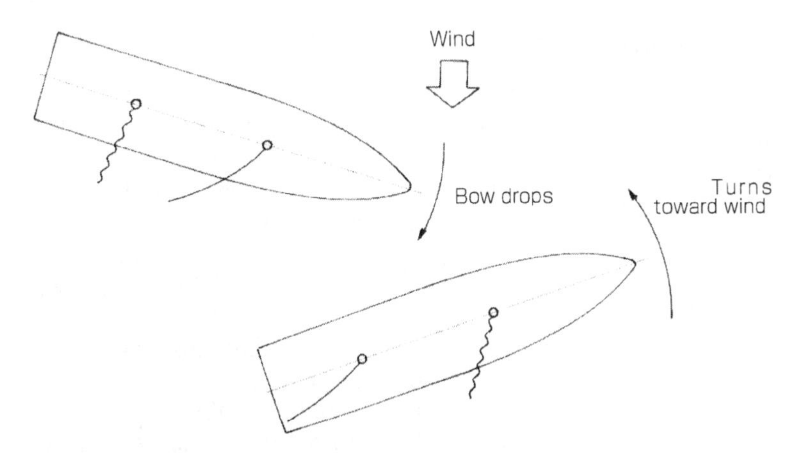

Fig -12 Windsurfer-like system

design.

While your boat is planing you must pull up the keel, otherwise you are likely to experience a knockdown of the boat during a turn. How about a hydraulic system for the retractable keel? Would it work reliably over a long term? Providing a pull-up system for a large rudder was another issue. I could not get a clear picture of the ideal arrangement and was worried.

After racking my brains on the problem, I finally decided on the method used by windsurfers to steer their boards without using a rudder. (Fig - 12). The rider steers the surfboard by inclining the sail forward and aft causing the center of pressure to move longitudinally. In my boat, I had sails fitted forward and aft, and intended to steer the boat making use of the difference in tautness of the forward and aft sails (keeping one sail taut and easing the other sail). I thought that the leeway (drift) of the boat could be controlled within permissible limits since the boat had a large bottom deadrise angle, so I designed Hateruma without a keel.

My first sailing experience on the Hateruma was terrible. The boat could barely sail from abeam (side wind) to 30 degrees downwind, and it could be held on its course and the drift could be stopped only when the boat was accelerated by running the main motors. The speed when the sails were up was inadequate; this was very different from the fun sailing that I had anticipated earlier.

To hoist two sails and to trim their sheets single-handedly is hard work. Moreover, the aft sail would shiver above the table on the deck and I could not fix the awning. I became disgruntled, stopped sailing for some time, and ran the boat only on the main outboard motors.

I think I pondered over this problem for nearly half a year. I thought of using a 10 hp four - stroke outboard motor intended for sailboats, which would provide the steering ability. I was aware that leeway (drift) could be prevented by ahead speed, and this problem would also be resolved by using a small outboard motor. I could fix the awning too since the aft sail need not be used. (Fig - 5).

I bought a US-made outboard motor bracket for the 10 hp motor to help raise and lower the outboard motor, and also connected it with one of the main outboard motors using a connecting rod to enable interlocked steering. This system worked very well. The boat attained a speed of about 4 knots with the small motor. At half - power, the four-stroke motor was practically noiseless, and yet the boat speed was 3.5 knots. By using the forward sail in an 8 m/s wind, I could feel the thrust increase strongly. The boat made 5.7 knots, not a very good speed, but adequately gave the feel of sailing.

The boat's heel was less than 5 degrees while sailing; there was no chance to test its range of stability of 110 degrees. The boat could be steered satisfactorily using the outboard motor, regardless of other conditions. Thus, tacking and jibing (tacking is to change the windward side of the boat through head wind, while jibing is to change the windward side of the boat through tail wind) could be controlled and the drift could be stopped.

In a slight wind condition, the outboard motor was dependable. In a strong wind, the sail was effective in spite of its small size, and was also dependable. Finally, we were able to enjoy barbecue and beer parties under the awning while sailing the Hateruma.

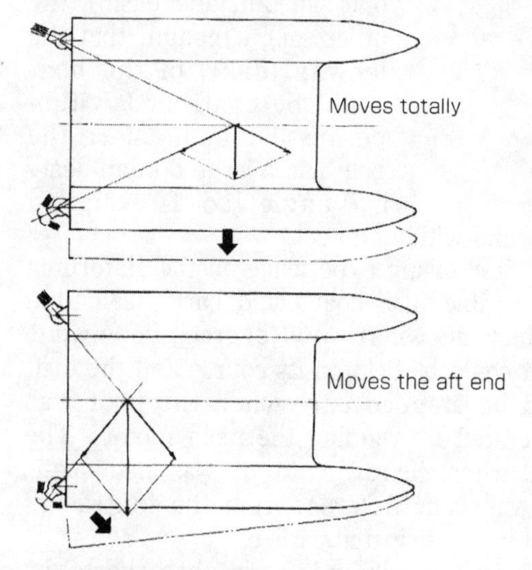

Fig -13 Australian catamaran

I invited Mr. Eguchi, the President (then) of Yamaha Motor, and other executives on board the boat. They enjoyed the cruises and the boat gradually began to build up a good name for itself. Mr. Eguchi loved sailing, and enjoyed a quiet boat but he did not like to handle a sailing boat by himself. The outboard motors of this boat were well aft of the cockpit and behind the transom, and were reasonably noiseless. Later, I tested the boat for easy sailing (auto-sailing).

6) Sideways movement

Normal sailboats have a keel and a rudder with large lateral area, and can travel on their intended course by inertia even in the presence of a strong wind. Therefore, it isn't very difficult to bring a sailboat alongside a pier, but it is difficult to bring a powerboat, light in weight and with large lateral area facing the wind, alongside a windward pier. Hateruma, in addition, is a large boat, and it was not easy to bring this boat alongside a windward pier single handedly.

Usually my wife goes to the forward deck to hook a looped rope over a cleat on the pier to prevent the boat from drifting, and I run the outboard motors in reverse to bring the boat's stern alongside the pier. My wife could never hook the rope on the cleat skillfully, and I was always uneasy at the thought of her falling overboard while performing this work, especially when only the two of us were on board. A single person should be able to bring the boat alongside, and I decided to find a solution.

A catamaran sightseeing boat in Australia does this in a very interesting way (Fig - 13). If the thrust from each of its jet propulsion nozzles is directed as shown in Fig - 13 with one nozzle exerting thrust ahead and the other in the reverse direction as shown, then the line of action of the two thrusts intersect forward and the resultant thrust acts to move the boat sideways, as shown in the figure.

By changing the direction of thrust of each jet nozzle, the point of application of the resultant thrust can be moved longitudinally, and the catamaran can be moved sideways while adjusting the direction of the boat, and the boat can be brought alongside the pier. I liked this method and considered adopting it in Hateruma. If the length of steering rod connecting the two outboard motors is shortened, then the same resultant force mentioned above should be obtained (Fig - 14). In case of the said catamaran, the distance between two jet nozzles was large, so both lines of thrust were made to intersect nearly at midship, a resultant force was generated at this point of intersection and the boat was moved sideways. In this situation, the direction of the jet nozzles is 30 degrees inward, which means nearly half the jet thrust contributes to the side force.

On the other hand, in case of Hateruma, the distance between the two outboard motors is 720 mm, which is extremely small. The direction of the outboard motors can be changed only by a small angle, therefore, the resultant side force obtained is very small (Fig - 14).

If the direction of the angle made by the outboard motors with the longitudinal centerline of the boat is increased, and a large side force is generated further aft, but the situation does not improve. As I didn't have a better idea, I planned to move the bow close to the pier before applying the side force so as to take the boat parallel to the pier. Even if I could not move the boat parallel to the pier, I expected to force the boat alongside using the side force after the boat's aft end touched the pier. I practiced this method.

The hydraulic steering cylinder controls the outboard motor on the starboard side in Hateruma. The outboard motor on the port side is connected to the starboard motor by a connecting rod and moves simultaneously. If I shortened the connecting rod, I could probably achieve my target. In the initial stages, I would adjust the rod using screws, which is an inconvenient method, but later on, I planned to introduce a hydraulic system.

When I tried out this system actually, the side force was generated. Although the boat turned on the spot around the bow, the force was not powerful enough to bring the boat alongside. I found also that it was too troublesome to shorten the connecting rod before arriving at the pier. I was also a bit uneasy to abandon my position at the steering wheel and go aft to shorten the connecting rod. The method was interesting, but there were too many practical issues, and I was forced to give it up finally.

Next, I thought of using a bow thruster (a

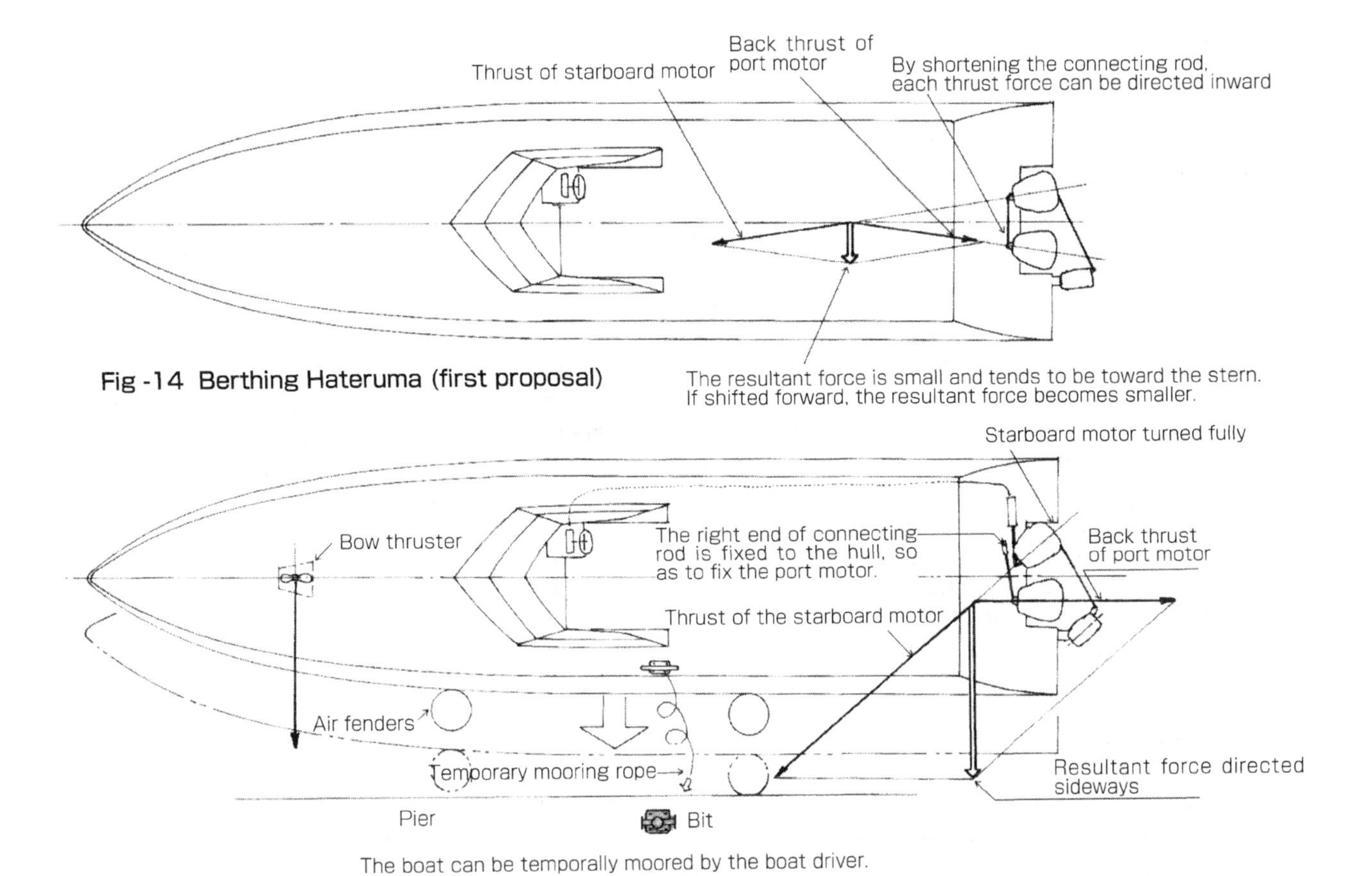

Fig -14 Berthing Hateruma (first proposal)

Fig - 15 Bringing Hateruma alongside the pier (final proposal)

Fig -16 Bow thruster

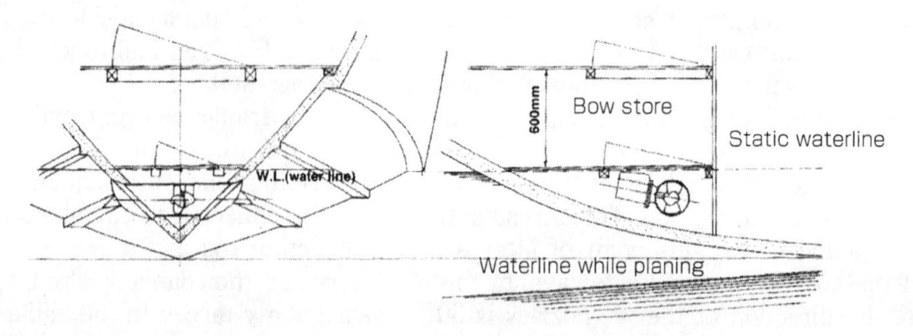

device that generates side force at the bow) to supply the bow force, and rely only on the outboard motors for the required aft thrust. This would be a more effective method.

To eliminate preparatory work before arriving at the pier, I thought of disconnecting the right end of the connecting rod from the starboard motor and fixing it to the boat (Fig - 15). Only the starboard outboard motor would be used for steering in this case.

When the starboard motor is turned fully and operated in the ahead condition, and the port motor is maintained in the fore-aft direction and operated in the reverse condition to cancel the thrust in the fore and aft direction, only the lateral or side force remains. By using various combinations of this lateral force and the thrust provided by bow thruster, both sideways movement and turning on the spot by reversing the direction of the fore-aft thrust are possible, enabling the boat to be maneuvered freely and as desired when it arrives at the pier. When cruising normally with this steering system, the steering performance drops a little, but sharp turns are not necessary when cruising in Lake Hamana. When navigating the sea in rough weather conditions, the connecting rod could always be restored to its original position.

I decided to adopt this method. Bow thrusters were very expensive those days, but I found one with a 50 kg thrust for about US $ 2500 in West Marine (US mail-order house) catalog, and ordered it. Initially, the FRP pipe for installing the thruster did not arrive. I re-ordered it from the Netherlands through the USA. It took two to three months but finally everything was at hand. The bow thruster and other accessories cost me approximately 300,000 yen.

The bow thruster I bought was very compact, and could be installed below the bottom plating of the store under the bow berth (Fig - 16). The source of electric power for the thruster, however, was a headache. As the bow thruster needed 200 amperes, we had to use electric wires with adequate capacity and use power from the batteries at the stern. The wiring would weigh as much as I did, and laying such cables after completion of the boat was a tremendous problem.

Finally, I decided to install the battery on the bow deck exclusively for the thruster. To charge the battery, I decided to use a 5-W solar battery installed nearby. The time to operate the bow thruster is 10 seconds on the average, and 20 seconds at the maximum, which amounts to only 1% to 2% of the capacity of an 80 AH battery, which can be recharged in 2 to 3 hours by the solar battery. Therefore, I need not worry about power for the next berthing of the boat. After carefully contemplating this solution, and knowing that it would work, I felt happy (Fig - 5).

The system actually worked very well. The boat could be moved sideways easily. When approaching the pier, I stop the boat parallel to the pier, and after adjusting the fore and aft side forces, move the boat sideways. Just before touching the pier, I switch off both outboard motors (thrust) and bow thruster as the boat moves by inertia and I get on the pier to moor the boat amidships while the boat's air fenders (pneumatic devices protecting the gunwale) push against the pier. The entire opera-

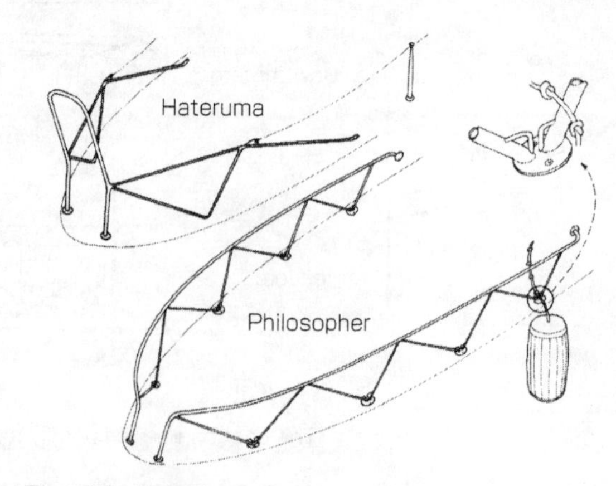

Fig -17 Pulpit

tion can be done very easily.

The bow thruster is powerful and can be used against wind speeds of 7 to 8 m/s from the pier. When departing, I move the boat laterally from the pier and then reverse the direction of thrust at the bow and stern, and the boat turns on the spot. During the trials at the company, people were impressed with the berthing and de-berthing of the boat, and remarked that it was similar to that of a ferryboat. After fitting the bow thruster, I have enjoyed operating the boat single handedly.

7) Pulpit

A gate (Fig - 17) is fitted at a convenient position at the center of the pulpit. To get on to the boat moored by the bow from a high pier, we need to hold on to something to steady ourselves. Gripping the top of the gate makes the process of getting on board very easy and safe. Conversely, if the pier is low, it is not easy to straddle the pulpit carrying heavy provisions, water, and fuel. A gate enables persons to pass through easily, and affords a handhold also.

Pulpits adjoining this gate are made of trusses. Normally, guardrails at the bow are connected and are rigid, but as we approach the aft, the pulpit tends to be less rigid, with excessive forces acting on the installed bases. However, if stays are arranged in a zigzag fashion similar to a truss structure in a bridge and since the boat edge is curved in the plan view, the pulpit become three dimensional truss structures supporting each other, and they prevent excessive bending moments from occurring at the bases. The structure of the pulpit is light and strong, similar to a steel pipe - welded structure often used in the frames of racing cars and light aircraft. Thanks to this structure, the pulpit is rigid all around. Even after making the pulpit discontinuous at the bow or installing a gate for persons to pass through, overall the pulpit remains strong and rigid. I wonder why everyone does not adopt this style.

Satouchi Kazuhiko and Tazura Mitsuharu, who were in charge of the production design of Philosopher thought of an interesting idea. They fitted two angled bars on the pipes near the base plate as shown in the figure, to enable fenders to be suspended at the knots made on the rope attached to the fenders. This method enabled us to conveniently adjust the vertical position of fenders by making knots at desired locations. I had purchased more than ten different kinds of fender adjusters and tried them out, but none were as useful as the method that we devised ourselves. Of course, this method is exclusive to the truss-type pulpit, and the fender ropes may suffer from wear and tear in the long run, but it is a very reliable and extremely convenient fender adjusting method in my experience.

Fig -18 Fuel oil tank

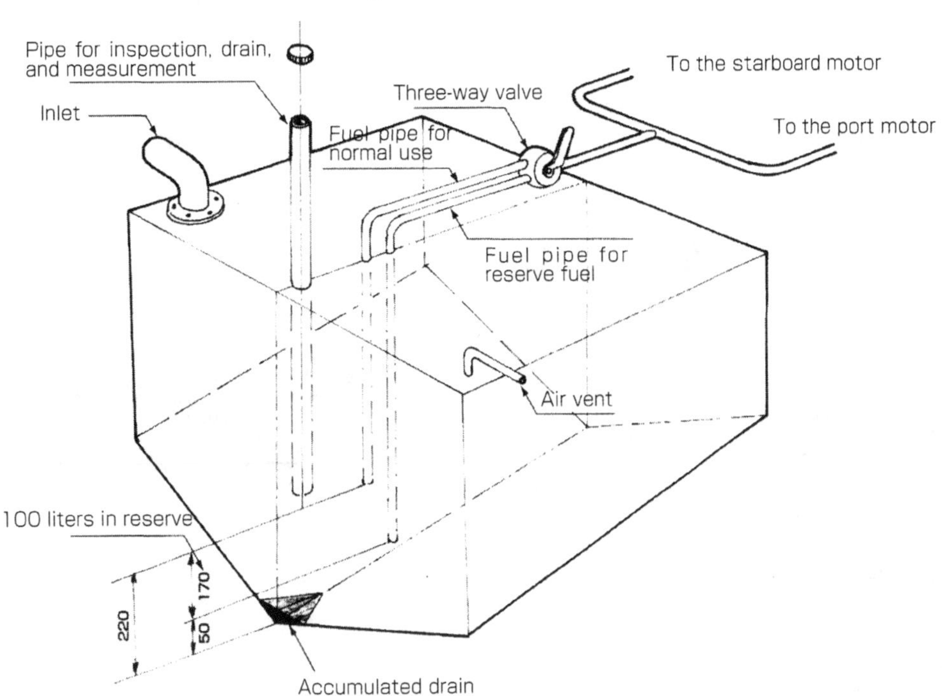

8) Fuel oil tank

This boat has a 600 ltr FRP fuel oil tank positioned outside the cabin in front of the main engines. As the fuel oil tank is integral with the hull, the expanded urethane form surrounding it insulates the tank from heat. The tank, however, is inaccessible from the bottom, and drain plugs cannot be fitted. We tried out various ideas, and some of them gave good results. (Fig - 18).

A pipe for inspection/draining fuel oil was erected at the deepest point of the tank. If a measuring scale is inserted through this pipe, amount of fuel remaining can be measured to an accuracy of a few liters. I also installed an electric fuel gauge beside it, but its accuracy was of the order of 100 liters. Accurate measurement is naturally important to know the correct fuel consumption and to know whether one can continue cruising even with very little fuel remaining in the tank.

Usually the instruction manuals recommend topping up fuel oil tanks but I hate carrying 600 liters of fuel every time I set out. I like to run the boat in the light condition as far as possible. Since I did not always top up the fuel tank, I sometimes checked for water accumulation but could not find any traces. For some time, I did not understand why there was no water in the tank. Finally, I figured it out. If the tank has voids, large amounts of air move in and out frequently because of temperature changes within the tank, and at low temperature, water condenses in the tank. However, in a tank insulated from heat, both these disadvantages are eliminated and water does not condense in the tank. Thus, a heat-insulated tank has advantages.

Another arrangement we used was to have two fuel outlet pipes on the tank. The intake point of one pipe was 20 cm and of the other was 5 cm above the lowest point of the tank. I normally use the former to prevent suction of dirt accumulated at the tank bottom. After the fuel is used up, I turn the cock to switch over to the other pipe and use the remaining 100 liters or so. The arrangement is a kind of reserve fuel valve, but it could be improved upon. The main outboard motors have been working very well since the last thirteen years.

9) Dealer meeting

We presented Hateruma at the meeting of Yamaha dealers at the Yamaha Marina, Lake Hamana, in the autumn of 1989, where new boats were demonstrated.

Many salesmen who boarded and went for a trial cruise on the Hateruma found the boat unconventional, but the presidents of important dealers did express interest in buying the boat.

Fig -19 Bridge of Philosopher

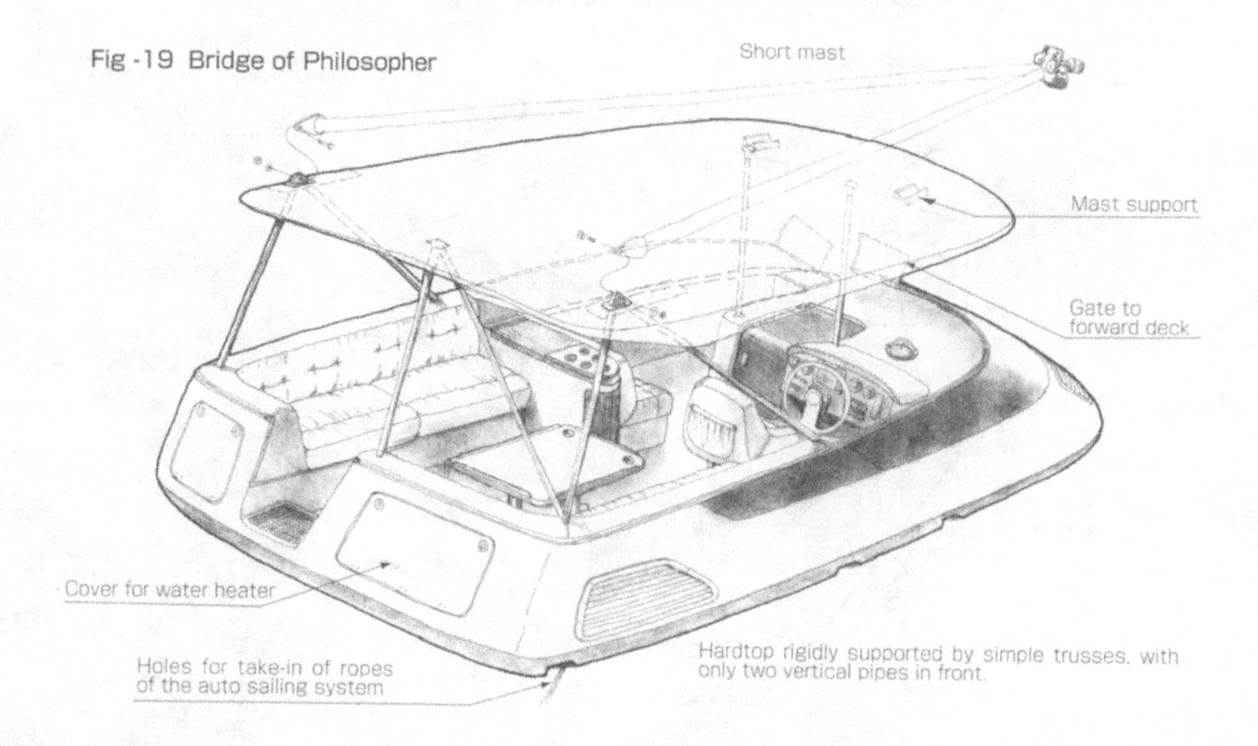

Fig - 20 Philosopher on sail

Fig - 21 Owner's room of Philosopher

Fig - 22 Dining room of Philosopher

The boat was initially designed and built for myself, I had no intention of selling it. Those who evinced interest wanted to own the boat themselves, after seeing the comforts, the spacious owner's room, and the bathtub. I was pleased to find so many people with the same feelings as mine.

Later, based on the opinions of the sales people, the Company decided to go for production of the boat. I spoke about my experiences with Hateruma in detail to the young engineers, whose job it was to commercialize Hateruma.

To eliminate the awning installation, I redesigned the layout so that the bridge and the barbecue zones were combined into one area and covered all around by a transparent enclosure during the winter season and on rainy days. I also redesigned the hard top arrangement supporting it by simple trusses and enabling the mast to be erected on the hard top, thereby eliminating the aft mast (Fig - 19). Compared to my own boat, which was built with careful cost considerations, the production model gradually assumed luxurious specifications and became more expensive (Fig - 20, Fig - 21, Fig - 22). During this period, the bubble economy burst, and the sales of boats dropped steeply. The presidents of dealers, who had initially wanted to buy Hateruma gave up the idea because they were hard pressed to manage their own companies. Under such circumstances, the boat went into production. The name of the production model was Philosopher, and my boat Hateruma became a prototype of the Philosopher.

10) Auto sailing system

President Eguchi had often mentioned in the past that sailing should be simple and performed automatically. I decided to demonstrate the auto sailing system that we developed when exhibiting the Philosopher at the Tokyo Boat Show.

Tsuide Yanagihara of the Horiuchi Laboratory and Atsushi Uchiyama of the Electric Engineering Department, developed a fairly good auto sailing system, which was actually demonstrated at the boat show. We also tested the system at Lake Hamana and found it to work satisfactorily, but to my regret, its performance did not reach the level required for actual sales. We found that the specially designed winch for auto sailing system had to cope with an unexpectedly wide range of speeds and rope winding forces, and its development would require considerable time. Thus, the auto sailing system project was shelved after a brief demonstration at the Tokyo Boat Show.

The auto sailing system worked as follows: firstly, the skipper runs the boat with the 10 hp outboard motor at half power and sets the course to 45 degrees or more from windward. He pushes a button, and a rolled-up front sail (like a furling jib) unfurls, with its sheets (ropes for sail control) trimmed to the optimum length automatically. When the boat is steered and it changes course, the sheets are adjusted automatically. If the wind increases in intensity, the

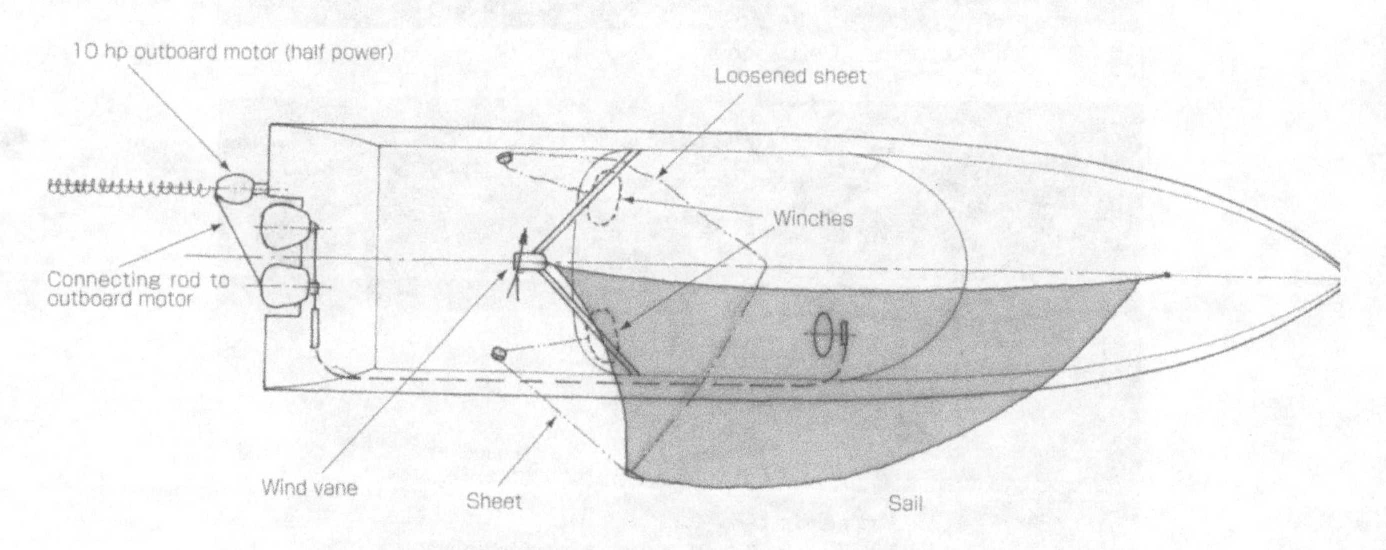

Fig - 23 Auto sailing system

sail is reefed (sail area reduced) automatically. A computer controls the system after reading the wind direction and wind speed (Fig - 23).

When the boat tacks, the sail is temporarily wound up so that it does not come in contact with the bridge top. The sail is fully spread out by the system after the boat turns into the wind coming in at an angle of 45 degrees.

Since the outboard motor is running all the time and is used to steer the boat, tacking will not fail. The skipper's job is to merely press the button of the auto sailing system and to steer the boat using the outboard motors. In addition, the skipper can use an autopilot. His only job would then be to watch the course to avoid obstacles, and leave sailing to the auto sailing system !

The main issue was the development of a sheet - winding device that loosens the sheet smoothly in response to a slight tension when the boat turns to leeward under the breeze. On the other hand, when the boat is running to windward in a strong wind, the device should fully and strongly tighten the sheet. The range of forces and speeds to be handled by the specially designed winch is excessively large, as also the large length of rope required for winding up. A single-stage transmission is inadequate, and another winding-up reel is necessary. We needed a winch that would satisfy the conditions mentioned above.

We looked through various catalogs, and found a rather nice one used in mega-class yachts (yachts of over 100 feet). I did not anticipate this winch to fulfill all our expectations, but by studying it, we could develop one for smaller sailboats that suited our requirements.

Another issue is to provide a power source of adequate capacity. If these two issues are resolved, a perfect auto sailing system is not too far off, since we have powerful computers available to us these days.

Auto sailing is a welcome feature for those who wish to enjoy sailing but do not want to learn how to sail or handle sails themselves. Auto sailing can also pave the way to popularizing sailboats, which account for only about one-tenth the number of powerboats in the world.

Having a beer or enjoying a meal with friends in the quiet and peace of the open sea with the waves gently lapping on the hull is the feature of a sailboat, and is an experience that cannot be had in a powerboat. The preparatory work that goes before one sets out for sailing is cumbersome. When you invite guests onboard, you become very busy with the preparations and get irritated at guests who have no inclination to assist you! I still remember my happiness when I first installed the furling jib, and kept the main sail on the boom with a cover on it rather than taking it off. As we grow older, we look for ways to make sailing easier.

11) Epilogue

Only a few Philosophers were sold to persons who appreciated my design concepts. I invited those people onboard who wanted to hear more about this boat or wished to take a ride on the Hateruma. I was never happier than when talking to them about the boat.

A boat as large as 15.3 m looks difficult to sell especially after the economic recession. I retired from Yamaha Motor in 1996 and moved to Kamakura to spend the rest of my life. The charges for mooring the boat at a marina near my house are so high that I still keep the Hateruma at Lake Hamana.

When I built my boat, boats over 15 m in length were exempted from luxury tax, while boats below 15 m in length were subjected to a 30% luxury tax. Nowadays, there is no luxury tax but a consumption tax of 5% is levied. I have a desire these days to build a boat of a smaller size yet possessing all the advantages of Hateruma. If the boat is for my personal use, the length would probably be around 8 m, but if it is to be more like the Hateruma, then maybe a length of about 10 m would be ideal.

A layout in which the entire length below the deck is taken up by cabin space and the space on deck used for barbecue and other leisure activities may not be possible as the center of gravity of the boat may rise. It will give me great pleasure to work out afresh the plans of a boat to be used exclusively by a married couple, has a comfortable accommodation space, and is ideal for both powered and sailing cruises.

Fifty years have elapsed since I first entered this field of work. During this period, Shiro Uchida (after resigning from Yokohama Yacht Works, he joined Yamaha Motor and participated in the Pacific Ocean 1000 km powerboat marathon) has been a trusted colleague and friend. I would like to acknowledge with gratitude his invaluable assistance in the design of Hateruma.

SMALL HIGH SPEED SAILBOAT TWIN DUCKS

After watching human - powered boat races every year where boats are run at speeds of nearly 20 knots, I began to think that their efficient hydrofoil systems could be used in sailboats. In the 1980s, we developed a fast hydrofoil sailboat named Avocet in cooperation with Mr. Greg Ketterman, but this boat was large, and with a speed of 40 knots, it was too fast. I dreamed of designing a sailboat with a speed of 20±5 knots that would be fun to sail, but this was not an easy task. I studied the design for almost three years, and one fine morning the solution came to me.

1) Avocet

In 1985, Yamaha Motor established an R&D Center (Research and Development Center) in the head office, and also established R&D California in Los Angeles and R&D Minnesota in Minneapolis for research and development of products to be sold in the US market.

As I was appointed in charge of the R&D Center, I visited both R&D California and R&D Minnesota every two months for discussions. The R&Ds in the USA were given a free hand to find research topics through interaction with friends and acquaintances, or through participation in leisure activities in the marine environment with the aim of developing new products.

In 1983, Toshiharu Yamada of R&D California met a young American named Greg Ketterman, who was enthusiastically carrying out research on a hydrofoil sailboat, which had already achieved a maximum speed of 24 knots.

Yamada persuaded Greg to start joint research work for future production with Nick Larson, who was Yamada's colleague. Nick named the project Avocet and worked on it earnestly. The dictionary describes avocet as a kind of snipe. He probably thought of avocet after visualizing a hydrofoil in the water with its thin legs.

The sailboat developed with Greg was a trimaran type hydrofoil boat with two windsurfing sails arranged in parallel. (Fig - 1).

A large diameter beam made of aluminum tube is set in front of the center hull and extends on both left and right sides. At the end of the beam on each side, there is a surfboard. An arm extends forward from each surfboard at the forward end of which a planing plate is attached. The planing plate works as a sensor that senses the water surface. A curved hydrofoil is fitted on the outer side of each surfboard.

When the boat is at rest, both the planing plate and surfboard float flat on the water and remain statically stable (Fig - 2a). In this condition, the hydrofoil has an angle of attack of about ten degrees, which is adequate to lift the surfboard out of the water when the speed of the boat increases.

As the planing plate traces the water surface and the surfboard lifts above the water surface, the front end of the arm drops relatively, the angle of attack of the connected hydrofoil decreases, and the lift decreases. Likewise, the surfboard can continue its foilborne running condition maintaining the foilborne height when the lift becomes equal to the weight supported by the hydrofoil (Fig - 2b).

The skipper sits in the cockpit at the rear end of the center hull. The rudder is located just aft of the cockpit, and the hydrofoil is fit-

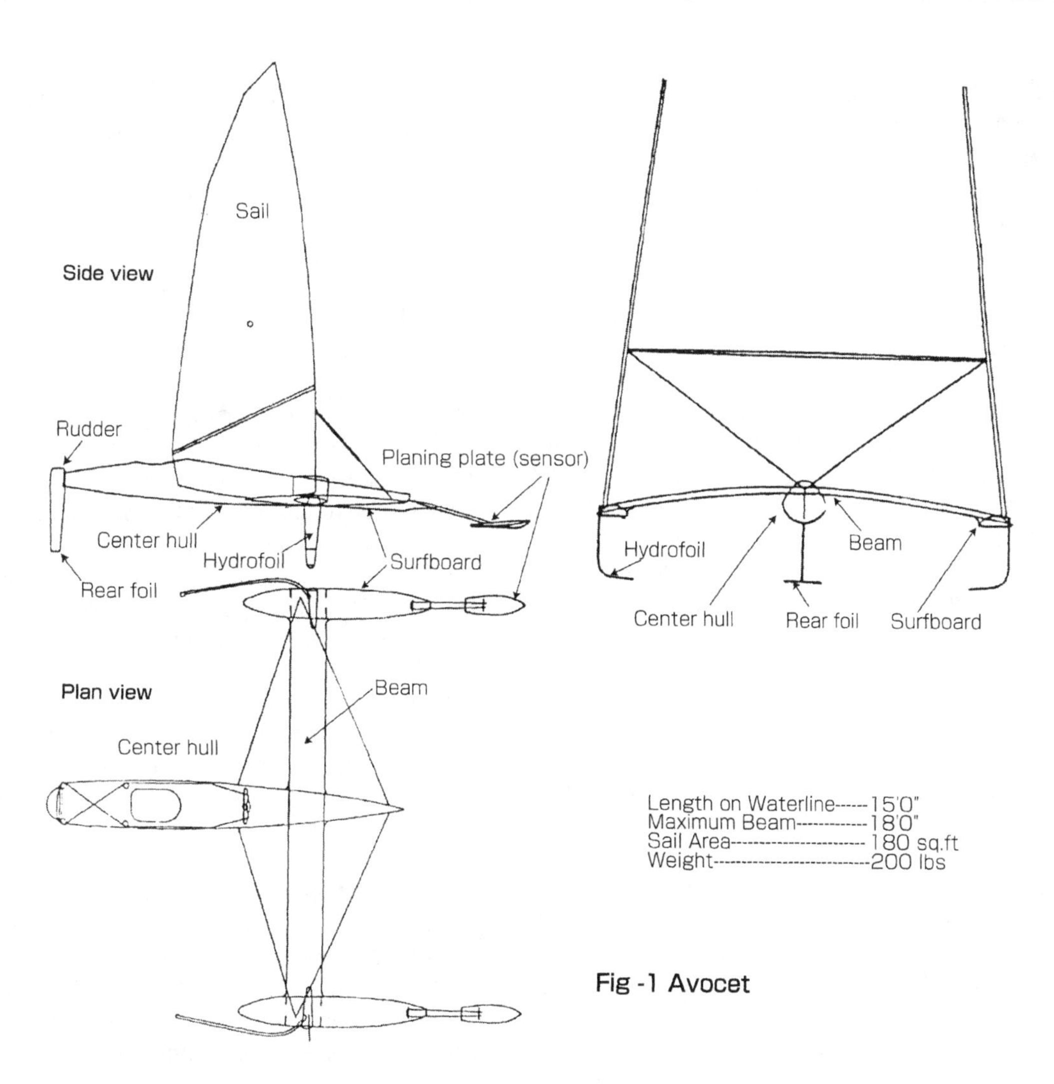

Fig -1 Avocet

Fig - 2 Transition to foilborne condition

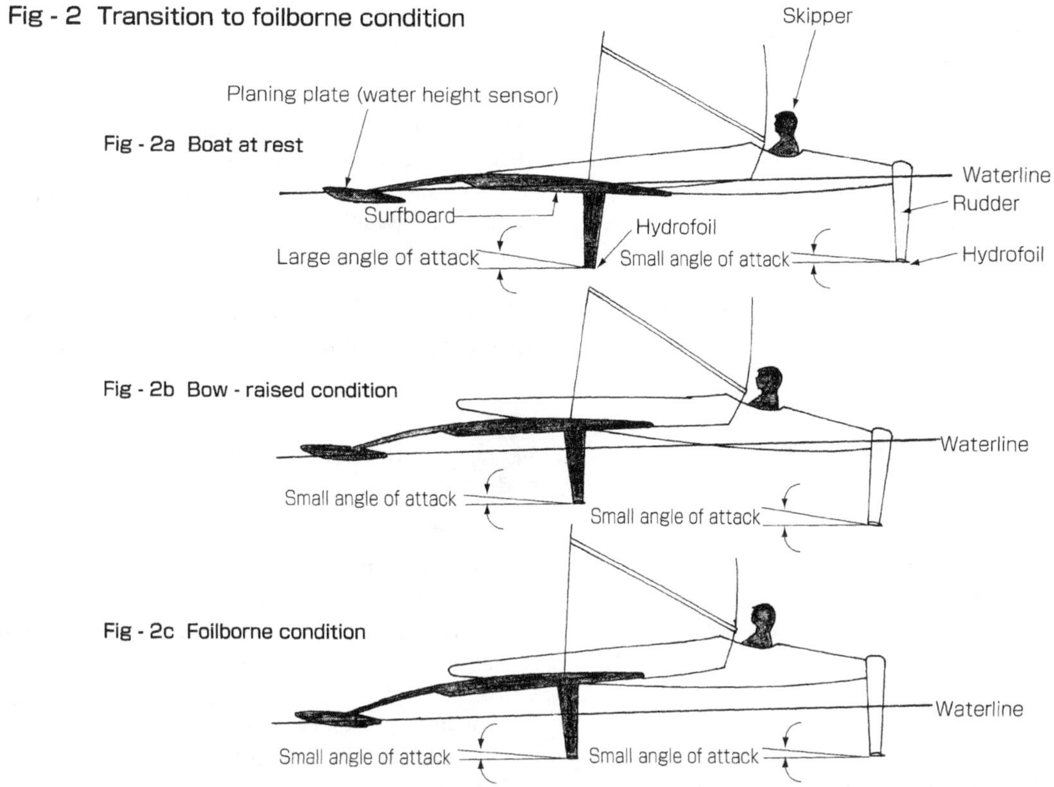

ted at the bottom of the rudder. When the surfboards begin to become foilborne, the bow of the main hull rises, and the angle of attack of the rear hydrofoil increases, which increases the lift and raises the stern also. In this case too, with the rise of the stern, the angle of attack of the rear hydrofoil decreases, and the boat runs with the stern at a height at which the lift and the load are in stable equilibrium (Fig-2c).

Normally, sailboats have to withstand a heeling moment (lateral heeling force). In the Avocet, when the leeward surfboard is pushed down by the heeling moment, the angle of attack of the hydrofoil under the surfboard increases and the lift increases, while the lift on the windward side decreases. Thus, the heeling moment is balanced automatically. When a strong wind or a gust is encountered, the lift of the hydrofoil on the windward side may sometimes become negative.

Greg dreamed of breaking the sailing speed world record. We, on the other hand, wished to market a sailboat that was easy to operate, low in cost, and with a reasonably high speed. Yamada and Nick tried to persuade Greg to develop Avocet as a product acceptable to the market, but Greg's intention and ours did not match. He did not wish to modify the initial layout, or to modify Avocet to make it smaller, or to change the layout.

During the development stage, I had three to four opportunities to try out the Avocet. I was possibly too heavy, since the boat did not get

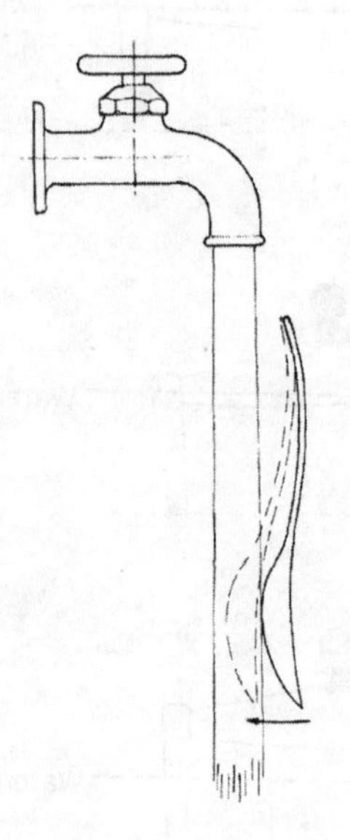

Fig - 3 Spoon sucked into the water flow

foilborne, except in the presence of a very strong wind. On the other hand, once the boat was foilborne, the subsequent acceleration was so high that the boat sprinted violently. Tacking was also difficult, but jibing when foilborne was very easy.

While I tried out the boat, the speed was probably about 25 knots. My impressions from the trial runs were as follows:

a) I was scared that if the boat speed exceeded 30 knots, it would turn over, break into pieces, and cause bodily injuries. Anyhow, I was glad that the boat didn't go any faster.

b) The boat should preferably take off at a lower speed. Since the difference between the speeds before and after it became foilborne seemed too large, it gave me a feeling that the resistance while taking off was too large. If we could reduce the wind speed at take off to 5 to 6 m/s, the foilborne running of the boat would become very enjoyable. Also, as soon as the wind drops and the hull touches the water slightly, the condition returns almost immediately to the condition before becoming foilborne, which is not very pleasant.

c) The boat is too big, heavy, and difficult to set up, and these take away the enjoyment of sailing. The present size is acceptable if the intention is only to break the speed record, but to enjoy sailing, and naturally, as a marketable product also, a smaller and lighter boat with simple structure is preferable.

Finally, we gave up on the idea of bringing Avocet to the market but supported Greg's research according to his will for several years. In this period, he achieved the world record of 43.55 knots with a record-breaking boat named Long Shot, which was similar to Avocet.

I saw the photo of Long Shot making a tremendous splash while running for the world record, and felt once again the fear of sailing at speeds of around 40 knots.

2) Avocet in Japan

After the contract with Greg terminated, we brought an Avocet to Japan. The members of the YAMAHA sailboat development team had already made several trial runs. It appeared that Avocet would not become foilborne except in the presence of a strong wind. I suggested introducing a step at the stern.

The bottom of the center hull rises up in a

Fig - 4 Step of Avocet photo by Masaaki Ozawa

circular arc at the aft part, and along this surface, the water flow is sucked up. The counter force sucks the hull downward, and this was probably the reason that the take-off resistance increased. This is exactly the same phenomenon as that of the round surface of a spoon sucked into the water flowing down from a tap (Fig - 3).

I thought the problem might be solved by arranging a step at the stern to separate the water flow from the hull (Fig - 4).

Isobe, a crew member of the Nippon Challenge team that participated in America's Cup and in charge of sailboat design, measured the take - off resistance of Avocet by towing the actual boat. The result showed that the take - off resistance, which was 135 kg in the original model, was reduced to 30 kg after providing the step. Naturally, it took off easily and I estimate that the wind speed for take-off would have decreased drastically from 8 m/s to 5 m/s.

Subsequently, the boat was victorious in the first All - Japan Sailing Speed Trial Event held at Matsumigaura in Lake Hamana in November 1992 (Fig-5). The speed recorded was 21.1 knots in a 12 knot (6m/s) wind.

After the event, the boat has never been on water. The difficulties in setting and launching it might be the reasons. I myself have hesitated to ask my staff to launch the boat.

Since then, I have supported the human powered boat race that takes place annually. In this race, human powered hydrofoil boats became foilborne very easily. Particularly from 1996, a race has been introduced that has seven tight turns over a 1000 m course to be covered in the foilborne condition (Fig-6). Furthermore, their speeds were two or three times that of displacement boats. Considering these circumstances, I have concluded that it is unreasonable not to use hydrofoils in sailboats,

Fig - 5 Avocet -- the Winner
photo by Masaaki Ozawa

which make use of the propulsive force of very strong winds compared to human power.

There are many reasons why hydrofoils have not been used in sailboats. For instance, the wind direction is not fixed; at times, it's not there, at times, it becomes a gust. As sails are tall, the wind forces work on the upper part and generate a large heeling moment, causing the boat to pitch pole (longitudinal knock down). Also, sailboats have to withstand side forces. There are yet some items that cannot be resolved easily, compared to human powered boats.

Recently, the world sailing speed record was set by hydrofoil boats, which are playing a major role in this field. If we can find a design of a sailboat for popular use that resolves various issues well, we will come up with an interesting boat.

3) My favorite sailboat

I have no intention of challenging the world record. Instead, I would prefer to be on board a sailboat and sail from Enoshima Island to Sagami - Bay, swiftly winding in and out among crowded dinghies at 20 ± 5 knots.

If the sailboat is able to tack, that would be fine; but even if not, I will be happy with jibing only. The most important point of the sailboat should be the ability to launch and stow it by just one or two persons.

If wind speed is 8 to 10m/s, the sailboat's speed may reach 30 knots, but such speeds are for young people. My favorite sailboat will be one that is carried on top of the car to the marina, assembled, and launched all by a single person.

Such expectations together with my impressions after seeing many hydrofoil human powered boats, gave me the feeling that it should be possible to design and build a small hydrofoil sailboat. However, I could not arrive at concepts that would clear the design thresholds, and days went by.

During an interview in 1997 for a boat magazine, I talked to Mr. Masahiro Mino about my dream of a 30 knot sailboat. But when asked for details, I told him that I could certainly design such a sailboat, but at that moment, I only had its concept in mind.

4) Conception

The conversation with Mr. Masahiro Mino might have stimulated me. While thinking about sailboats lying in my bed one early morning in July, an interesting plan suddenly occurred to me.

The base was already there. In the 1960s, I had become interested in the layout of a simple and small catamaran in the USA named Aqua

Fig - 6 Super Phoenix 1997.8.23, Ran 100m in 9.99 sec. Recorded 19.45 knots (not recognized by IHPVA)

Fig - 7 Aqua Cat
drawn by T.Tadami

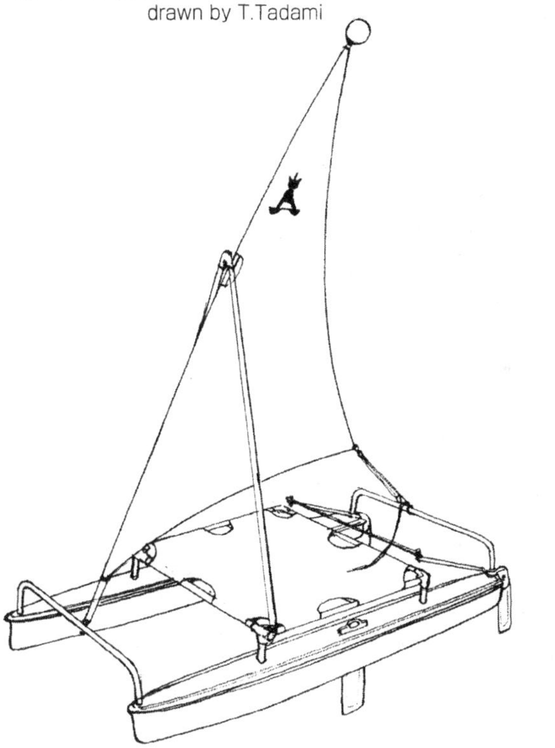

Fig - 8 SC-14

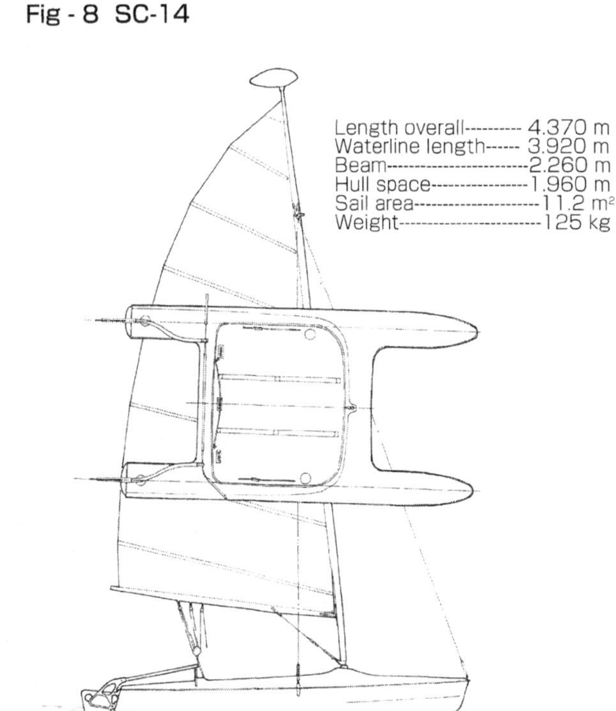

Cat, especially in its sailing rig (Fig - 7).

The Aqua Cat had become very popular with its sail number having reached 4 digits, and coming very close to 5 digits. Later, Yamaha Motor developed a catamaran dinghy named SC - 14 (Fig - 8). Both these boats made me feel uneasy because the beams connecting their hulls had a tendency to twist.

When a thrust acts on the sail of the catamaran boat, its slender leeward bow easily tends to plunge into the water. The skipper must place his weight on the aftermost deck on the windward side and try to bring the bow up, but since torsional rigidity of the beam is inadequate, only the bow of the windward hull comes up, while the leeward hull remains submerged. As a result, the boat tends to pitch pole during a strong blow.

I had been planning on a catamaran hull for my new boat, so while I lay in bed, I considering how the forces would work while thinking of how to solve the issue of torsional rigidity. I thought about a layout that would make the boat torsion free, and arrived at a very interesting solution.

The layout plan is as follows : left and right hulls would have the layout of the hydrofoil in the human powered boat, and each hull would independently maintain longitudinal stability and foilborne height. To ensure that the motion of each hull was not disturbed, they would be connected by an aluminum tube which enables the two hulls to twist freely and also possess transverse stability (Fig - 9). This plan may be regarded as the layout of only the front part of Avocet.

The human powered boat (Fig - 6 or Fig - 10) has a front hydrofoil with a height sensor, which keeps the foilborne height at the bow constant. Instead, if we consider a large planing plate sensor used in Avocet, the combination of the planing plates, surfboards, and attached hydrofoils corresponds to the layout of the human powered boat. Therefore, the center hull of Avocet is not really necessary to maintain the attitude, and it may be considered similar to towing an unnecessary wheeled cart.

Also, Avocet accounts for the heeling moment by the difference of lift in the left and right hydrofoils and does not fully use the windward lift, whereas the new layout enables the loads on the left and right hydrofoils to be equalized by the balance of the skipper's weight and enables lifts of the foils to be utilized equally.

This means that the efficiency of hydrofoil can be increased, and at the same time, the foil area required for take-off can be reduced, thereby reducing the resistance at high speeds and enabling the boat to go faster. If the hydro-

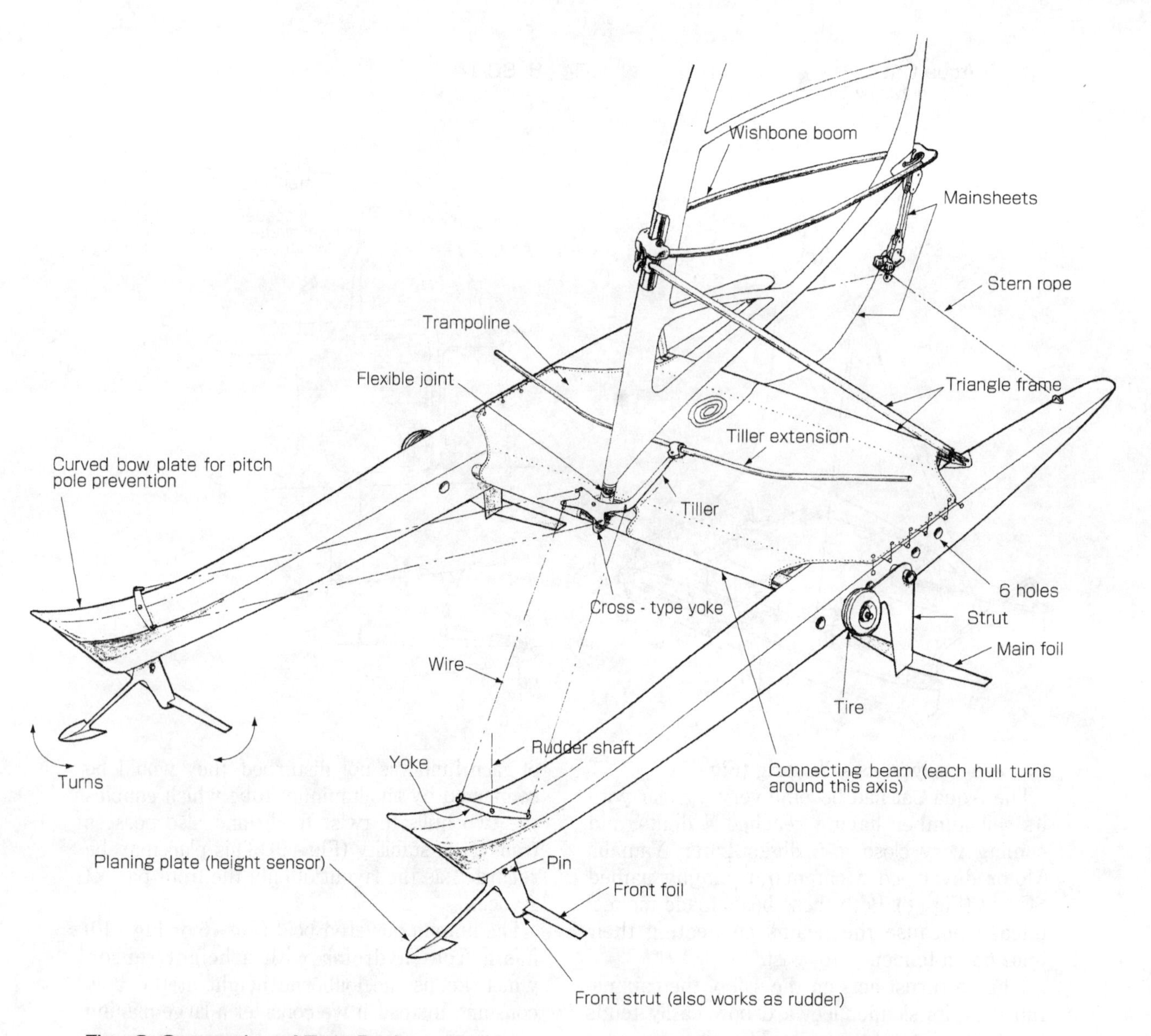

Fig - 9 Conception of Twin Ducks

foil area is kept constant, the boat can take off easily. When its transverse stability is compromised by a gust of wind, the boat twists and a difference arises in the angles of attack of the left and right hydrofoils. This difference in lift between the left and right hydrofoils resists the heeling moment automatically, which is reassuring.

I was worried however, that if a gust blew during a slow speed run, the lift of the front hydrofoil would not counter the plunging down of the bow, even if the skipper moved to the aftermost position at the stern. If that happened, the boat may not become foilborne and may pitch pole. To prevent such an occurrence, the front foils should be made a little larger.

5) Structures of each area

Hull

Considering the boat to fit on top of a car, its length may be taken as 4.5 m. At the largest section, its beam may be taken as 28 cm and its depth as 24 cm. The hull form may be basically similar to that of a normal catamaran sailboat.

Normally, the hull of a catamaran sailboat has adequate reserve buoyancy since such boats are designed to run on one hull. This boat may also experience a sudden gust of wind at slow speeds where hydrofoils are not effective, so the boat must possess adequate reserve buoyancy to ensure transverse stability.

At the bow, the plate can be shaped with the upward curve of an Australian surf ski to prevent pitch poling (Fig-9). Since the plate shapes resembled the bills of two ducks, I have named the project Twin Ducks.

The hull shape is symmetric, so only one female mold is required. The structure should preferably be light considering the ease of handling and ensuring good performance during tests. I adopted sandwich construction as in a rowing shell eight although it seemed too expensive for mass production.

This sandwich construction has 4 to 6 mm honeycomb core or structural plastic foam, bonded on both sides with 0.1 to 0.5 mm Carbon Fiber Reinforced Plastic (CFRP) providing a very lightweight structural material weighing only 0.7 to 2 kg/m^2. The surface area of one hull of this boat is about 3 m^2. If the unit weight is 1.2 kg/m^2, the weight of one hull will be about 4 kg.

At the front end of the hull is a vertical rudder shaft with a strut, front foil, and planing plate sensor fitted with pins at the bottom of the shaft. Around the midship part of both hulls, there are six 70 mm diameter holes facing the transverse direction, and a connecting pipe (beam) through the hole connecting the left and right hulls. The second or third hole from the forwardmost hole is used for supporting the hydrofoil strut. These six holes are provided for adjusting the longitudinal position of the connecting beam or hydrofoil

Hydrofoil

The positioning of the hydrofoils follows the positioning in the human powered boat Cogito (Fig-10). The only difference from Cogito is the strong lateral force that acts on the struts (of the hydrofoil).

The load on the front hydrofoil increases steeply during a gust. The bow sinks and main foil also sinks due to the decrease in the angle of attack. If the motion becomes severe, pitch - poling occurs. Therefore, it will be safer to provide the front foil with a slightly larger area to compensate for the lack of area and to prevent stalling. When the extra front foil area is used to start the foilborne run in a light wind, and the skipper shifts forward as far as possible to apply a load on the front hydrofoils corresponding to their areas, the loads on the main hydrofoils will decrease by the same amount and the boat will become foilborne faster. However, as the front hydrofoil area increases, the top speed of the boat decreases.

Cogito has a front hydrofoil are of 0.03 m^2 and main hydrofoil area of 0.1 m^2 supporting 180 kg. This boat has two sets of front hydrofoil area of 0.04 m^2 and main hydrofoil area of 0.08 m^2 supporting 120 kg. The calculated take - off speed is 5 to 6 knots, which is half that of Cogito's take-off speed.

If the buoyancy of stern is excessive, it will hinder the take off because the bow does not lift soon and the foilborne running is delayed. The arrangement should also be such that the main hydrofoil has an adequate angle of attack when only the bow is foilborne.

Fig-10 Cogito Won the 200m - speed race for 6 years consecutively. Ran 100 m 10.57 sec.

Fig - 11 General arrangement of Twin Ducks

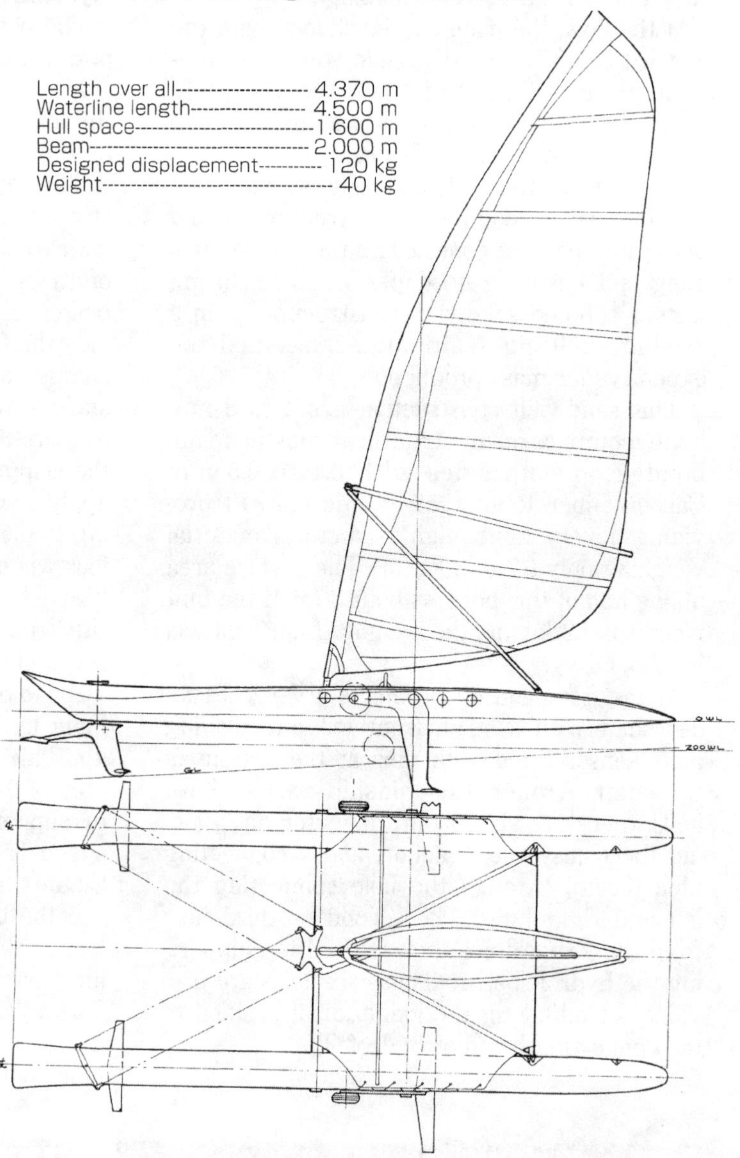

The load on front hydrofoil varies considerably according to skipper's longitudinal position in relation to the main hydrofoil. Adjusting the initially set angle of the main hydrofoil is a delicate job and we have to find the best angle by trial and error. Therefore, the hydrofoils should be temporarily attached from the outside so that their longitudinal positions and initially set angles can be adjusted using the six holes described above (Fig - 11).

The construction should be such that the main hydrofoils can be folded back before launching and then restored to their original positions after floating the boat. Also, an arrangement should be provided such that if a large object strikes the main hydrofoil or strut during foilborne running, the foils and struts swing up backward, just like a circuit breaker after it trips.

This structure must be strong because the strut of the main hydrofoil is tall and is affected by large side forces. Later, when the complete layout is finalized, the strut can be extended from the keel of the boat. Then the strut can be made short and light, and its resistance decreases, which is very welcome. The layout of the front hydrofoil should be almost the same as that of Cogito, and no other changes need be made except reinforcing it to adequately resist the side force.

Rudder

The struts of the front hydrofoils work as rudders. The main struts take up the major part of the side force, but the front struts are also anticipated to take up impact side force. The front struts should withstand the side force, and at the same time, the struts must be exchangeable in case they are damaged due to collision. Since the left and the right hull each rotate around the connecting pipe, the left and right tillers cannot be connected directly to each other.

The cross type yoke fitted at the middle of the connecting beam is connected to the left and right yokes independently with yoke wires. The left and right yokes are fitted at the upper end of each rudder shaft. The cross - type yoke is part of the tiller connected to a long tiller extension, both ends of which can be controlled by the skipper.

Sail and its support

I intend to use the windsurfing sail as - is, since I have heard that windsurfing sails have made tremendous improvements in the past several years.

Windsurfing sails, particularly, can be selected from sail areas ranging from 4 m² to 9 m² and from a wide range of characteristics, such as sails for high speed or for maneuverability. That is the main reason why I would like to use fittings and parts that enable the windsurfing sail to be set up as - is.

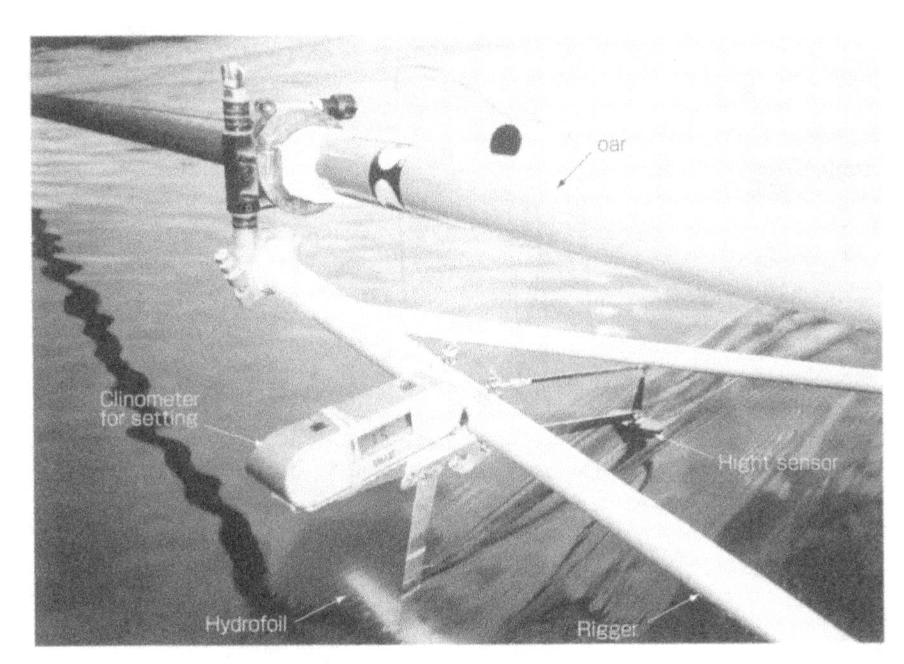

Fig - 12 Stabilizer for a rowing shell

When this side of boat becomes low, a height sensor activates to increases the angle of attack of the hydrofoil, and the lift works to raise this side.

The flexible joint at the mast base is used as - is. To keep the mast erect as in Fig - 9, the front edge of wishbone boom is fixed on top of the triangle frame. The bottom ends of the triangle frame are fixed to each hull to keep the mast erect. To enable the left and right hulls to twist freely without constraints, the bottom part of triangle frame could be made extendable. Or, the bending of the connecting beam may even absorb the twist, so the bottom part of the frame may not be required to be extendable. Both extendable and fixed frames should be practically tried out before taking a decision.

The main sheet connects the aft end of the wishbone to the stern rope. The stern rope tends to pull the left and right hulls at the stern toward each other, and the bottom triangle frame must take up the compression. As the sail is fitted set within the triangle frame, it restricts the sail angle.

I do not think this will hinder sailing at high speeds. If the bottom ends of the triangle frame are moved aft, the sail angle decreases; if moved forward, the space required for the skipper to go through during tacking and jibing becomes narrower. I think the position may be decided after a test run.

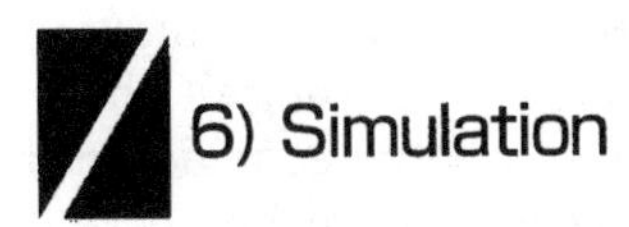

6) Simulation

When I design a boat that has no precedent, I always try to visualize the controls and the corresponding motions of the boat in my mind. Such a mental exercise enables me to find defects in operation and structure that might occur, and I proceed with the design after adopting measures to eliminate such defects. If I cannot find any solution, I wait for the next opportunity to solve the problem.

At the time of the final design of the boat, probably about two or three points may remain unsolved, but these are minor issues, and practically the entire design of the boat has been simulated.

For this project, as the motion of the boat is extremely complicated, I plan to build a scale model and run it. The size of the model will be one fifth the actual size, and the model will probably have an overall length of 90 cm.

The chord (length of hydrofoil section) of the hydrofoil at less than 2 cm is too small, and its Reynolds Number[1)] differs too much from that of the actual boat to accurately estimate the performance, but there should not be any problems in confirming its motion. Since I have tested a small hydrofoil stabilizing system for a rowing shell, I have a good idea on how to make a small hydrofoil with a sensor (Fig - 12).

It is not easy to make a soft sail for the scale model, so I will make an asymmetric sail, which has a fixed curvature for wind blowing from one side only, and is made of paper glued on a wooden frame, which is like the wing of a model airplane.

The hull is whittled down from a solid balsa wood and painted. Its course is kept by the front rudder, which is operated either by a wireless signal or a wind vane. The wind vane will be made of a feather and attached on top of one of the rudder shafts. I also have success-

1) Reynolds Number is a number that is proportional to the product of length, such as boat length or hydrofoil chord, and speed. The smaller this number, the greater is the influence of fluid viscosity. Generally, both speed and length of model are smaller than the actual boat.

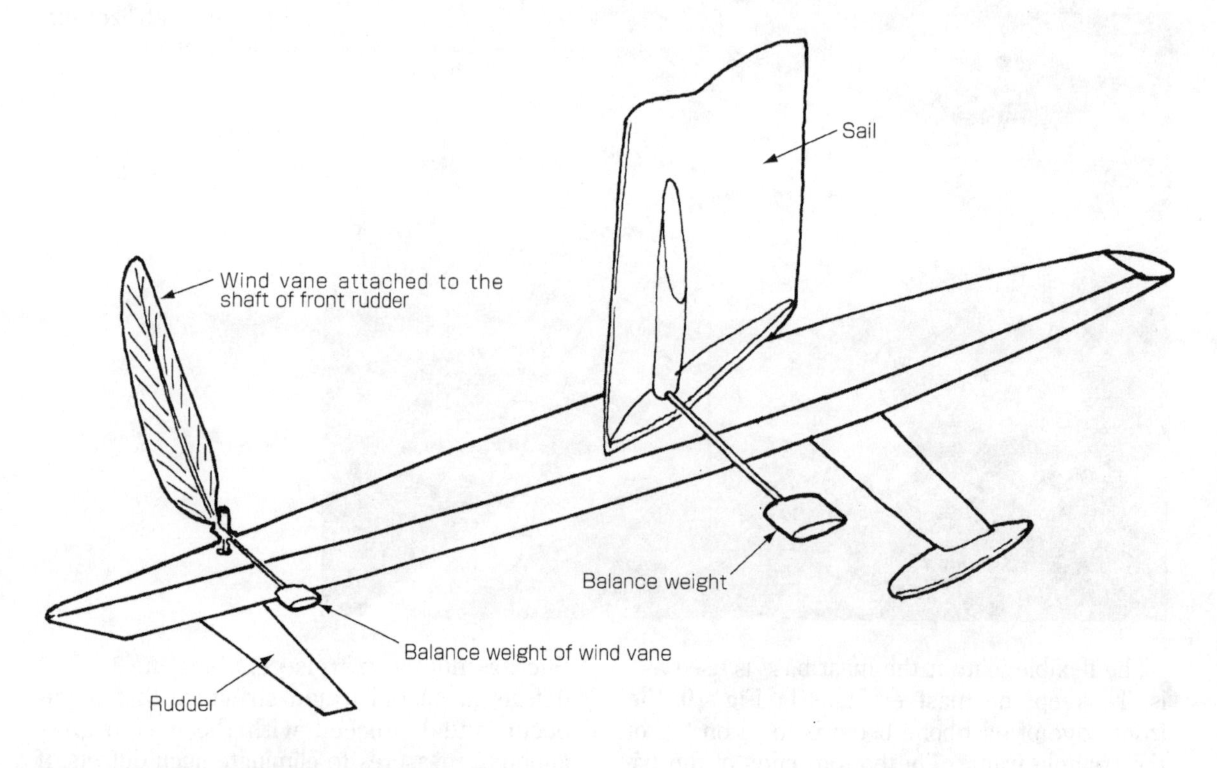

Fig - 13 Wind vane of model sailboat

Wind vane made of feather is attached to the front rudder shaft. If you align the bow in the direction of advance and the feather to windward, the sailboat runs keeping the course

fully tried out the wind vane on the model of a small sailboat (Fig - 13), so perhaps the wireless control may not be necessary.

As I wish to minimize the twist and run the model, I will set the weight corresponding to the skipper's weight on the windward hull. If I perform tests by the trial and error method adjusting various points, I should be able to find a condition that permits stable running of the model.

If the project is to be abandoned because of a basic defect, it is preferable that this happens before building the actual boat. To decide the limit of twist angle between the hulls will become a very important point, but better results are likely to be obtained by basing the design on the model test data obtained.

Four important issues in the simulation should be resolved in addition to the model test. If I do so smoothly, I can start building the actual boat.

The first issue is the take-off. Initially, I considered getting the boat foilborne in a gentle breeze, if possible. First, the skipper while accelerating the boat in an abeam wind, places his body weight aft so that the boat's bow rises. After the bow rises, he moves his weight forward such that the bow does not drop, and reduces the load on the main hydrofoil so that the stern rises and the boat is foilborne. During this stage, the skipper moves his weight to the side that has risen to equalize the loads of the left and right hydrofoils and to ensure that there is no twist between the left and right hulls. When the wind speed decreases during foilborne running, it is preferable to reduce the loads on the main hydrofoils and make them uniform in a similar manner.

The second issue is foilborne running. At high speeds, the boat will be very stable, and there is no need for the skipper to balance the loads. Since the boat twists to balance automatically, if the rudders and main sheets are fixed, the skipper may walk around the deck, or do whatever he likes.

However, it is desirable for the skipper to continue to move around, balance the loads, and minimize the torsion. In a 5 to 6 m/s wind, the skipper just takes position on the windward hull. If the wind speed increases, he has to hike out (to lean backward and over the side of the boat). By balancing loads using the body weight in this way, the foilborne condition can be maintained regardless of wind speed, and adequate margin for stability and foilborne running can be ensured. As mentioned earlier, in addition to balance in the lateral direction, the skipper has to move forward to reduce the load on the main hydrofoil if the wind drops, and maintain the foilborne condition until the next blow. He also takes a position toward the aft to prevent a bow-down attitude, which is the same for a normal sailboat also.

The third issue is tacking/jibing. Since the boat weight is small, it has little momentum and it may be difficult to tack. But if it is already running at a reasonable speed before tacking, it can turn by inertia. Even if it drops from the foilborne condition after tacking, the boat will complete its intended motion because of its inertia. In such a case, the boat should be accelerated by turning it to the abeam condition.

In this condition, the skipper has to move around the mast, which does not pose any problem. But while changing sides, the mainsheets and rudders should be fixed. The main sheets can be fixed in one action through a cam cleat (cleat that can fix or loosen a rope in one action according to the rope-pulling direction), but the fixing of the rudders needs to be considered henceforth.

Jibing must be possible in the foilborne condition, but if the boat speed has dropped, the moment the sail has filled out, the bow will go down into the water and the skipper may not be able to move to the other side from the front of the mast. Therefore, he must cross over from under or behind the triangle frame. Naturally, in this case also, the rudders and main sheets must be fixed.

The last issue is the launching/landing of the boat. Since the boat is designed for operation by one person, the launching/landing of the boat should preferably be done by one person. Since the main hydrofoils are positioned deep under the hull, this work becomes difficult. Therefore, it needs to be considered separately according to the condition at the waterside. The main hydrofoils are to be folded up toward the aft so as to have as shallow a draft as possible. The front hydrofoils are to kept at their original position because they cannot be folded up as easily as the main hydrofoils.

At slopes, such as in marinas, the boat is generally placed on a trolley and launched into the water. If the boat needs to be alongside a pier, a floating pier that has no obstacles below is necessary since the hydrofoils protrude outside the boat. Or, guard pipes should be provided on the boat. The guard pipes can be easily attached using the unused holes of the connecting beam. In the future, the front hydrofoils may be protected by using pitch pole prevention boards. In spite of these modifications, it may be difficult for one person to launch and pull up the boat using a trolley. Assistance from people at the marina or from close friends will be required.

Another case needs to be considered. Before launching the boat from a sandy beach, some preparations are necessary. If we arrange lightweight pneumatic tires when taking the boat down with the main hydrofoils pulled up, a trolley is not necessary. We can also erect the mast on the beach and launch the boat. After the boat floats on the water, we can lower the main hydrofoils and the tires will come up automatically.

When returning to the beach, it is easy to tilt up the hydrofoils and lower the wheels, so one person can pull up the boat if the conditions are good. The pneumatic tire system may be useful for keeping the boat at the marina too, as a trolley is not necessary.

7) Realization of the project

After thinking so much and arriving at this stage, I do want to make this project come true. I will be very happy if I succeed in adding a new enjoyment to the field of marine sports

Initially, I thought of building the boat in my backyard. This is not an impossible task. After simulation however, I realized that it would be very difficult to go ahead with the project by myself. Many trial runs are needed for tuning the hydrofoils, and I would also need an escort boat. If the test boat capsizes, we would need help.

If the model test goes well, I may team up with some company to build the boat. Since I cannot foresee whether the boat would sell well and contribute to the company, I cannot say anything at this moment.

Every time I visualize this boat running noiselessly in a fine breeze at a speed of over 20 knots, I become happy.

Nobody has considered the idea of a twisting catamaran, and thanks to this twist, the boat is stable and runs fast without the need for balancing it. I am also happy that the techniques used in human powered hydrofoil boats are contributing to this project. I am looking forward to the project with so much pleasure and anticipation that I am sure it will be realized some day in the near future.

3

SEVEN IDEAS FOR ROWING SHELLS

The history of rowing shells is one of tremendous effort and enthusiasm in winning races and maintaining pride, as these shells are the fastest human powered boats. Three years before the Atlanta Olympic Games (1996) I had the opportunity to study how to improve the speed of rowing shells and win the race.

With the cooperation of a team of proficient researchers, we used advanced Computational Fluid Dynamics (CFD), tank tests, and wind tunnel tests to study rowing shells. From hull shape and structure to air resistance, we studied and tested every item that might contribute to increasing their speed, and arrived at interesting results.

1) History of rowing shells

It has been reported that rowing boat races began in 1810 in England. Naturally, slender hull with small resistance is required if racing in a competition is the main object. In 1828, a boat having outriggers with outboard supports for oars appeared (Fig - 1). Until then, rowlocks were installed on one side on the gunwale and the crew positioned close to the other side to derive adequate inboard length for the oar from the rowlock. So, it was not possible to reduce the boat's beam. By adopting outriggers (hereafter call riggers), the beam of rowing boats was quickly reduced to half, which greatly contributed to the increase in boat speed.

In 1857, a boat with no bar keel (which had been attached outside the hull until then) appeared in a race. In 1870, sliding seats made their appearance in the USA. But initially, the range of movement of seats was inadequate and these seats were ineffective. It took 55 years for the fixed seat to be replaced by the sliding seat of the style used currently. The boat's beam continues to decrease, the hull form is becoming smoother, and the weight smaller.

Also, the oar has changed gradually from a bar - like oar to a wide, short blade. In recent years, it has become very light, with a shaft made of CFRP[1] and an asymmetric blade (Fig - 2).

The rules for races were also generated spontaneously. Initially, the sterns of all boats were clamped so that they were all in the same line when seen sideways at the start of the race. At the finishing line, the outcome of the race was judged when the bow crossed the line first. This led to the trend of building longer boats, and to prevent this, the present rule was framed by which the bows of boats are aligned at the start line, and the outcome is judged when the bows of boats cross the finish line.

There is also a fine anecdote of the cox diving into the water during the last spurt to make the boat weigh lesser after he ensured that the boat was on course, which makes for an interesting story.

It may be because of such historical reasons that the restrictions on boats and oars were very simple until the 1980s.

For example, an eight-oars shell must use 8 oars and if the weight of the cox is less than 50 kg, he must carry a ballast weight to make up the total weight to 50 kg, which were the only restrictions. There were no restrictions on boat and oar dimensions, structure, and materials. After going through a long history, the high - speed rowing boat has developed to a shape

1) Carbon Fiber Reinforced Plastic: Instead of glass fiber in FRP, carbon fiber is used. As CFRP is light and very rigid, it is used for airplanes, golf shafts, etc.

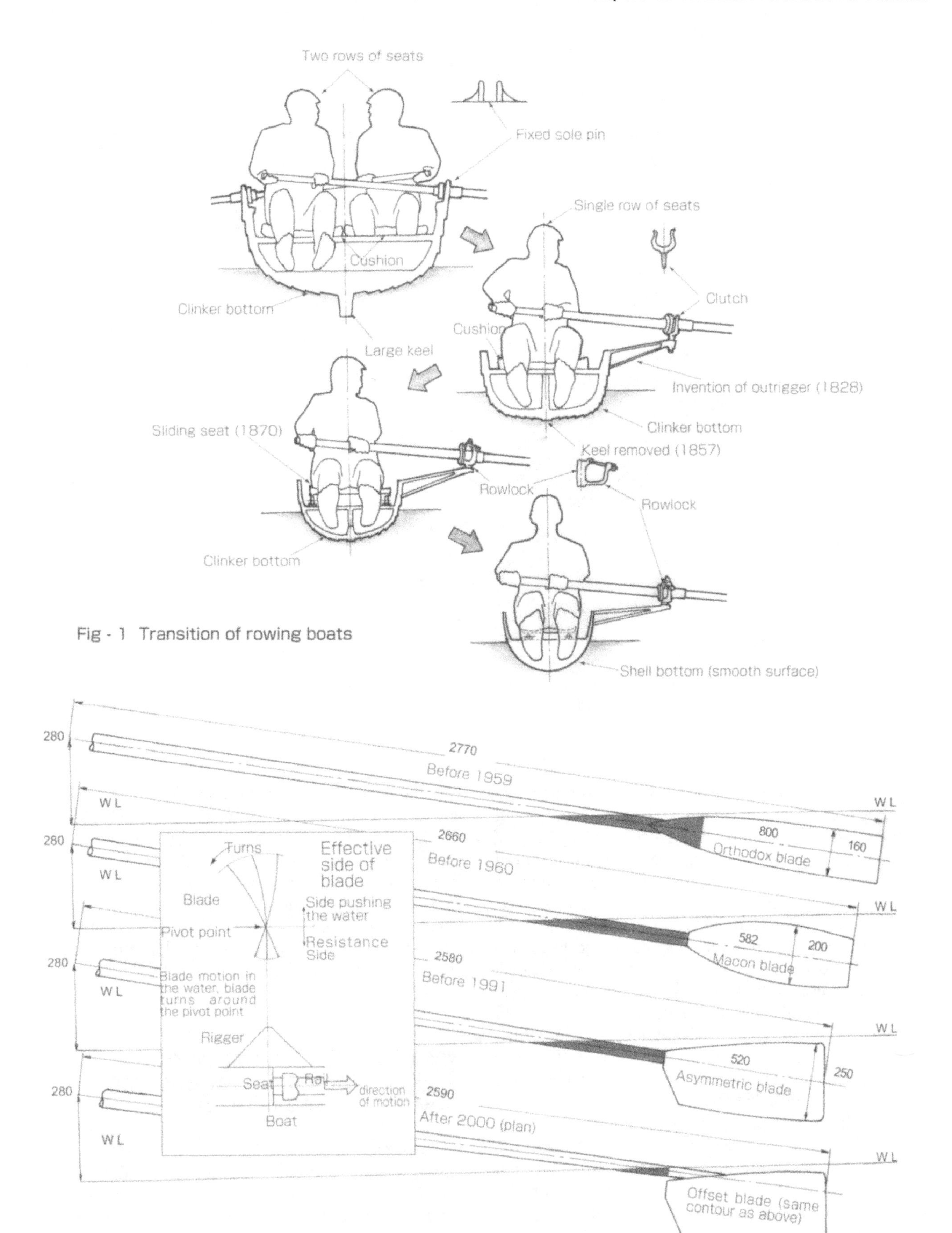

Fig - 1 Transition of rowing boats

Fig - 2 Transition of oars

The blades have become shorter and wider with time. The effective width of the asymmetric blade has become larger and its length has decreased, without any increase in the time (angle) for submerging the blade in the water. The black colored areas in the figure are parts that become resistance as they move forward in the water. The lowermost figure shows the oar in which this resistance is minimized.

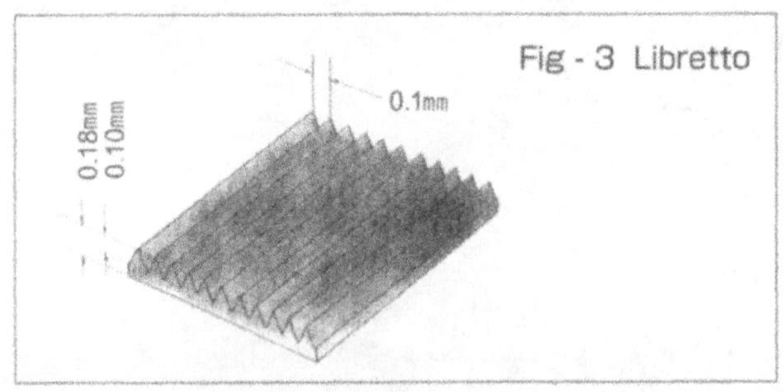

Fig - 3 Libretto

Fig - 4 Sliding rigger
On normal rowing boats, shoes are fixed on the boat and rower's body moves forward and aft, causing the boat to pitch and surge (fore and aft motion). Such motion leads to a 2 to 3% increase in boat resistance. The Sliding Rigger was developed to prevent such boat motion. FISA (international rowing federation) prohibited this system because of the increase in cost.

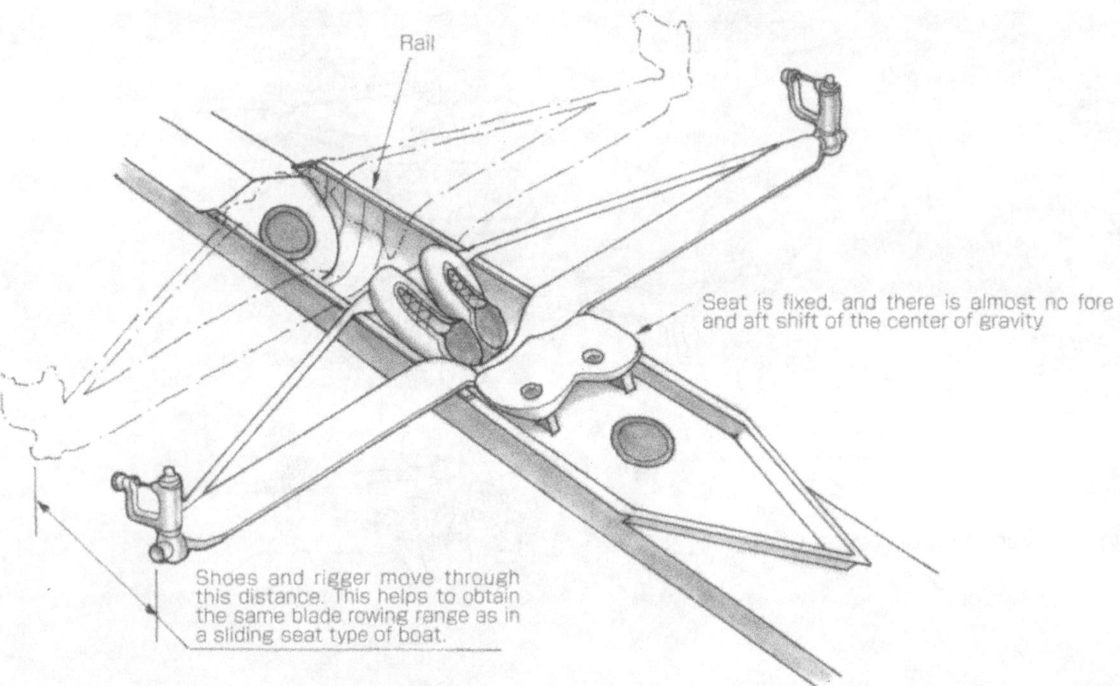

with the best balance between resistance, longitudinal and lateral stability, buoyancy, and rowing ease. Therefore, restrictions by way of rules may not really be necessary. I considered this as the true shape of the fastest human powered boat.

I was really fond of such boats. At some stage, I had designed and calculated the performance of every type of shell eight including hydrofoil and semi - submersible boats, and had come to the conclusion that faster shell eights would not appear in the market for a while, which made me revere the present style all the more (Chapter 3 of my previous book LOCUS OF A BOAT DESIGNER).

In recent years, the situation has changed. The rules have become complicated. Like sailboat races, attempts to reduce the frictional resistance by sticking Libretto (Fig-3) on the shell or pouring a polymer liquid along the hull surface, and so on, are prohibited. Also, a type of rigger called the Sliding Rigger (Fig-4) is prohibited. In the Sliding Rigger, the shoes and rigger form one unit, which slides on the rail, but the seat is fixed to the boat. As there was no movement of the crew's weight, single sculls with the Sliding Rigger very frequently won races. I expected considerable progress in this type of rigger, but I was disappointed.

Other rules included the thickness of oar blade edge to be over 3 to 5 mm for safety, depending on the boat type, and minimum boat weight prescribed according to the boat type.

2) R&D Committee of Rowing Equipments for Atlanta (RCRE)

I resigned in 1993 as Director of YAMAHA Motor Company and became a regular advisor. As I had some spare time, I thought I might work to strengthen the Japanese rowing team for the next Olympics. The Barcelona Olympics had just finished, but the ardor and enthusiasm of the games were still in the air. In Barcelona especially, the shape of the oar blade changed

2) Libretto: A film NASA developed that has fine longitudinal stripes and delays transition from laminar flow to turbulent flow to reduce the frictional resistance by several points. The US boat "America" won the America's Cup in 1986. This boat used film on its outer surface.

from symmetric to asymmetric (Fig-2) and with the arrival of CFRP as the boat material, a new era of developments in this field began.

As I had been engaged in America's Cup activities since the Nippon Challenge started in 1986, and I had assisted in the setting up the technical team, I was interested in competing with new technologies that were special to the America's Cup race. I had also developed close relationships with naval architects and professors, and with the assistance of these persons, I hoped to improve Japanese rowing boat technology so as to aim for better results at the next Olympics.

In the summer of 1993, I consulted with Mr. Matsui of Akishima Laboratory of Mitsui Shipbuilding Co., Ltd. and Professor Kinoshita of IIS (Institute of Industrial Science), University of Tokyo, to organize the R&D Committee of Rowing Equipment for Atlanta. This Committee consisted of about 15 engineers and researchers from various universities, research laboratories, and companies. The aim of the Committee was to improve the equipment in the rowing boat with the final aim of raising the Japanese flag at the winners' podium at the Atlanta Olympics. The hope of success was based on the following:

a) With light and rigid CFRP effectively used as the new material and with the arrival of new processing methods, very high performance boats but with high costs that had never been built earlier, came to be built.
b) I predicted new calculation methods and experimental techniques would open up new aspects based on my activities with the America's Cup technical team. I also anticipated assistance from the members of the said team.
c) I thought that we could get a medal if we came a few seconds faster in the two lightweight classes to be introduced for the first time at the Atlanta Olympics, since the rowing team had become strong in recent years. This was the target of the Japanese rowing association for a long time, and if this target was achieved, I expected it to trigger further progress and developments in Japanese rowing activities.

3) Activities and overview of the Committee

Two classes were chosen for the target: lightweight coxless four and lightweight double scull.

The present rowing boat has been finished to a fairly high degree of perfection. Therefore, we should not aim for radical improvements in specific areas but use conventional methods to improve hull lines, in addition to reduction in air resistance, improved ease of rowing, and ten other items, which seemed to be advantageous. We began to study these items.

As improvement of hull form was the top priority, we put all our efforts in it. We can change the hull form as we like; no other country can object to it. An excellent hull form would also assist domestic boat builders who suffered on account of imported boats.

Committee members included specialists in hull form and aerodynamics. Japanese tank tests are highly accurate and reliable. The level of Computational Fluid Dynamics is also said to be high.

We borrowed three coxless fours and three double sculls from domestic owners. These boats were the best boats in the world at that time. We transported them to the Ship Research Institute (currently the National Marine Research Institute), Ministry of Transport (currently the Ministry of Land, Infrastructure and Transport). With the assistance of Yamaguchi, the Head of Tank Testing Department, we conducted resistance measurements to assess the latest performance of the boats and to prepare the basic data for the project.

Kouichi Matsumiya, at that time, was a student in the Kinoshita Laboratory, and he selected this project as his master's thesis and worked actively on it. From the analysis of the test results, we understood the approach to take to improve the hull lines. We made full scale models of the coxless four and double scull to measure their resistance.

Unfortunately, the resistance did not reduce as much as we expected. It was time to build the new boats but we could not come up with ideas to reduce resistance of the hull adequately using the conventional method. For the coxless four, we thought of an idea to reduce the resistance by about 0.1 kgf out of the total resistance of 23 kgf. But the reduction was almost the same as the measurement error, and it was difficult to confirm the effectiveness. The design of rowing boat is accomplished by delicately balancing resistance and stability. If you make the beam narrower, the resistance is reduced but the boat rolls making it difficult for the crew to row with their full power. However, if the resistance can be reduced ade-

Fig - 5 Stabilizer
This is a hydrofoil system that maintains the lateral stability of the rowing boat. The right side black planing plate in the photo always planes on the water surface. If boat leans on one side, the planing plate rises relative to the rigger, and this motion increases the angle of attack of the hydrofoil to generate a lift and to restore the lowered side. In the photo, the instrument on the left side held in position by rubber bands is a digital clinometer, which is removed after the system is installed. The weight of the system was 250 g and the resistance was 0.1 kgf at 5.5 m/s speed.

quately by reducing the beam, we could think of new devices that would increase the transverse stability in the range of the resistance reduction.

With this consideration, I used the hydrofoil system called a stabilizer, which had given good results at the human powered boat race, and arranged it at the far end of the riggers to retain good stability (Fig - 5). I installed the stabilizer on my scull to confirm its effects and also measured its resistance in the water tank to prepare for its intended use.

However, even after installing the system, the boat resistance did not reduce as we anticipated. As half of the reduced resistance was accounted for by the additional resistance due to the stabilizer system, we did not get any noticeable improvement.

After struggling for two years with the hull form, we did not find a good solution and ran out of time. In view of such circumstances, we decided to give up building a new boat because there was little improvement in hull resistance, and not sufficient time to confirm the reliability of the structure. Regretfully, we had to give up the idea of building a new boat .

On the other hand, we got some interesting results for oar improvements. We improved the effect of the oar by offsetting the blade on the shaft and decreased the resistance area (Fig - 2 bottom).

Professor Doi of Hiroshima University measured the effect of this type of oar with a precise instrument designed by Mr. Ikeda, the Managing Director of Electronics Industry Limited, and found the oar to be 3% more effective than the conventional oar. The result was so encouraging that we made the oars for the coxless four and for the scull for rowing trials (Fig-6), but the crew required some time to become accustomed to the new oars. There

Fig - 6 Offset blade
The photo shows a sculler rowing with offset blades. The center of the pressure of the offset blade is below the shaft. The shaft is bent to an S - shape near the grip so as to balance the torsional moment of the offset blade. Measurement by Hiroshima University showed a 3% increase in efficiency, that is an advantage of 20 m in a 2000 m course, but on the other hand, further technical improvements were required to ensure torsional rigidity.
(The sculler is Sahara of Yamaha Motor Co., Ltd.)

was no fundamental problem with the oars, but it still needed something more - some small improvement that we were not sure of - to satisfy the crew. However, we did not get the time to make that small improvement. Unfortunately, the time was up before we could resolve the problem and satisfy the crew's feelings, and we had to give up the research. We concentrated on reducing the air resistance during the last stage. It was October 1995, nine months before the Olympics.

4) Details of the resistance and plans for their reduction

Fig - 7 shows the breakdown of resistance for lightweight coxless four and lightweight double scull, which were the targets of this research. The same figure also shows the methods that we developed to reduce the resistance, and the estimated advantages (how many meters ahead).

Looking at the table (Fig - 7), total air resistance accounts for about 12% of the total resistance. This percentage is not very large. However, although we had thoroughly studied water resistance and there was little scope of reducing it, we had not made any efforts to reduce the air resistance.

For example, a 3% resistance reduction corresponds to a 1% speed increase, that is, this reduction in resistance gives a 20 m advantage over a 2000 m course, which is substantial; thus, reducing the air resistance should give a beneficial effect.

I visited the Olympic course in Atlanta in the summer of 1995 and found considerable head winds at the course, which suggested that reduction of air resistance would be effective.

In the breakdown of air resistance, oars come first accounting for 6%, while crew comes second accounting for 3%, then riggers at 2%. To reduce each kind of resistance, we developed oar fairings, body fairings, wing riggers and rigger fairings, and then tried to reduce the water resistance of individual parts such as rudder and fin. We went to the Olympics with all the improvements that we made.

These are explained in detail below.

5) Oar fairing (streamlined cover over the oar shaft)

Let us consider the oar in two parts. One is

Men's lightweight coxless four

items | reduction method | reduced amount

kgf
23
22
21
20
19
0
air resistance(12.2%)
water resistance(87.8%)
resistance(kgf)
crew
0.90kgf
body fairing (BF)
reduced amount of 0.67kgf is corresponds to 17.8m advantage in 2000m course
oar
0.77kgf
oar fairing(OF)
1.45kgf
1.04kgf
rigger 0.39kgf
rigger fairing(RF)
0.27kgf
increased resistance by moving 0.59kgf
hull 0.14kgf
fin, rudder 0.36kgf
same as left
hull 19.77kgf

Men's lightweight coxless four
Total weight of the boat is 340.4 kg. Each resistance is at a boat speed of 5.5m/s. Total resistance is 23.62 kgf. The water resistance is at water temperature of 25℃.

Men's lightweight double scull

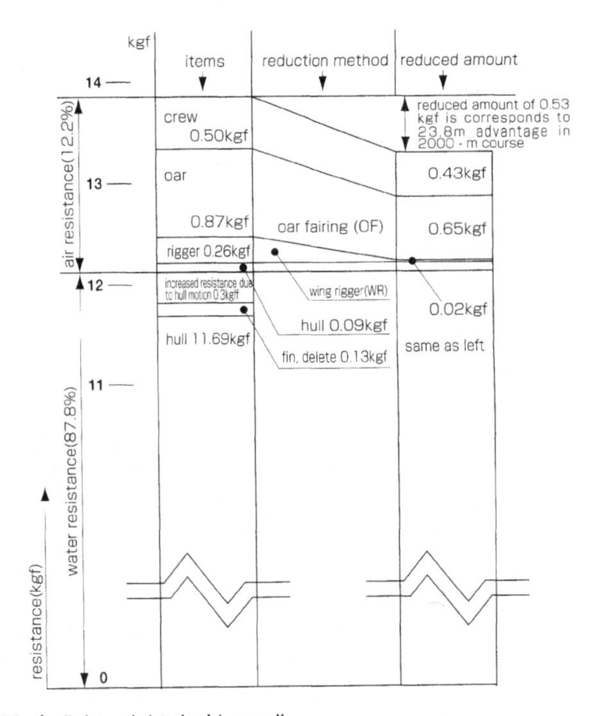

Men's lightweight double scull
Total weight of the boat is 172.6 kg. Each resistance is at boat speed of 5.25 m/s. Total resistance is 13.85 kgf. The water resistance is at water temperature of 25℃.

Fig - 7 Breakdown of resistance and its reduction

Fig - 8 Oar fairing tried out first
Oar fairing of wing section made of 2 to 3 mm thick Styrofoam covered by thin polyester film. It is made of two separate pieces (inboard and outboard) to facilitate its manufacture and assembly.

the shaft, which is a round bar, and the other is the blade, which is a flat plate. The blade pushes the water when in the vertical position, is brought back after pushing the water, and turned to the horizontal position to minimize air drag. That means most of its air drag is due to the shaft. As the shaft is made of round bar, fitting a small streamlined fairing over the shaft should reduce the air resistance.

If the diameter of the bar is small and the speed is low, the Reynolds Number is small, and a small fairing has a very small effect. To reduce the air resistance to one fifth or more, the streamlined cover should have a section in which the length is four times the thickness (Fig - 8).

In case of the scull, we needed to fair its 42 mm diameter shaft. The fairing has a streamlined section 5 cm thick and 20 cm in length. The fairing should cover a 1.5 m length of the shaft. If the oar becomes heavy because of the fairing, the sculler will soon get tired and it will become difficult to get a high pitch at the start or at the starting dash.

Since I had doubts, I consulted Naoto Suzuki, a representative of the YAMAHA team for human powered airplanes. He immediately suggested cutting out a 2 to 3 mm thick streamlined outer shell from a large Styrofoam block using electrically heated nichrome wire, covering the outer shell with one - micron film, and assembling it. The complete oar fairing had adequate rigidity and weighed 100 g per oar, which was as planned (Fig - 8). We could not think of any other structure because the weight of the shell was only 100 g/m^2.

The next stage was to make and fit a mechanism to keep the fairing horizontal even when the shaft was rotated. If such a mechanism were not made, the air resistance would increase as opposed to our aim of reducing the resistance. I fitted the mechanism that I had made myself to my scull and made improvements by trial and error, by rowing, modifying, and improving it. Finally, after trying out two systems that did not work, I successfully made one that did and connected the fairing and the rowlock (Fig - 9, Fig - 12).

Since a clearance of several mm remained between the oar and the rowlock, I attached a long lever to both the oar fairing and rowlock, and connected the top parts of both non - rigidly with Velcro tapes, by which I could keep the oar fairing horizontal. The mechanism was simple and very practical.

During the summer of 1994, I rowed the scull with the oar fairing every morning (Fig - 10). Initially the fairing broke many times but I repaired, improved, and modified it so that it worked. The Styrofoam structure was stronger than it looked and even if it broke, I could repair it easily using a special adhesive,

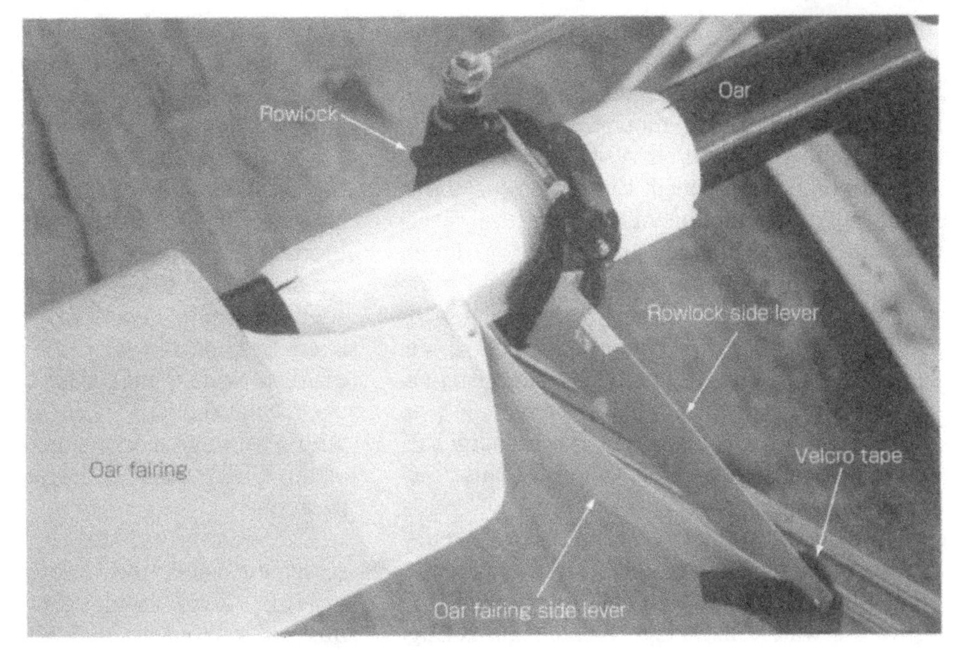

Fig - 9 Mechanism to keep the oar fairing horizontal
Rowlock and oar fairing are connected non - rigidly at the end of long aluminum levers. For the connection, a Velcro tape is used, which enables the oar to be taken out easily.

Fig - 10 Author conducting trials with the oar and body fairings
The first body fairing slides on three rails attached to the boat. It looks large but weighs only 250 g. I rowed every morning, and made improvements by trial and error.

so I used only one set of oar fairings throughout the summer season.

Unexpected gain

In the spring of 1995, we performed the wind tunnel test. The wind tunnel had a turning table that could be used to change the direction of wind. We placed a round bar that had the same shape as the oar shaft having a diameter of 42 mm and a length (height) of 740 mm. The air resistance of the shaft alone at a 30 km/h wind speed was 0.15 kgf.

The resistance of the round bar covered with the fairing was only 0.03 kgf, indicating a large reduction. As seen in Fig-11, 0.03 kgf is the resistance when the wind comes from straight ahead on to the streamlined fairing and the air resistance decreases when the wind direction changes by a few degrees. When the wind direction exceeds 3 degrees, the resistance in

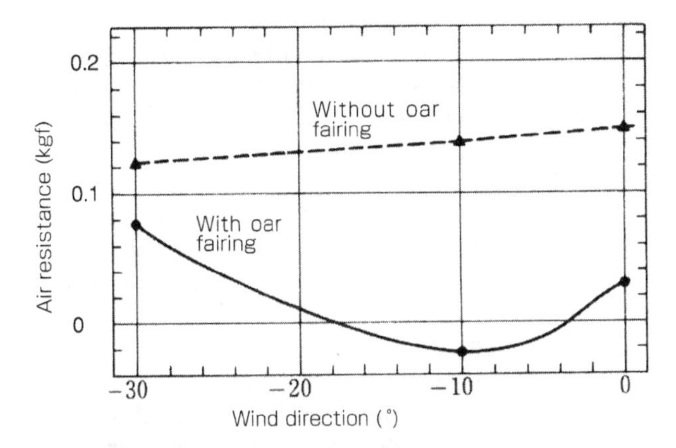

Fig - 11 Effect of oar fairing
The graph above shows the air resistance when a part of the oar (74 cm) was exposed to a 30 km/h wind, which is the average wind speed, and corresponds to the stage when the oar swings back at 10km/h on a boat running at 20 km/h speed. The resistance without oar fairing was 0.15 kgf. With the oar fairing, it reduced to 0.03 kgf, that is, by more than one-fifth the original resistance. Furthermore, when the oar is moved up or down during the forward stroke so that the angle of attack becomes 10 degrees, a thrust of about 0.025 kgf is generated.

the direction of the axis of the section becomes negative.
For example, if you move the oar at a constant height during forward motion, the oar fairing remains horizontal and air resistance is 0.03 kgf at a 30 km/h wind speed. But if there is some slight up/down motion of the oar, the resistance decreases further, and if the up/down motion exceeds 3 degrees, the resistance becomes negative. When the angle reaches 10, a thrust of 0.025 kgf is generated! That is, we can expect considerable reduction in resistance by fitting the oar fairing, and moreover if you add an up/down motion when returning the oar, a thrust is generated, which means you are rowing the air.

The structure, however, is not very strong, and if the oar is plunged deep into the water, it may break, making it difficult to pull the oar out of the water. Therefore, to keep the fairing out of the water while in motion, I had to restrict its length, and I pondered for a long time before deciding the length.

Since January 1996, I requested our national team to make trial runs many times, and the length of the oar fairing became shorter after each trial run. In rough conditions, during high pitch rowing, starting dash, and combinations of these, the blade sometimes became submerged deeper in the water, therefore, the length of the fairing became shorter after every trial.

One day in February when I asked the crew their general opinions, they told me that they disliked the color (sky blue) of the fairing. Initially, I considered coloring the Styrofoam or the film but later I felt that they would not be satisfied by such qualitative impressions. So, I individually collected opinions from several crewmembers and found that many of them liked the feel of CFRP, which was used as the material of the oar shaft.

I had considered using CFRP at the initial stage of this project and had given up because the weight was not light enough. But when I heard these opinions, I felt that the concern of the crew over the fragility of Styrofoam had translated into dislike of the color of the material (Styrofoam color, sky blue). Moreover, their opinion that they did not care about the increase in weight due to CFRP used probably encouraged me to reconsider the use of CFRP.

From that time onward, I consulted and learnt from Mr. Atsuo Terada of Iwata City, an expert in model airplanes, who had represented Japan in the world championship meet. Even the sandwich construction used in the wings of model airplanes was too heavy, too rigid and fragile. Fragile structures tend to break easily. Mr. Terada suggested ideas one after another.

For the oar fairing, there is an oar shaft at the position of spar (main beam or backbone) and the oar shaft supports the main force, therefore, we need not worry about bending stresses. Also, since we do not require lift, the section profile need not be accurate. So we need to consider only the following: Can we make it with a thin shell of CFRP (not sandwich construction)? Can we reinforce rigidity with a thin Styrofoam sheet glued on the inside of the CFRP shell? I decided to go ahead and make it.

The date was March 10, and 130 days remained before the Olympics. I began to look urgently for a mold shop but they were all busy. I started to make a wooden plug by myself, but due to my unfamiliarity with the work, it took longer than expected, and the female mold was ready only after April 20.

Developments in manufacturing techniques

After overcoming many difficulties, I made one oar fairing and handed it over to the national crew training at a river in Gifu Prefecture so they could try it out. They liked the color of CFRP but I did not like the wavy surface. After it was fitted to on the oar, I found that it had a slight twist, which was not good. After the trials, the crew got accustomed to the fairing, but found the oar a little heavy. Since the first prototype weighed 360 g, which was twice the planned weight, it was natural that the oar felt heavier (Fig - 12).
From this point onward, we began to build up on our manufacturing techniques. In our prototype shop (at YAMAHA Motor), there was a person named Hattori, who was a skilled worker in all fields from metal work to CFRP work. Thanks to the assistance and advice from Mr. Terada and Hattori, even I, unfamiliar with this sort of work, made rapid improvements in the manufacturing method.

We made steady progress by making jigs, changing materials, improving the structure, and changing manufacturing processes, through trial and error. Hiroshi Kobayashi, a Tokyo University student, who assisted us since April, picked up skills, and worked together with me. During the Golden Week holidays in May, we continued to manufacture the fairings for the coxless four, and at the same time, we also worked on the wooden

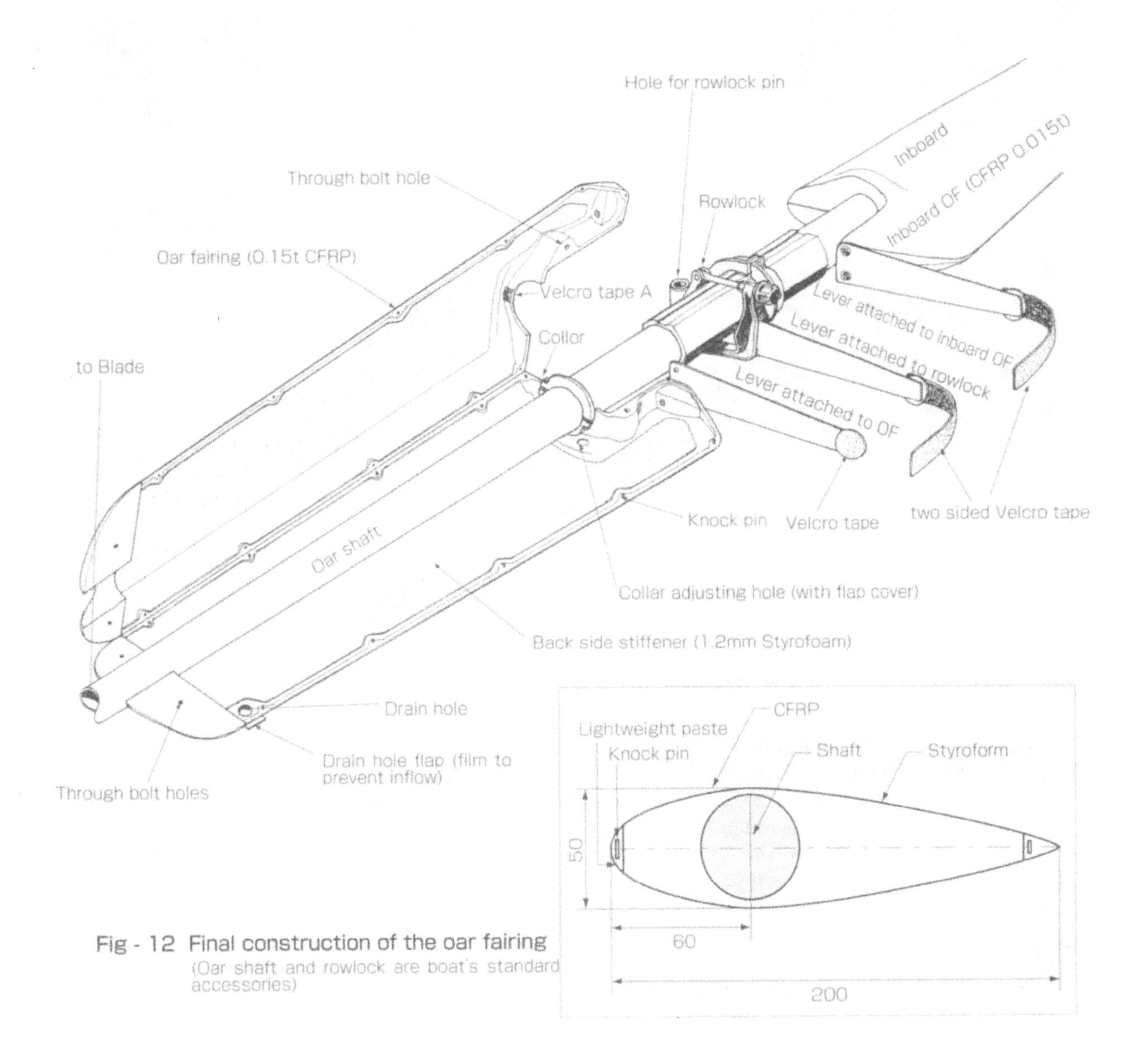

Fig - 12 Final construction of the oar fairing
(Oar shaft and rowlock are boat's standard accessories)

plug of the oar fairing for the double scull.

The National Rowing Championship Regatta took place from June 6 to 10 in 1996. This was also the race in which members of the Olympic team representing Asia would be decided. After this race, Japan won the right to field representatives in the following events: the men's lightweight coxless four and men's lightweight double scull, women's lightweight double scull, men's heavyweight pair and men's single scull.

We were delighted with the results because we had anticipated selection in only two or three events, but it was not easy to prepare new devices for all five categories. We did our best, and prepared oar fairings for all categories before the departure of the crew on July 26.

After manufacturing the parts, Kobayashi and I followed the crew to the Olympics. Since the Olympic courses would open only a week before the race, the crew worked out at a lake near a small town named Elberton, 150 km east of Atlanta. We occupied a corner of the boathouse, used it as a workshop, and continued to work to attach the oar fairings to the oars or detach them for maintenance, while making adjustments to satisfy the crew.

Finally, only the men's coxless four used the fairings, which were also cut several times and shortened at Atlanta. At last, they became so short that they were even attached inside the

Fig - 13 Final oar fairing finally they were attached on both side of the rowlock

rowlocks at the crew's suggestion, and in that state, they were used in the race. The length of the oar fairing in the race was about 1.2 m including the part inside the rowlock, and the calculated advantage after rowing 2000 m was about 2 seconds, or about 10 m (Fig - 13).

6) Body fairing (Streamlined cover attached on the bowman's back)

The air resistance of the crew's body is probably large. Especially, the wind pressure that the bowman feels on his back is said to be rather high. What would be the air resistance including the side force component? By how much does the air resistance change if a streamlined cover is used? For these questions, we had no data at all. We planned to estimate the body fairing weight and develop a body fairing mechanism that moved with the bowman's upper part of the body, before the end of 1994. From 1995 onward, our plan was to find effective shapes by wind tunnel tests.

The initial body fairing was large with a length of 70 cm, height of 45 cm, and shell weight of 250 g. We mounted it on three rails fitted on the boat, and connected it to the sliding seat so that it moved forward and aft with the movement of the bowman's body.

There were many restrictions for carrying out tests on the boat itself, so I firstly assembled frames around my rowing machine in the corridor of my house, and installed rails on which I set the body fairing. I rowed sitting on the machine actually and captured the motion on a video camera, and then watched the movie with the intention of making improvements and confirming them myself before installing the fairing on the scull. Since adequate tests were carried out on the rowing machine, no problems in particular occurred after installation on the actual boat. Both oar and body fairing were installed on the boat, and all tests were completed by the end of 1994 (Fig - 10).

Confirming the effects

From the beginning of the next year, wind tunnel tests began. Mr. Matsui of Akijima laboratory, Mitsui Zosen, as the leader, associate professor Yasuaki Kohama, student Hidemi Oguri of Tohoku University, and members of the R&D Center, Mazda Motor, in Yokohama, worked hard to progress smoothly with the measurements. I made 1: 4 scale models of the boat and the crew, and then prepared six types of model body fairings and one full - sized body fairing. If the wind speed is 4 times (for four, 5.5m/s x 4 = 22 m/s) the speed of the actual boat, the resistance would also be same

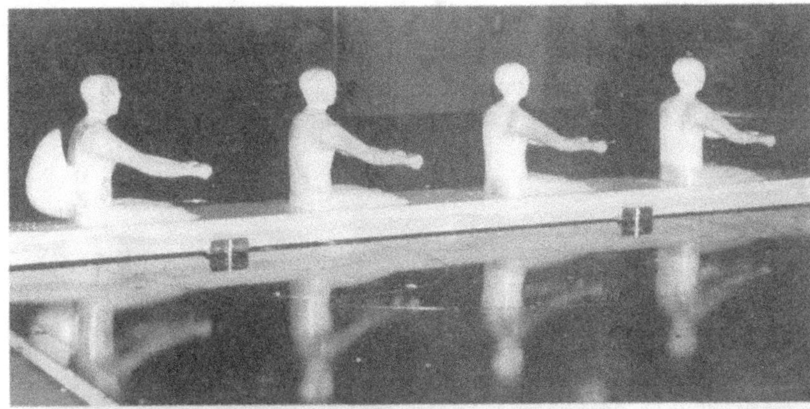

Fig - 14 Scale model of wind tunnel test

This is a one - fourth model of coxless four. Tests with one, two persons can be conducted by taking off the remaining lifelike models. The body fairing is attached only to the bowman. The resistance in oblique winds can be easily measured as the platform can be rotated.

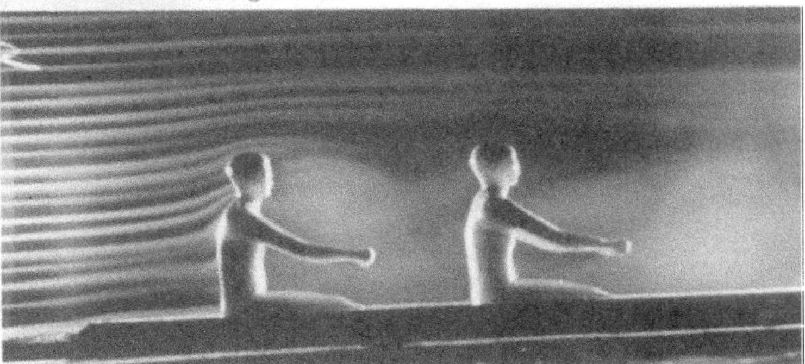

Fig - 15 Body fairing and air flow

It can be observed clearly that the airflow does not strike the second crewmember when the body fairing is used.

as that of the actual boat, which is easily understandable.

During the experiments, the actual value of resistance of the body in wind from various directions, the rolling moment (the force causing heel) and the yawing moment (the force causing the bow to swing to the side) when 1, 2 and 4 crew sat in the boat became clear for the first time. We found that the body fairing was effective in all cases. It was interesting that even the smallest body fairing gave the same effective result as the initially planned size, which indicated the feasibility of using smaller body fairings.

However, the body fairing was considered an integral part of the design of the boat hull, and in those days, since a good hull form could not be found, construction of the new boat was abandoned, and development of body fairings was suspended, which was depressing since the effect of the fairing was significant. The project was suspended for more than half a year.

From the hull to the seat

During January and February 1996, we struggled with the oar fairing, rudder, and wing rigger (mentioned later). Considering the effectiveness of the body fairing, it was sad that the project was abandoned. I thought that body fairing could be put to use without building a boat. I considered a body fairing separate from the boat, which also makes the fairing mechanically easier to produce. The question was its effectiveness. To confirm the effectiveness, we had to conduct wind tunnel tests again. I requested various concerned persons to perform the tests, and this opportunity presented itself in April (Fig - 14, 15).

I chose a body fairing that had the least resistance and smaller size in the previous experiment, then rounded off the lower half of the fairing, and took this fairing independent of the deck as the base model. We prepared three sizes of this model-large (same size as base model), medium, and small.

Downsizing and weight reduction

The experiments showed that the effect did not change appreciably from the one connected to the boat. To my delight, there was no difference in effect among large, medium and small body fairings (Fig - 16). Compared to the initial body fairing, the size difference was large - like parent and child. As the results were so good, I wanted to make the body fairing immediately. I soon had a mould made and manufactured the body fairing from CFRP. Its size was about the same as a helmet and its weight was less than 100 g.

Since the body fairing became smaller, I eliminated the rails on the deck side and modified

Fig - 16 Results of wind tunnel test for body fairings

The graph shows the resistance of crew and boat at a speed of 30 km/h, which is assumed to consist of 20 km/h (about 5.5 m/s) of boat speed, speed at which the crew moves forward, and about 1 m/s of head wind speed. The shaded areas show the effects of body fairings, seen to be effective for all cases. The dotted lines show the case when the body fairing was inserted under the crew's wear.

the body fairing so that it could be attached to the crew's seat directly, which could swing fore and aft according to the motion of the bowman's back. In addition, to harmonize the motion with the bowman's upper body, the upper part of the body fairing was held close to the bowman's body by a rubber string, and a roller was attached at the area where the fairing touched the bowman's back so as to slide smoothly.

After I arrived in Atlanta, I found the situation had changed. The crew of the lightweight four, who had been eagerly testing the fairing suggested that the roller touching the back be replaced by an easy-to-slide flat, and instead of rubber string holding the fairing to the bowman's body, the body fairing should be pressed gently on the back with a spring. So I attached a plate glued with an easy-to-slide film inside the body fairing, used a spring, and completed the desired structure (Fig - 17a, b).

Later, some small modifications were made and the final body fairing with these specifications was ready for the coxless four race.

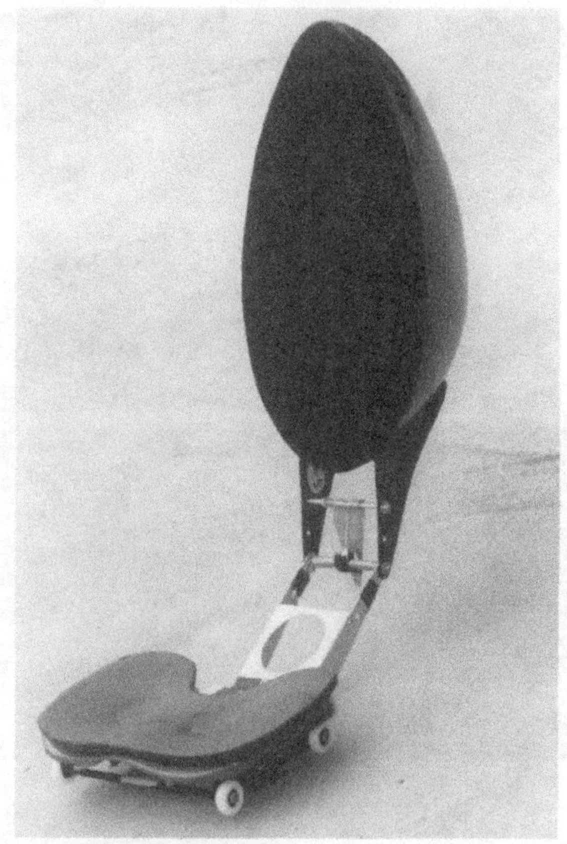

Fig - 17a Body fairing

Upper part of body fairing presses gently on crew's back by two springs encircling the hinge shaft. A cover plate is fitted to prevent the springs from coming off by accident.

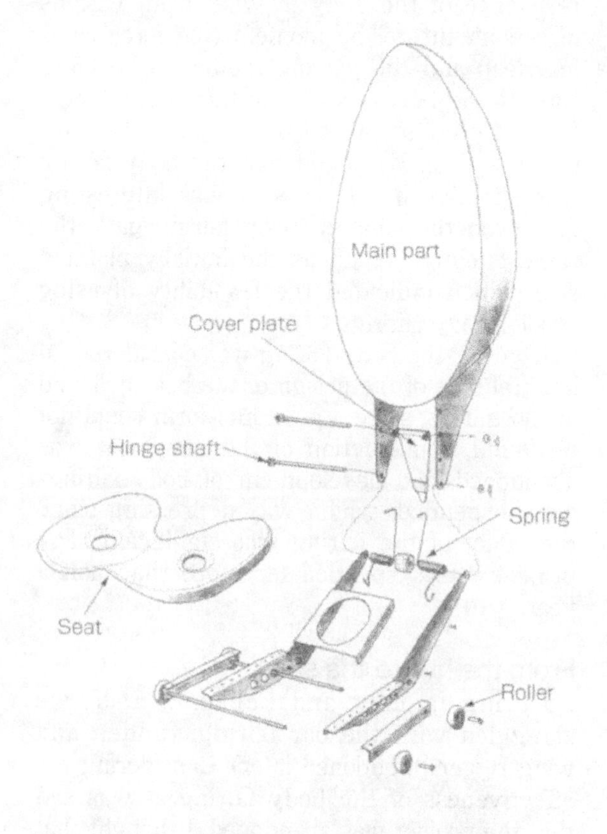

Fig - 17b Exploded view of body fairing

Fig - 17c Wing rigger & body fairing
The wing riggers were used in the women's lightweight double scull but not used in the men's lightweight double scull, because their adjustments could not be completed in time, which was a pity.

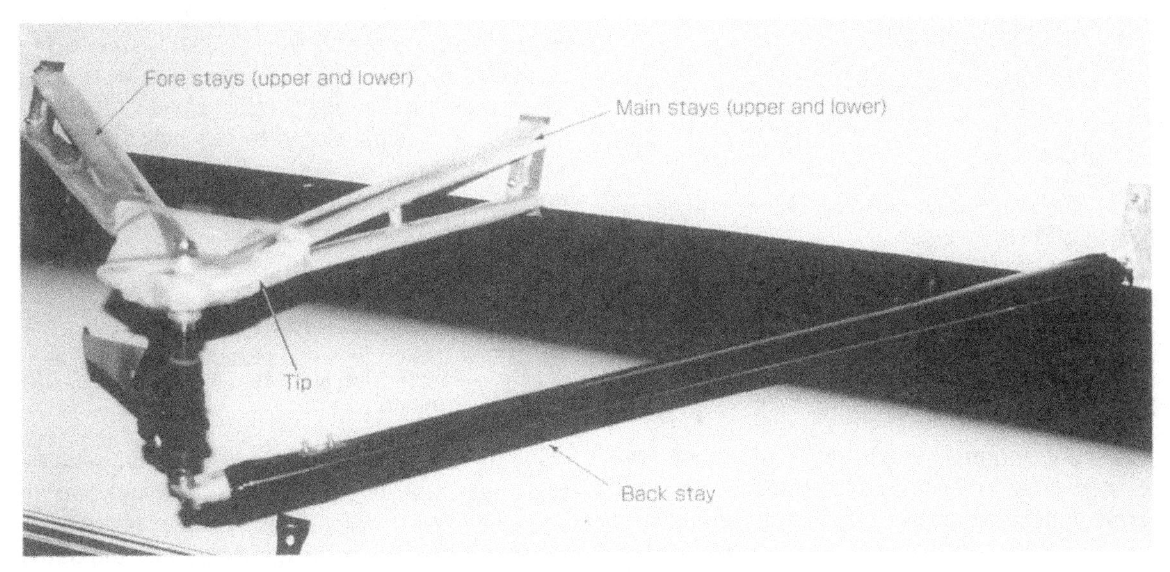

Fig - 18 Rigger fairing
The rigger fairing could reduce air resistance by 0.124 kgf, but we could not use it because of lack of time.

During the preliminary race, close-ups of the body fairing that we had made were shown on television and attracted considerable attention.

7) Wing rigger and rigger fairing (Measures to reduce the rig ger resistance)

Rowing shells have riggers that support rowlocks and extend outboard on both sides so that the pivots of oars can be located adequately outboard (Fig-1). To reduce the air resistance of this part, which was about 2%, we developed a wing rigger (Fig-17c).

This is not a new idea, and we can see some in foreign countries, which are mainly used in touring sculls. We do not think they made the wing riggers with the aim of reducing the air resistance after analyzing the results of wind tunnel tests.

We asked GH craft at Gotenba City to manufacture the wing rigger. GH craft is a well - known engineering company that manufactures CFRP structures used in racing car bodies and aircraft parts. The same company also manufactured the offset blade shown in Fig - 2.

The beautiful finish of the wing rigger obtained after making good use of Computer Aided Design (CAD), Finite Element Method (FEM) and Computer Aided Manufacturing (CAM) techniques, had a favorable impression on the crew. It was difficult, however, to fix the wing rigger using fittings on a boat that was originally not designed for wing riggers.

We made wing riggers only for the scull. Although we planned to manufacture wing riggers for the coxless four also, we could not do so because of lack of time. Therefore, for the coxless four, we made a fairing attachment for

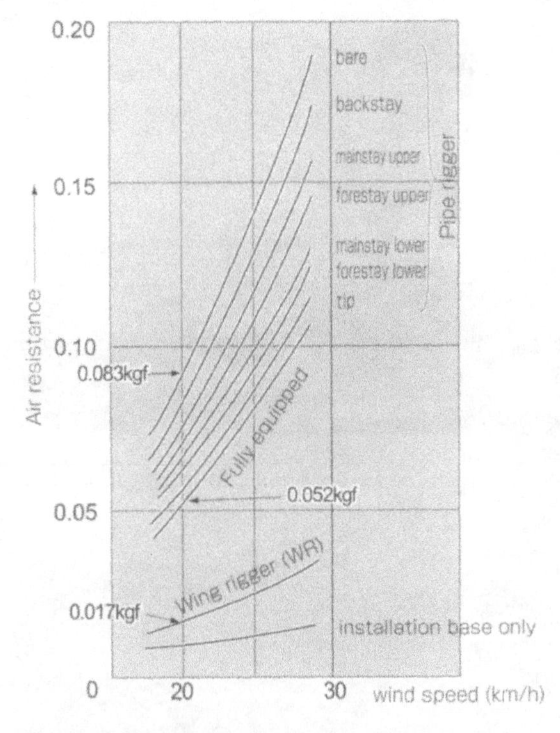

Fig - 19 Wind tunnel test results for rigger fairing and wing rigger

The figure shows data that we measured first with a set of normal rigger (bare) fitted with all fairings (fully equipped), and took off fairings one by one to measure each resistance. You can see that the resistance of a normal rigger in a 20 km/h wind speed was 0.083 kgf; after fairing it decreased to 0.052 kgf; with wing rigger it decreased to 0.017 kgf, which is one fifth the original resistance (all data include the resistance of installed base).

each conventional rigger pipe to reduce its resistance as far as possible, as desired by the crew, which became the rigger fairing (Fig-17c, Fig-18).

Fig-19 shows the results of wind tunnel tests in 1995. The resistance of the wing rigger is extremely small. Women's lightweight double scull used the wing riggers during the race, but men's lightweight double scull did not because there was no time for adjustments. Riggers look like simple frames that support oars. However, it is not easy to make the rigger rigid enough to enable free movement of oar, allow no deformation to occur due to large forces from every direction, and also maintain correctly the angle of blade in the water.

Considering these points, wing riggers are advantageous because of their large sectional areas. The resistance of the wing rigger is extremely small and as it is positioned high, waves do not strike it as in a normal rigger even in rough conditions, therefore, the increase in resistance is small. In addition, since the wing rigger contributes to the rigidity of the boat, enables the structure to be made simple, and reduces costs, it is likely to come into the mainstream in racing boats of the future.

On the other hand, a rigger fairing that has 40% less resistance is also advantageous. But at Atlanta, we did not get the opportunity to install the rigger fairing because the tests for oar fairings and body fairings had priority. Since the rigger fairing need only be fixed to the rigger, we thought we could install it anytime. The rigger fairing has not been used because of the coaching staff's policy to not use devices that had no prior results.

The rigger fairing is a device that reduces the resistance of the existing rigger. Without modifying the mechanical characteristics such as rigidity of the rigger, the rigger fairing reduces 40% of the air resistance with a little increase in weight (100 g in case of four).

Fixing the rigger fairing on each pipe requires very patient work. Also, it has to be handled carefully when taking it off the rigger. The procedure we used was to push a pair of molded Styrofoam pieces to the pipe from forward and back, and lap over the outside with 5 micron polyester film. A simpler handling procedure may be preferable.

8) Rudder neutralizer

(Device to keep the rudder neutral and prevent steering to avoid excessive resistance)

There is no cox on the coxless four or coxless pair, but these boats have rudders and one of the crew has to steer the rudder by moving the toe to the left or to the right. The toe naturally moves a little when rowing the boat or when balancing the boat.

The resistance of the rudder is about 0.11 kgf at the neutral position. Once the rudder is turned to 10 degrees, its resistance shoots up to nearly 0.60 kgf. The rudder neutralizer is meant to prevent this sudden increase in resistance. The rudder neutralizer was not considered initially but is a product of a so called spin-off. Beside the rudder, the fin is also an appendage in the coxless four. The resistance of the fin was 0.25 kgf and adding it to that of the rudder resulted in a total resistance of 0.36 kgf, which was rather high.

The rudder turns the boat and the fin keeps the course straight. Each has a role to play. I thought that if the rudder was fixed at the neutral position, it could work as a fin at the same time. While the fin is positioned far from the stern, the rudder is fixed at the aftermost position; therefore, I thought that if the original rudder area was made a little larger, it should work effectively as a fin also.

Fig - 20 Combined fin and rudder
The rudder is meant for steering and the fin for course stability, with a resistance of 0.11 kgf and 0.25 kgf respectively. This photo shows the rudder in a coxless four that I designed and built with the intention of making the rudder area a little larger, and made the shape oblong to increase efficiency. If the rudder can be fixed at a neutral position when not steering, its resistance can be reduced by about 0.20 kgf, since it would work as a fin as well. In practice, the course stability was inadequate and only the rudder neutralizer (Fig - 21) was used.

To realize the idea, the rudder should always be fixed in the neutral position and should work as a normal rudder only when the crew wants to steer the boat. Even if some pressure acts on the rudder surface, such as side wind or unbalance in the left and right rowing forces, the rudder should maintain its neutral position and work as a fin.

I designed the rudder shape and installed a mechanism on top of rudder shaft that kept the rudder at the neutral position. We carried out the first tests on a coxless pair, which was available then (Fig - 20).

It was a very cold season in January and February. If the usual fin is positioned at the normal position, the boat's motion would probably be moderate, but without the fin, a small movement of the rudder would soon cause the bow to swing to the left and the right. While rowing, the crew was apt to move the toe unintentionally, and the rudder could not be kept at the neutral position.

After giving much thought to stopping the meandering of the boat, we designed and installed a mechanism called rudder neutralizer (Fig-21) at the toe. The mechanism was designed to hold the rudder firmly at neutral position and prevented motion to the left or right unless some force was exerted by the toe. With the rudder neutralizer, the boat could now maintain a straight course.

However, there were still some problems for the national team when the fin was taken off the boat. The stability of the boat was unsatisfactory when the boat encountered a strong side wind or a diagonal wave created by other boats. The boat was also reported to be hard to control when running against a mild side wind.

To resolve these problems, we made a fine

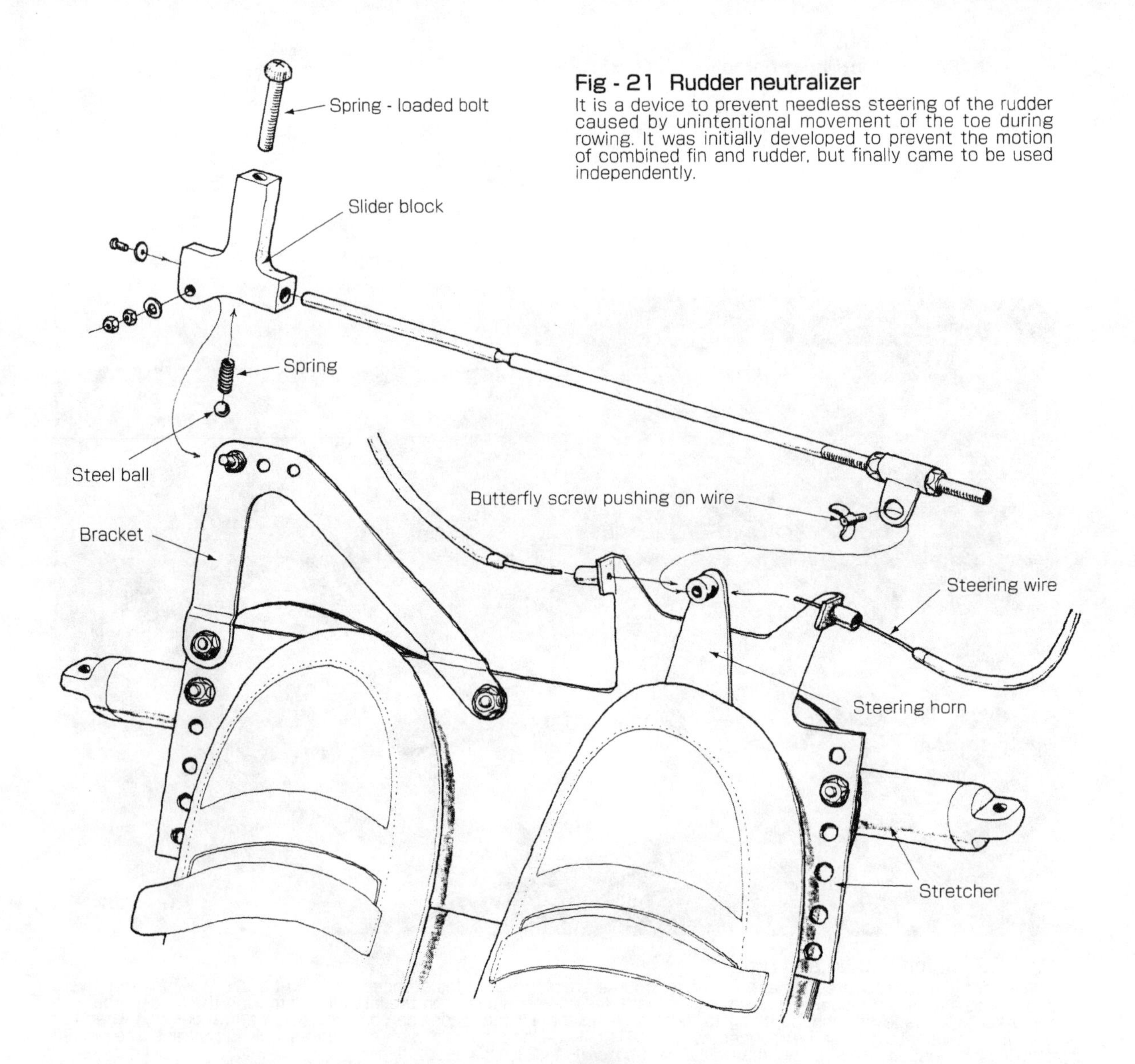

Fig - 21 Rudder neutralizer
It is a device to prevent needless steering of the rudder caused by unintentional movement of the toe during rowing. It was initially developed to prevent the motion of combined fin and rudder, but finally came to be used independently.

adjusting device for the rudder against a side wind and also made a mechanism that the crew could adjust while rowing. The crew on their part too, tried to solve the problem by trying out the boat by changing over to a fin having half the original area.

Finally, we could not fulfill all the requirements, and as time also caught up with us, we had to give up the idea of using the combined fin and rudder. The mechanism to keep the rudder at neutral position using the toe was well received, and became an indispensable part of the boat .

Later, several rudder neutralizers were manufactured at the request of the crew and used in all the men's and women's coxless boats, and those boats participated in the meet to select teams to represent Asia at the Olympics. Since then, the mechanism that we made has become an indispensable item for the national team.

9) Results at the Atlanta Olympics

The results at the Atlanta Olympics were very unexpected. After time analysis of men's lightweight coxless four and double scull for which we had hopes, we found the time difference from the leaders to be about 10 seconds, which this was close to our initial plan, and one likely to produce good results. In practice however, the final ranking was 16th and 15th, which was not good. The reason was that the number of events for the lightweight class was restricted to coxless four and double scull only. Thus, almost all outstanding lightweight oarsmen from other classes switched over to these classes, raising the level of the race and making it highly competitive.

Although the new devices we developed were used, their completion was delayed and

Fig - 22 Preliminary race at Atlanta Olympic Games
Men's lightweight coxless four participated in the elimination race using oar fairings and body fairings. Soon after the race, however, the use of body fairings was prohibited.

only a few boats used them. Therefore, their effects were probably small and we could not judge their effectiveness.

We pursued too many items within a short period, and in the final stages, we worked hard to develop and make the devices for all five teams. As a result, we probably diluted our capacity to develop the devices to a higher level of completion and finish. However, it was later verified from a series of tests that the devices we had made in preparation for the Atlanta Olympics did have the effects for which they were developed. Had adequate time been available for the crew to become fully accustomed to each of these devices, their effects would have been amply demonstrated.

Another unfortunate incident was that the body fairing (Fig - 22) used during the preliminary (elimination) rounds of the race was abruptly prohibited from use in subsequent races. We demanded explanations from the international rowing association (FISA) for the prohibition.

In 1998, FISA established a new rule that laid down conditions for using new devices in international races and in the Olympic games. One of the conditions was the actual use of the relevant device for more than one year in international races after obtaining the approval of FISA. Other conditions were fair use of the device by all crew, inexpensive, and so on. For these reasons, it has become difficult to participate in races with loaded advantages offered by new devices.

In human powered boat races at Lake Hamana, a pedal boat ran faster than an eight - oars shell recently (Chapter 5). In the sailboat events of the Olympics, there is a class of sailboat named 49er (forty niner), which is very fast and highly sophisticated.

Now we are shifting to an era in which people's interest lie in the high-speed boat, which mobilizes speed and technique. Therefore, I feel a little sad sometimes when I think that such restrictions imposed on devices to increase the speed of rowing boats will move people away from rowing boats to other sports.

DEVELOPMENT OF THE MARINE JET

I could not forget the excellent maneuverability of the hydrofoil boat that I had developed in 1955. Twenty years later, I decided to develop a water bike reproducing the maneuverability of the same hydrofoil boat. I went through various twists and turns during the course of development of the water bike, failing several times and becoming discouraged, suspecting sometimes that somebody, somewhere in the world had already worked on the same idea and had developed it. During those years however, we learnt from mistakes, accumulated expertise, and finally, after thorough market research, we successfully brought the Marine Jet to the market.

1) The glamor of airplanes

In 1985, I went to Oshkosh, Wisconsin, to see an air show, where the builders of about 1,200 self - made airplanes display their handiwork. Owners of about 12,000 planes fly in from all over the United States to participate in this show. During the one-week long show, the plane owners pitch their tents under the wings of their airplanes and camp there. It is truly a spectacular airplane show. Throughout the week, visitors enjoy watching acrobatic stunts, vertical take-offs and landings of Harriers, and shop around for materials, parts, and new products. Airplane construction is demonstrated at some booths. Visitors check up on prices of used airplanes, some look around for opportunities to renew their airplanes, and others try to trade their planes.

You could buy an airplane for about $3000 in the 1930s, so they were very popular with the people in this era. An airfield that can accommodate as many as 13,000 airplanes is amazing, but the fact that nearly 800,000 visitors come to see this show boggles the imagination. What attracts so many people to airplanes?

In 1952, while employed with Okamura Manufacturing Company, I worked on the design of the first of the domestic-built aircrafts after the World War II, namely Nippon University's N52 and Tokyo University's LBS-1, a two-seater soarer. At that time, I was surprised to know that many of the well-known senior designers loved airplanes, and they loved flying more than designing the airplanes. I suspect that their love of flying airplanes had much to do with their attraction to Gravity (G). Gravity normally works vertically on the human body, but when in motion, acceleration or centrifugal force gives the feel of increasing gravity force acting on the body horizontally. This unusual but pleasurable feeling of G attracts people to sports that involve gravity. This glamour of G and the glamour of speed possibly attract people to skiing, skating, and motor sports. Flying, I believe comes at the top of this list.

2) Building a hydrofoil craft

In 1954, I started work as a designer in the Design Department of Yokohama Yacht Works. All the time while designing ordinary commercial boats, I was dreaming of building a hydrofoil craft.

Hydrofoil craft technology made rapid advances during World War II. This technology was utilized in the commercial arena immediately after the war. High-speed hydrofoil ferryboats were popular in Switzerland (see Fig-

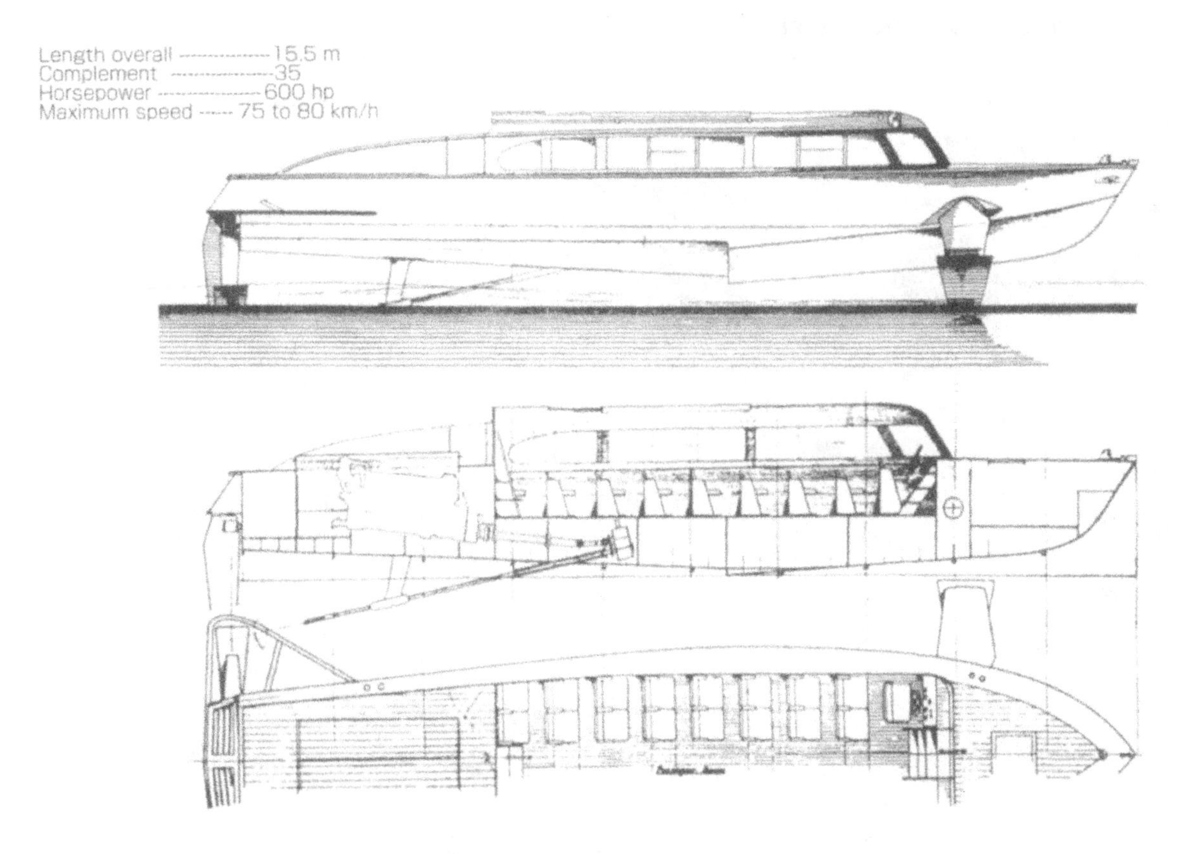

Fig - 1 Supramar hydrofoil craft, Switzerland

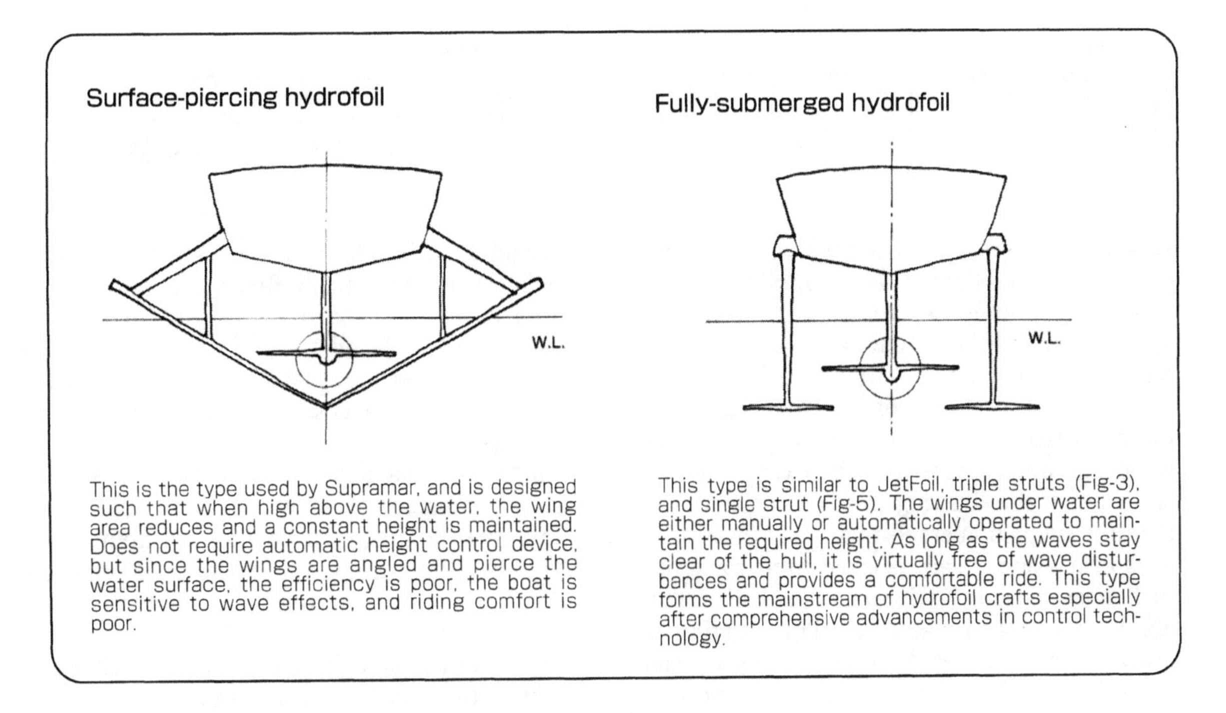

Fig - 2 Types of hydrofoil craft

Fig - 3 Three - strut hydrofoil craft

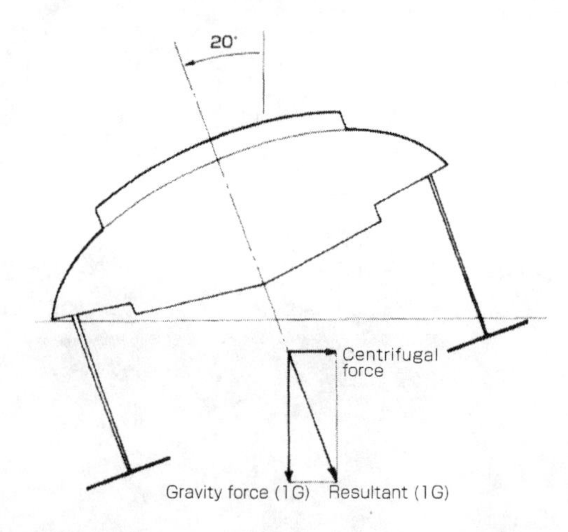

The boat cannot bank beyond 20° because at this angle, either the hull on the inside touches the water, or the outer underwater wing protrudes out of the water.
At a 20°- bank, the gravity force on the body is 1.06 G, which is not apparent because it is almost the same as 1 G.

Fig - 4 Explanation of the bank

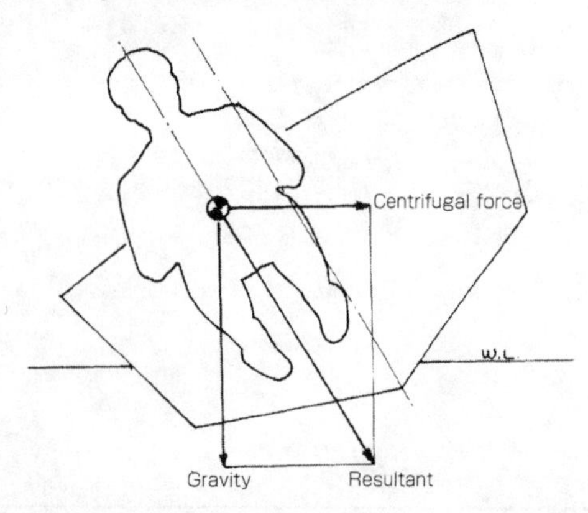

Perfect bank
If the resultant of the centrifugal force and the force of gravity is in the boat's symmetric plane, the rider will not be thrown sideways. The "perfect bank" is when a boat while turning maintains a proper bank (inside) toward the center of the turn.

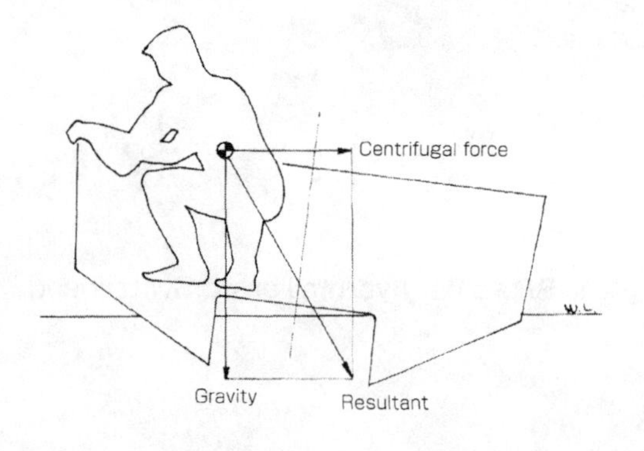

Inverse bank
An "inverse bank" occurs when a boat heels outward during a turn. The sketch above shows a catamaran in which the centrifugal force combines with outward heel to throw the rider outward during a turn. The rider is holding on to the inside part of the gunwale to avoid being thrown out.

1). I was convinced at that time that hydrofoil craft held the key to the company's future.

One day, I happened to see a drawing of a hydrofoil craft in an American boat magazine. At that time, surface-piercing hydrofoils formed the mainstream of hydrofoil crafts, but the main issue with these craft was that they were extremely sensitive to waves. On the other hand, the drawing in the magazine was that of a fully submerged hydrofoil (see Fig-2). This boat was virtually isolated from wave disturbances as long as it maintained a certain height above the waterline. This particular drawing showed that the boat had a stick and a foot bar just like those in a light airplane, and it seemed that driving it was almost similar to driving an airplane. I thought that an airplane pilot must have designed this boat.

My request to the President of my company to build a hydrofoil craft was granted. After building one and actually driving it, I was surprised to discover that it was similar to flying an airplane. I enjoyed the feeling of being foilborne. I noticed, however, that the hydrofoil craft required an excessively long time to turn (see Fig-3).

Since this boat had a wide beam and short struts in comparison to its beam, it could not bank over a large angle like an airplane. Excessive banking caused the hull on the inside to touch the water, or the outer foil to protrude out of the water.

On the other hand, if you try to turn without banking, the thin hydrofoil struts will break. If you turn the boat maintaining a proper inward bank such that the direction of the resultant of the force of gravity and the centrifugal force lie in a symmetric plane of the craft, you will

Fig - 5 Two photos of a foilborne single - strut hydrofoil in action

The boat with this arrangement could bank up to 45 degrees and make a tight turn. The force of gravity acting on the rider was as high as 1.4G.

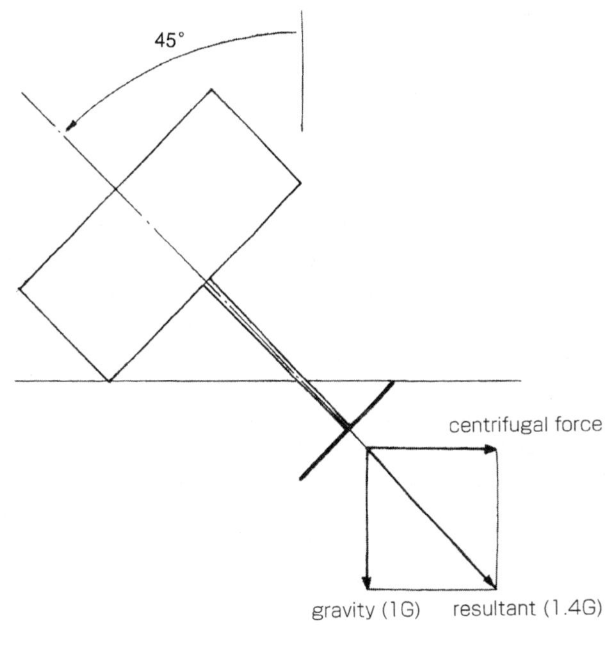

neither be thrown sideways nor will the struts of the boat break (see Fig - 4).

The boat, however, was unable to bank adequately during a sharp turn. An airplane will slip sideways during a sharp turn, and a bicycle or a motorcycle will overturn sideways. It is obvious that good maneuverability requires an adequate banking angle.

At that time, I was working in the tank testing laboratory. During my spare time, I drew up plans of a new hydrofoil craft that could bank up to 45 degrees. At this banking angle, the rider experiences 1.4 times the force of gravity (1.4G), which naturally is not as large as the gravity force during a 60 degrees turn (2G) of an airplane, but certainly cannot be experienced while skiing or riding a motor - cycle (see Fig - 5).

Again, getting approval to build a prototype boat that might not sell was not an easy task. However, I really wanted to build this boat because I was confident that it would perform well. After overcame many hurdles, we finally built the boat, and I was delighted with its successful run. It had excellent maneuverability and the G that acted on my body during a turn was identical to that of an airplane.

This prototype boat would probably never sell in post-war Japan, but I began to dream more than ever before of building a boat that would offer superior maneuverability, and I set aside my initial dream of designing a high - speed ferryboat.

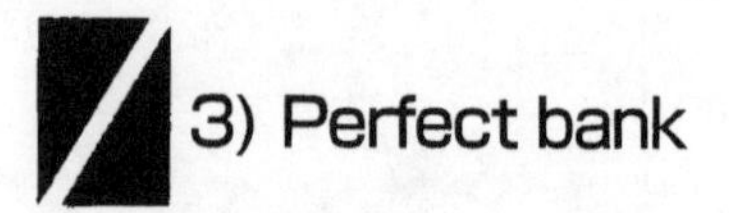

3) Perfect bank

This boat had a successful run in 1955. Since that time, I have believed that vehicles should turn with a correct inside bank that prevents the centrifugal force from pushing the rider outward. When a boat turns, it is difficult to maintain an ideal banking angle irrespective of rudder angle, speed, and weight distribution. Inside banking is not a major issue for an airplane or a motorcycle, but it has to be artificially created during the hull design of the boat. In view of the difficulties in doing so, I have named this condition Perfect Bank.

The words Perfect Bank do not exist in flying terminology. Sideslip occurs when an inexperienced pilot is unable to maintain a Perfect Bank. Excessive sideslip can be dangerous, but pilots sometimes intentionally allow sideslip to occur. This is a special skill used by them to compensate for windage on landing, or to dupe enemy pilots during an air battle. Maintaining a Perfect Bank is part of the normal operation of an aircraft.

Fig - 6 SEA DOO introduced in 1967

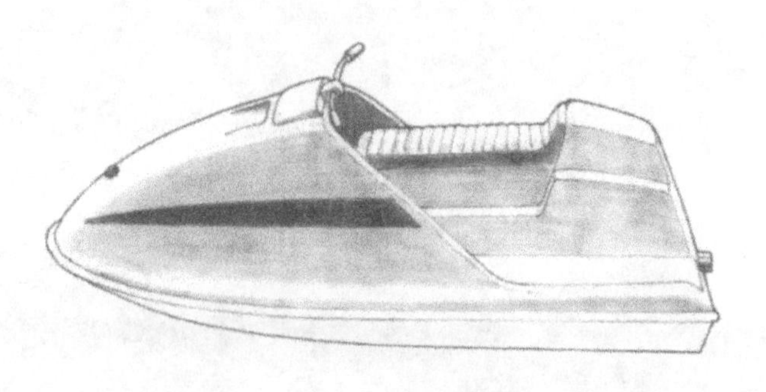

Later I designed air-propelled boats (propeller boats) and racing boats where I pursued the Perfect Bank concept. After I changed jobs and moved to Yamaha Motor Co., I more or less perfected the Perfect Bank concept with round bilge hulls and deep-V hulls, but was helpless when it came to catamarans and flat - bottom runabouts, which turned almost flat on the water, and looked unnatural to me. Automobiles that heel outward during a turn also seem unnatural to me, but this seems a natural thing to do in motorcycles.

4) SEA DOO

I think it was probably in 1967 that Bombardier of Canada introduced an interesting boat named SEA DOO. This was a small boat in which the rider sat astride a seat and drove the boat; the name Water Scooter would describe it appropriately. This boat had no propellers, but was equipped with a jet propulsion system with its inherent safety features (see Fig - 6).

It came as a bit of shock to me to find that somebody had come up with the kind of highly maneuverable boat I had been dreaming of. However, when I rode one of these, it was quite different from what I had imagined. Since it had a flat bottom, it ran all right, but its turning ability was poor and it did not even come close to my concept of the dream boat. It was well received in the United States for a while, but later developed engine problems, which was followed by numerous claims. I heard that the manufacturer recalled all the 1,500 boats sold, and later withdrew the boat from the market.

SEA DOO revived my dream. I thought that with some improvements to the hull design, I could design a water scooter with excellent maneuverability and with Perfect Bank. I also felt at that time that the era of water scooters was not too far off in the future.

How does one carry such a heavy boat as SEA DOO ashore? It was too heavy to carry from the truck to the shore. I decided after studying SEA DOO to develop a light, maneuverable, water-jet propelled water scooter. This opportunity did not arrive any time soon because we needed an appropriate engine and a suitable water jet system.

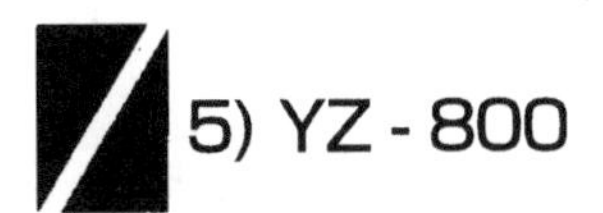

5) YZ - 800

After about five years, I got hold of a small UA - Jet made in England. I thought I could use a 25 HP outboard engine with its head turned horizontally, modify the carburetor and the exhaust, and give shape to a jet system. I then began to consider the design of the hull. I remembered the maneuverability of the single - strut hydrofoil craft, and wanted somehow to realize the Perfect Bank of that craft in the water scooter. The boat must have a front rudder. An aft rudder would not help to achieve the Perfect Bank. However, with this layout and form, the stability of the hull would limit its banking ability. To minimize that effect and to reduce the righting force of stability, the hull has to take the form of a deep V.

Minoru Sugasawa was in charge of this boat. He was involved mainly with the designs of large boats including a 15 m boat in Yamaha Motor Co. He showed a keen interest in my project because of the radical change from his usual work of designing large boats to the development of a small boat. Since the profile around the saddle was complicated, we built a 1:10 model first, then constructed a polyurethane foam mockup for the male mold. We actually sat on it and cut out the parts to confirm the arrangement. The finished prototype YZ - 800 (see Fig-7(a)(b)) looked like a snazzy offshore racer. It was much more compact than SEA DOO and weighed only 60 kg.

It was very difficult to ride the YZ - 800 just after starting the engine because the narrow deep-V hull made the boat very unstable. After a while, when the boat picked up speed and started planing, the stability improved. As I anticipated, the stability was too large for the boat to heel inward. We were very far off from the Perfect Bank concept. We tried to steer the front rudder to heel the hull inward, but the water pressure pushed the inside of the hull up and wouldn't permit the boat to heel, which was quite natural.

I don't recall performing simulation of the trial run before constructing this boat. When I built the hydrofoil craft earlier, I successfully simulated the trial run, but I must have missed it this time. I think I was too busy, or I might have given up simulation entirely because of the complexity of the boat that required considerations of stability of both rudder and hull, or I did perform the simulation, but the results were inconclusive. I just don't remember now.

Sugasawa was a young man who worked very hard to improve the performance of the boat, but he could do nothing with the front rudder configuration. I must have given him a hard time with my strong prejudices. Finally, after trial and error over a prolonged period, we installed the UA-Jet rudder at the jet outlet. The result was the same as the rudder mounting system of any normal jet boat.

Kokichi Takahashi, who now runs a marina on Lake Hamana, continued with the trials. I had still not regained my confidence since the failure of my front rudder concept. While I remained low in spirits with my personal problems, Sugasawa and Takahashi steadily learned to maneuver this boat and tamed it in the process. They showed that that it was quite a fun boat to ride.

Initially, I thought I had given too much of a deadrise which caused the boat to lose out its lateral stability. Nevertheless, I found that once you get used to riding the boat, you can control it well. The boat turned out to be a soft

Fig - 7a Sugasawa on a trial run of YZ - 800

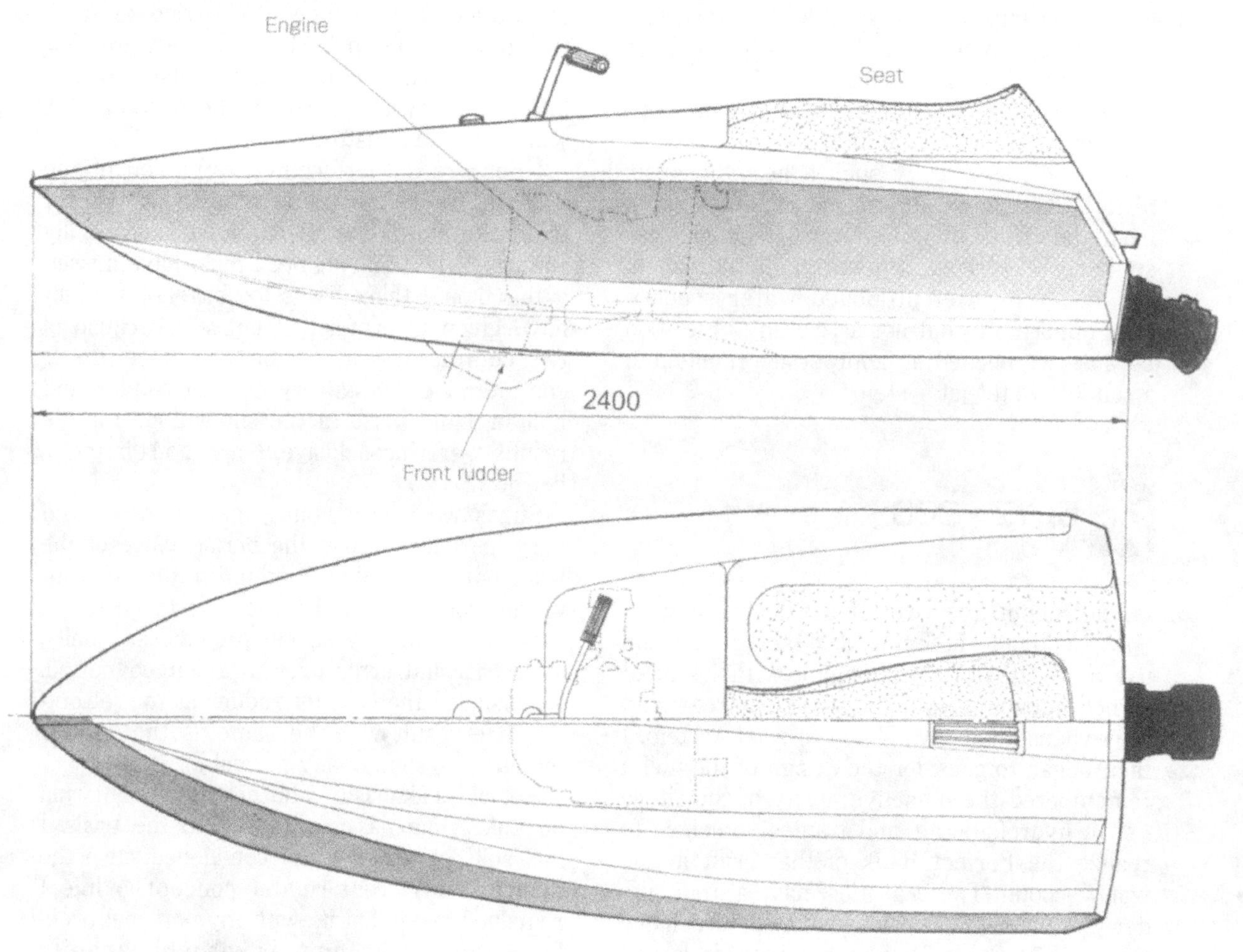

Fig - 7b YZ - 800

riding and comfortable one; it was compact and fun to maneuver, and felt as if one was riding a motorcycle. Though we did not achieve the Perfect Bank as in the hydrofoil craft, we could perform the banked turns of a regular boat by shifting our body. With some redesigning of the hull, this boat should become much easier to ride. Over time, I gave up on my pursuit of the boat with the Perfect Bank, and instead focused on and became confident of turning this boat into a marketable product.

Until now, none of the projects we had worked on was a recognized, official development project, but were developments in secret. To realize this boat as a product, we had to have this project officially recognized and placed among other normal development projects. I was the Senior General Manager (Marine Operations) at that time, but I had to obtain the President's approval before I could embark work on a new product. Mr. Hisao Koike was the President of Yamaha Motor Co. at the time, and he didn't give me the nod to go ahead. Unfortunately, prior to my meeting with him, Kawasaki Heavy Industries had introduced the Jet Ski in the market. During their demonstration, a woman was hurt and the accident made the front pages of newspapers. The injury itself was not serious, but the incident occurred amidst a sudden interest on safety and product liability issues in the society.

I came to know later that Kawasaki Heavy Industries had given up its plan to sell the Jet Ski in the Japanese market and had moved its operation to the United States (currently, Kawasaki Motors Japan, deals with the Jet Ski). I had witnessed a typical case of a new technology facing barriers in this country. Mr. Koike said: I am sorry, but the timing is not right. Give it up for now.

I understood that this was the consequence

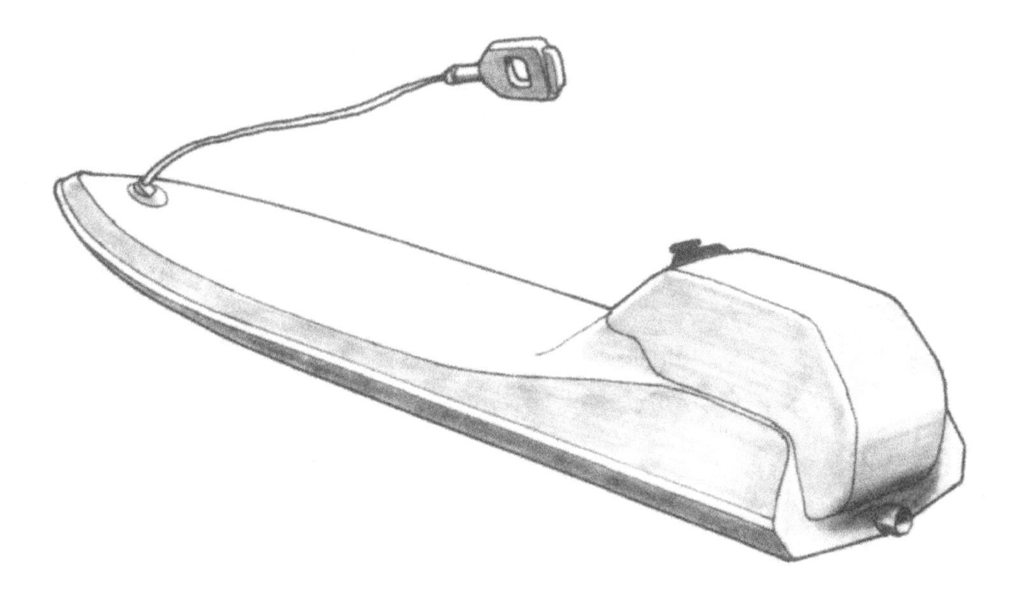

Fig - 8 SURF JET

of the Kawasaki Jet Ski accident, and just had to accept it. I apologized to Sugasawa and Takahashi, who found it very difficult to digest the news. That was the end of summer, 1973.

6) Wing

Eleven years elapsed since we gave up on the development of the YZ - 800. In 1984, as soon as the Horiuchi Laboratory was established within Yamaha Motor Co., I listed the water scooter as a research project. We were not involved directly in the development, but we provided support to Noboru Kobayashi, who was the senior test engineer, Boat Operations. He had been very enthusiastic in the development of the water scooter. The Horiuchi Laboratory at the Head Office supplied Kobayashi with the jet components and research funds, and offered to coordinate work with the Outboard Division and other departments. We offered him technical guidance also.

We were trying to develop a watercraft that would be small and light enough to be carried on top of a car and yet offer good maneuverability. It had to be lighter and less expensive than the YZ - 800. The reason for this change in the approach was that a powered surfboard called SURF JET had arrived in the American market (see Fig - 8) at that time, and its performance was surprisingly good. This boat weighed only about 50 kg. We wanted to develop a product that would offer features that are more attractive.

The design of SURF JET was very simple. Its body was like that of a surfboard and it was equipped with a 15 hp engine and a centrifugal pump[1] manufactured by Fuji Heavy Industries. It was manufactured in Minnesota, USA, which is known for innovative vehicles such as the SURF JET and the WETBIKE. We decided to use the SURF JET engine and pump initially, and considered designing a body that would dramatically increase the maneuverability. We would think of the motor later. This kind of vehicle had to be fast, and to achieve high speeds, the boat must plane. However, a large planing surface is necessary if the boat is to plane with a small horsepower engine. At the same weight, the smaller the engine horsepower, the longer and wider is the planing surface required. Since we needed a lightweight and small - horsepower craft, we had no choice but to adopt a board type layout such as that of the SURF JET to make maximum use of a

1) Centrifugal pump: Water pump operated by centrifugal force. Recent jet pumps are mostly axial flow pumps.

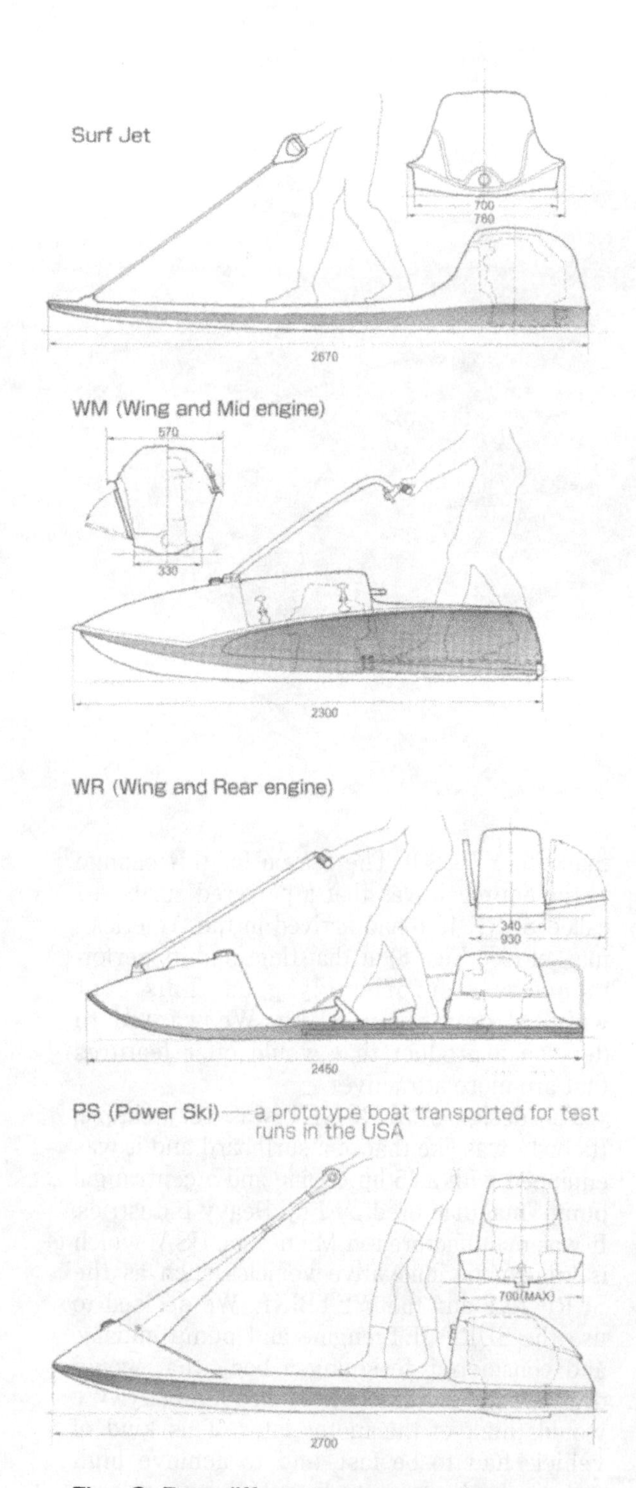

Fig - 9 Four different arrangements

wider planing surface. At high speeds, the planing surface should be narrow, which means smaller wetted surface of the boat. I came to the conclusion that if the boat was designed to use a wide planing surface as it entered the planing mode, and if the side planing surfaces were raised as soon as the boat entered the high speed running mode so that it was supported only on the central narrow planing surface, it would go faster. This was the concept of the Wing (see Fig - 9 WM, WR).

Kobayashi worked very hard at it and in addition to my Wing concept, he tried various layouts one after another, testing the boat by moving the wing, engine, jet, and intake to different positions. Yoshiki Hirahara, in the same test team, also assisted eagerly. Throughout these modifications, certain features remained unchanged. These included the surfboard type hull, operating the boat while standing, jet propulsion, and tilting handle that moves only in the vertical plane.

A SURF JET rider must hold the end of the cable coming out of the bow at the handle grip while operating the throttle. The rider must also control the hull with his feet just like a regular surfboard. If a tilting handle is fitted to this, the rider can operate the rudder at the jet outlet to steer the boat and support his body. The rider can also control pitching and rolling by pushing the handle in the fore/aft and left/right directions.

We believed such functionality was important in making the ride more interesting and fun even for beginners compared to the SURF JET, and was also necessary for enhanced maneuverability. Such controls would go a long way in expanding the scope of boating activities, especially the feel of G during sharp turns and the joy of riding on the plunging breakers.

Kobayashi tried to modify the Wing in various ways, but none of his methods proved successful. No major effect was achieved by using an excessively large structure when the boat was fitted with a small horsepower engine, which restricted the speed. This project was my second failure following the YZ-800. Meanwhile, Kobayashi tried various layouts, evaluated the results of each layout, and studied what should be avoided, both in theory and in practice. He finally came up with a layout (see Fig - 9 PS) similar to the one with a tilting handle installed on the SURF JET. This boat offered improved sharp turns and better surfing in plunging breakers compared to the SURF JET (see Fig - 10).

Fig - 10 PS in a turn

Fig - 11 Prototype boat sent to the USA

7) Presentation in the USA

We learned from our experience with the XY-800 and the JET SKI that this type of product would sell better in the US market than in Japan. We held discussions with Yamaha Motor Corporation USA (YMUS) about marketing this product. YMUS was involved with the sales of outboard engines, and showed quick understanding and interest in the product. Ryozo Okita extended his full support and did everything possible for its presentation and commercialization. Thanks to Okita's efforts, three of our prototypes (see Fig-11) were transported to Long Beach and offered for test rides to many YMUS employees in the summer of 1984.

Once the boat began to run, it more or less satisfied the rider, but its static stability was unsatisfactory, and it was difficult for riders in the USA to climb on board. This was mainly due to the difference in the body weight of Americans and Japanese. We realized for the first time how heavy Americans were. Overall, our prototypes did not win approval.

8) The turning point - Marine Jets

YMUS studied this subject in detail for a whole year. Denis Stephanni of the Product Planning Department, in particular, conducted extensive market research and presented to us a simple and straightforward conclusion. I received his report while I was in Los Angeles on business. According to his report, the users wanted a powerful two-seater boat; they would not mind even a heavier boat so long as the boat was a two-seater and had power. That certainly made sense in the USA where most people used boat trailers. In short, the Stephanni report suggested a two-seater SEA DOO with a highly reliable engine and other components for the US market promptly. The report ended with a remark that suggested that the market would probably need a lightweight boat such as the prototype in the future. It appeared that this remark was meant to console me.

We were using Japanese specifications of a boat to be sold in the United States. We didn't see any engineering obstacles to satisfying the YMUS requirements. After all, during the last two years, we had learned many lessons on what not to do, and we had attained a fair degree of confidence in our engineering skills. I personally thought that we could develop the required model immediately.

After receiving the report, Kobayashi officially started work on this project. Since he had worked as a service engineer, he had been in close contact with marine customers. He knew the kind of fun the customers wanted. He had also handled many complaints during that period and had acquired first-hand knowledge about product liability (PL) issues. His knowledge proved to be very valuable for this project. Although the project was not very demanding in terms of technology, it required much thought and consideration related to PL issues. Should the engine stop automatically when the boat has capsized? Should the boat slowly circle around at idling speed and return to the fallen rider? Should the boat right itself automatically? If it does not right itself automatically, how much force would be required for the user to turn over the capsized boat and upright it? How should we handle collision? Throttle lever? Kill switch? These and many other issues that we had never imagined arose successively and demanded solutions. Kobayashi resolved all these issues very well. Finally, in November 1986, just one year after the YMUS report, we introduced the marine jet MJ - 500 satisfying almost all PL issues and meeting all the requirements of YMUS (see Fig - 12).

9) Popularity

MJ - 500 immediately captured the hearts of marine sport lovers and sold quickly. Soon, many American manufacturers including Bombardier came up with similar water scooters one after another. The sit-down water scooters quickly caught up with and exceeded the ten-year old Jet Ski in number. Water scooters began to carve out a niche in the world of the marine sports. This product must have given rise to business of over $10 billion by now. (as written in 1998)

Success arrived after a long time; I first wanted to build a hydrofoil craft some 32 years ago. It has not been a straight road, but our experiences over these years have contributed to the successful development of the Marine Jet. A few years ago, that Bombardier reportedly

2) Product Liability

Fig - 12 Successfu Marine Jet 500

advertised that this product was originally developed by them. I do admit that they came up with SEA DOO first, but they withdrew from the market twenty years ago, so the Marine Jet is a product developed by Yamaha Motor Co., Ltd. I also would like to tell the world that the success of the Marine Jet is entirely due to the sincere efforts and hard work of dedicated persons like Kobayashi and many others.

5

HIGH SPEED HUMAN POWERED BOATS

In the 1980s, the enthusiasm for human powered boat races in the USA led to a human powered hydrofoil boat attaining a speed of over 10 knots. The Dupont Award of $25,000 also stimulated interest for achieving 20 - knot speeds. Even in Japan, more than 200 human powered boats participated in the exciting Dream Boat Contest. Two engineers of the Horiuchi Laboratory have been continuously winning this Contest since it started in 1991. I present herewith the history and techniques of human powered boats. The record achieved in 2000 was certified as a world record.

1) IHPVA

There is an organization called the International Human Power Vehicle Association (IHPVA) in the United States. It is headed by Paul MacCready, who holds the record for first 8 letter flight on a human powered airplane and who developed the Gossamer Series that flew across the Dover channel. Beside human powered airplanes, this association has been organizing many bicycle races and also officially recognizes records. These bicycles are not ordinary. The rider lies down on his back to reduce the front profile to a minimum. A streamline cover minimizes air resistance. In other words, these bicycles are designed to maximize the utilization of human power. Enthusiasts enjoy competing with these bicycles in the short-distance dash and the long distance endurance race (Fig - 1).

In 1985, the IHPVA decided to organize a speed race for human powered boats. The race was called Hydro Challenge. Since I was a member of IHPVA at that time, I learned about this event from the IHPVA newsletter.

Mr. Jack Lambie, one of the founding members of IHPVA, provided technical information to the Yamaha Development Division. When Yamaha was investigating the business possibilities of the hang glider and the ultra-light plane, he gave us details about the American market and introduced us to some related people. His specialty was bicycles, light airplanes, and gliders. He has written some books too.

At that time, I had been thinking of developing a hydrofoil scull with the capability of 15 knots speed and I shared my dream with Jack Lambie in a party. Though it was only a concept, he insisted that I should send him an illustration so that he might insert it in the IHPVA newsletter. I finally drew a conceptual sketch and sent it to him (Fig - 2). I learned later that he wrote an article about the scull and the single-strut hydrofoil that I had built a long time ago (1955). Unfortunately, I missed this newsletter, but I suppose that he used that story to boost Hydro Challenge.

The concept of the hydrofoil scull unfortunately didn't go too far, but the mechanical water surface sensor of this boat, however, was something I can be proud of even today. This sensor maintains the height of the boat in motion as well as its transverse stability. Dragging an arrow-feather-like skid on the water surface, the scull can rise above the water when the speed increases because of its hydrofoil. When one side sinks, this skid is pushed up relatively and the aileron, which is located at the rear edge of the hydrofoil and connected to the skid through the link mecha-

Fig - 1 High - speed bicycle with minimal air resistance
Source: Scientific American: Dec. 1983

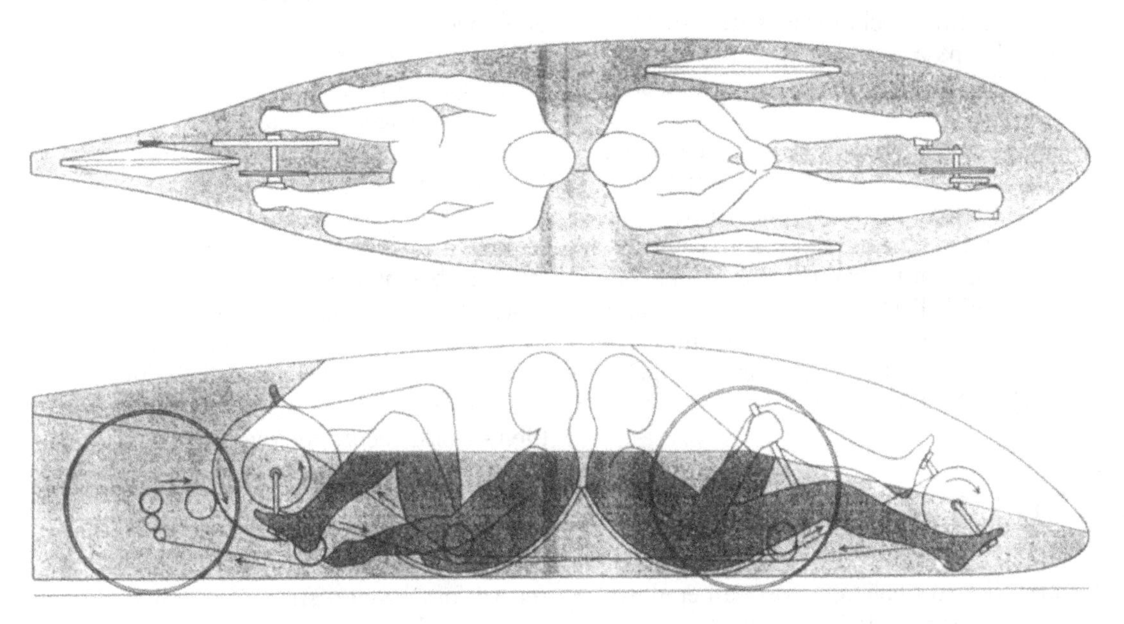

Vector tandem
Recorded 62.5 mph (about 101km/h) on a 200 m course in 1980. Later, in California, it ran on the interstate highway for 40 miles (about 64 km) at an average speed of 50.5 mph (about 81 km/h).

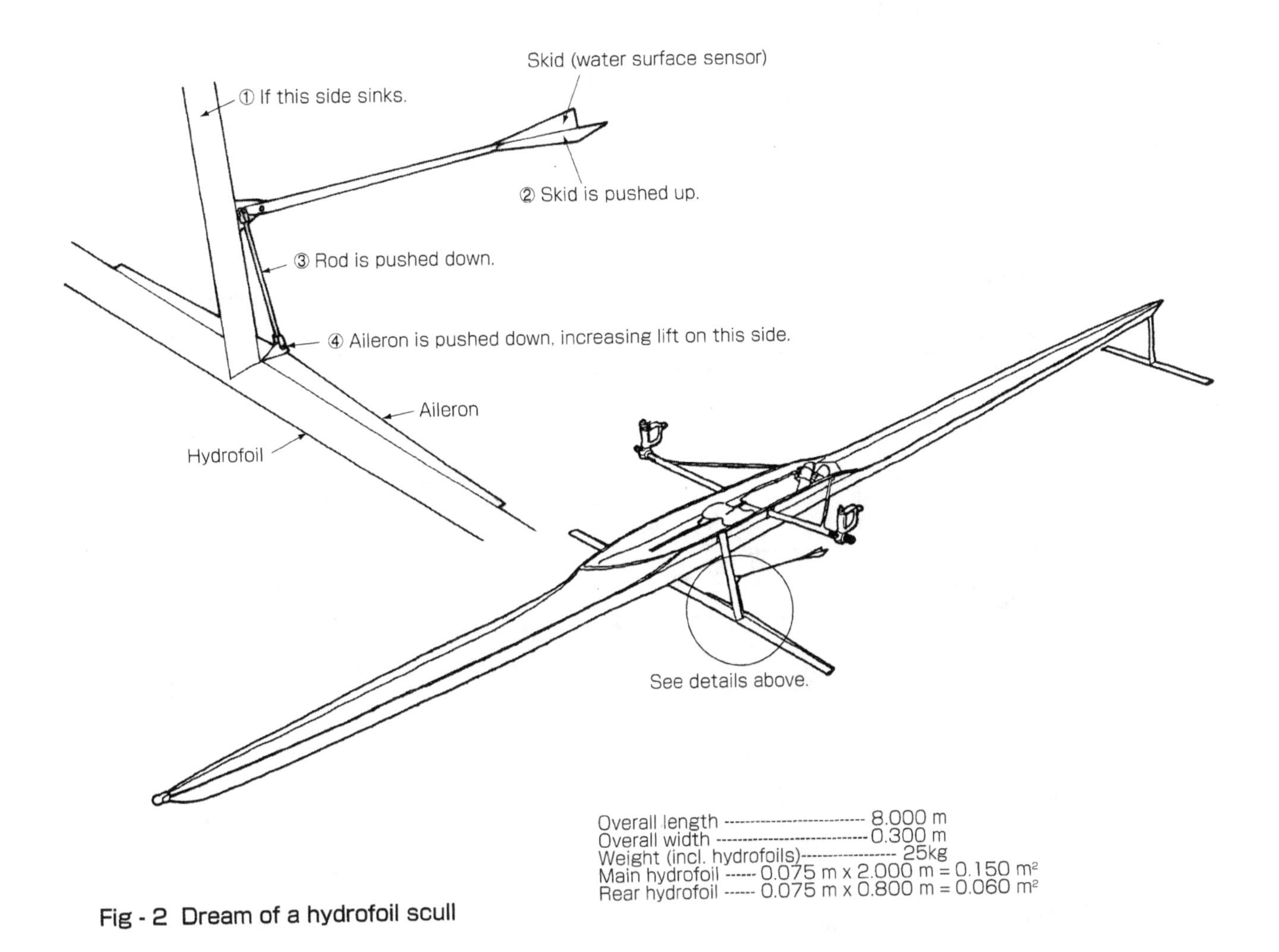

Overall length ------------------------- 8.000 m
Overall width --------------------------- 0.300 m
Weight (incl. hydrofoils)----------------- 25kg
Main hydrofoil ------ 0.075 m x 2.000 m = 0.150 m²
Rear hydrofoil ------ 0.075 m x 0.800 m = 0.060 m²

Fig - 2 Dream of a hydrofoil scull

nism, is pushed down. As a result, the side of the scull rises to the horizontal position. This is the theory of the mechanical water surface sensor that keeps a boat (scull) in motion at a certain height and maintains the transverse balance.

The single-strut hydrofoil, on the other hand, adopts a system where the rider operates the front foil by moving the handle bar of the boat forward and aft, adjusting and maintaining the boat height when in motion (Fig-3). I built this boat around 1953. This boat was a success and showed a high level of maneuverability. I wrote an article about this boat in the magazine Kazi (July 1981) (Fig - 4).

2) FLYING FISH

The first Hydro Challenge was rather desolate. A newsletter reported that nobody attended the event and that they asked a rowing boat (four-oars shell) that had happened to pass by to allow them to measure its speed. Some years later, an interesting boat called Flying Fish appeared (Fig-5). This boat did not have a hull and wouldn't float properly, but was equipped with flotation that would prevent it from sinking to the bottom of the sea.

One has to load this boat onto a cart equipped with roller board wheels, and let it roll into the water by gravity on rails installed on a slope. In this way, the boat enters the water at a speed that is sufficient to allow the boat to run on its hydrofoil. The rider maintains the speed to travel on water. Should the rider slow down, the main hydrofoil sinks, the front hydrofoil springs up above the water, and the boat begins to sink from the rear (Fig - 6). I imagine that it would have been a difficult job to recover the boat from the water. This boat ran a course of 2,000 m at 6 minutes 38 seconds, which was 11 seconds faster than the then - world record for the single scull.

Flying Fish was developed in 1984 by Allan Abbot and Alec Brooks. Allan Abbot was a medical doctor and professor at the University of Southern California, who held the speed record for bicycles. Alec Brooks held a doctorate in engineering, specializing in theoretical hydrodynamics, and worked for Aero Environment Company run by Paul MacCready. Both gentlemen had a profound interest in human powered vehicles.

Later, they installed pontoons on the modified Flying Fish so that it would float on the water, could be started on the water and become foil-

Steering is like that of motorcycle.

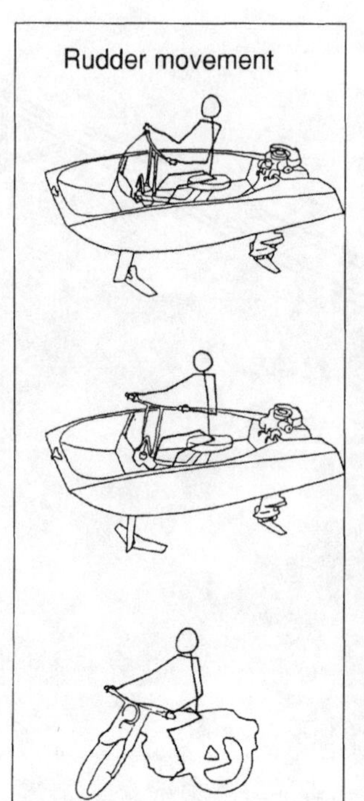

Pull towards you to raise the front. Push away from you to lower it.

Fig - 3 Operation of single - strut hydrofoil boat

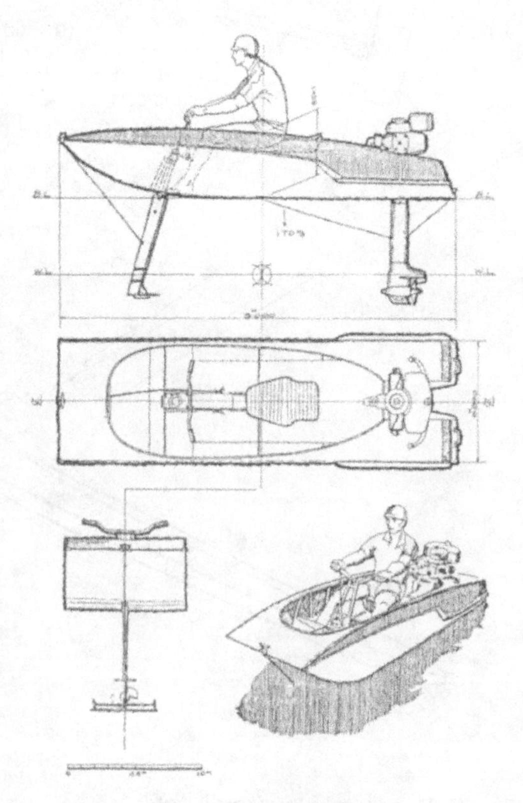

Fig - 4 Single - strut hydrofoil boat

Fig - 5 Flying Fish on a run

Fig - 6 Flying Fish in a stalll

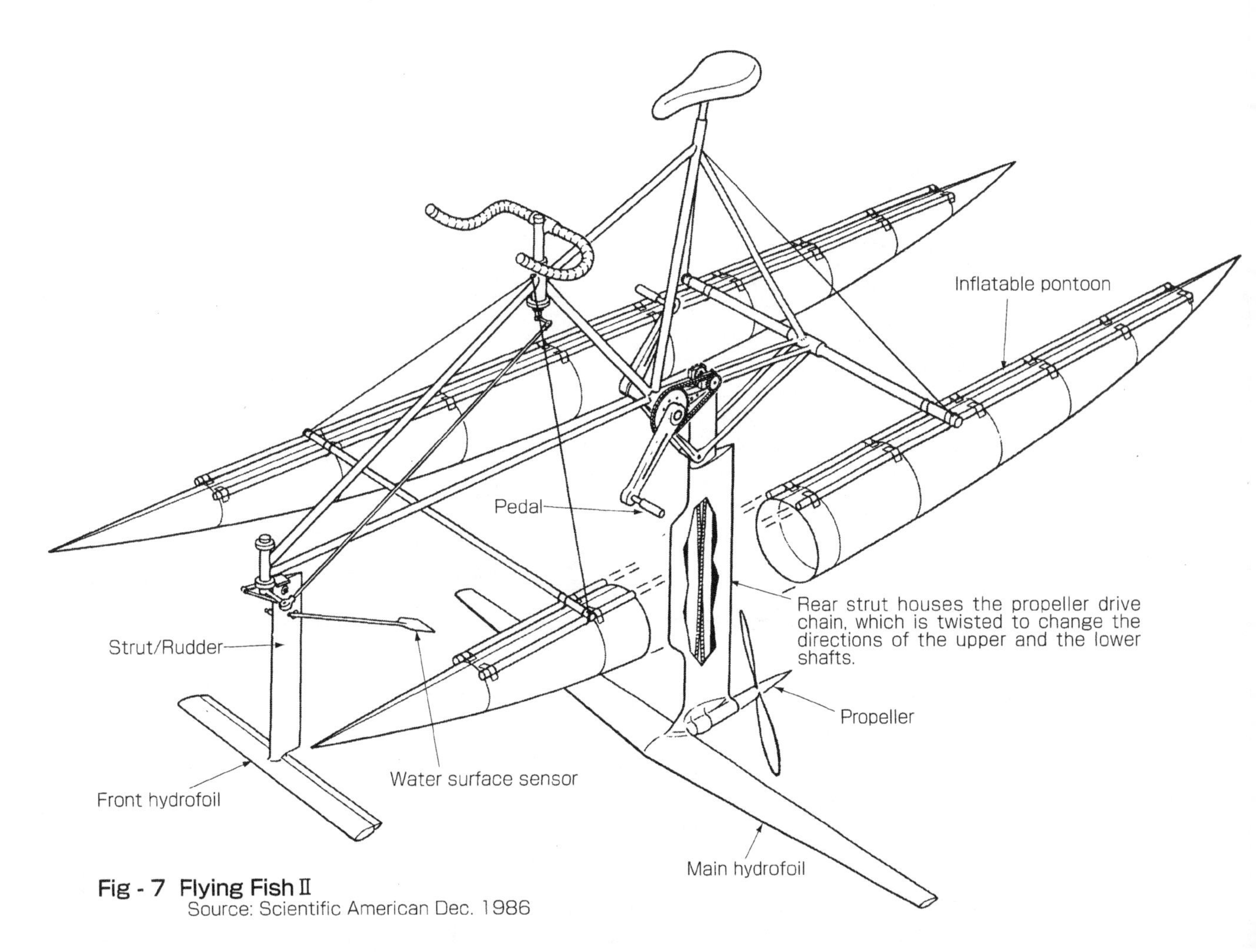

Fig - 7 Flying Fish II
Source: Scientific American Dec. 1986

borne without rails, and thus completed the world's first human power hydrofoil Flying Fish II (Fig - 7). This boat attained a time of 5 minutes 48 seconds on a 2000 m course, which had never been achieved by the eight-oars shell. They presented the method of building this boat in the newsletter and shared their dream of achieving 20 knots in a separate article. That was truly exciting.

Let me explain the construction of the Flying Fish. The main hydrofoil of this boat is located aft. The front hydrofoil is equipped with a water surface sensor, same as our hydrofoil scull, and it could adjust the angle of elevation. The water surface sensor is so installed that it would always be in contact with the water surface. If the rise of the boat above the water surface is small, it would increase the angle of attack of the front hydrofoil. If the boat rises excessively, this device works to reduce that angle and to bring the front hydrofoil down. The sensor activates to maintain the height of the bow at a nearly constant level irrespective of its speed.

Fig - 8 Displacement type Hydropet
Source: HUMAN POWER 7/4

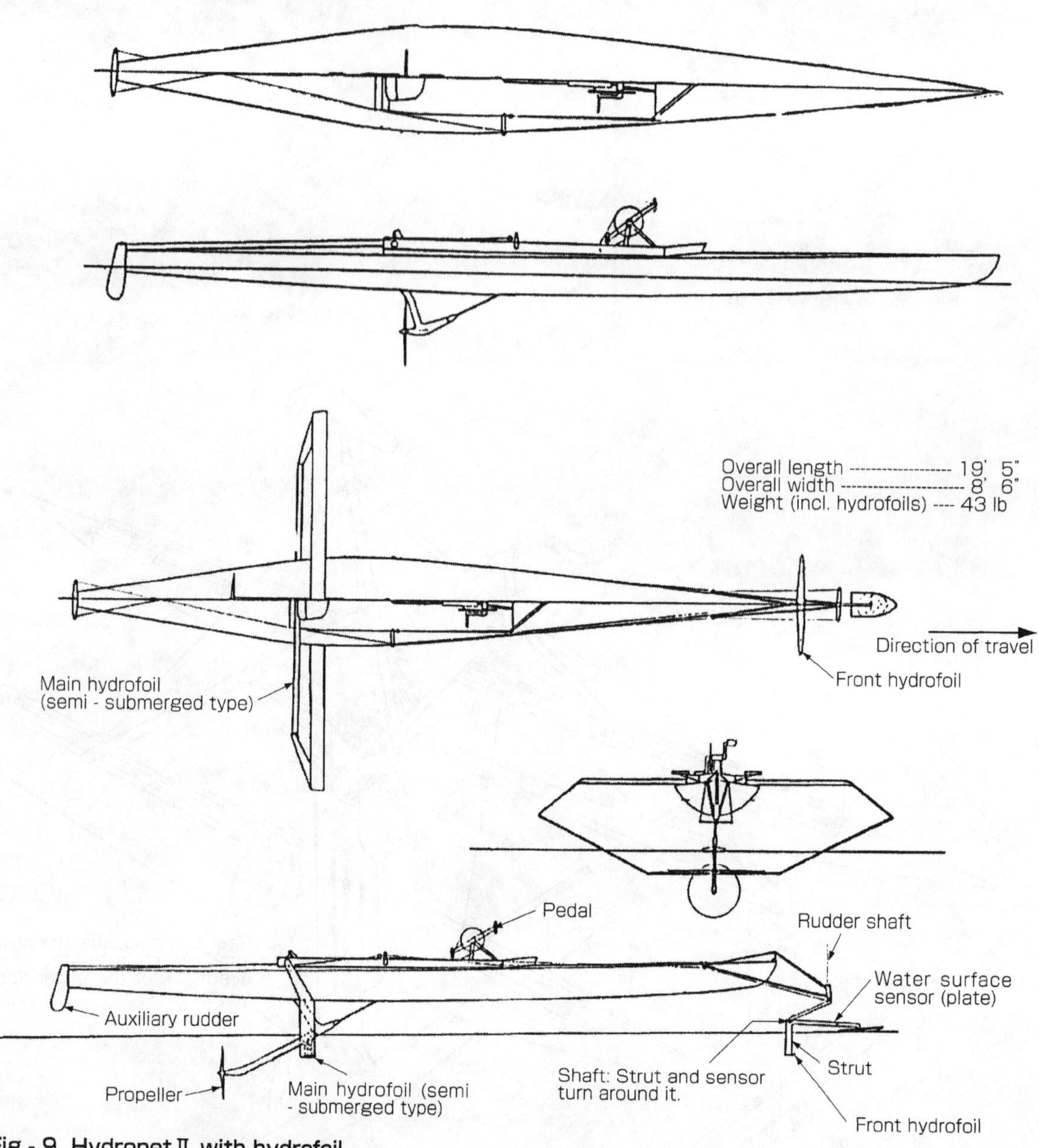

Fig - 9 Hydropet II with hydrofoil
Source: HUMAN POWER 7/4

The angle of incidence of the main hydrofoil is fixed with respect to hull. The rise of the bow naturally increases the angle of main hydrofoil, thus increasing the lift to raise the stern. When the boat comes up above the water surface and becomes level, the angle of attack declines relatively until the lift on the main hydrofoil and the load are in equilibrium. When this point is reached, the boat can run in a stable condition. Its transverse stability is similar to that of the single strut hydrofoil boat (Fig - 4). My idea of turning the front strut as in a bicycle to maintain transverse stability was adopted in this boat.

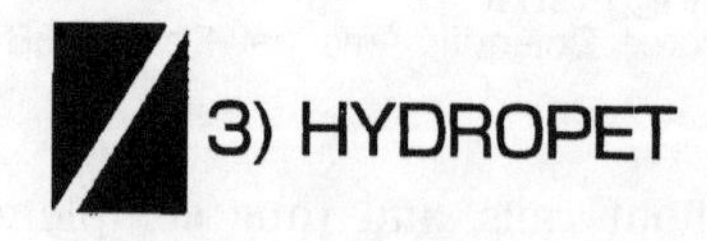

3) HYDROPET

Following the Flying Fish, Hydropet Series appeared on the scene (Fig-8). The designer was Sid Shatt, a Californian aviation engineer who had retired from Boeing Co. This boat used the hull of a commercial racing kayak and was designed for the so-called recumbent position, where a rider sat and pedaled with a low profile for transverse stability. Even without a hydrofoil, it attained a speed of about 10 knots. In 1985, he installed a hydrofoil, called it Hydropet II, and improved its performance progressively (Fig - 9).

The layout of Hydropet II and Flying Fish was Canard-type or Ente (main hydrofoil is located aft). The Hydropet II adopted a V-shaped main hydrofoil. This type was also known as the wave-piercing type or the semi-submerged type, and it was used in hydrofoil ferryboats built by Hitachi and Mitsubishi around the 1980s.

The submerged surface area of the V-shaped hydrofoil increases as the boat sinks. The hydrofoil increases the lift of the boat so that the original above-water height is maintained. If one side sinks, the submerged surface of the hydrofoil on that side increases, and a restoring force is generated. Thus, this layout automatically maintains the above-water height of the boat as well as its transverse stability.

A plate is directly installed on an arm that extends out of the small front strut and it acts as a water surface sensor. In this type, as compared to the original Flying Fish and my trailing arm (dragging a plate behind), the sensor movement is free of the surface agitation caused by the strut. The water surface sensor touches the undisturbed water surface, thus increasing the stability of the boat. This type of sensor is also simpler to construct.

4) DUPONT AWARD

In 1988, an American chemical company, Dupont, established the Dupont Award to promote progress in human powered boats. This contest required a 100 m run under a wind velocity of less than 3.22 knots (1.67 m/s) and a current of less than 0.2 knots (0.1037 m/s) in the presence of the IHPVA staff. The contest period was for three years starting from January 1, 1989 and it offered $25,000 as reward to the first boat that attained a speed of over 20 knots.

Until then, human powered boats could run as fast as 10 knots without hydrofoils, and at 12 to 13 knots with hydrofoils. Therefore, the target set in this contest was reasonably high and I personally thought that it was an excellent target. Naturally, the US human powered boating circles were excited by this challenge and Hydropet and Flying Fish renewed their own records. In Japan, Fumitaka Yokoyama, who worked at my laboratory in Yamaha, participated in the 1991 Dream Boat Contest (a large-scale human power boat race sponsored by the Japan Marine Vessel Promotion Association (currently, Japan Foundation) in the '91 to '93 period) with his Hydropet-type hydrofoil Phoenix, and was victorious (Fig - 10).

In the summer of 1992, Yokoyama aggressively worked on completing his advanced Phoenix II (Fig-11). I thought of sending him to the Dupont race scheduled for August if his boat was completed by that time. I began making the preparations. Yamaha had its R&D California in Los Angeles, so I began checking the racing site and making transportation arrangements for the boat. Despite all these efforts on this side, Yokoyama struggled to increase the speed of his boat. Phoenix had demonstrated a very stable run the previous year, but the new Phoenix II might have been too ambitious. In the beginning, we didn't think 20 knots was a distant dream, but it did not turn out to be easy. We could finally achieve only 12 knots. We decided to give up participa-

Fig - 10 Victorious Phoenix

Fig - 11 Phoenix II (1991 model)
Black board at bow is a reflector of the electronic speed-measuring device.

tion in the US race, cancelled all the arrangements, and focused on Japanese races only.

5) VICTORY OF DECAVITATOR

Decavitator was introduced in 1991. This boat was developed by a team of MIT professors and students headed by Mark Drella, a professor of space and aeronautics engineering. Its structure appeared to be very fragile compared to that of Flying Fish or Hydropet (Fig - 12).

This team, in fact, was one of the world's best engineering teams for human powered vehicles. They built Dydallos, a human powered airplane in 1988 and flew over the Mediterranean Sea in April the same year, starting at Crete and finishing at Santorini. It was an incredibly long distance of 119 km. They took advantage of this engineering legacy and utilized the above-water propeller on their human powered boat. Furthermore, they installed a two-phase hydrofoil system to achieve higher speeds. This system works such that when the boat reaches a given speed, the rider manually raises the main hydrofoil forward, using the very small hydrofoil only in the high-speed range, thereby reducing resistance to a minimum (Fig - 13).

The human powered boat race held at Milwaukee in 1991 where Decavitator collapsed during the race, clearly indicated the weakness of large machines constructed of long members. The above-water propeller is vulnerable to winds, and the two-phase hydrofoil system might have been less desirable on a water surface with waves. Nevertheless, in October, this boat set a record of 18.50 knots in Boston, the home of MIT, probably during one of the many trial runs performed in the morning mist over the River Charles. In the summer of 1992, Flying Fish recorded 17.96 knots in the final official run for the Dupont Award, but this record did not surpass that of Decavitator. Finally, none of the contestants reached 20 knots, but the Drella team was awarded the prize money of $25,000. This marked the end of the enthusiastic speed competition for human powered water vehicles.

Decavitator was brought to the Boston Museum of Science after this event for display along with Dydallos and it never returned to the waters. I haven't heard of Flying Fish being operated since 1993 either.

As I reviewed the IHPVA newsletters, I noticed that recent records in human powered boat contests in the United States were below 10 knots. I suspect that Hydropet too must have lost interest. I have not seen any article on such races after 1995. It seems that the end of the Dupont award practically discouraged the interested handful of people in the United States and they lost their passion for human powered boats.

The following article describes human powered boats in Japan with the focus on the Dream Boat Contest.

6) HUMAN POWERED BOAT RACE IN JAPAN

On one fine day in autumn in 1990, the NIPPON CHALLENGE engineering team of which I was a member, visited the Japanese team for the 1991 America's Cup at their camp in Gamagori, Aichi prefecture. Despite the strong winds, the sailboat ran very smoothly. We felt very excited as we watched it from an escort motorboat. We stayed that night at a hotel in Katahara, a hot spring resort.

Soaking myself in the hot spring prior to the meal, I enjoyed a pleasant conversation with Professor Terao of Tokai University and Professor Masuyama of Kanazawa Institute of Technology. We began to talk about speed trials of human powered boats and sailboats. The two professors earnestly wished to give their students an opportunity to experience the spirit and passion of engineers through actually building the boats for those events. I fully agreed with this idea. There was another reason I wanted to hold these events. At that time, I was responsible for the Marine Division at Yamaha Motor and I was under pressure from Mr. Kawakami, the then-chairman of Yamaha, to hold many events at the marina. I had been looking for a suitable event. The different motivations of the three people were combined and ignited by the NIPPON CHALLENGE boat. By the time we left the hot spring, we all had agreed to hold the event.

I proceeded with the plans and in 1991 Yamaha held the All Japan Human Powered Boat Championship at the Yamaha Marina on Lake Hamana. I made the 100-m time trial the main event since I had known about the Dupont Award. In addition, I included a race in which the boats had to turn back after reaching a buoy at the 200-m mark, to add more fun to the event. Naturally, the core entries were Tokai University, Kanazawa Institute of

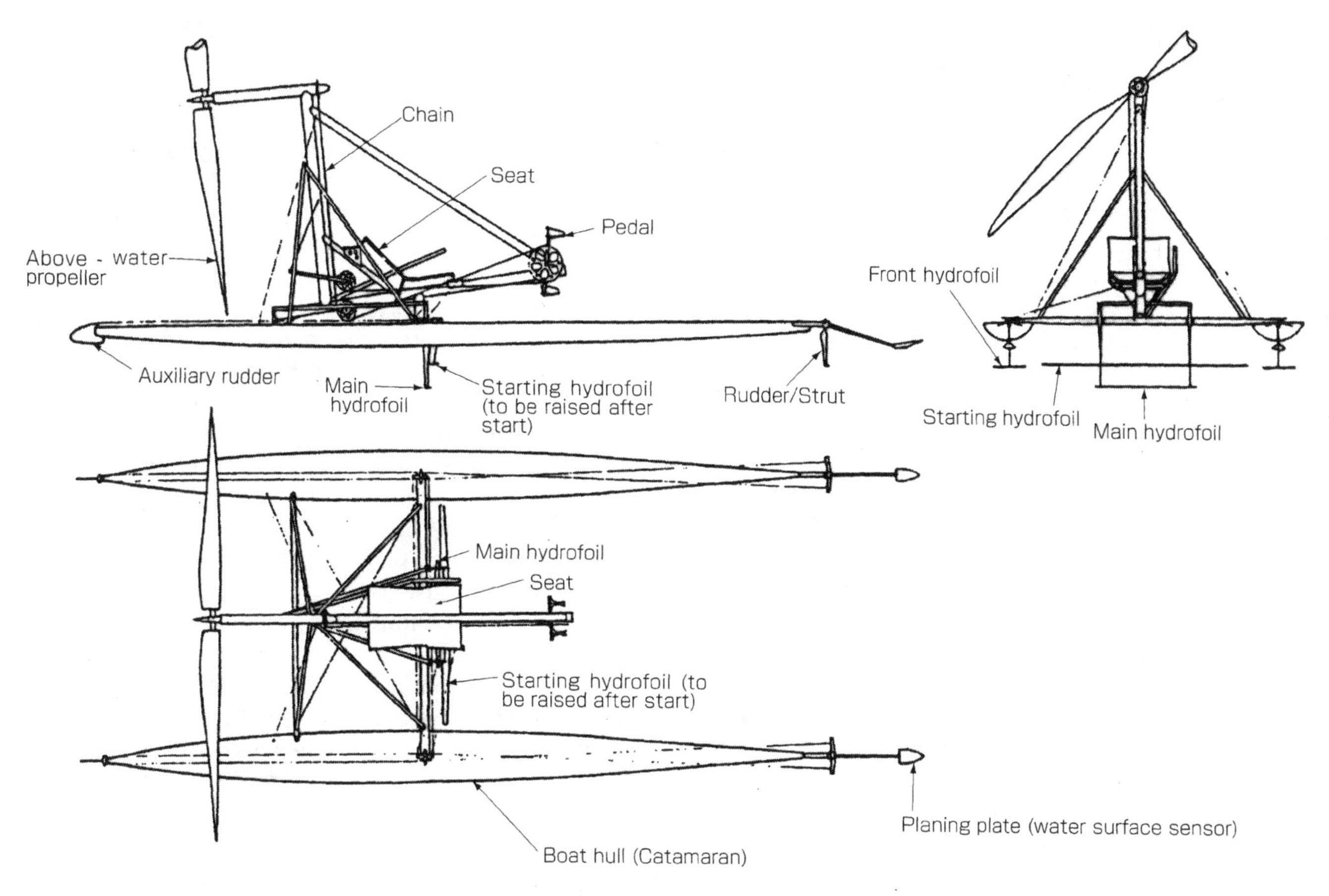

Fig - 12 Decavitator Source: HUMAN POWER, fall & winter, 1991 - 92

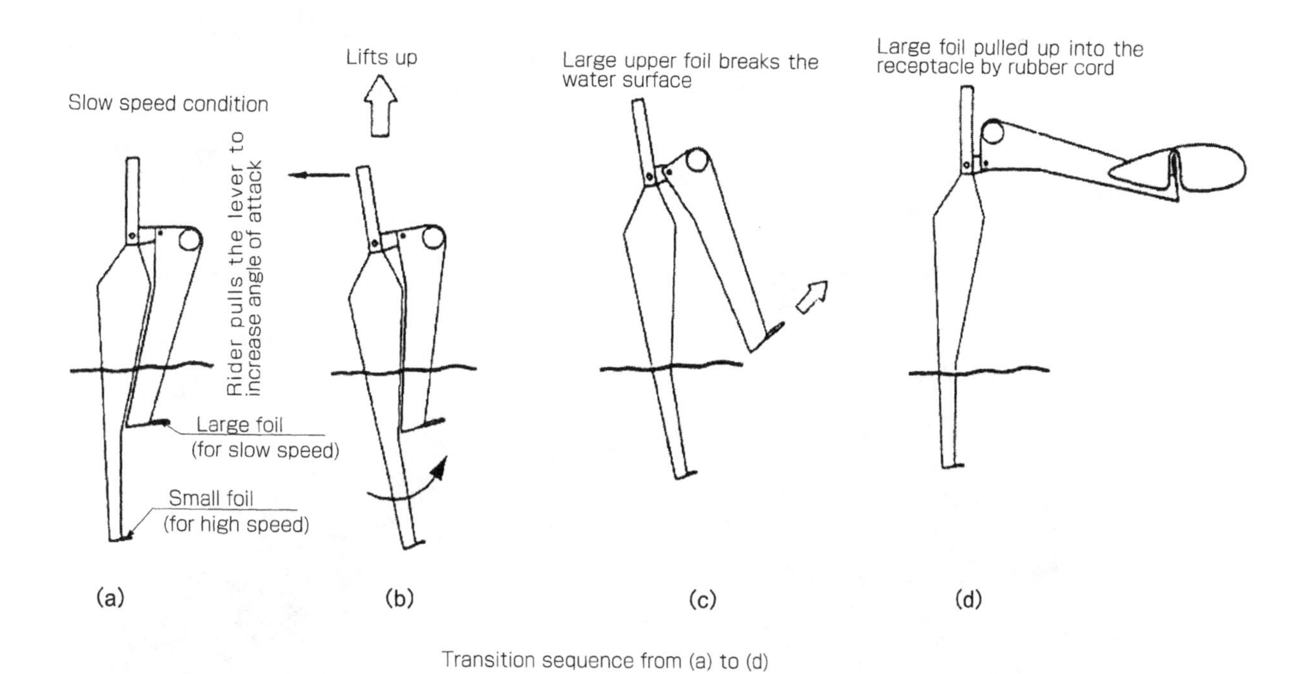

Fig - 13 Sequence of raising main hydrofoil - Decavitator
Source: HUMAN POWER, fall & winter, 1991 - 2, vol. 9/3 & 9/4, p.7

Technology, and Yamaha.

Fumitaka Yokoyama of Yamaha won the first championship with 11.75 knots (Phoenix). Shunsuke Horiuchi came second with the single scull. Horiuchi was the all-Japan scull champion five times and he participated in the Olympics in Los Angeles. The second event took place in 1992 and focused on creating awareness of the Dupont Award. I had Tsuide Yanagihara of my laboratory prepare the speed - measuring device that was used in the contest. Two piles were installed beyond the course so that the infrared rays could be directed between the piles and the pier. When a boat interrupted the infrared ray, time was recorded. I set up this arrangement because I thought that if our record was good, we could send it for consideration for the Dupont Award. Unfortunately, the fastest boat that year recorded only 12.36 knots (Phoenix II), followed by Cogito built by Yanagihara's group (Fig - 14).

Since this event, the All Japan Human Powered Boat Championship became a battleground between Phoenix and Cogito. They continued this keen competition in the time trials for the next seven years until 1998. As a result, Japan has been successful in developing world-class human powered boat technologies. I will explain each boat in more detail later.

At about the same time, the Japan Ship Building Industry Foundation (hereafter referred to as JSBI Foundation) and the Ship and Ocean Foundation had been planning a large-scale human powered boat race, based on the reports of a prior survey mission to the United States. When the plan was officially announced as the Dream Boat Contest, it tuned out to be a huge event involving the entire organization and all large and medium - sized shipbuilding firms in Japan. In comparison, our All Japan Human Powered Boat Championship at Lake Hamana was like a hand-made event on a small scale. I asked myself as to how the All Japan Human Powered Boat Championship should position itself. I concluded that this race at Lake Hamana should be in line with the philosophy of the Dupont Award and the IHPVA, and should continue to be a time trial race. That was why I made Yanagihara prepare the time - measuring device in 1992 that would satisfy

Fig - 14 Cogito

the Dupont Award rules and used it.

Both the race management and the development of the measuring device were very demanding propositions. In spite of only a few entries, the race required a lot of work. It was unfortunate that the race demanded everything from each individual who only wished to build a boat. It appeared that it was more fun entering the Dream Boat Contest, and it was also apparent the Dupont Award was approaching its end. The All Japan Human Powered Boat Championship was terminated after being held twice – 1991 and 1992.

7) DREAM BOAT CONTEST 1991

The first Dream Boat Contest in 1991 was divided into three categories: Human Powered Speed Boat, Human Powered Idea Boat, and Idea Boat. The Human Powered Speed Boat category was the only racing contest, while the other two were competitions to judge the quality of ideas. The race was carried out with a standing start of six boats in a group for the 200 m course in tournament style.

The first Contest was held in 1991, the same year when our first All Japan Human Power Boat Championship was held. The Contest's preliminary elimination time trial was held at three boat racing locations in Japan, namely Miyajima, Lake Hamana, and Tamagawa, and 24 boats were chosen to compete in August at the Heiwajima boat racing arena. Phoenix (Fig - 15) won the preliminary, semi-final, and final races at exceptional speed. Since the team had planned the race itself and had good experience, included Takahiro Sahara, who was an all - Japan class oarsman, and moreover, with the proven Phoenix design, it definitely had a lead over all other participants. The team recorded 43 seconds in the 200 m standing start race, maintaining a comfortable lead over the second place contestant, which recorded 60 seconds. Yokoyama and his team walked away with one million yen in their hands with pride and sense of accomplishment.

The Phoenix had a Canard type layout. It was equipped with a semi-submerged type main hydrofoil, and a front hydrofoil with a water surface sensor. The hull was a 5.5 m Ocean scull prototype. To meet the rules of the Dream Boat Contest, the hull was shortened to 5 m. The boat width was only 46 cm and it

Fig - 15 Phoenix

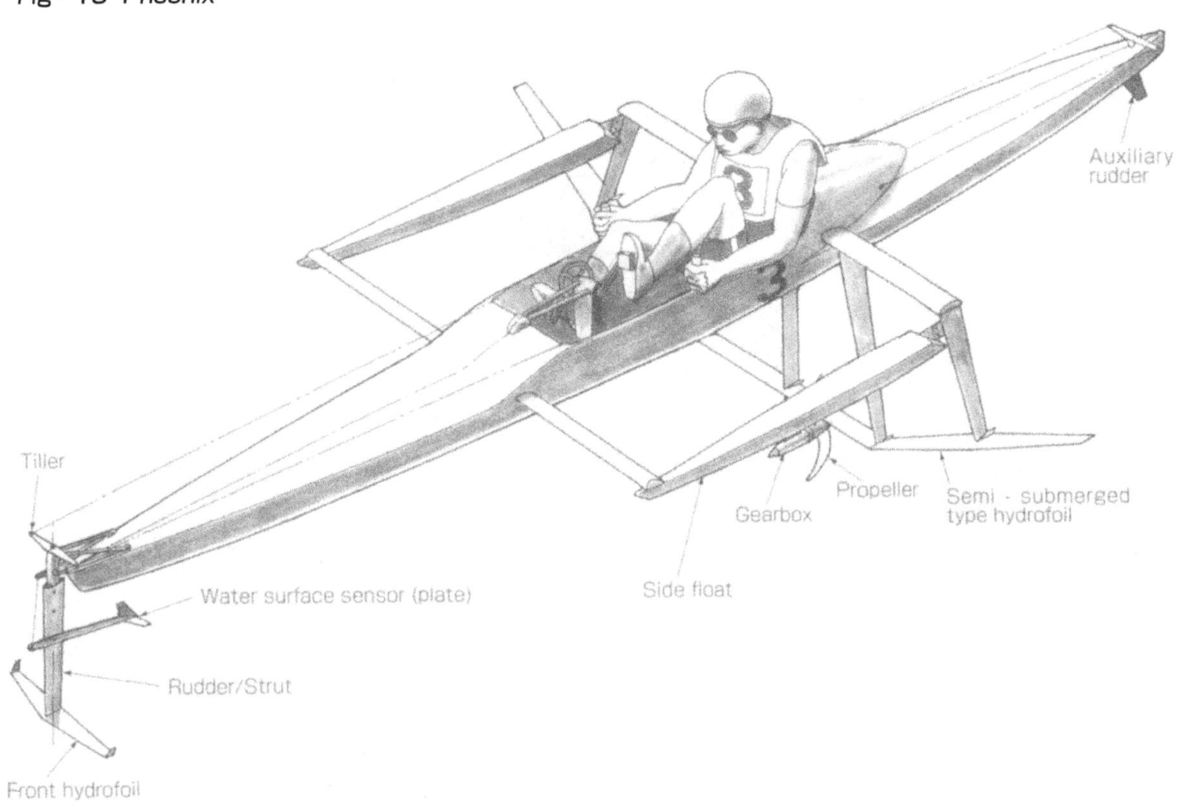

was very unstable even if run with a low profile. For this reason, this boat was equipped with a side float on each side. The sensor was a trailing type (sensor pivot shaft positioned in front of its dragging feather) sensor, which is different from that used in the Hydropet, but similar to that in the Flying Fish. This trailing type sensor was chosen by Yokoyama, the designer of the boat, because he had actually used it and gained confidence in it during the course of development of a motor-powered hydrofoil boat. In this sense, the boat was soundly designed and showed good stability from the beginning.

8) DREAM BOAT CONTEST 1992

The second Dream Boat Contest attracted 154 entries. The preliminary elimination process was held at three locations toward the end of June. The championship event took place with the selected 24 boats on August 2 at the Heiwajima boat racing arena. Having gained confidence with Phoenix the previous year, Yokoyama created a record-breaking Phoenix II (Fig-16). As mentioned earlier, we had to give up our plan of participation in the US race. Nevertheless, with a speed of 12 knots, we were optimistic about winning this race.

To aim for a speed of 20 knots, Phoenix II was equipped with a complex hydrofoil system that would accommodate two modes of travel; low and high speed hydrofoil operations. Let me explain why it was necessary to implement such a system.

Using a large hydrofoil, the boat is able to take off and become foilborne at low speeds. It reduces the hull resistance and allows foilborne state to be reached at a small horsepower. If you try to increase the speed in this condition, however, the large hydrofoil itself offers resistance. For example, a boat becomes foilborne

Fig - 16 Phoenix II ('93 model)

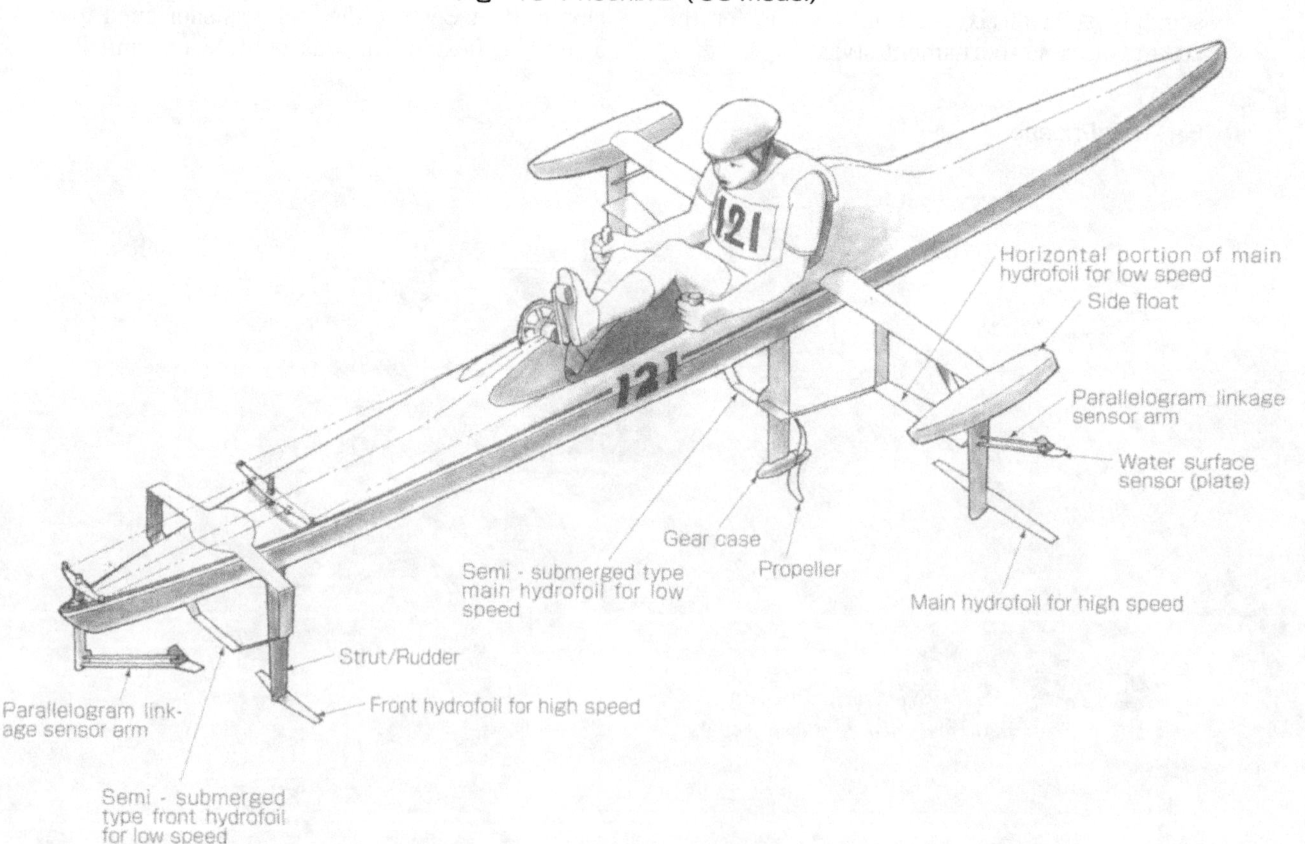

at 10 knots, and you want to double the speed to 20 knots. Since resistance is proportional to the square of the speed, the resistance becomes four times in this case. That means that the lift also becomes four times greater. The boat under such conditions would jump out of the water. You need to reduce the lift down to one-fourth by reducing the angle of attack, but it hardly reduces the resistance. In other words, three-fourth of the hydrofoil area makes no contribution to the boat running at 20 knots. Since it doesn't make sense to keep a white elephant (large hydrofoil) that increases the resistance by four times, you should take away three-fourths of the hydrofoil area, that is, reduce its area to one-fourth. This is what every designer of a hydrofoil boat wished.

An airplane extends a flap behind its small wing upon take-off and landing to generate lift of magnitude equivalent to that of a large wing. The density of water is 800 times that of air, which means that you need a wing size with an area of 1/800 in. Such a small hydrofoil has only an average section of 6 cm chord and 5 to 6 mm in thickness, which makes it difficult to obtain the minimum strength and rigidity. If you install a flap without evaluating it carefully, the hinge may produce so large a resistance that it would prevent this airplane solution from working effectively.

To resolve this problem, Phoenix II was equipped with a pair of large hydrofoils for low speeds and a pair of small hydrofoils for high speeds at the boat's forward and midship parts. At low speeds, the boat received primary lift generated by the large hydrofoils. At high speeds, the large hydrofoils were raised above the water level and only the small hydrofoils were used. Furthermore, it was designed such that the angle of elevation could be adjusted while in motion, which made the structure even more complex. The adjustments and the operation were quite difficult.

This boat presented various problems. The semi - submerged hydrofoil did not rise adequately above the water surface, and it caused water splashes, increasing the resistance, as a result of which the speed did not increase. Unless the boat took off cleanly, much energy was lost. This meant that the run could be rather unpredictable. In the worst case, the boat may not take off. The team participated in the race with these problems unresolved.

Hideaki Fukamura was the rider. He too was a rowing athlete and an engineer, but had tremendous energy. He once recorded a maximum of 2.4 horsepower during 40 second full - power pedaling. His recorded average was 1.6 horsepower.

Meanwhile, Yanagihara's group was working

Fig - 17 Cogito

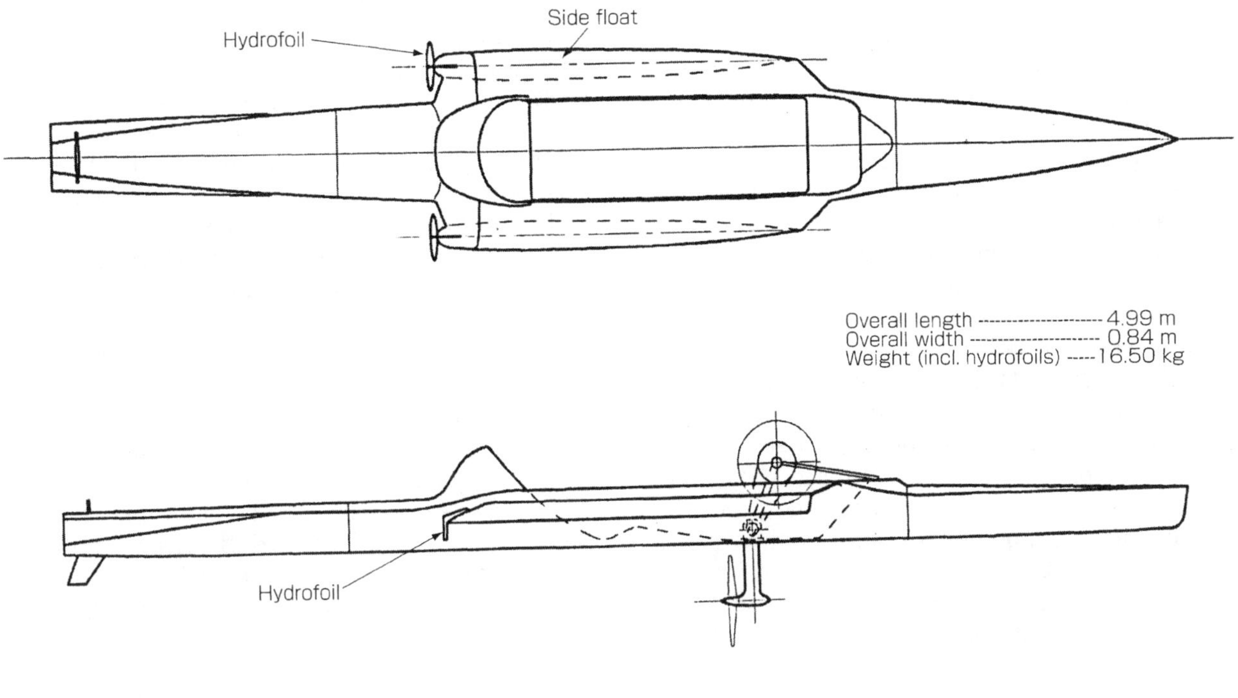

Overall length ---------------------- 4.99 m
Overall width ----------------------- 0.84 m
Weight (incl. hydrofoils) -----16.50 kg

Fig - 18 Cogito winning the 1992 Dream Boat Contest (Final race)

hard to finish their displacement-type racing boat, Cogito (Fig - 17). The hull was built from the same mold that had been used for Phoenix, and to compensate for rolling stability, the boat was equipped with short-arm side floats. However, the arm was so short that the boat sometimes rolled, and the speed reduced. Yanagihara and his team members installed small hydrofoils each with 3 cm chord x 10 cm span on both the side floats. They were installed at a fixed angle and served as a sort of damper to control rolling while the boat was in a high-speed run. Obviously, the boat had disadvantages compared to a hydrofoil boat.

The group modified the stern so that the overall length was 5 m and tried to improve water separation as part of their desperate efforts to increase speed. This boat was to be operated by Shunsuke Horiuchi. An unexpected incident occurred at the Heiwajima boat racing arena. Phoenix, the boat, which was predicted as most-likely-to-win, failed to take off. In the semi-final event, Phoenix first appeared to have taken off, but lost its balance and never got the chance to take off again. Once you fail to take off, you have already used up all your energy and there is no more energy left for another attempt to take off. A hydrofoil boat that has failed to take off cannot beat even a displacement-type boat. Phoenix did not make it to the final championship race.

On the other hand, Cogito demonstrated its potential in this race (Fig-18). Against all odds, Cogito with Shunsuke Horiuchi beat all other boats including hydrofoil boats, and won the race. Their engineering efforts at maximizing the speed of the displacement type boat brought them unexpected rewards.

9) DREAM BOAT CONTEST 1993

The third Dream Boat Contest in 1993, particularly the human power speedboat category, attracted considerable attention and the entries went up to 213. Yokoyama refined Phoenix II to create Phoenix III with considerable improvement in the stability. By this time, Fukamura was well accustomed to the boat, which was painted with a new color.

The Cogito team retained its confidence from the previous year, and developed an entirely new boat. Besides Yanagihara, who was a good aviation engineer, this team consisted of Takashi Motoyama, who was an experienced sailor and familiar with advanced composite materials (ACM), and Tokuzo Fukamachi, who was an expert in fabrication, including aluminum fabrication. There was one other person in this group - Masayuki Hattori, a prototyping expert nicknamed Master in the NIPPON CHALLENGE camp of America's Cup team. As such, the group was very talented, and they were able to build a very reliable boat after a thorough evaluation of different perspectives. The new boat was named Cogito

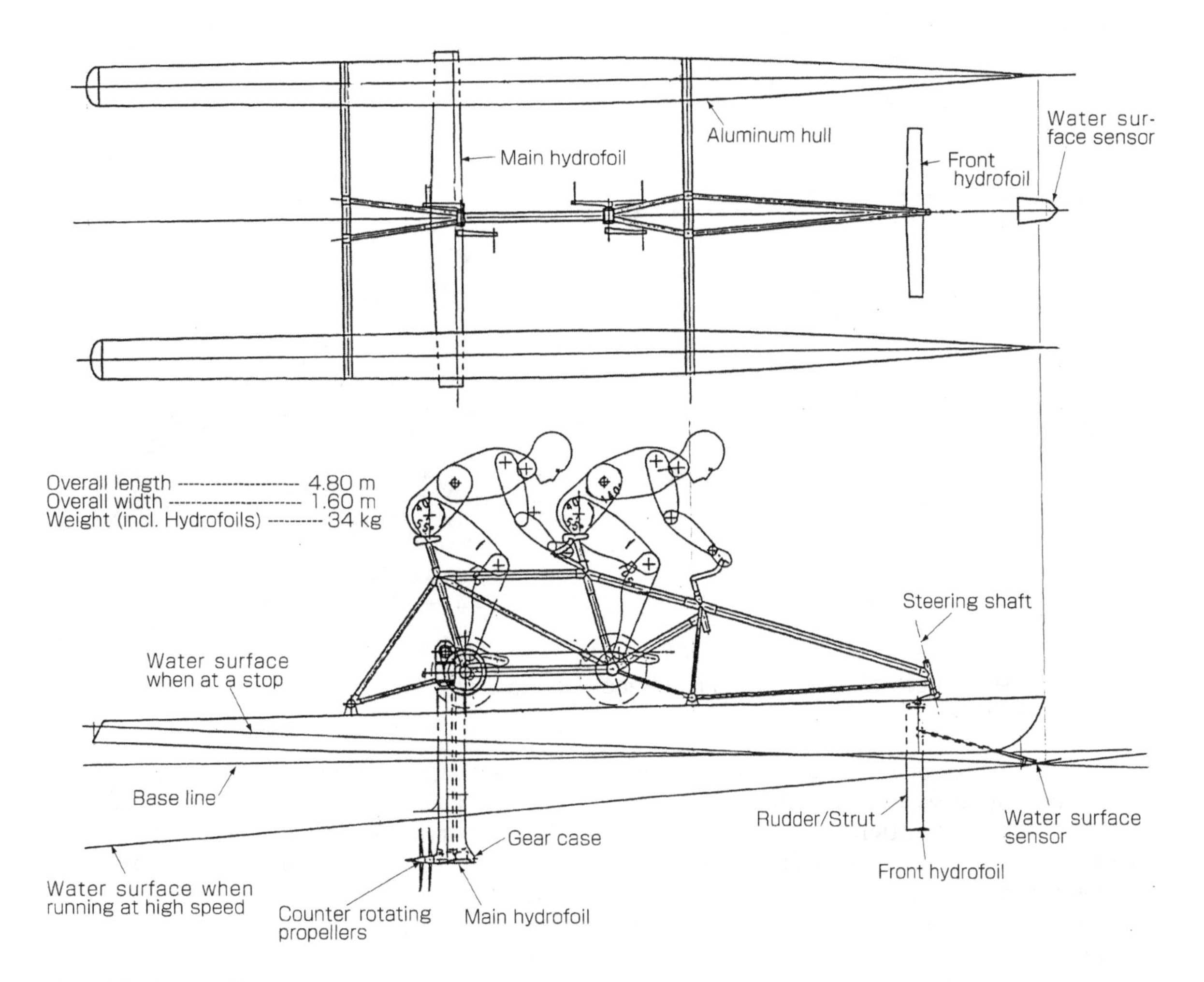

Fig - 19 Cogito II

II (Fig - 19).

The rules of the Dream Boat Contest required one or two riders for all events. In the past races, many other teams used two - seaters, but the Yamaha team was consistently making single-seat boats since we always had the Dupont Award in mind. Two-riders meant double the horsepower. In addition, a slight shift in the pedal angles of each rider could make the torque on the propeller uniform and favorably affect the propeller efficiency. Furthermore, two riders do not cause the resistance to double. Since the rear rider is behind the front rider, the air resistance should be about 1.2 times that of a single-seat boat. For the same reason, the water resistance as well as the air resistance of the hull would be about 1.6 times. The two-seater configuration would increase the hydrofoil and propeller sizes, which should be an additional advantage. The weight of the boat would be less than double. All these pointed to the advantages of using a two - seater boat.

The layout of Cogito II was almost the same as that of the Flying Fish, having a bicycle type structure with a rolling stability device and Hydropet type water surface sensor. When they tried to make the hull a catamaran, they found that the Ocean Scull female mold was too large for the new boat. Because it would take lots of work and time to build a new mold, they built an aluminum hull. The hull would not require much durability and for this reason they used 0.5 mm ultra-duralumin (A2024) sheet, which was bent and attached to the hull without three-dimensional forming. On the deck, they used a thin CFRP (Carbon Fiber Reinforced Plastic) sheet. The frame was assembled with thin aluminum alloy tubes using rivets. This boat was designed very well. Except for the repair of the water surface sensor in the beginning, it had good integrity. To ensure human power, they reserved Shunsuke Horiuchi, Masaki Kamimura, who was a bicy-

Fig - 20 1993 Dream Boat Contest Final Race : Cogito II and Phoenix III fighting for the lead

cle racer and engineer, and Daisuke Chiba, who was a professional bicycle racer.

The race was anticipated to be a fight between Phoenix III and Cogito II, and both teams still had reserve energy after they won through the preliminaries. I had no idea which team would win this race. If Phoenix III with its small hydrofoil for high speeds ran well, Cogito II would have no chance. The rider was the powerful Fukamura. A concern was that if he exhausted his energy before take off, it might be difficult for him to continue the acceleration. In other words, there were some uncertainties in the performance of Phoenix III.

Cogito II, on the other hand, was consistent. It would take off easily with its large main hydrofoil and the power of the two riders. The race was very close, but finally Cogito II won (Fig - 20) at 30"21 and Phoenix III came second with 30"88, followed by the third-place boat that came in four seconds behind. Cogito II reduced its time by 14"48 as compared to its previous year's time of 44"69, which was 48% faster than a year ago with its speed increasing from 8.7 to 12.87 knots. I was very happy to see the engineers of Horiuchi Lab winning the first and second places.

With this victory, Cogito II became the engineering benchmark for human powered hydrofoil boats. Through continuous improvements, this boat won the 200 m speed races for five consecutive years without any change in the basic design. In 1997, it marked an impressive 22.91 seconds (16.97 knots), which is admirable.

The Dream Boat Contest too came to an end. The popularity of the human powered boats in Japan seems to have faded. The next article introduces the Lake Hamana Solar & Human Power Boat Race that replaced the Dream Boat Contest.

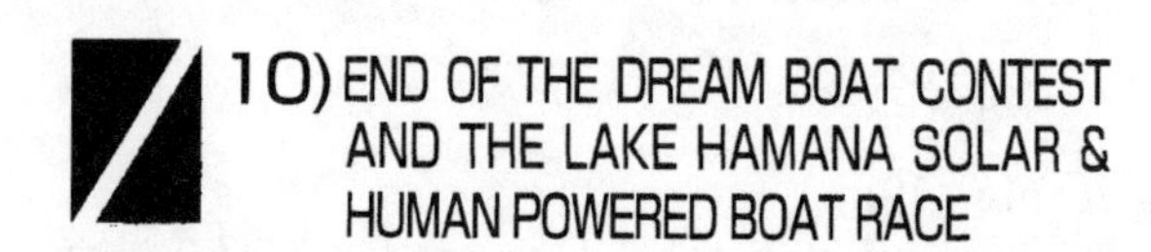

10) END OF THE DREAM BOAT CONTEST AND THE LAKE HAMANA SOLAR & HUMAN POWERED BOAT RACE

I heard a rumor that the Dream Boat Contest was to be terminated at the end of 1993. It appeared to be true. I was caught by surprise because I had withdrawn from the All Japan Human Power Boat Race to focus on this race. Professor Miyata of Tokyo University, Department of Naval Architecture, was also very apprehensive, because he had already included the building of a human powered boat in his regular curriculum for his students.

Professor Miyata had his students design miniature boats and compete for speed in the testing tank in 1989, and he was very impressed with his students' devotion and involvement in this assignment and with the efforts they put into this project. He was so moved by their enthusiasm and passion that he wrote an article about it in the Bulletin of the Society of Naval Architects of Japan (May 25, 1990). Like Professors Masuyama and

Terao, he believed in the merits of students engrossed in actually building something.

After conversing with them, I said to myself that I must do something about it. These professors were obviously in trouble and the Dream Boat Contest had once attracted nearly 200 entries. Those who became interested in the building of human powered boats might also be in trouble not knowing their goals. I personally was sorry to see the end of development of human powered boats . Everybody would stop building the boats and human powered boats may disappear in due course unless something was done. How could I revive such a big event that cost several hundred millions of yen by myself? A thought occurred to me - the human power boat race could be combined with the Solar Boat Race that was held annually on Lake Hamana.

The Solar Boat Race was held on a summer day when the sun was at its strongest and precipitation was minimal. It attracted about 20 boats, all of which were less than 6 m in length and were equipped with solar batteries of up to 480 W, and supplied batteries of about 24 AH. They raced in a 200 m flying start run and a one-hour endurance run. Since I had been involved with the management of these races, I knew the organization and the details of the races. I saw no reason why they couldn't be combined, although there were some differences in the rules and the race procedures. A positive aspect was that the human powered boat race would be an added attraction to the Solar Boat Race.

Combining it with the Solar Boat Race would naturally require additional operating expenses. The Solar Boat Race had always been managed by minimizing cost, and combining the human powered boat race would require money.

As soon as the New Year began, I made use of every possible opportunity to meet with people in the JSBI Foundation, the Ship & Ocean Foundation, and the Ministry of Transport that had financially supported the Dream Boat Contest. Many of the people I met felt sorry that the Dream Boat Contest had terminated and were supportive of my idea. Aside from the financial needs, I was concerned about how many teams would participate in this event. I sent out questionnaires to the 170 teams that had participated in races in the past to find out if they would be interested in participating in future races, and if they would attend the rules meeting at a hotel in Hamamatsu in February. I received 149 long-awaited responses; 77 indicated a positive response and possible participation. I felt very encouraged. On February 19, fifty people came together for the meeting, and we discussed and agreed upon the rules and how to run the races.

This idea was received favorably by many people. The JSBI Foundation promised a five million-yen assistance, and that made it possible for the human powered boat racing event to combine with the solar boat racing event. In 1994, the Lake Hamana Solar Boat & Human Powered Boat Race took place at the Lake Hamana boat-racing arena, in which 22 solar boats and 50 human powered boats participated.

The race maintained the same standing - start method as in the Dream Boat Contest. It cost us money to build a pit for six boats to start simultaneously, and the pit was in the way during the circling race. The starting method also needed many people. The steering committee found a collapsible pier for a relatively small sum and used it that year. From the following year, the human powered boat race adopted the flying start just as in the Solar Boat Races, which made the race much easier to hold. The flying start, however, was more difficult for the participants. To make it easier for the participants, Masao Yagi, the Chief Judge, along with others asked their business partners to develop an F - 1 type starting machine that would light lamps in succession. It was ready the day before the race for us to inspect, try, and practice the flying start. This problem was almost resolved.

From 1995 onward, the 100 m time trial for human powered boats was added to the event so that we could compare the times with the records of the Dupont Award. In this event, an adequate approach run was allowed before the boat traveled the 100 m full course enabling us to measure the speed for the full 100 m. However, we had to do this time trial during a predetermined time of the day, and because of this, we could not comply with the Dupont Award criteria for the wind and water velocities.

In 1995, we tried a new race of about 700 m run encircling two buoys 300 m apart. This race was designed for both solar and human powered boats. To our surprise, most of the participants, numbering about 60 boats took part in this race. All the participants started simultaneously. We were a little worried about collision, but the race went smoothly without any incident. After this race, we realized that the participants wanted to take part in as many events as possible since they were already there. This trial revealed that for this distance, solar boats held a definite advantage compared to human powered boats. This mixed race was not offered the following year. Instead, two new official races were added : one was a circling slalom race on a course of about 1000 m using seven buoys to see how many trips a solar boat could do in an hour, and the other was a one-circle time trial for human powered boats using the same setting. These races had all the participating boats start at the same time, so they were spectacular. Moreover, the race challenged the riders' operating skills and the boat's turning performance since seven buoys were used, which made the race more fun. I felt that this kind of race would promote wholesome development of the boat - building culture. I wished more boats would run the entire course foilborne using their hydrofoils.

In 1996, Yokoyama built a new boat called Super Phoenix, which was the culmination of all his engineering experience and knowledge. This boat, however, probably needed some more time to mature. It definitely ran faster than Cogito at one time, but capsized at other times and didn't win any race.

11) THE ALL JAPAN SOLAR & HUMAN POWERED BOAT CHAMPIONSHIP, 1997

The core sponsor of the event, Lake Hamana Solar Association disbanded at the end of 1996. Soon after, the members of the steering committee and their respective organizations formed the Japan Solar & Human Powered Boat Association in January, 1997. With this change, the Lake Hamana Solar Boat & Human Powered Boat Race was renamed as the All Japan Solar & Human Powered Boat Championship.

In the All Japan Solar & Human Powered Boat Championship 1997, the battle was once again between Cogito 3.8 (Fig-21) and Super Phoenix (Fig-22). Exhibiting superior starting technique, Cogito 3.8 won the 200 m speed race. In the slalom, it held the lead up to the last corner, but unfortunately lost the race to Super Phoenix finally. Super Phoenix set an excellent record of 19.45 knots in the 100 m dash. Cogito 3.8 registered 18.39 knots, which was faster than the previous year by 2.36 knots, and came in second. This record of Super Phoenix was accomplished with a tail wind of 5 m/s. Nonetheless, it was well beyond

Fig - 21 Cogito 3.8 (1997)
The hull was made lighter and the stern shortened by adopting the CFRP sandwich hull construction. As a result, the boat's bow rose easily, making the take - off easier. A wider hull with smaller depth now allows greater banking angle and superior maneuverability.

Fig - 22a Super Phoenix (1997)
Note the very small wake even at the maximum speed.

Fig. - 22b Main hydrofoil of Super Phoenix (1997)
The long and slender wing with aspect ratio of 20 is made of solid CFRP

Fig - 22c Front hydrofoil of Super Phoenix (1997)
Water surface sensor (right) equipped with a parallelogram linkage arm and small front hydrofoil

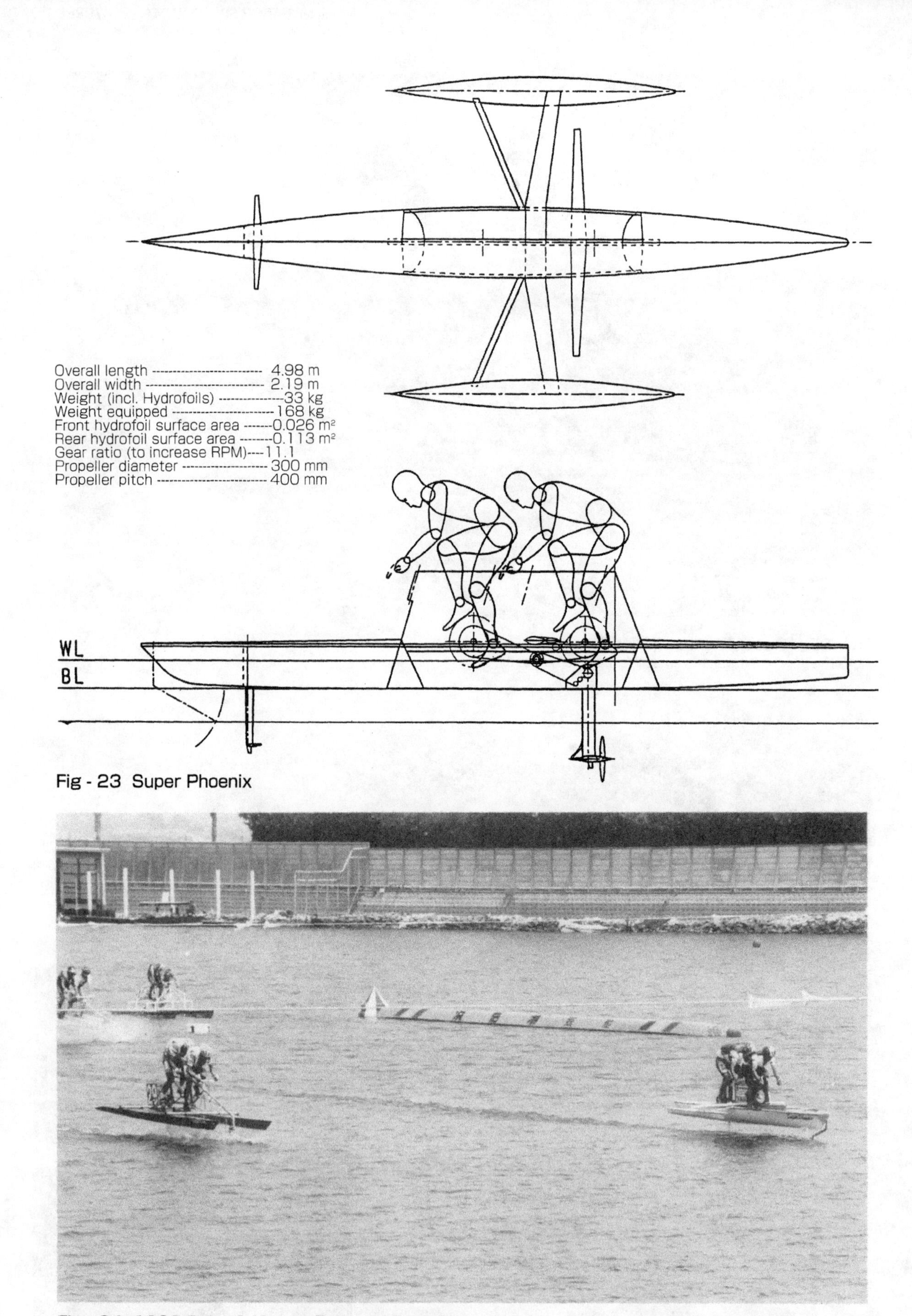

Fig - 23 Super Phoenix

Fig - 24 1998 Solar & Human Powered Boat Championship, 200 m time trial, final race

1st place - Super Phoenix, right and 2nd place - Cogito '98, left

Fig - 25 Contra rotating propeller and main hydrofoil of Super Phoenix
Black gap at the center of the main hydrofoil edge is for high - speed runs when the angle of attack reduces and the trailing edge of the main hydrofoil rises so that the gap is filled.

the official world record of 18.50 knots established by Decavitator. That made us very happy. Compared to the one-seat Decavitator, Super Phoenix was a two-seat boat and we were assisted by the tail wind. Yet, the result was superior to anything earlier. Yamaha Motor Company awarded them ¥500,000 (Fig - 23) in recognition of their good work.

Super Phoenix built by Yokoyama, is possibly the most advanced human powered boat today. For this reason, I would like to tell you about the design of this boat. If you are interested in the details, you should refer to his article in the Bulletin of the Society of Naval Architects of Japan, July 1998 (in Japanese).

To reduce the resistance in the high-speed range, it is necessary to use a small hydrofoil as mentioned earlier. To take off easily with the small hydrofoil, it is very important to drastically reduce the resistance immediately before and after the take off. Super Phoenix adopted the following measures:

(1) Reduce the water resistance by about 30% by using a mono hull rather than a catamaran.
(2) Minimize[1] induced drag by making the main hydrofoil long and slender (aspect ratio of about 20).
(3) Make the strut shorter and thinner by installing it directly on the keel. The strut supports the hydrofoil and the propeller.
(4) Maintain a consistent angle of attack that ensures the least resistance by supporting the water sensor plate using a parallelogram linkage arm. The arm also moves freely around vertical axis to prevent damage by unexpected side force. This design makes it lighter
(5) Reduce the angle of elevation automatically at high speeds by flexibly mounting the main hydrofoil. This structure minimizes changes in the boat attitude and allows installation of a shorter and thinner strut.
(6) Allow the rider to sit low and reduce air resistance since this is a mono hull. A low profile provides for better stability.

Yokoyama had a support team consisting of Fukamura, a powerful rider and an engineer, and woodworking and CFRP experts such as Mitsuo Yamade, Tatsuyoshi Ishihara, and Masayuki Fujita. Yokoyama himself studied very hard, designed, and built what he had designed. It was impressive seeing the energy they expended even after the normal working hours. They were fearless as they took up new challenges, yet sometimes they were beaten by the superior stability of Cogito.

Super Phoenix improved remarkably. It was expected to exceed 20 knots in the 1998 race. The steering committee prepared for official approval of the world record. However, the boat failed our expectations (Fig-24) at the race. It registered 9.96 seconds in the 100 m dash with tail wind, which was not much different from the 9.99 seconds the previous year. The average speed for the round trip fell short

1) If the maximum lift is generated by having a large wing angle upon take - off, the wing edge generates vortex increasing the drag. Having a thin and long wing creates smaller edges, thus reducing both the vortex and drag.

Fig - 26 Speed records of the Cogito series and the Phoenix series for eight years

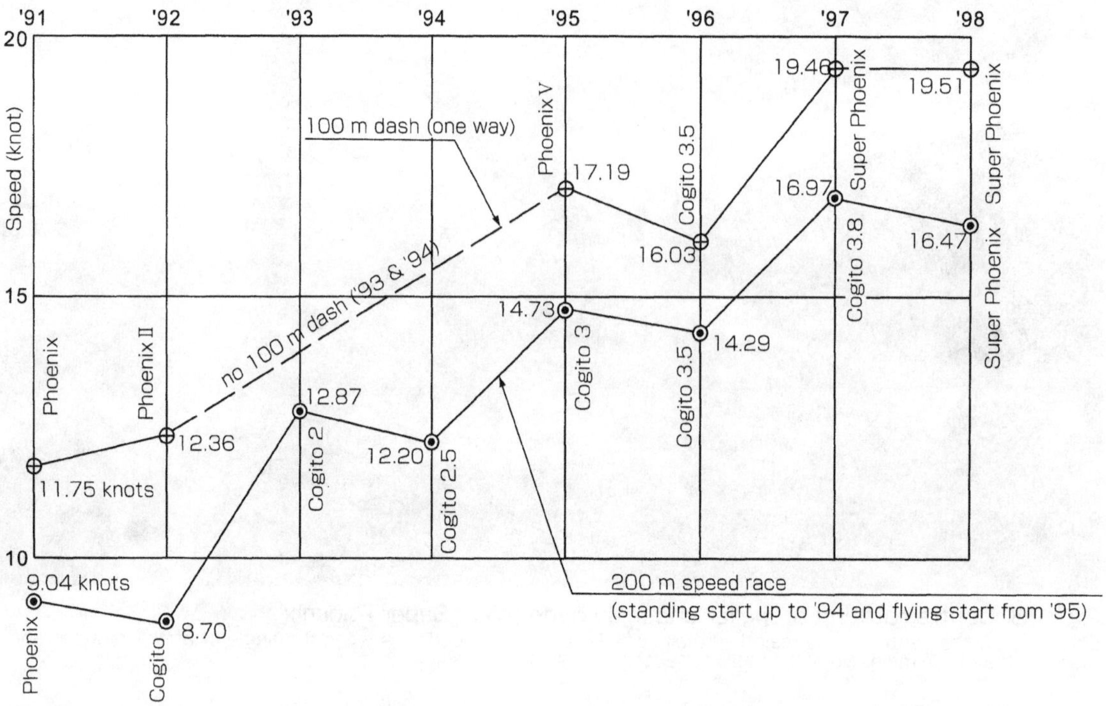

of 20 knots, and the boat made only 18.16 knots. Nevertheless, both the winning Super Phoenix and the runner-up Cogito 98 clearly demonstrated the benefits of repeated improvements (Fig-25). They gave us great hope for the next year.

12) EXPECTATIONS HENCEFORTH

Fig-26 shows the speed records of human powered boats for the past eight years. During this period, there has been an improvement of 7.76 knots in the 100 m dash and 7.36 knots in the 200 m speed race. This means an annual average increase of 1 knot, which suggests that we should exceed 20 knots in 1999 or 2000.

What is the next target? Calculations show that you can achieve up to 24 knots by increasing the rider's horsepower by 15% and reducing the drag of the boat by 35%. The former is very possible. What about the latter, reduction in the resistance? We could use a fairing on the rider to reduce air resistance. We could create ideal high-speed hydrofoil runs by switching over to the smaller hydrofoil upon reaching adequate speed. However, we know from experience that riders dislike the fairing because it makes them susceptible to winds. On the other hand, switching over to the high-speed hydrofoil run has many delicate issues that troubled us earlier, and this wouldn't be easy in adverse water surface conditions.

Such a 'delicate' boat may be able to set the record, as demonstrated by Decavitator in the past, through repeated runs in calm water in the mornings. To win our races, however, is an entirely different matter. Since our races are held even under strong winds, the boats must withstand waves and winds, and they must have good turning ability. Under these circumstances, I would say that reaching 20 knots is a matter of time, but it would take considerable time to reach 24 knots and it would require us to set up especially favorable conditions. The key question here is how to motivate ourselves and prevent loss of interest so that we can continue with our efforts.

The popularity of the human powered boat race in the United States seems to have diminished now. The builders of fast boats are still young, and they will return given the opportunity. There are quite a few enthusiasts in Europe, too. It is time we revitalize the human power boat race and upgrade it to a worldwide event.

If one can purchase ready-to-cut high-performance CFRP and readymade driving systems, one can easily build a single-seat or a double-seat boat that travels at twice the speed of the America's Cup sailboat or at 1.5

INTERNATIONAL HUMAN POWERED VEHICLE ASSOCIATION

Records Committee Awards this

Certificate of Record

to

Masayuki Akasaka & Shigeru Takayanagi

For Their

100 METER FLYING START SPEED TRIAL

A Men's Water Record, for an

Elapsed Time of 10.409 seconds

18.67 Kt (21.49 Mph)

In Fumitaka Yokoyma's Super Phoenix

At Shizuoka, Japan on the 27th Day of August 2000

Beth Wichers Schreur, President

William Gaines, Records Chairman

Fig - 27 Certificate of World Record

Over all length	4.98m
Over all breadth	2.19m
Height	1.46m
Weight	33.0kg
Front wing span	680 mm
Front wing area	0.0231m²
Main wing span	1100 mm
Main wing area	0.0649m²
Sub wing span	2000 mm
Sub wing area	0.169m²
Propeller diameter (front)	230 mm
Propeller diameter (rear)	215 mm
Gear ratio	11.1

Fig - 28 Specifications of Super Phoenix

Fig - 29 Record run of Super Phoenix

times the speed of the eight oars shell. That would be wonderful.

Since I came to know IHPVA closely while working with them on the official approval of records, I would like to discuss with them ways to make any activity related to human powered boats an enjoyable international activity. These days, I am considering how I can remove bottlenecks in boat-building processes so as to encourage more people to take part in this pastime.

Postscript:

The HUMAN POWERED BOAT RACE has continued even after 1999. As of 2001, regrettably, a speed of over 20 knots has not yet been attained. One important reason is that the progress of such boats has slowed down; another reason is that the method of measuring the world speed record has changed, and only speeds of boats measured in the windward state are recognized. In such conditions, we submitted the record that Super Phoenix attained in the windward condition in 1999 to IHPVA as application for the world record. The record of 18.67 knots was certified for the multi rider class. This record is certainly not over 20 knots, but it is the best speed on water attained by human power.

Note : Author is the Chairman of the JAPAN SOLAR & HUMAN POWERED BOAT ASSOCIATION.

REMOTE-CONTROL HELICOPTER R-50

Professor Washizu in the Department of Aeronautics of Tokyo University had told me once that YAMAHA should develop an inexpensive helicopter. The chance for development of a remote controlled pilotless helicopter came up suddenly. But it was very difficult to develop a remote controlled helicopter with two contra-rotating rotors. Worried that we might not succeed, I developed in parallel, another tail rotor type helicopter that could be flown by manual control with an engineer. Over time, I changed the initial project of the committee to the latter, and the production of the tail rotor model has now reached 3000 units. Nowadays, these helicopters are used for spraying agricultural chemicals, rice seeding, and for taking photographs, and they have also led to an increase in jobs for women who operate them.

1) Professor Washizu

I was Prof. Washizu's student just after World War II, when the present Department of Aeronautics of Tokyo University was called the Department of Applied Mathematics. Professor Washizu was well known as an expert in analytical engineering, and the theory of elasticity. I, on the other hand, was poor in this subject. I had spent most of my time rowing in the boathouse at Mukojima and producing model airplanes. Our personalities (Professor Washizu and I) were very different.

I used to visit him frequently when he ordered me to make a drafting machine (hereafter called Perspector) for perspective projection drawings. Around 1953, Mrs. Washizu was an illustrator and she was involved with the preparing the parts list for Hino Automobile Manufacturing Co. Professor Washizu wanted to produce a drafting machine for perspective drawings that would replace his wife's work. He wanted me to produce a working model that would operate exactly the same as the manual perspective projection drawing that students studied at that time.

I used a 90 cm-square plywood board and

Fig - 1 PERSPECTOR (original)

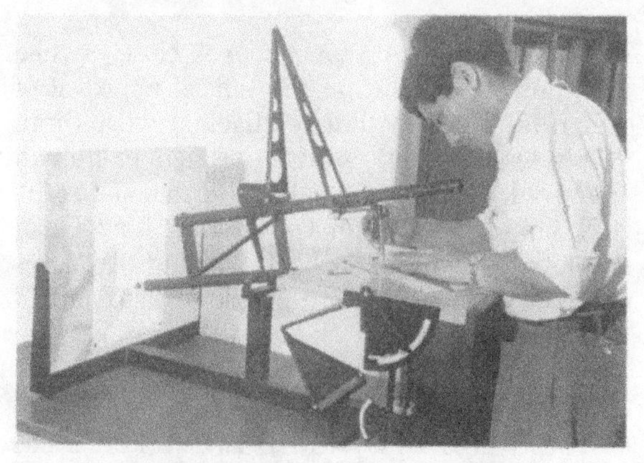

Fig - 2 PERSPECTOR (working model)
As the tracer tip traces the plan view, the pencil draws the perspective projection on the paper set vertically. The pencil can be raised by pressing on the trigger.

built the mechanism on it (Fig-1) with model airplane wood material and a celluloid sheet. You could place the top view on the right hand side of the plywood, trace the drawing with the tracer needle that projected out to the right, and then the pencil attached to this device would draw its perspective drawing underneath. It did work, but to my disappointment, the perspective projection drawn by this machine was very small because of excessive interference between the mechanical components.

I learned of the limitations of mechanical conversion of the drafting process, but at the same time, I thought of substituting the direct mechanism with a light beam that connected the 3D object and its perspective projection. Instead of the light beam, I used a straight bar, which was supported by gimbals (permitting free yaw and pitching). As the device traced a 3D object, the other end of the device drew its perspective projection on a vertical surface. In this device, firstly a sheet with a plan view was laid on the board, and a perspective drawing could be made from the plan view on the board by adjusting its height and by tracing the horizontal cross section that corresponded to the board height. It was necessary for the bar to freely extend and retract on both sides of the fulcrum. I brought the fulcrum of the bar to the point where the hole of a pinhole camera would be, and substituted the bar for the light beam.

The small model I made at that time worked very well. Later, we received a scientific research grant to make a demonstration model, and my device was introduced to the public in the Scientific Asahi magazine (Fig - 2). We also applied for and obtained a U.S. patent, and had two manufactures prototype it. I worked with Prof. Washizu for nearly ten years and we developed a strong relationship based on mutual trust.

One day, when I visited Prof. Washizu at his office. He told me that he wished that motorcycle manufacturers would build helicopters. Prof. Washizu was involved in the development of a helicopter at that time, but the high cost and difficulties of commercialization apparently frustrated him. As I was working for Yamaha Motor Company at that time, I took his comment as a suggestion. I was convinced that small helicopters and motorcycles for a manufacturer would be a good combination.

The light airplanes Cessna or Piper[1] built by US manufacturers would not sell well in Japan since the topography and the population density would not allow for safe emergency landings. In this respect, small helicopters seemed to be more suitable and had good potential for commercialization in Japan only if safety as well as cost issues were resolved. I was a young boat designer at that time. I was convinced by his words, but that didn't mean that I could move in the direction he suggested right away.

2) Helicopter YHX

In 1975, Yamaha Motor began the development of the YHX helicopter. The project started when Koji Ogawa, who had been involved with the helicopters at Kawasaki Aircraft Company, suggested to Yamaha to build a small and easy-to-fly helicopter. Ogawa's idea was to install a weight at the tips of the rotors to gain flight stability. His concept was like that of a low-cost flying motorcycle. The airframe planned at that time was supposed to have a dry weight of 300 kg, equipped with a 115 horsepower engine, and to have a cruising

Fig - 3 YHX prototype in flight test

Empty weight -------------- 300 kg
Total weight ---------------- 515 kg
Seating capacity ---------- 1 or 2
Engine -------------- 115 horsepower
-------------- (Lycoming)
Cruising speed ---------160 km/h

1) Manufacturer of Piper Super Cub and other fine airplanes

Fig - 4 Yamaha's motor hanglider
Author at extreme left with Takehiko Hasegawa, second from left, the - then President of Yamaha Motor

speed of 160 km/h. The target price for this helicopter was about $20,000 (about ¥6 million). Compared to the US-made Robinson R-22 with similar performance, the target price was quite good (Fig - 3).

The project leader of the YHX helicopter was Yuichi Imani, who was once a rowing athlete at Tohoku University. When he visited me at my house, he often asked me for advice on helicopters. I was happy to hear that the project was moving in the direction that Prof. Washizu had hoped for, but unfortunately, this project was cancelled after two years.

3) Dreams of Chairman Kawakami

From the beginning of 1980, Genichi Kawakami, Chairman of Yamaha Motor Company told us frequently to develop light airplanes. Hangliders and ultra-light[2] planes had become explosively popular in the United States. The free and unrestricted design of these light planes seemed very attractive to us. We wanted to develop them, but we were intimidated at the same time by product liability issues[3] that had already become very complicated. By that time, the total claim amount for damage brought against Yamaha had reached several tens of billions of yen mostly involving motorcycles.

We wondered why anybody would build hang-gliders and ultra-light planes in a country where product liability (PL) lawsuits were so common. After we visited the United States, we found out that accidents would not usually lead to PL lawsuits against small manufacturers because legal costs were typically higher than the awarded damage relief. On the other hand, if Yamaha built these products, people would look at us (Yamaha) as having deep pockets and will try to sue us. Nevertheless, we had to do something. We surveyed the US market in 1981, and we bought an ultra-light plane and made a demonstration flight in the Hanglider World Championship at Beppu (Fig-4). I was very interested in this project, but it never took off. I learned that most of the ultra-light plane manufactures had gone out of business due to PL issues in the past. Our judgment might have been correct, but we failed to realize Mr. Kawakami's dream.

4) Remote - control helicopters

Ever since Yamaha began producing motorcycles, Honda had always held the largest market share. Around 1980, the President of Yamaha, Hisao Koike, tried to take the top

2) Ultra - light plane is the ultra small aircraft under 115kg weight and few legal controls are applied.
3) Product Liability. Manufacturer's responsibilities were greatly pursued in the United States then.

market share by deploying a vigorous marketing and development offensive. In the industry, it was called the HY (Honda versus Yamaha) war. Unfortunately, Yamaha lost this battle and its business fell upon hard times. There was an air of disappointment and distress in the company.

In March the same year, Shigemitsu Aoki, who was a graduate of Tokyo University, Department of Aeronautics, visited Professor Hiroshi Nakaguchi of the same department and brought some interesting information. The Industrial Remote-control Helicopter Development Committee, chaired by Prof. Nakaguchi, was about to lose one member, namely a company that manufactured airframes, and the committee was trying to locate a replacement. Prof. Nakaguchi told Aoki that Yamaha could be a new member.

The parent body of this committee was the Japan Agricultural Aviation Association (hereafter called JAAA). Within this organization was the RCASS Development Committee[4], headed by Prof. Nakaguchi. The members included Prof. Akira Azuma of Tokyo University, who was well known as the Chairman of the Birdman Rally (Human Powered Aeroplane Contest) and for his study on dragonflies, and many other experts in this area. Prof. Nakaguchi and I had been close friends for a long time. In addition, my uncle had served as a chairman of this organization (JAAA) until about 10 years ago, which helped us become better acquainted with this organization. The process of assignment went smoothly.

The business at Yamaha declined since we had lost the HY war, and it was obvious that the company had to go through some restructuring. The company wanted to do it without losing engineers. Under such circumstances, this project with a grant was attractive. What was even more attractive was that this project would involve a remote control helicopter. That meant that we could somehow bypass the PL issues. The project was not exactly in line with Mr. Kawakami's wishes, but it could be considered a beginning. Soon, a meeting was held and this project was approved.

In April, the management of the company was restructured. I resigned as director the following month and was appointed as manager for this project. We were right in the middle of a sea of changes.

The division in charge of air operations in the company performed the actual day-to-day tasks, which were run by Mr. Jiro Oguma. This group was responsible for repair and maintenance of reciprocating type engines as well as turbo engines that were being used on the target aircraft of the Self-Defense Forces. The group was also involved in the manufacture and sales of engines for ultra-light airplanes. In other words, this group was engaged

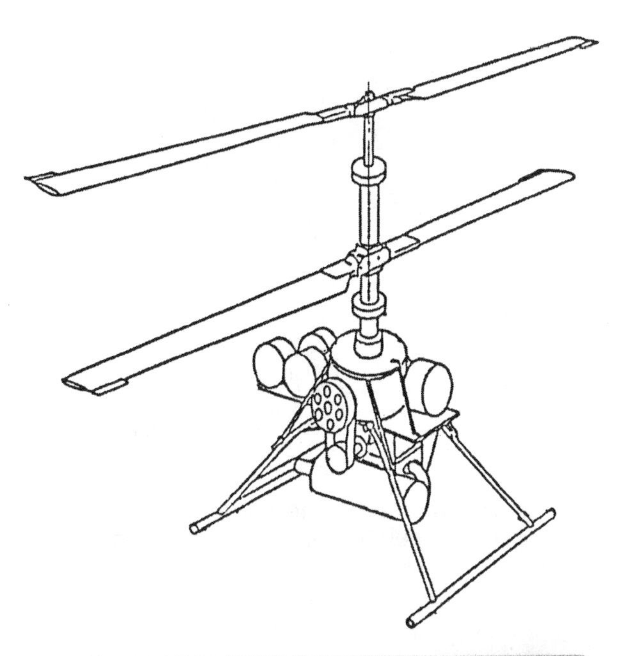

Fig - 5 Airframe (photo) transferred from Kobe Kiko Co.

4) RCASS: Remote Controlled Air Spray System

in business with the Self Defense Agency related to aviation. It was also responsible for the development of a wind - powered generator.

5) Transfer of Project

This project began after its transfer from Kobe Kiko Co. to YAMAHA. We visited their workshop to view the airframe under construction. Mr. Furuichi, managing director of Kobe Kiko Co., explained the progress of the project to us. At the end, we asked him to ship the airframe to us and we left for Yamaha (Fig - 5).

This airframe was patterned after the US Navy's on-board remote control torpedo helicopter (Fig-6) called Dash. It was made to a very compact size. This helicopter was equipped with contra-rotating rotors, and the airframe's front and rear could not be distinguished. Therefore, it had to be flown by automatic control. There was no other choice. As we didn't have much knowledge about helicopters, we were just impressed by its layout.

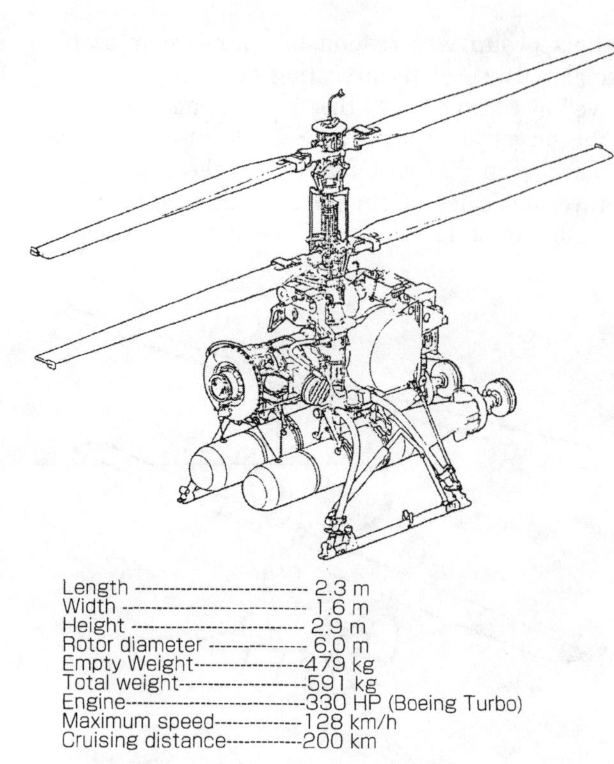

Length ------------------------------- 2.3 m
Width --------------------------------- 1.6 m
Height -------------------------------- 2.9 m
Rotor diameter ------------------ 6.0 m
Empty Weight-------------------479 kg
Total weight---------------------591 kg
Engine---------------------------------330 HP (Boeing Turbo)
Maximum speed---------------128 km/h
Cruising distance-------------200 km

Fig - 6 Remote control helicopter Dash developed by the US armed forces

In Canada, a remote control reconnaissance helicopter called Peanuts (Fig-7) was developed for military use. I read in a newspaper that the US Armed Forces and NATO were very much interested in it. For some reason, the military was quite interested in the remote control helicopters equipped with contra-rotating rotors during this period.

Height---1.64 m
Body width----------------------------------0.64 m
Rotor diameter---------------------------2.80 m
Maximum take-off weight---------190 kg
Payload-- 88 kg
Speed--130 km/h
Cruising time---------------------------- 4 hours

Fig - 7 Peanuts
Canada Air CL - 227 Sentinel

As soon as the airframe built by Kobe Kiko Co. arrived at Yamaha, we began the performance tests. The engine produced 12.2 horsepower with the carburetor at the maximum setting, but soon we came to know that it would not fly because of its 140 kg weight. A rotor of this size would produce a lift of 6 to 7 kg per horsepower, but the horsepower was not adequate to lift its own weight even without its payload. We were at a loss as to what we should do. By this time, we had already gone beyond the point of no return. We replaced this engine with a hydraulic motor that would produce up to 20 horsepower, and studied the feasibility of using this system (Fig - 8). Finally, we realized that some issues were unsolvable and our only option left was to redesign the helicopter.

Let us review some of the regulations relative to helicopters of this kind. If the total weight including the airframe, fuel, oil, and payload exceeds 100 kg, it is an airplane. If we must meet the regulations for an airplane, the cost

Fig - 8 Testing with hydraulic motor
The engine that came with the airframe from Kobe Kiko Co. was replaced with a hydraulic motor, and the rotors were tested.

would increase beyond our control. On the other hand, if the craft's total weight stays under 100 kg , it would be considered a model airplane and no regulations would apply. In other words, it was absolutely necessary for this helicopter to weigh less than 100 kg, inclusive of the payload. This was the only area where motorcycle manufacturers had a design advantage in building a helicopter

6) RCASS

The development of RCASS was contracted to Yamaha Motor Company in the spring of 1983 by JAAA. We were too busy that year with the analysis of the aforementioned airframe, so the new design began the following year. Our plan was to use a 20 horsepower Yamaha engine, to realize a total weight under 100 kg, and to carry over 20 kg as payload (Fig - 9). We worked on the driving system, airframe, control system, and the software. The final product had to be completed by the end of 1984.

Since an entirely new control system was

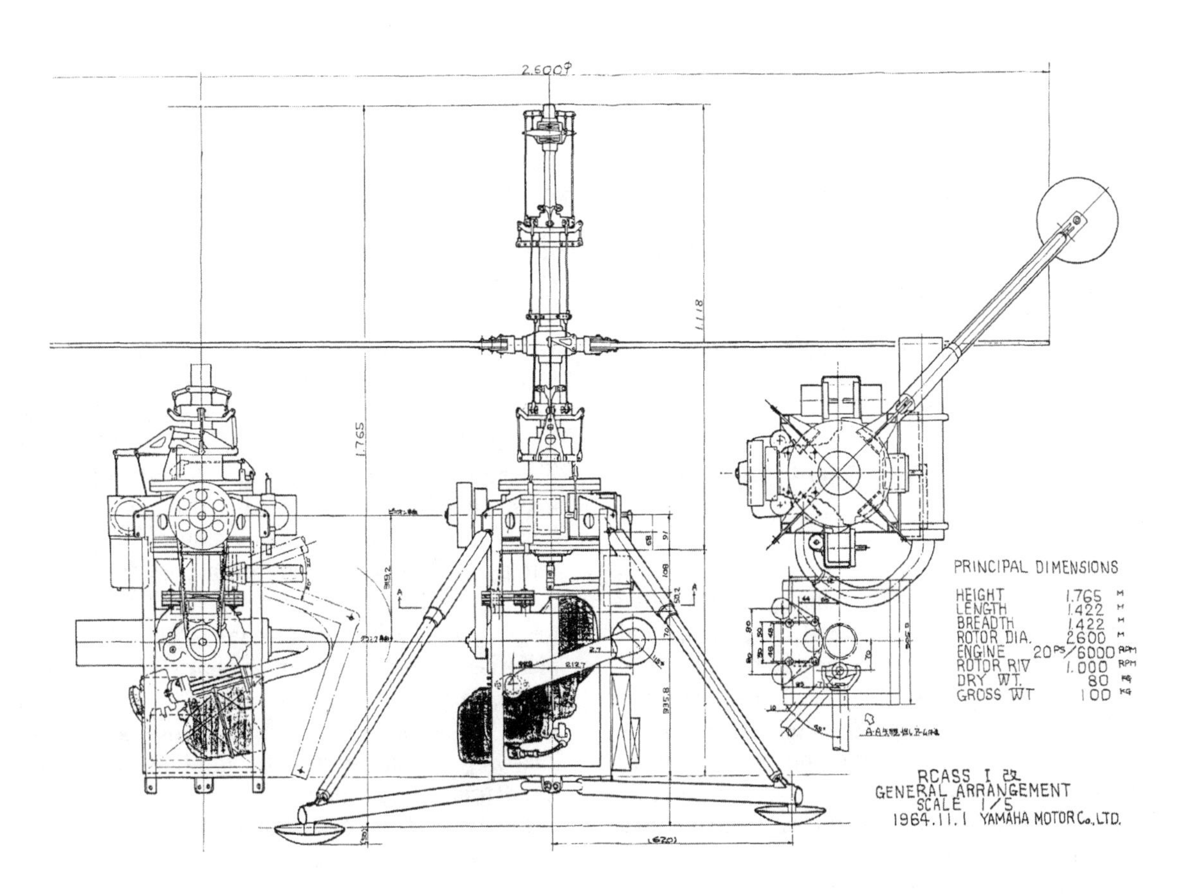

Fig - 9 RCASS (Yamaha model)

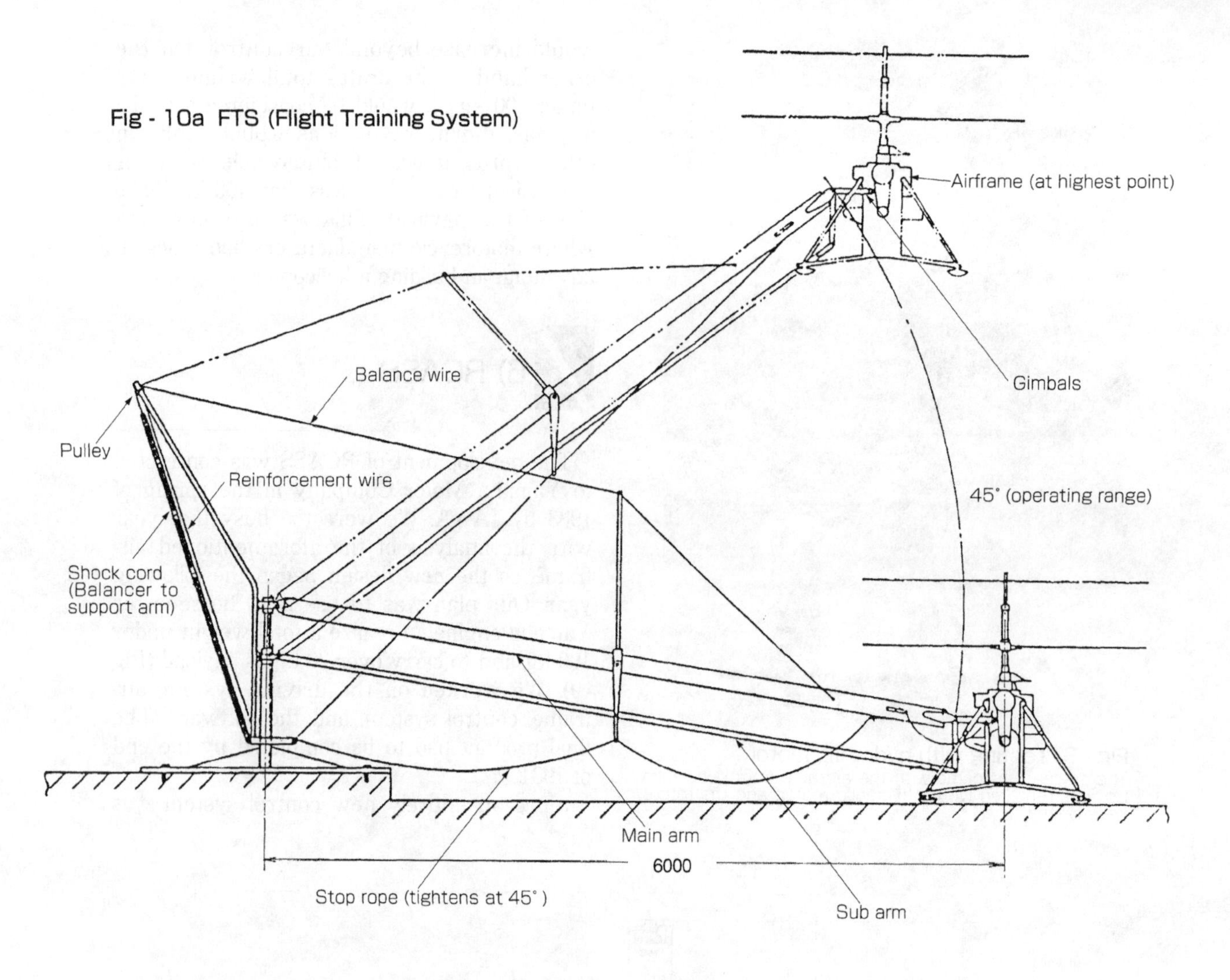

Fig - 10a FTS (Flight Training System)

used, we anticipated that the helicopter would not fly well from the beginning. For this reason, we built a flight training system (FTS) to restrict the freedom somewhat in order to prevent major damage to the helicopter. This system would allow semi-unrestricted flight within a metal net cage.

7) Flight Training System (FTS)

The FTS had a 6 meter main arm (Fig - 10a) that was supported at one end by gimbals and the other end connected to nearly the center of gravity of the airframe, allowing the airframe to fly over a hemisphere 6 meters in radius. A device was installed on one end of the arm to prevent the airframe from tilting more than 20 degrees, thus forcing the airframe to always land on the landing gear that was equipped with a damping system even in the worst situation. This device prevented damage to the helicopter.

The arm had to be strong enough to support the RCASS should it overturn unexpectedly. On the other hand, if it was too heavy and affected the flight, the purpose of the test would not be fulfilled. Selection of the size was important. After extensive discussions, we decided to use aluminum tube measuring 130 mm in diameter and 1.2 mm in wall thickness. To reduce the impact of the weight, we installed a balancer that utilized tension of rubber. As to inertial resistance[5], we had to close our eyes. Whenever I create a very new product, I would normally draw a sketch on a graph paper with various ideas about balance of forces, strength, weight, local structure, and so on, in the empty space. This way, you can show the entire construction on one sheet and it is easy to find errors. In addition, when you must make corrections, you can easily find the influence of the corrections on other areas so that you will not omit the necessary modifications in related areas.

Finally, the FTS test went well and subse-

5) Resistance that occurs when a heavy item is to be moved suddenly. The movement becomes slower due to addition of portion of the arm weight to the airframe weight.

Fig - 10b RCASS on FTS

quently a series of flight tests were conducted on this equipment (Fig - 10b). The goal of these tests was to ensure free flight stability using a gyro and to move the airframe to given locations using a remote control device. Without any prior experience, it was not an easy task for this group to realize the goal. Everything was new to us. The biggest headache was that no good gyro was available at that time. The gyro for precise measurement available at that time would cost around $40,000, and moreover, it was delicate against impact. We had to wait for the introduction of an optical fiber gyro or the next-generation gyro. These devices were available but not fully ready. None of them had acceptable performance.

Another difficulty was the absence of a suitable actuator (driving device to change the rotor angle). An actuator for model airplanes was too small, and one for general use was too heavy for our helicopter. The actuator we were looking for had to be compact, light-weight, resistant to strong vibration, highly reliable, and inexpensive. The constraint on inexpensive components of motorcycle manufacturers prevented us from acquiring a suitable actuator. In the end, we had to keep going with a combination of equipment that failed to meet our expectations.

8) Serious concern

RCASS began to take shape and the concept of FTS grew more mature. Yet, I was unable to overcome my concern of actually building a self-stabilizing airframe successfully.

Even though the flights on the FTS were successful, there was a large technical hurdle between the FTS flight and a free flight. Any minor failure could bring down the helicopter, and crashes and damages would be inevitable, once we began free flights. If such failures required repairs over a long period, how many more years would it take to complete and market the product? It was obvious that this project would be terminated if it required many more years. Fully automatic control was one of the main concepts of RCASS. But I began to feel that the concept would not be appropriate in situations where effective components were difficult to procure. If success was the goal, we

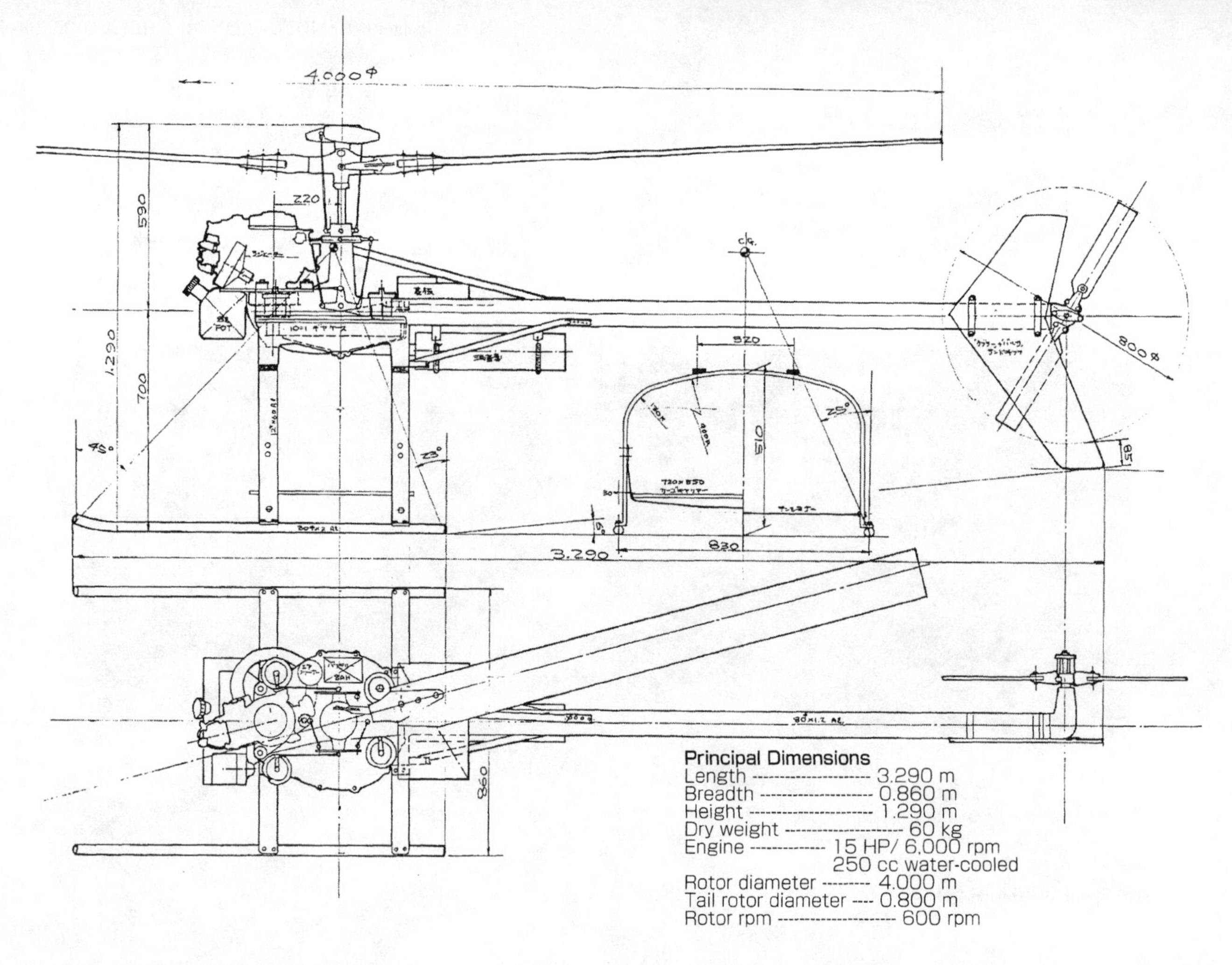

Fig - 11 R - 100 Planning drawing

should have introduced the advantages of model helicopter technology that was at the near-perfect level, starting with a manually controlled helicopter. Once it reached the commercial level, the control should be simplified in phases adopting automatic stabilizer functions in altitude, direction, and so on. Finally, in the future, when a good gyro became available, we should adopt it to complete the fully automatic stabilizer.

9) Professor Hiroshi Nakaguchi

My concerns about the project continued and I thought about them all the time. This project was not Yamaha's own project. It was a project of the RCASS committee lead by Professor Hiroshi Nakaguchi and Yamaha was given a grant. It was not easy to change the current development course. I mentioned my concerns to Ryohei Ichikawa, the secretary of JAAA and the person responsible for the RCASS project, then visited Prof. Nakaguchi at his home on one hot summer day in 1985. I asked for his approval for the development of a tail rotor model in parallel to RCASS. Prof. Nakaguchi himself had been concerned with the delay in the automatic stabilizer, and he sounded as if he were encouraging me to begin the project in case the RCASS was stuck.

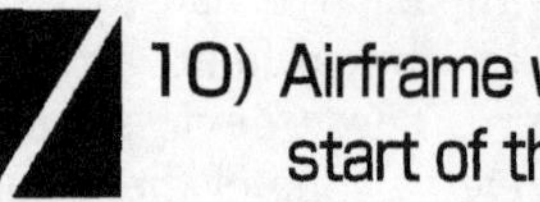

10) Airframe with tail rotor and start of the R-100 project

At Yamaha, Director Shunji Tanaka was responsible for the remote control helicopter. He understood my concerns and a consensus was reached for parallel development of a tail rotor model. However, since all the design engineers were working on RCASS, no one was available for the tail rotor model. I had to do the preliminary plans by myself with support from Jiro Oguma, Director of Operations. I began planning immediately.

As I mentioned earlier, the total weight must

be less than 100 kg and the helicopter was to be equipped with a tail rotor so that it could be flown like a model helicopter. This model was powered by 250 cc Yamaha outboard engine. See the airframe of R - 100 (Fig - 11).

Because this would be a working helicopter, I made it a skeleton type craft for simplicity and ease of serviceability. In addition, I used the leaf spring type landing gear that produces a large damping stroke and enables rough landing. In the past, when I had designed a light airplane, I investigated this type of landing gear made of vanadium steel[6] for use on a Cessna. I was very impressed then with the performance. This time I used FRP instead. These landing gears offer a high damping capacity by opening each leg outside until the bottom of the fuselage touches the ground.

The next step was to contact a model helicopter expert. I requested the assistance of Hirobo Co., the leader in the industry. I explained the situation, everything went smoothly, and they promised me to meet Mr. Oguma and me in early December. We prepared the layout drawings, plans, and ideas for each allocation. As we arrived at the Head Office of Hirobo in Fukuyama, the young President, Keitaro Matsuzaka welcomed us. We were very delighted to know that he was also interested in building a working remote control helicopter.

Hirobo would develop the rotor system, assemble, and test it. Yamaha would prepare the master plan and develop the engine and the drive train. I felt that we could assist each other and develop a good airframe. While we were there, President Matsuzaka strongly recommended that I should personally experience flying a model helicopter

11) Flying a model helicopter

As soon as I returned, I bought a model helicopter and began to fly it. Due to the confidential nature of this model helicopter, I had to learn by myself how to fly this helicopter for some time without a coach. I was fully engrossed in flying this model helicopter through the year-end and into the New Year. I learned a lot from flying it.

Once a sudden wind blows and changes the helicopter's flight path by 180 degree, you will lose the sense of directional control completely and the helicopter will crash. One day, the on-board battery ran out during the flight and I lost control of the helicopter. I was really scared by the thought that it might crash in the town. On January 15, 1985, the helicopter crashed and lost the rotor. The same night, I broke my Achilles tendon while playing badminton and had to stay in a hospital. My flying lesson was interrupted for a while. Still, I learned a great deal about helicopter operation

Fig - 12 R-50 skeleton

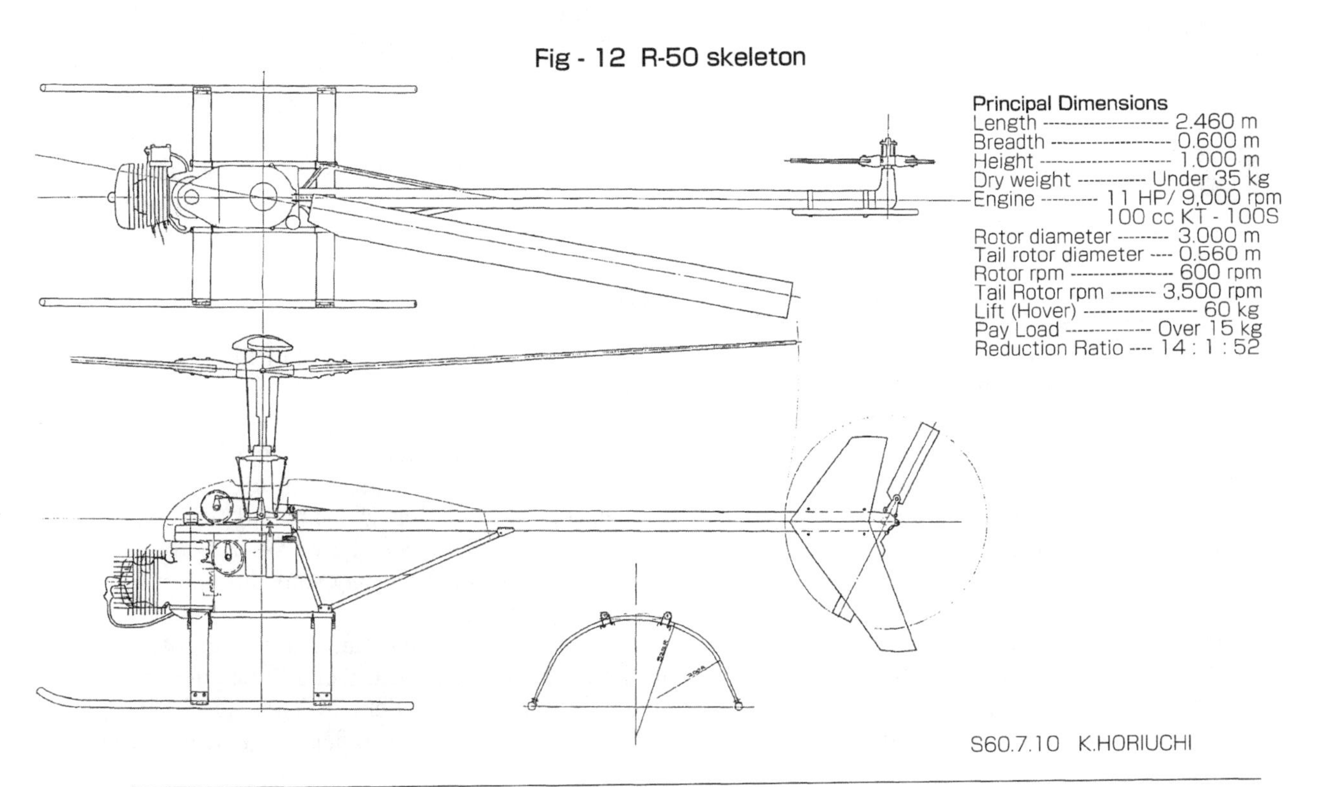

6) Vanadium steel: Special steel that is very tough and used for springs

and some of the difficulties in designing one. I also learned about uncontrollable disasters that could have far-reaching consequences. I said to myself that the project would likely be terminated if we met with any accident during the development stage.

12) Transition from R-100 to R-50

When I visited Hirobo in July, Mr. Matsuzaka told me that the R-100 was too large. I suspected that his comment came from his research on the commercial world. Though I didn't ask for explanations, I quickly responded to his comment. Shortly after that visit, I drew and sent him a sketch of the R-50 (Fig-12). This time, I used 11 HP power, 100 cc Yamaha kart (go-cart) engine and designed this helicopter such that it had a weight of 35 kg with a payload of 15 kg. So, the total weight was estimated as 50 kg. There was not much reduction in the payload although the body was smaller. I viewed this as progress in engineering design.

The length of the airframe was reduced from 3.3 m to 2.46 m, and the rotor diameter was significantly reduced from 4 m to 3 m. Though we were not aware of it then, the reduction in these dimensions helped us a great deal later. The helicopter shouldn't be larger than this size otherwise more than one person would be required to load it onto a truck or to carry it through narrow paths between rice fields

13) Monocoque airframe

With the framework of the project defined, both the Yamaha and the Hirobo staff began to meet more often. I met with Mr. Mitsugu Matsumoto, Manager of R&D Department at Horobo, and discussed the details of the specifications. He had had extensive experience. On the other hand, I was a layman and could utilize some of my experience in designing an airplane. I learned a lot from the discussions. I was glad that I had practiced flying a model helicopter. That helped me to keep up with him.

When we were working on the details, Mr. Yoshihide Fujita, General Manager of Model Operations at Hirobo, visited Yamaha and made some comments. He would found it difficult to sell skeleton bodies. He wanted us to design a good-looking fuselage. He was looking at the product from the sales point of view. Again, I quickly responded to his comments. A monocoque[7] airframe might help reduce the weight. I personally preferred an attractive fuselage. If it didn't have to be a skeleton, I would rather use a monocoque construction. This assessment concluded that the monocoque frame had no particular adverse impact. We decided to go along with the monocoque fuselage. We knew nothing more than that at that time, but later we would find a monocoque structure was inevitable.

Since the helicopter was a remote control type, the operator controlled it by watching it. In other words, you couldn't fly the helicopter beyond a point where you could no longer tell which direction the aircraft was heading. Therefore, the airframe had to be large enough and colorful enough to ensure proper control of the helicopter, to be able to fly farther away from the operator. This issue, however, assumed top priority only after the helicopter was built. If we had continued the project with the skeleton model in mind, we would have had to accept a completely revised plan.

A major model change was made for the first time in 1997, but until that time for 12 years, little had changed in the basic dimensions or the structure. In that sense, we can say that the ideal size and structure were defined at the very beginning, by which developments in the later stage were performed much more efficiently than we expected. You could keep most of the airframe as-is, and you could make localized improvements. The ideal size and structure made it possible for us to make progressive improvements. I am very thankful to President Matsuzaka for his foresight and to the relevant advice of the Hirobo staff.

14) Contracts with Hirobo

In July 1985, I returned to Yamaha Motor to my previous position as a director. I was responsible for the Horiuchi lab, and was also the Assistant Director of Marine Operations, as before. Nonetheless, I did pay some attention to the helicopter project.

Mr. Matsuzaka visited Yamaha on October 8 in the same year, and at the end of his meeting with Mr. Eguchi, President of Yamaha, signed a contract, marking the official kickoff of the

---

7) Monocoque: Reinforced skin. Stressed skin construction is synonymous.

Fig - 13a R-50 Monocoque structure

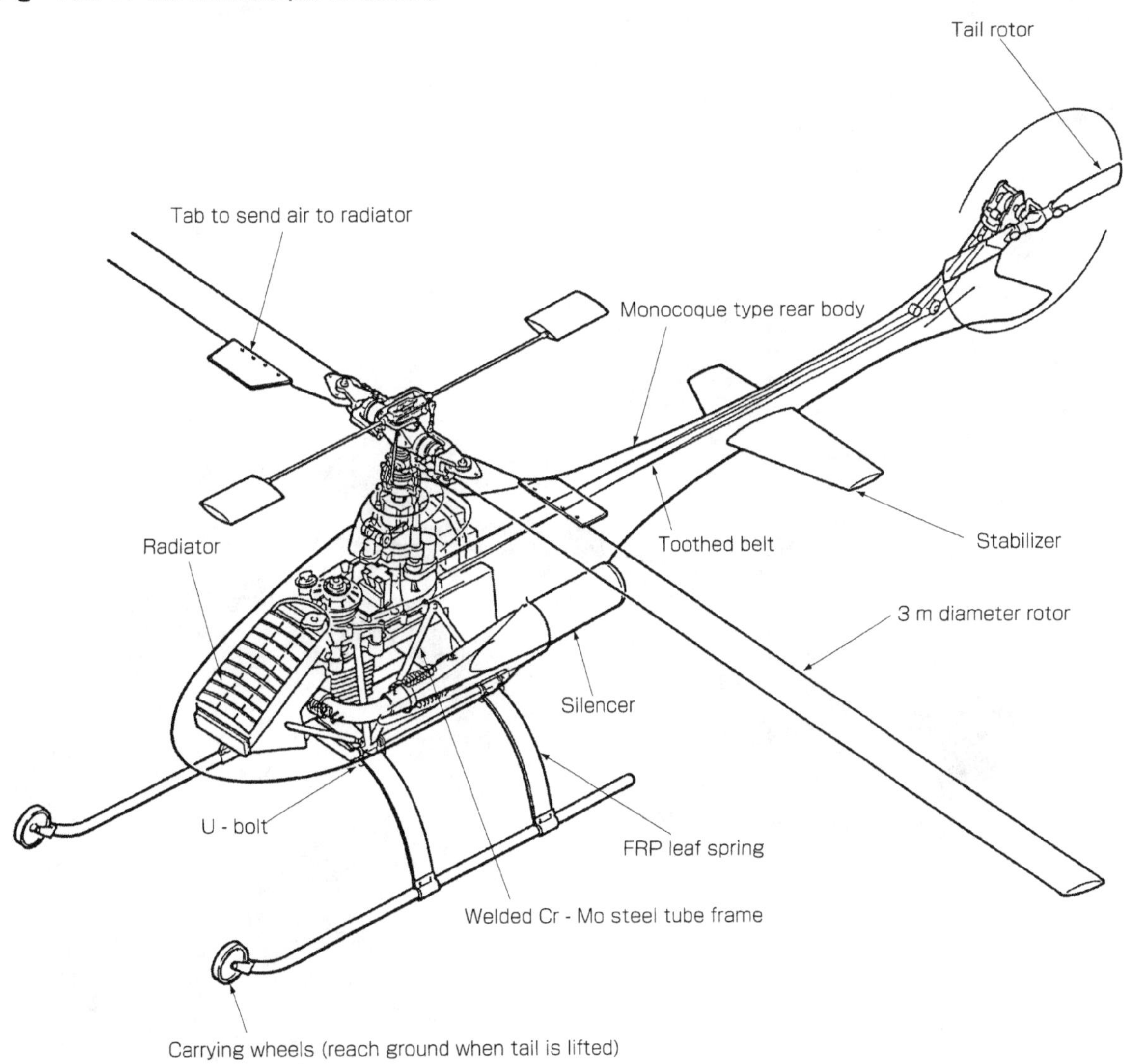

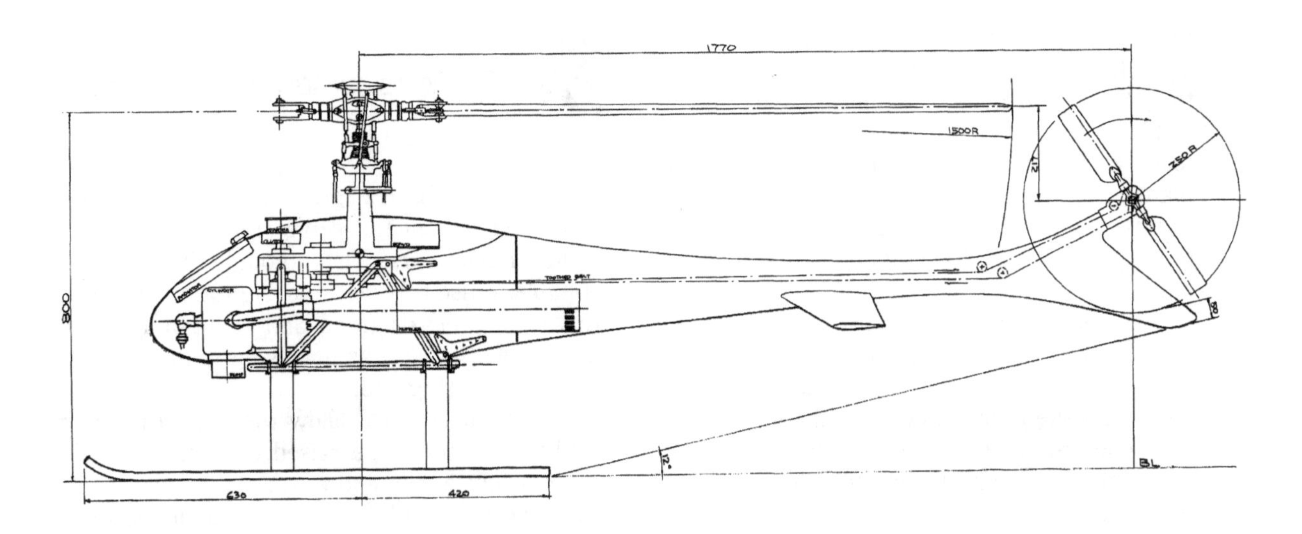

Fig - 13b R-50 Monocoque, Plan

Fig - 14 First flight of R - 50

(Above) Two prototypes
(Right) Front wheels were very handy.

project. All the helicopter design team members, however, were still fully involved with the RCASS project, and had no extra time available for the R-50. I had expected this. I told Mr. Itaru Hisatomi, who was department manager and an expert in engine development, of this situation and we began to work together to produce the drawings outside the working hours.

Our plan was to complete the two prototypes by October of the following year. We were supposed to complete the assembly in the first six months, make improvements, and conduct a 50 hour endurance test in the next six months. To accomplish this plan, we had to finish almost all the necessary drawings by the end of the year.

15) Development of the R - 50

In 1984, Hirobo had great success in utilizing a toothed belt in the drive train of radio controlled model cars. They thought it would be suitable to use this belt to drive the tail rotor of this helicopter. I agreed and thought this the best choice as long as it was light, more silent, and efficient (Fig - 13).

Use of a belt would allow the axis of rotation of the tail rotor to be raised close to the main rotor height at the end of the tail boom. In this way, the rolling of the airframe could be minimized when changing the flight direction. Any new design such as this belt, however, could result in a large change to the schedule. If problems occurred with the new component, we would have to modify the master plan. So we had to test this structure beforehand and if it caused a major problem for the project, we would have to switch to a proven system at an early date in order to stick to the master schedule.

Mr. Hisatomi immediately built a test device that had a pulley installed just like it would be on a real airframe and began the endurance test. He also devised a water-cooling system on the cylinder and the head of KT-100 kart engine to prevent it from suffering thermal wear[8] in an extended flight. The gear case design was also making progress. Since the engine rpm was 8,000, the gear case was supposed to reduce it down to 620 rpm for the rotor. A clutch was installed for auto-rotation[9] and a pulley for driving the tail rotor.

As for me, I was working on deciding the overall arrangement and the structures in each section. I designed a double structure front airframe consisting of a truss frame and skins to bear the concentrated loads caused by the landing gears and the engine, and for serviceability. A truss structure frame constructed of Cr-Mo steel tubes was installed under a large aluminum gear case. This structure was covered with an FRP shell provided with a latch for opening. The rear body was an FRP monocoque structure.

For the landing gears, we adopted the structure of our original concept in which an FRP leaf spring was mounted on the airframe using U-bolts. Later on, we learned that this structure was desirable since it would absorb the upward impact during a rough landing and also it would absorb the lateral impact by allowing the leaf spring to move at the U-bolts.

The toothed belt Mr. Hisatomi had been testing was successful and was adopted without problems. The belt started from a pulley on the vertical shaft and after being twisted through 90 degrees, was inclined to 40 degrees upward through the guide pulleys on the horizontal shaft, and then reached the tail rotor pulley. It was a multi-functional belt that changed the direction of the axis and transferred the driving force to the shaft at the rear end 1.8 m from the main rotor shaft. The monocoque airframe that housed all these components was slim and nice-looking. I liked it.

As only two people were working on the design, there was very good communication. We worked as if we were one. There was no misunderstanding at all. I really enjoyed working with Mr. Hisatomi.

From 1986, we began the partial assembly and testing of the R-50. Shortage of staff forced Mr. Hisatomi to frequently run between the test site, Hirobo, and the vendors. The prototypes made steady progress and they completed the first flight at Hirobo without incident exactly on schedule. That was on June 7. Compared to a model helicopter, the prototype moved gently and was easy to operate. I had no concerns about the test flight because the helicopters were in the hands of the best pilots of Hirobo. I was told later that R-50 completed its first free and unconstrained flight successfully.

July 1 was the 30th anniversary of Yamaha Motor. The R-50 made its first flight in the

Fig.15 Free flight of RCASS

8) Thermal wear : Decrease in engine output because of continuous combustion heat
9) Auto - rotation : Technique that reduces the rotor pitch during a mid - air engine stop and causes the rotor to accelerate due to the descending speed of the airframe. The pitch of the blades is increased just before it contacts the ground, thereby ensuring a safe landing.

presence of the directors of Yamaha (Fig - 14). All those concerned with its development shared joy as it flew steadily and freely. Following this event, our job was to continue improvements and at the same time, to extend the flight hours from 10 hours, to 50 hours, and so on. At about this time, Yamaha appointed a project team for the R-50 and secured the services of an operator with the assistance of Hirobo.

The flying of the R-50 by both the parties helped in the progress of the project for commercialization and at the same time, we shared problems.

16) Free flight of RCASS

As for the RCASS project, which began in the spring of 1985, there was improvement in the control technique of RCASS due to the flight restraint by FTS. Also, operators became accustomed to its control, and the flight became more stable. The demonstration flights by FTS for the committee members and for the Ministry of Agriculture, Forestry, and Fishery, were successful. RCASS was scheduled to demonstrate a free flight at the end of the year (Fig - 15).

When the committee visited the test site in November, we attempted a free flight. However, it failed during landing due to wind and rolled over, minor damage occurring to the airframe. We repaired the rotor and the skid[10], and then tried another free flight indoors on December 24. It was successful.

The free flights revealed some control issues that had not been observed in the FTS tests. The response to rolling was too slow to control the flight, and it was difficult to maintain the flight direction. We repeated the free flights in 1986, some of which were for the RCASS committee. There were quite a few issues that were difficult to resolve. One of the issues was that the helicopter was unable to maintain altitude and was difficult to fly in the wind even within an area of the size of two tennis courts. Once the helicopter began to move, it had to be stopped by flaring[11], This was not easily accomplished in our case because the helicopter landed with some ground speed instead of zero ground speed, which caused it to roll over.

The causes were complex: poor maneuverability, immature control technology, lack of operator skills, and difficulty in identifying the front and the rear of the airframe. A combination of these factors made the flight difficult. It appeared that it wouldn't be an easy task to resolve the issue of sudden winds. We had to simultaneously resolve multiple major problems that we had never experienced before. We were facing a very difficult situation technically. Another large concern was that any major damage to the helicopter would set us off by three to six months, or even longer, compared to the schedule.

My concerns of two years ago were gradually turning to reality. Furthermore, the airframe of RCASS was heavy, and the payload was small. It would be of no commercial value unless it was redesigned. To resolve these problems, the RCASS type R-100 project went back to the planning stage by the end of the year. By this time, it became obvious that an effective guidance system was required to put RCASS into commercial use. It would be very difficult to operate this slow-reacting RCASS steadily along the predetermined spray rute under various wind conditions. A fully automatic control system would be necessary if the helicopter was to be commercially successful.

The committee began to discuss the guidance system, and a budget was allotted for its development. But that was the beginning of the end of the RCASS project. Besides, the R-50 was about to complete its endurance test and it had sufficient payload.

17) Official recognition of the R-50

The R-50 project, even with its advanced achievements, was not an officially recognized project within the RCASS committee. On March 11, 1986, when the key committee members gathered at the Yamaha headquarters in Iwata city, Shizuoka Prefecture, we demonstrated the R-50 for their understanding. In January 1987, Prof. Nakaguchi agreed to include R-50 in the mission of the RCASS committee. The committee was happy to include the R-50 project, which was Yamaha's own development, as one of the committee's business achievements. Later during a regular committee meeting, it was officially approved and the R-50 project started being discussed openly along with the RCASS project in all committee meetings.

In January 1987, Mr. Shunji Tanaka, Director

10) Skid : A type of landing gear like sleigh.
11) Flare : Since a helicopter is not equipped with a side - ways propulsion device, it must tilt the airframe to move forward or stop. To zero the ground speed, the helicopter must tilt the airframe in the opposite direction from the flight direction, raising the nose and apply the brake. This operation is called 'flaring'.

Fig - 16 Production model R - 50
New coloring that is highly visible

of Yamaha Motor, approved the transfer of production of the R-50. In February, a production plan was formulated. Although the plan was more like a pilot production plan, ten monitor helicopters were to be completed by the end of April and delivered in August (Fig - 16).

The RCASS project added a new plan with R-100 for commercial use. However, many engineers had to move to the R-50 project with the designated active production plan. In 1988, the R-100 plan came to a standstill, leaving research on the control system to which a grant had been awarded. On the other hand, as we looked to the future, we believed R - 50 should be made easier to operate by gradually implementing automation. The previous studies on the RCASS will become useful.

The last RCASS R&D conference was held in March 1989 when this project was terminated.

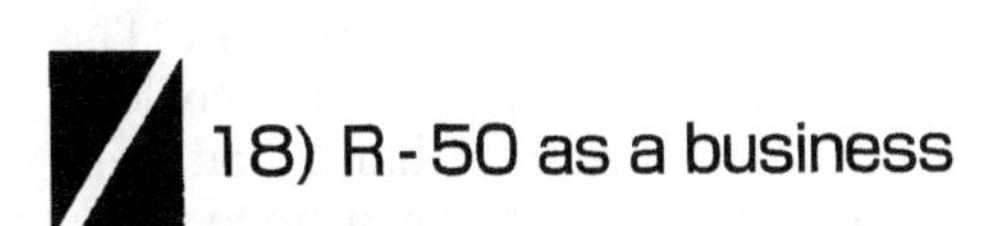

18) R - 50 as a business

Started as contracted research, the remote control helicopter project had reached the point of whether or not it could establish itself as a business.

Applications of the remote control helicopter were many. We started with the development of this helicopter at the request of JAAA, whose goal was to spray agricultural chemicals. For this reason, Yamaha concentrated 99% of its energy on the agricultural application of the helicopter.

By then, other manufacturers had been announcing agricultural remote control helicopters one after another. They were slightly larger than model helicopters. In November 1987, the manufacturers brought their industrial remote control helicopters to Uchihara, Ibaragi prefecture, for a demonstration flight. At that time, the future of the remote control helicopters was viewed with optimism as a tool for modernization of the nation's agriculture. The implementation of a ban on the spraying of agricultural chemicals by manned helicopter had already started in some prefectures.

With regard to agriculture during this period, direct seeding of rice was attracting the attention of many people since this type of rice growing did not require nurseries or transplantation.

Operation of a remote control helicopter was generally viewed as a difficult task. In this respect, the R-50 was easy to operate with its gentle movement compared to that of a model helicopter. Yet, the operator had a major responsibility not to crash this expensive helicopter. An organized effort was necessary to teach and improve operation skills that would be adequate to operate helicopters on a commercial basis. For this purpose, Yamaha established an organization called Yamaha SkyTech to sell and promote the R-50 The organization had the SkyTech Academy where they taught how to operate helicopters. Yamaha had established piano classes and boat license classes earlier, so organizing the SkyTech Academy was nothing new.

In December 1989, six years after Yamaha succeeded in the RCASS research project, the R-50 went into official production and sales.

It has been ten years since then. During this period, a minor change was made to the tail

Fig - 17 Minor change to R - 50
Tail body is now molded by filament winding and made straight

body for better productivity and reliability (Fig - 17). The remote control helicopters sold rapidly during this period probably because people considered this method of spraying agricultural chemicals as a key tool for next-generation agriculture. By 1997, over 4,000 people were certified as licensed operators. The large number of women operators was noteworthy. It was nice to know that young women and housewives looked at it as a new and attractive work. They are active in this field today (Fig - 18).

Probably, the number of R-50 in the fields would have exceeded 1,000 and the spray area expanded to 200 thousand hectares by 1999. The crops subject to this type of spraying have also expanded to cover rice, wheat, soybeans, lotus roots, radish, chestnuts, and citrus. I have been away from this business since 1988, but now the R-50 has made significant progress. Altitude and direction control technologies have matured, high-performance gyros have become available, and in 1995, the long-awaited YACS[12] appeared on the market. YACS utilized three optical fiber gyros and three accelerometers, making helicopter control extremely easy.

About that time, I heard that Mr. Hasegawa, President of Yamaha Motor, flew the R-50 and was surprised. I learned later that he operated one equipped with YACS, which allows recovery of the hovering position by itself when the operator releases the control stick, and which had been our dream for many years.

19) Evolution to the RMAX

R - 50 made significant progress in 1997 to become RMAX (Fig-19). I was invited to observe a demo flight immediately before its sales, and I was amazed by the changes. It was quite natural since the design team was full of airplane enthusiasts. Mr. Makoto Sugimoto, Manager of the Engineering Department, was a long-time expert in model helicopters, and the core engineers were the Heavenly Dragonfly team who won an overwhelming victory by flying all the way to the northern end of Lake Biwa in the 1998 Birdman Rally (human powered airplane contest).

No doubt, this helicopter was excellent. This new design team did everything they could to improve it. The helicopter that I had been involved with didn't go beyond a multi - purpose remote control helicopter. RMAX, on the other hand, realized every idea that evolved from chemical spraying experiences during those years. Its horsepower as well as the payload doubled. It is equipped with a starter motor and YACS and yet the weight has not increased much. The cassette type chemical tank is very convenient, and the bent skids provided to facilitate lifting are superb. The helicopter is also equipped with a self-diagnosis device and a low fuel/radio jam warning device. Everything was so impressive. Despite all these changes, little has changed in the size or the basic structure. Just the right size for

12) YACS: YAMAHA Attitude Control System

Fig - 18 Women operators who competed in All Japan Remote Helicopter Operation Competition

handling has been retained. It has retained the small wheels in the original design that I was proud of. I was very happy to see them. The good features have been maintained and additional improvements have been made.

Excellence in their job made me aware of another point. The Heavenly Dragonfly team members are good friends, and they have always been good friends even after their marriages. Now that their children have grown up, the team continues to compete in the Bird Man Rally. These fellows must have acquired a state of mind that keeps them happy, allows them to pursue what they want to accomplish, and encourages them to make the best of their engineering expertise to attain the goals. Their ideas and their spirit must have given shape to the RMAX.

RMAX itself must be happy that it was designed by this team, and at the same time, I was very happy that they took over my project and made it so successful.

I must not forget to thank Yamaha Motor for their support and understanding. I heard that some other companies tried to build a remote control helicopter, but withdrew from the project. Those companies probably were not patient. Yamaha, on the other hand, has been supporting this project for more than 20 years now. Takehiko Hasegawa (ex-President of Yamaha), Shunji Tanaka (ex-Director), Oguma (ex-General Manager of Operations), who worked with me from the very beginning and some others who retired after completion of the RMAX, and all those people involved in this project loved their work dearly. I believe that the devotion of these people gave strength to Yamaha, led to its success in business, and nurtured the RMAX team.

Fig - 19 RMAX and old model R - 50

WATER SCOOTER DOLPHIN

Tremendous efforts by a young man who wanted to escape from East Germany to the western democratic world led to the development of a remarkable product. This was the underwater scooter, the propulsion means of which was a submerged reciprocating engine. He was arrested by the local coast guard during his escape through the Baltic Sea on the underwater scooter, and imprisoned for three months.

A year later, he improved the water scooter and succeeded in escaping from East Germany to West Germany, where he worked to develop it and bring it to the production stage. Unfortunately, he died two years later. In this article, I would like to introduce the underwater scooter and similar leisure products that have appeared subsequently

1) Birth of the underwater scooter[1)]

Was there ever a submerged combustion engine in our long history? I have never seen such an engine except in the one mounted in an underwater scooter, which is the subject of this article. How the water scooter was conceived and developed makes for a very interesting story.

In the1960s, many persons tried to escape from East Germany and lost their lives because the escapes were always dangerous. However, a young man, who was convinced of his own genius, decided to develop an underwater scooter with which he intended to escape. I admired his strong self-confidence, excellent ideas, and his activities supported by technologies of the highest quality.

I have become fascinated by both this wonderful underwater scooter and its inventor, and I take off my hat off to this inventor. His name is Bernd Boettger. He was a chemical engineer employed in a large company in East Germany, and was a good swimmer, and an experienced lifeguard. At that time, he considered escaping from East Germany to Denmark by swimming across the Baltic Sea, but gave up the idea because the Baltic Sea was rough, cold, and the distance was too large to cross since he had to swim 24 km.

After considerable thought, he developed an underwater scooter. During the first trial, he

Fig - 1 Inventor Bernd Boettger and the underwater scooter he designed and used for his escape

(excerpt from referred article)

1) Reference : From Escape to Escapism, The Aquascooter Story, by M.L. Jones and John Donovan, [Soldier of Fortune] magazine.

was discovered and caught by the local coast guard, arrested, and imprisoned for three months. Finally, he was released because he worked for an important company in East Germany.

Soon after his release from prison, he started to develop the second prototype. The first failure gave him the essential requirements for success of the second prototype. The requirements were a silent engine and continuous underwater towing ability to tow him over a long distance. A long range and durable engine was an indispensable requirement, and of course, his body had to withstand the long voyage. He had to be successful on the first and only attempt at crossing the sea without any trials.

It took two years for the second prototype to be completed. It was made with a 1.5 HP scooter engine and other easily available components. This so-called submarine had a weight of 10 kg inclusive of FRP tank, engine, propeller, and snorkel. It was estimated to travel for 5 hours at a speed of 5 km per hour.

At 11:30 on a starry night in September 1968, he dived 1.5 ft beneath the water surface and departed from a lonely beach at Bad Warnemuende, not far from Gral-Mueritz in northwestern East Germany. The place was just 300 meters from the coast guard watchtower but he was not discovered because only a slender snorkel protruded above the water.

An hour after diving and convinced that he was not being pursued, he threw away the 6 kg lead waist belt needed for submerged travel, and continued traveling on the water surface. In another waist belt, he had food such as vitamin C tablets, sugar, milk, chocolate, drinking water and repair tools, and even an inflatable mattress for floating. His careful preparations, including those for the water scooter, brought him success after a long voyage (see Fig - 1). At last, after traveling for 5 hours, he came upon a Danish light ship and was rescued from the chilly waters.

2) Progress of the underwater scooter

Bernd Boettger of East Germany invented the underwater scooter in 1968. Using a prototype, he escaped from East Germany. The news was broadcast all over Europe. One of executives in Rockwell International (a large US company) became interested in both the news and the scooter. The executive consulted with Rockwell's subsidiary in West Germany to commercialize the water scooter. Eventually, Bernd Boettger was employed to develop the water scooter he invented to a marketable product.

Unfortunately, Bernd Boettger died in the early 1970s while diving along the Spanish coast without seeing the progress on the underwater scooter that he invented.

In 1974, Rockwell sold its patents and production license to James Taylor, who established Aquascooter Incorporated in the USA, and the first production model was completed in January 1978. The flow of progress is shown in Fig - 2. Subsequently, the product was called Aquascooter.

In 1978, Arcos Co. in Italy also started production of the Aquascooter under license, and has continued to do so even today. On the other hand, in the USA, New Yorker Robert Steven Witkoff acquired many of the rights to the Aquascooter and Aquascooter Inc. as well. Witkoff shared worldwide selling rights of Aquascooter with Arcos Co.

I am not sure who named Aquascooter - Boettger or Taylor - and when it was named.

However, the name sounds natural and polished. The name Aquascooter is probably a combination of aqua and scooter taken from Vespa scooter, which was used in the film ROMAN HOLIDAY in 1954, starring Audrey Hepburn, and had a cheerful and pleasing image.

According to Witkoff, 60,000 units of Aquascooter were manufactured by Arcos Co. for eight years until 1985. This showed that Aquascooter was acknowledged as a superb

Western year	Events
1968	Invented by Bernd Boettger. He escaped to Denmark
1969	Bernd Boettger took job in Rockwell co. to complete the products into production. Patents
1970	Bernd Boettger died during diving
1971	
1972	It was said that 3000 units were produced during 4 years(70-73)
1973	
1974	Rockwell co. sold out its patents and license for production.
1975	American James Taylor got every right and estabished Aquascooter inc..
1976	
1977	
1978	January, Aquascooter inc. got success in prototype trial. Italian Arcos co. began production of Aquascooter under license.
1979	It was said that 60,000 units were sold during these years (Witkoff)
1980	
1981	Patents application in Japan
1982	American Robert Witkoff bought both patents and license for production
1983	
1984	
1985	YAMAHA MOTOR co. developed Dolphin D-1
1986	YAMAHA MOTOR co. developed Dolphin 1.5. Pilot production for trial sale
1987	
1988	Production in Arcos co. has continued reaching
1989	total units of about 200thousand
1990	
1991	
1992	Patents of Aquqscooter inc. have expired.

Fig - 2 History of Aquqscooter

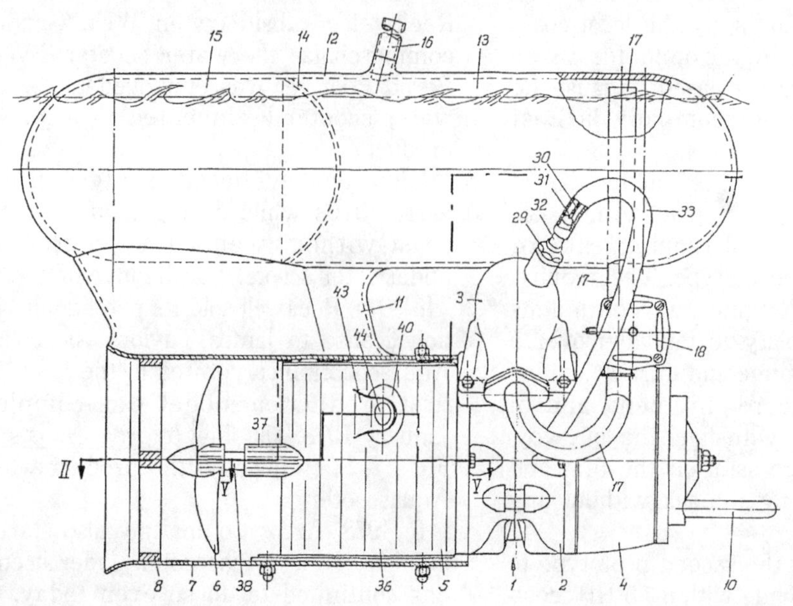

Fig - 3 Underwater scooter
(Excerpt from patent applied for in 1969)

product.

Arcos still continues to produce Aquascooter, and made major modifications to it around 1996. Perhaps until today, the total sales of units worldwide would be about 200,000. It is thus established and has a large market.

3) Excellent mechanism of Aquascooter

Fig - 3 shows the drawing for the patent applied for Aquascooter in West Germany in 1969. Comparing its design with that used for the escape from East Germany, some differences in the arrangement can be observed. Modifications were made to the design later, based on the experience of the voyage during the escape. However, the basic layout of the submerged engine was completed in the first stage, and Rockwell International applied for the patent for Aquascooter. Witkoff of Aquascooter Incorporated in the USA had held the rights from 1982 until 1992 when the major patents expired.

As the Aquascooter's engine is in direct contact with water, no special cooling device is necessary. The propeller connected to the crankshaft is covered by a duct[2] for safety and for increasing thrust. The air required for the carburetor is introduced through a pipe that extends to the upper part of the air chamber located on top of the engine, which means that water will not be sucked into the engine as long as the air chamber is not full of water.

The snorkel extends upward from the air chamber and prevents water from entering the air chamber. In addition, both the air chamber and the fuel tank serve as buoyancy tanks. The total weight of the machine is moderately balanced with buoyancy. Therefore, the Aquascooter neither sinks nor floats too much, just holding itself in the water. It can pull a person faster than when the person is swimming and can be used 60 cm below the water surface. Its inventor escaped and crossed the sea with the scooter.

Looking at the attached drawing, the driver's handgrip is positioned aft of the machine, compared to the present model in which the handgrip is forward, indicating some difference in the way of riding it. I came to know about the Aquascooter in 1980 or around that time. At that time, it had already become famous and its design had become very refined. It is not known who designed the present model.

This model (Fig - 4, 5, 6) was more polished than that of the drawing in the first patent (Fig - 3). Although the basic arrangement has remained unchanged, the shape is simple and beautiful, hydro - dynamically refined, and with no unnecessary parts. According to the patent applied for in the USA in 1973, the inventor was Eggert Bueik of East Germany. The relationship between Bueik and Boettger is not known.

The present model has a thicker propeller duct different from the one in the patent. The duct wall has a wing like section that sur-

2) Duct : A cylinder that shrouds the propeller.

Fig - 4 Aquascooter (excerpt from Arcos owner's manual)

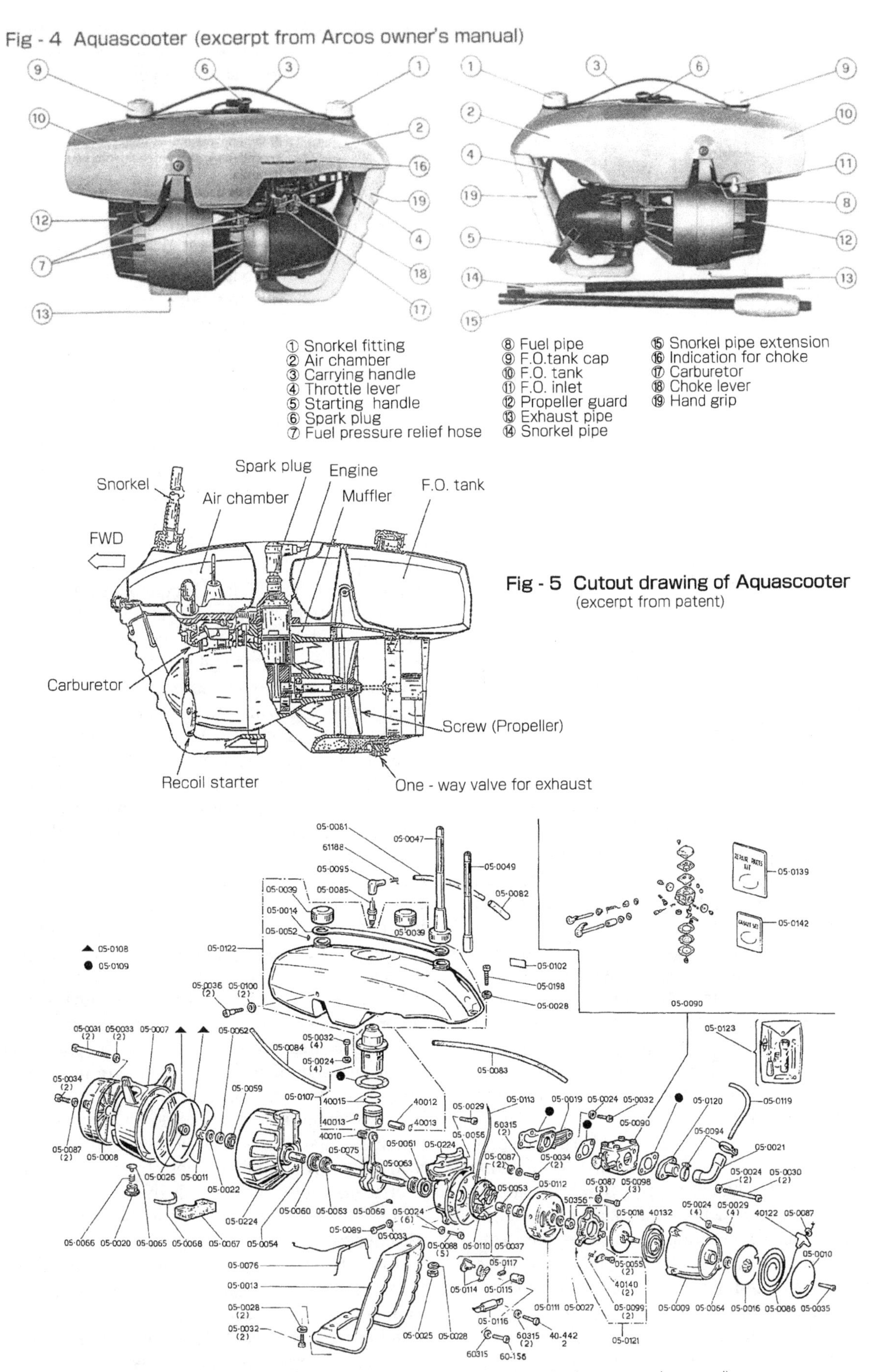

Fig - 5 Cutout drawing of Aquascooter
(excerpt from patent)

Fig - 6 Exploded view of Aquascooter (excerpt from the Arcos owner's manual)

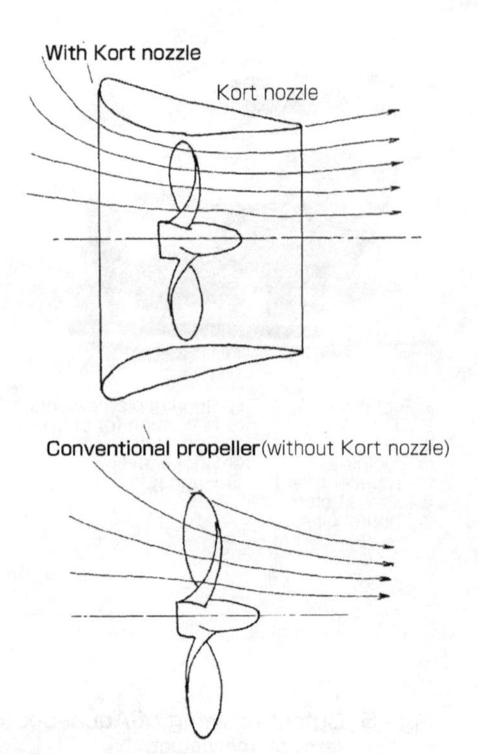

Fig - 7 What is a Kort nozzle?

In a conventional propulsion system, thrust is not fully generated at slow speeds because the after-flow of propeller becomes narrow (called shrink flow). A propeller with Kort nozzle prevents shrink flow (to generate large thrust), and the Kort nozzle itself generates thrust, thus the total thrust increases. However, a Kort nozzle has no advantage at high speeds because of its excessive shape drag.

rounds the propeller tip with very little clearance between duct and propeller. This system is called Kort nozzle (Fig - 7), which is usually fitted on tugs that need large thrust at low speeds. Compared to a normal propeller that has no Kort nozzle, the system increases thrust by 50%, which is sizable.

When you want a large thrust at low speeds, you have to rotate a propeller with large diameter at low rpm. But in case of the Aquascooter, the layout neither permitted a large propeller nor reduction gear, because the product had to be light in weight and of simple construction for easy portability. On the other hand, to obtain high output from a small engine, you have to use it at high rpm. A propeller shroud was essential to ensure safety of this product. That is to say, the Kort nozzle was the best solution to fulfill all requirements. In addition, the volume of thicker duct was used as a muffler, it served as a one-way valve that prevented backflow of water to the engine, and is thus a remarkably good design.

By incorporating such a good idea, Aquascooter when finished had a weight of only 6.8 kg, and could be carried in one hand. If you load a few Aquascooters on your pleasure boat, your cruising experience is sure to become more enjoyable than ever before

4) Other underwater leisure products

From 1985 to 1991, a large number of underwater or on-water leisure products were developed and sold both nationally and internationally. Many talented persons in the world developed such products during this period. But it was a boom period, and finally, only the Aquascooter survived.

Hereafter, I would like to show photos and a list of such products, and explain how these products have faded away (Fig - 8, 9). I regret that the list is not complete and some photos may be a bit ambiguous.

If we have a small tool that assists swimming, boating would be a lot more fun. If we have such equipment on a cruiser, the enjoyment multiplies several-fold. Once you think of a leisure product with engine, you will face some engine problems. Only a product that overcomes these problems will survive.

Especially, engines of compact products such as Aquascooter, which take in air and exhaust combustion gas while afloat on the water or while under the water, face harsh conditions and may not work well all the time without any problems. Two kinds of engines that are used in sea water and have overcome these problems are the outboard motor, which has surmounted many problems and stabilized its performance, and the engine mounted on personal watercraft, which has also overcome problems due to capsizing and swamping. However, I believe the Aquascooter is used in more severe conditions than the outboard motor or the engine on personal watercraft.

Another idea would be to use an electric motor instead of a combustion engine. The electric motor can be sealed off completely from water because it needs no air. But generally, it is heavy, has less power, and can be used only over small ranges. In general, the electric motor is only applicable to products with gentle and slow movements because of its low energy density.

Most of leisure products introduced here have faded away because of the harsh conditions. Only the Aquascooter has survived.

The seventeen models listed in Fig - 8, 9 are divided into three groups as follows:

Group A - Domestic model with jet propulsion

No.	Name	Dealer/Manufacturer	Length (m)	Breadth (m)	Weight (kg)	Type	Speed(km/h)	Propulsion	Price (10000yen)	Sold units / year
①	Jet Mini	Seibu marine/ Akashi Yacht	0.85	0.51	23.0	FRP boat	5	Suzuki jet	18.8	72/'89~'90
②	Jet Shark	Seibu marine/ Akashi Yacht	1.49	0.95	28.0	FRP boat	—	Suzuki jet	23	-/-
③	Jet Swim	Okamoto Outdoor Accessories	1.50	0.84	31.0	FRP boat	—	Suzuki jet	21.1	1000/'85~'91
④	Jet Swim	Okamoto Outdoor Accessories	1.65	0.80	32.0	FRP boat	13	Suzuki jet	18.8	incl. In above
⑤	Diver mate	Akashi Yacht / Akashi Yacht	1.00	074	25.0	FRP boat	6	Suzuki jet	13	130/'90~ '91
⑥	Jet Yuu	Akashi Yacht / Akashi Yacht	1.00	0.74	25.0	FRP boat	6	Suzuki jet	19	incl. above
⑦	Jet cCruiser	Akashi Yacht / Akashi Yacht	0.92	0.55	35.0	FRP boat	5	Suzuki jet	19	incl. above
⑧	Jet Iruka	Akashi Yacht / Akashi Yacht	1.60	0.74	30.0	FRP boat	7	Suzuki jet	13	incl. above
⑨	Mini Cruiser	Koizumi Industry	1.48	0.64	33.0	Torpedo type + Inflatable boat	5	Elc. Motor	16.8	—
⑩	Aqua Marine	Tohatsu/World Marine Tech	—	—	35.0	Torpedo type + Inflatable boat	11	Elec. Motor (500W)	—	—
⑪	Marine Drive Super 2000	Koizumi Computer	0.95	0.53	9.9	Torpedo type + Boat	3	Elec. Motor (1 hr)	7.5	3500/'90~'91
⑫	Underwater Scooter	Suzuki	1.30	Φ0.19	30.0	Torpedo type	4.7	Elec. Motor (Range 4km)	—	—
⑬	Aqua Hit	W.G.: S&F Weiskemper	1.84	0.48	23.0	FRP boat	12	Aquascooter	—	—
⑭	Water flip	Aquascooter Inc./Arcos	1.70	—	22.0	FRP boat	10	Aquascooter	Hull$1795 + Aquascooter$459	/'85~'86
⑮	Aquascooter	Aquascooter Incorporated	0.51	0.18	6.8	Aquascooter	—		$459	200,000/'76'98
⑯	TOTE - A - BOAT	Aquascooter Incorporated	—	—	—	Inflatable boat	—	Aquascooter	$1399	—
⑰	Aqua Skimmer	Aquascooter Incorporated	—	—	—	FRP boat	16	Aquascooter	$1795	—
⑱	Aqua Scuba Kit	Aquascooter Incorporated	—	—	—	Diver's kit	—	Aquascooter	—	not for sale
⑲	Dolhin D - 1	Yamaha Motor	2.40	0.54	17.8	Urethane form boat	11.3	Aquascooter	—	not for sale
⑳	Dolhin 1.5	Yamaha Motor	1.49	0.60	15.0	FRP boat	7.5	Aquascooter	—	not for sale

Fig - Self - propelled aqua leisure products of the world

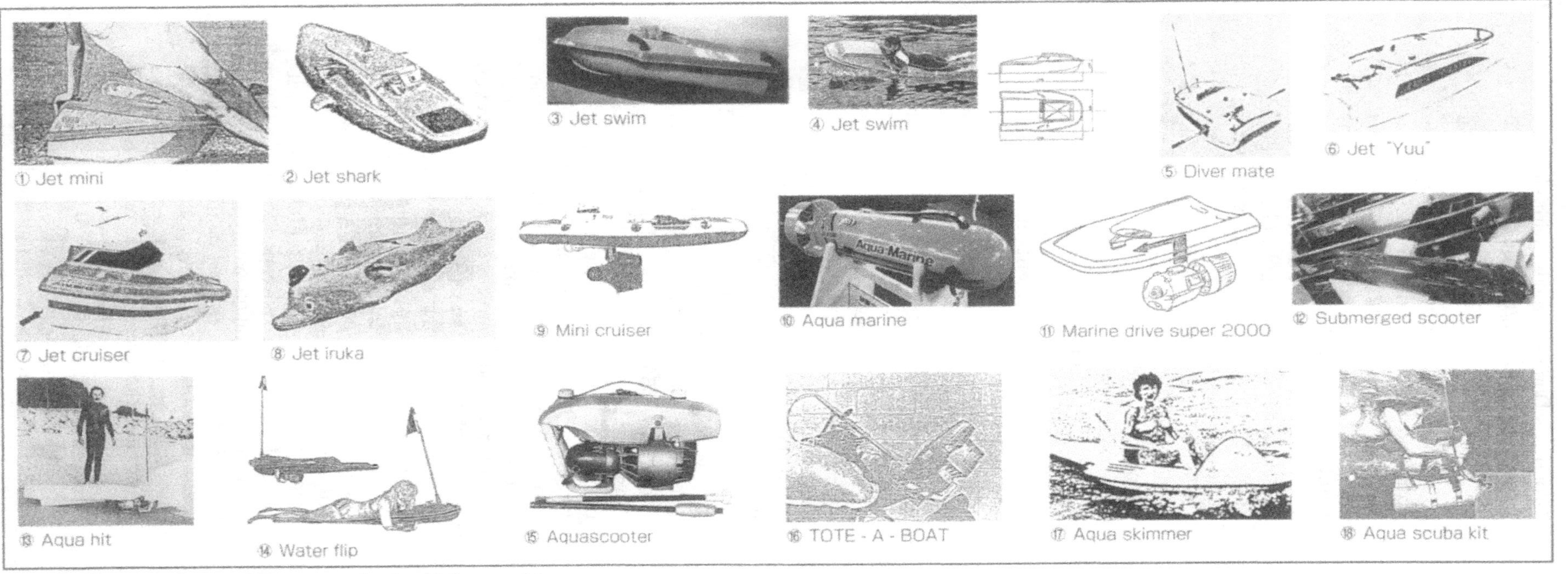

Fig - 9 Photos of aqua leisure products of the world

Group B - Domestic model with electric motor
Group C - Foreign model propelled by Aquascooter

Group A: All models in this group are equipped with Suzuki 100 cc engine and jet unit. The best selling product in this group was Jet Swim (3, 4), marketed by Okamoto Outdoor Accessories. It has a length of only 1.65 m and its speed is quoted as an unbelievable 13 km/hr. A thousand units were sold for several years, which I feel is a very good achievement.

Models 1, 2, 5 to 8 would really be called leisure products, and are appealing in terms of shape, but their heaviness and noise make them unattractive.

In my opinion, the engine with jet unit mounted in the Group A models was rather noisy and had insufficient thrust. Water jet propulsion has efficient thrust when its jet spouts out into the open air, but the jet unit of Group A models had only narrow jets spouting into the water and producing smaller thrusts.

Group B: Models of this group are propelled by electric motor. Both Aqua Marine (10) and underwater scooter (12) are similar types of submersible vehicle used by divers. They are more suitable as diving accessories rather than leisure products.

Model 9 Mini Cruiser has an inflatable hull and Model 11 Marine Drive has a rigid boat hull. Both are leisure products to be used on the water. 3500 units of Marine Drive are reported to have been sold, which is a measure of its huge success, possibly because of its light weight of 9.9 kg and economical price of 75,000 yen. The secret of its success was the selection of a suitable combination of less noise, small weight, and moderate performance. However, we cannot see Marine Drive anywhere today; I don't know why the product has faded away.

Group C: All models of this group have Aquascooter as their propulsion unit. The unclear photo of Model 13 Aqua Hit, makes it difficult to understand the intention of the designer. Model 14 Water Flip was a well-thought of product and reported to have been produced in cooperation with Aquascooter Co., but unfortunately, neither the number of units sold nor its present status is known. Its shape looks strange, but it is supposed to have less drag when the driver lies flat on the stomach on it. I suppose that is why Water Flip achieved such high speeds.

Model 15 Aquascooter has already been explained. The next, model 16 TOTE-A-BOAT, is a kind of outboard motor consisting of Aquascooter and attachment, which were intended to be installed on small boats such as inflatable boats. It was manufactured by Aquascooter Inc., but I have no idea how many units of TOTE-A-BOAT were sold.

Model 16 Aqua Scuba Kit was also developed at Aquascooter Inc. for scuba diving. With this, a diver can dive to a depth of 10 m below water. Air for the Aquascooter's engine is taken from a scuba air tank through regulator. This model basically seemed to work well and was interesting, but it was never marketed. I suspect they were afraid of product liability issues. I heard Witkoff was an industrial designer, so I presume such models as Aquascooter or Aqua Scuba Kit were produced as extensions of his ideas.

5) Development of aqua leisure products in YAMAHA

We at YAMAHA started developing aqua leisure products from February 1995. The idea evolved from a design company named YAC, which was a YAMAHA subsidiary responsible for designing and planning new products.
YAC extensively studied aqua leisure products suitable for development at YAMAHA and focused on a small aqua leisure product propelled by engine. In spring 1995, Yoshiaki Suzuki of YAC brought to me several kinds of hand-made models that he had made himself as preliminary ideas for a new aqua leisure product. He wished to find out if the models had a reasonable performance as boats and to discuss building a prototype with me.

The Horiuchi Laboratory had been established a year before in 1995, and one of my roles was that of a consultant for such technical matters. Suzuki brought with him five ideas. After comparison, I found that one had the ideal characteristics of a boat, such as speed, floatability, and stability. Others had some good characteristics. But once a model succeeds technically, there is a good possibility that others will be studied.

The model I had chosen had a length of 8 feet, a slender hull, and arrangement for a driver to lie flat on the stomach. It appeared to have less drag and suitable transverse stability. I felt intuitively that to proceed with this project promptly, it would be best to build a prototype and play with it.

I was so interested in the boat (idea) with

engine and jet in the hull that I visited Mr. Muroichi Shibata, who lived near the YAMAHA Head Office the same day, accompanied by Suzuki and Hiroshi Kubo of the Technical Administration Department, working on this project. I handed over to Mr. Shibata the selected model, and requested him to make patterns from it and produce a rough lines plan.

Mr. Shibata was an expert in building prototypes and had been creating molds, patterns, and full - scale lines plans for many years. It was not long after he had reached the retirement age as foreman of the prototype workshop in YAMAHA.

6) Spread of the Dolphin Vehicle

Besides the above-mentioned leisure products, I had wanted to confirm myself at an early stage as to how to popularize these leisure products. Figure 10 (Spread of the Dolphin Vehicle) illustrates the ideas that I had considered at that time. Suzuki's idea of a displacement model is shown in the middle of the illustration surrounded by other similar ideas.

I gave the name Dolphin to this project, to signify a machine that helps people to be like a dolphin, and also to swim and maneuver like one. I hoped that this name would bring success to this project.

Normally, it is not easy to fulfill all layout requirements in such a small vehicle. Speed and price are almost never compatible. Portability, payload, and speed are also trade - offs relative to each other. I drafted some preliminary designs around Suzuki's idea. Each had slight differences, according to leisure activities each was capable of. This process enabled me to make use of better ideas and to design without any omission.

In Fig - 10, you can see various calculations. In principle, I thought of using a 2 HP outboard motor with propeller positioned in a tunnel at the hull bottom as the engine. Initially, I had considered adopting water jet propulsion, the disadvantage of which I was not fully aware of; especially the poor performance when using the jet under water. I am not certain that I was aware of the Aquascooter at that time; but even supposing I was, I had missed the advantages of the Aquascooter.

Based on studies, the displacement type (center of Fig - 10) hull seemed to be the most orthodox design and the easiest to build. This boat was expected to reach a maximum speed of 12 to 13 km/hr using a semi displacement type hull shape.

The underwater airplane type was a model that could move about 1 m beneath the water at a 3 km/hr speed, inhaling air through a snorkel for both the driver and the engine. It did not have the same maneuverability as that of an airplane because depth was inadequate. The Aqua Scuba introduced previously, was better in this respect.

The inflatable hull type model had the same construction as an inflatable boat in the forward half of the hull. It could attain good speeds because of its large length, yet had good portability when deflated, but there were many items in its design that had never been tried out.

The planing type model had inadequate horsepower and was almost impossible to get on the plane. Therefore, the conclusion was that it would be interesting to use it to run about slowly and maneuver it with its 2 HP engine rather than trying to go fast with it, and enjoy wave surfing by accelerating with the available power. Otherwise, both boat and engine would become larger. Based on such studies, our first choice was the displacement type model.

7) Dolphin D - 1

I started the hull design based on Suzuki's plan. How should I choose the engine? According to Suzuki's idea, the engine system was mounted within the hull, but we did not have such an engine in either Yamaha or other companies. The fastest way to proceed for trials was to use the Aquascooter as the propulsive engine. An appropriate new engine and jet unit could be developed when the design for series production was decided. I concluded that we had to mount the engine and jet unit made in Yamaha in the production model somehow. I thought that if the upper half of the Aquascooter could be mounted within the hull and control parts, such as throttle lever, located near the driver's hands, then we would be able to achieve the first target. After deciding the above, the overall D - 1 project started. The participants of this project were Suzuki, who was the preliminary planner, Kubo of the Technical Administration Department, and myself.

Within a week, Mr. Shibata called us since he

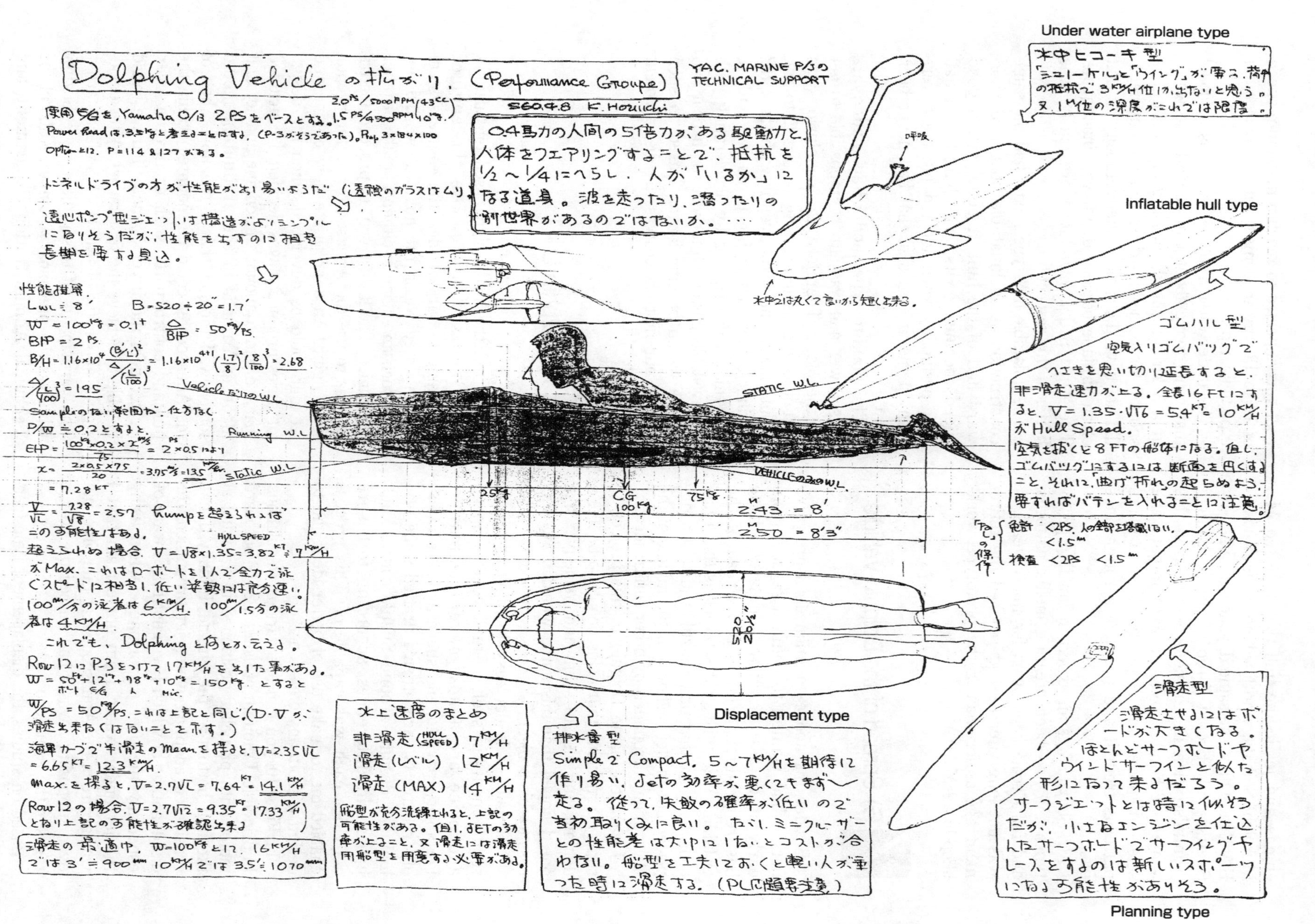

Fig - 10 The Dolphine Family

had finished the lines plan based on Suzuki's scale model. From this lines plan, I made the lines drawing of the prototype and sent it to Suzuki (see Fig - 11).

Suzuki brought a large block of urethane foam into the YAC workshop and began shaping it. Kubo placed the order for an Aquascooter, prepared parts for engine mounting, and also prepared for the trials. This was such an interesting project that everybody worked aggressively toward producing the prototype.

In June, we began the trials for which Kubo had made all preparations. Dolphin D - 1 ran smoothly. The wave from the bow to stern was satisfactory, and I was fully satisfied, except for the disturbance in flow at the stern. I felt that the length should have been a bit smaller. Since all three of us weighed nearly 80 kg each, with either one of us, the boat was too heavy during the trials. I requested Miss Keiko Furuhashi to try out the Dolphin D - 1, and during this trial, the waves around the boat passed by smoothly without any disturbance.
The steering was operated by foot (instep). When the driver extended a foot outboard on one side, the boat turned on that side steadily. If the driver wore fins, the stern waves were fairer and in addition, steering ability improved significantly. The speed at this stage was 11.3 km/hr, or about twice the speed of a good swimmer. The height of the driver's eyes were so close to the water surface that one felt its speed to be very high. Miss Furuhashi was very fond of the Dolphin (Fig - 12, 13).
All concerned persons felt good after the trials, and they thought this project should go into the product development stage. The Horiuchi Laboratory and the Technical Administration Department were part of the Yamaha Head Office, so the actual development of this project moved to the sales, planning, and design departments in the Marine Operations Division. Our group of three monitored future developments from the sidelines.

8) Dolphin D - 1.5

The problem that arose in the marine sales and planning departments was that people needed a license to drive this vehicle. If a license was required to enjoy such a slow and small leisure product, popularization of the product would be hindered considerably. We basically need a license to drive any powered water vehicle in Japan, but if the length of the vehicle is less than 1.5 m and half of the body of the driver is immersed in the water, no driving license is necessary, according to the annex of the relevant regulations (from 2004 onward, no license is required for watercraft below 3.0 m and less than 2 HP).

The length of Dolphin D - 1 was far more than 1.5 m. From the point of performance, even 2.4 m was inadequate, so poor performance was inevitable with a boat length of only 1.5 m. However, the decision to take the length as 1.5 m was adopted, and Aquascooter was chosen for power.

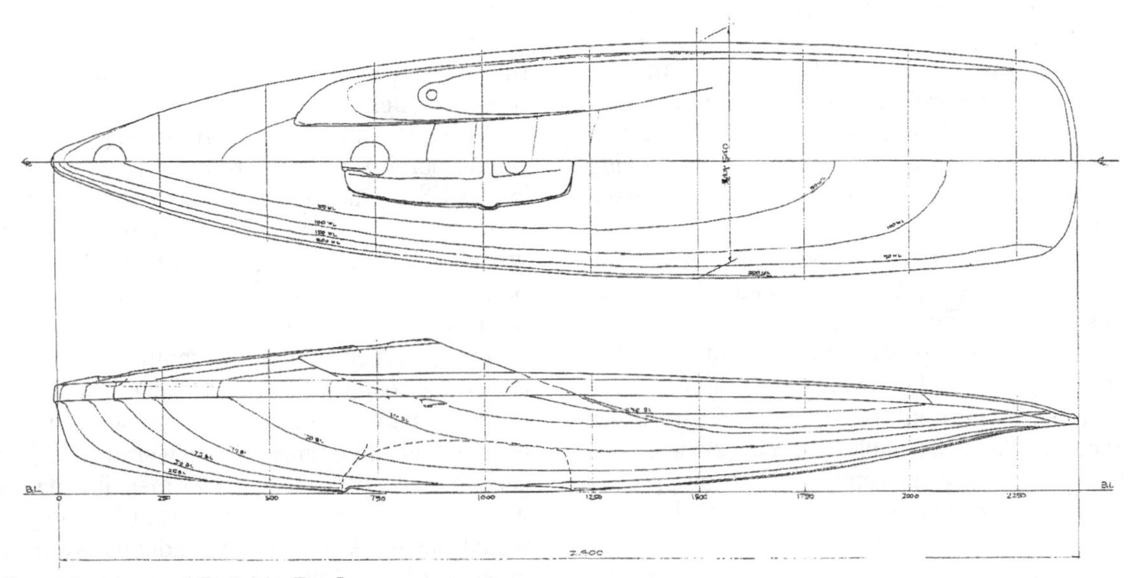

Fig - 11 Lines of Dolphin D - 1
DOLPHIN D - 1 LINES SCALE 1/5 S60.9.5
(TAKEN FROM ACTUAL MODEL MEASUREMENT) K. HORIUCHI

Fig - 12 Dolphin D - 1

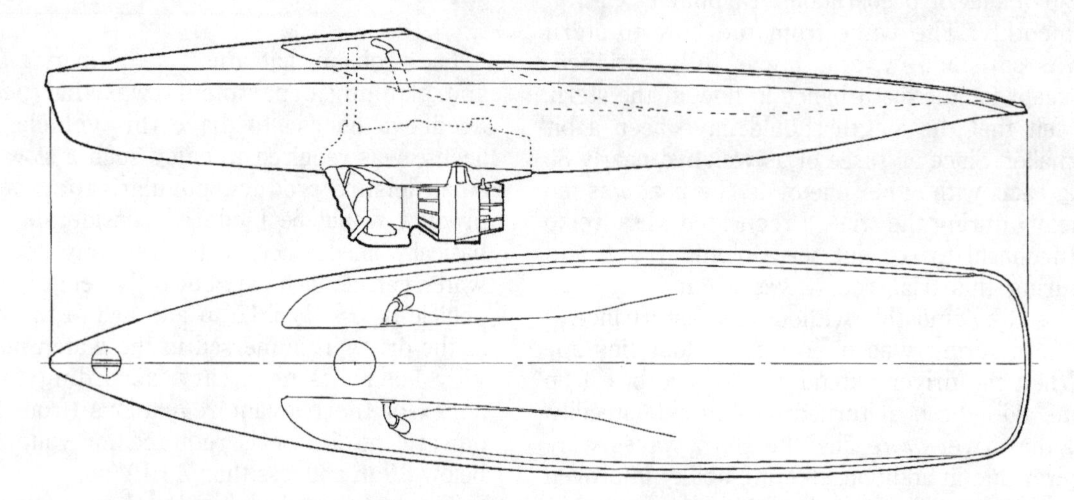

Fig - 13 General arrangement of Dolphin D - 1

I still liked the longer hull. I thought that both long and short hull moulds could be built since they were inexpensive, and people with driving license could enjoy higher speeds with the longer boat. However, my idea did not materialize; probably because I could not convincingly explain the differences in the speed that could be attained by each model. Moreover, the project was beyond my reach.

Kazuharu Fukushima was in charge of design for this project. He had been designing sailboats for a long time. It might have been hard for him to design a displacement hull with such a short length, nevertheless he completed the development of Dolphin D - 1.5 successfully (see Fig - 12, 13).

We had a pilot production plan and several of the Dolphin D - 1.5s were sent to Europe and USA for monitoring the market. We also held a race in which the public participated at a seaside resort near our company, and the product was well received. I had a feeling that official marketing of the product would begin soon.

Aquascooter had already been decided as the propulsion unit. I also felt that the design of Aquascooter had been perfected so much that Aquascooter itself would spread significantly. Since I wanted to have a product like the Aquascooter in Yamaha's product lineup, I agreed to incorporate it in Dolphin D - 1.5, but during the development of the Dolphin I became a bit worried about the reliability of the Aquascooter conpared with that of outboard engines.

At that time, I was in charge of the Minnesota R & D (Research and Development) in the USA. One of the employees, Jim Glinde, was an Aquascooter user. He enjoyed boating with friends on the Mississippi River on Sundays, and had participated in many races with the Aquascooter.

When I received this information, I thought the Aquascooter was reliable since it had been used in many races. I asked him what he did to maintain the engine. He answered that after each use, he personally performed its maintenance. It was also reported that the engine would not work if it was not adequately maintained during the off - season.

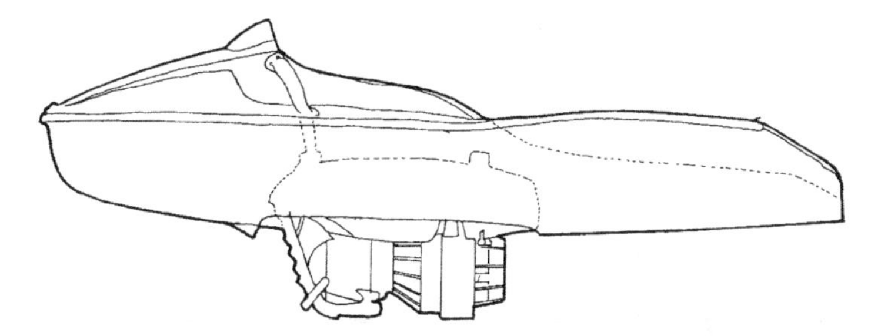

Fig - 14 General arrangement of Dolphin D - 1.5

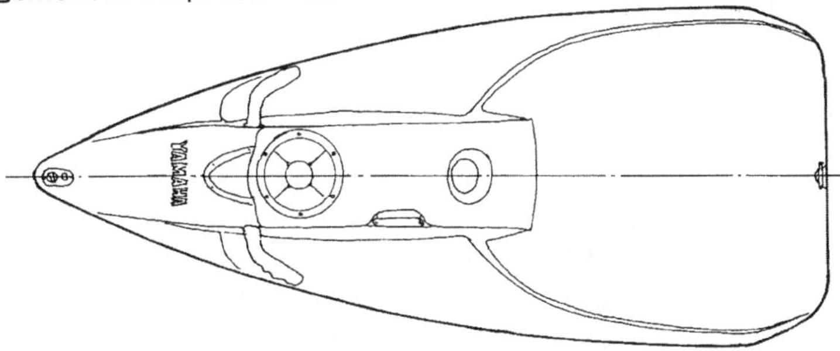

Using the Aquascooter for a long period might lead to some problems. The air chamber from where air is sent to the carburetor may be filled with water through the snorkel. Secondly, the exhaust outlet valve, located below the duct is sometimes not fully water sealed and this may cause water to penetrate the crankcase.

I did not take these issues very seriously. I thought that with some kind of care such problems would be resolved. However, the problems became serious as the development proceeded. Firstly, I estimated that the problems could be solved by the technology possessed by Yamaha Marine, but this was not so easy. Engineers from Yamaha Marine offered many ideas and tried to resolve the problems but could not find fully reliable solutions. According to the engineers, techniques underwater were quite different from those on - water.

On the other hand, negotiations with Mr. Witkoff to buy Aquascooter proceeded smoothly. Since we could not confirm whether its level of reliability was the same as that of an outboard engine, we had to give up the project. A few years later, many other powered aqua leisure products appeared in the market, but we could do nothing except watch from the side lines.

With the passage of time, the dream has faded out. There is only one underwater engine in the world, namely, the Aquascooter, and no other aqua leisure product is better than the Aquascooter.

Finally, I wish I could show the inventor, Mr. Bernd Boettger, the Aquascooter in its present form after 30 years, and congratulate him on his achievement.

Fig - 15 Photo of Dolphin D - 1.5

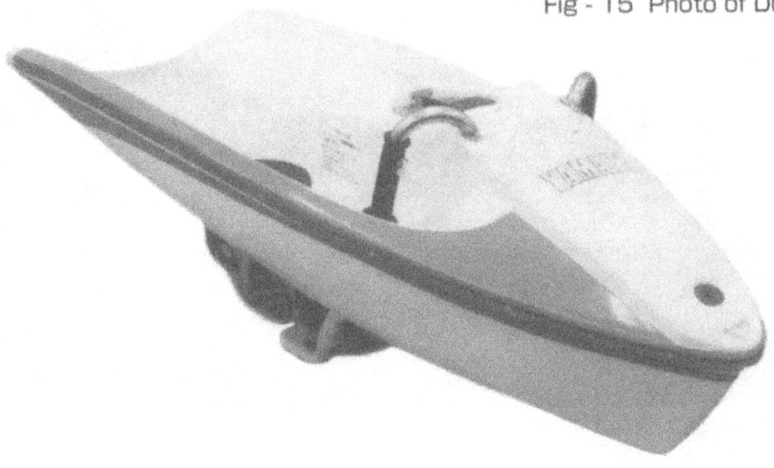

HYDROFOIL BOAT OR51

We can experience the most comfortable ride only onboard a fully submerged high - speed hydrofoil boat. I believed that if we could automatically control the up and down motions of a hydrofoil boat in high waves by sensing the waves in front of the boat, we could attain the dreamlike, soft riding characteristics even in small boats, and this is what this chapter is all about.

We had an inquiry for such a dreamboat from a boat racing federation. I thought I would rather build a hydrofoil boat suitable for boat races, and developed such a boat. However, just before completion, the order was canceled, so we converted the boat to an experimental one, from which we collected a large amount of technical data useful for designing hydrofoil boats.

1) Splendid, fully submerged hydrofoil boat

The small fully submerged hydrofoil boat that I developed in 1955 had a splendid performance. It had a tight turning ability along a steady course with a 45 degree inward bank (see Fig - 1). Such a turning ability cannot be achieved in a powerboat even with a fine hull shape. Even if such a turning ability could be realized in a powerboat for a specific condition, a small change in displacement, position of center of gravity, or speed, may cause a side force to act on the pilot or course deviation of the boat. However, to realize such a good turning performance, the layout of the boat needs to be

Fig-1 Single strut hydrofoil boat (1955)

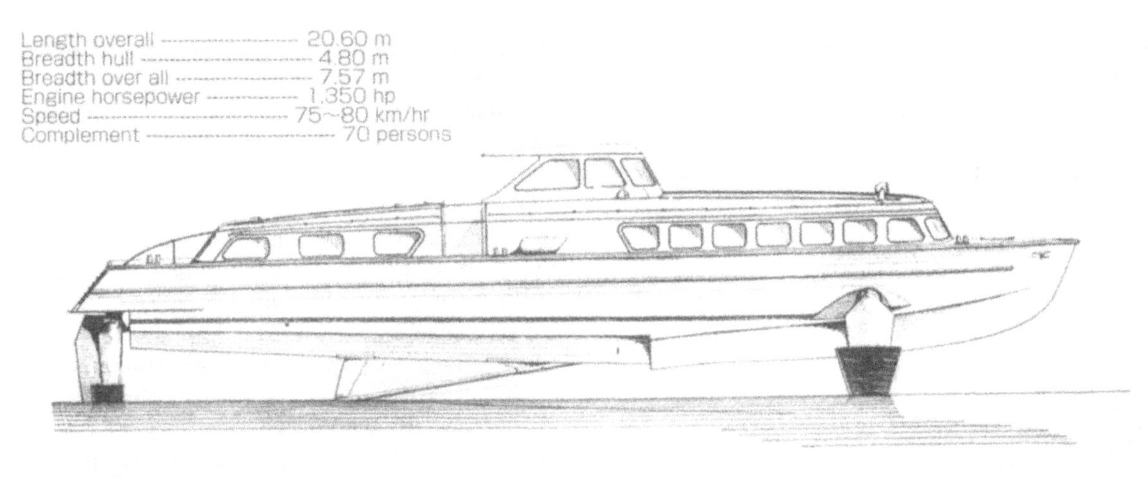

Fig - 2 Surface piercing hydrofoil boat (ship) PT 20

appropriate and skilled control technique is necessary. This type of boat could be used for sports or other special purposes, rather than as a normal high performance boat. Another advantage of a fully submerged hydrofoil boat is its excellent soft riding features. Because hydrofoils that support the total weight of boat are located deep below the water surface and connected only by thin struts, they are not disturbed by waves on the water surface. That is why the crew onboard get the feel of flying or gliding smoothly on air when riding a hydrofoil boat.

The feeling when riding a submerged hydrofoil boat is excellent soft riding characteristics, quite different from the hard feeling you experience in a surface piercing hydrofoil boat (Fig - 2). I believe no system other than the submerged hydrofoil gives soft riding characteristics to a high-speed passenger boat. To achieve this level, certain technical targets must be achieved.

When the altitude of the boat decreases in a surface piercing hydrofoil boat, its hydrofoils submerge to a greater depth and their effective areas increase. As a result, the hydrodynamic lift increases to restore the boat to its previous height. This system automatically works to maintain the altitude of the boat at the same level.

In a fully submerged hydrofoil boat (Fig - 3), there is no such automatic adjustment to keep the altitude of the boat at a certain level. In the first boat I developed, the pilot had to control the altitude of the boat by watching the water level in front of the boat, similar to controlling an airplane with the stick.

While watching the water surface far ahead in front of the boat and estimating the altitude of boat and submergence of hydrofoil, the pilot has to push or draw the control stick to maintain the boat at a constant altitude. This is a difficult technique to master. Compared to that of airplane, the pilot is required to have ten times more sensitivity.

I practised driving the hydrofoil boat for two or three days and was able to control the altitude to an accuracy of 5 cm. I was fully satisfied with the driving experience and the riding comfort.

However, it becomes difficult to maintain the altitude of the boat during a turn. The pilot finds it difficult to estimate the height, and in addition, he has to increase the lift force to control the stick according to the magnitude of the centrifugal force G.

Once when driving the boat, I failed to sense the height correctly and turned quickly at an excessively high altitude. This action caused the hydrofoil to emerge out of the water and resulted in a loss of control. The boat rolled over in the air and slammed into the water upside down. To prevent such accidents, you have to get into a turn with the hydrofoils deeply submerged.

Another problem is the difficulty of controlling the boat at high speeds. When the boat takes off, the speed is about 5 m/s (18 km/hr) and the angle of attack of the hydrofoil is about 10 degrees, which is its normal maximum. When the boat cruises at a speed of 10 m/s or twice the take off speed, the lift force becomes four times greater at the same angle of attack because lift is proportional to the square of speed. Accordingly, to prevent the hydrofoil from coming out of the water, you have to reduce the angle of attack to one fourth of the normal angle, or to 2.5 degrees.

With training, the pilot may be able to control the level. But once the speed reaches 15 m/s (54 km/hr), the pilot has to control the angle of attack to within one degree, which means that the control in the angle of attack at high speeds needs to be ten times more precise compared to that of a slow take off speed. In such high-speed conditions, humans cannot enforce such precise control, and the boat will move up and down considerably. Accordingly, the depth of the hydrofoil cannot be maintained at a constant level, and we cannot operate the boat at high speeds, even if it has more than adequate power.

2) Height controller

If an automatic control system for foilborne height of fully submerged hydrofoil boats, that is, a height controller can be realized, an ideal hydrofoil boat can be realized. This has been my dream since I developed my first hydrofoil boat in 1955. I now think I should have developed a mechanical height controller at that time. I did not actually make it; I had sketches of the same in my old notebook.

I think I was too busy at that time, since the first boat I developed was tested successfully in 1955, and Mr. Ryouichi Sasagawa, who funded the project, visited the factory to order an additional three boats, and I became busier. Even after receiving the order, I must have become so involved in many other interesting jobs that I did not find the time to complete the trials of the control system.

I moved to Yamaha Motor in 1960, and since then I was involved in developing production boats for nearly 20 years. We accounted for a large share of the market. One day, I visited Professor Kyuichiro Washizu of the Aeronautical Department in Tokyo University and I received some interesting information.

I heard one of my seniors, who graduated from the Aeronautical Department, had presented a scientific paper at a society meeting. The subject of the paper was control method to minimize motion of airplane due to gust on the premise that the condition of gust in front of the airplane was known.

Professor Washizu talked happily about the assumptions made and how they predicted the condition of the gust, which had been a rather interesting subject at the meeting. The same evening I realized that it was not easy to forecast a gust, although there were several methods to forecast sea waves such as by using ultrasonic waves, microwaves, laser, and so on. It might be more practical to use a beam of light as sea waves can be seen. Since I was not very familiar with these aspects, I was not certain these methods were actually usable, but I guessed it would be possible in the near future with technological progress.

The US Navy had been developing fully submerged hydrofoil war ships and a patent for Jet Foil had already been applied for in Japan. This showed that it was already possible to control the height of a fully submerged hydrofoil boat. I believed that we could develop a much more comfortable hydrofoil boat if we could realize an automatic height control system and forecast the conditions of the waves in front of the boat.

Unfortunately, I cannot recollect the name of my senior who presented the scientific paper and also the title of the paper itself. I think I once read the paper to confirm its effectiveness, but I am not sure.

Fig - 3 Three - strut hydrofoil boat (1954)

Fig - 4

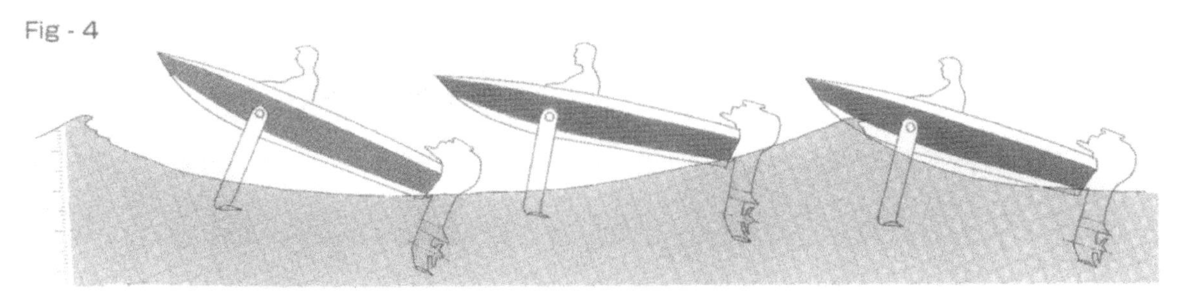

Three - strut hydrofoil boat ran comfortably in high waves (wave height 1 m, wave length 8 m).
Position of pilot1s head remained practically the same and did not move up or down.

My dream of realizing an automatic height control system intensified significantly because of one good experience of driving a hydrofoil boat in rough seas. I drove a three-strut hydrofoil boat that I had developed for the first time (Fig - 3) in one-meter waves. The boat's length was around 4 m, which means the wave height was 25% the boat length, but the ride on this boat was very comfortable.

If you head for waves at slow speed with the bottom of the stern dragging through the water, you can control the boat without the up and down motion at the pilot's position where the front struts are located (see Fig - 4). When the boat is on a wave crest, the hull almost touches the crest, and when in a trough, it runs as if the front hydrofoils are just about to fly out of the water. It is better that the boat breaks through higher crests.
In any case, since you can see each wave coming at you, it is easy to control the boat. Naturally, the bottom of the stern moved up and down considerably while touching the water, and bounced on each wave crest. But I was pleased with the comfort of the ride at the control position, and became convinced that this was the essential technology for a small boat when running in a rough sea.

The so-called platforming, which means running flat on a choppy sea without any up and down motion within the range of height of the hydrofoil strut, is the real advantage of a fully submerged hydrofoil boat. In addition, even if the wave is larger than the hydrofoil strut, you can skillfully control the attitude following the wave (contouring), and you can run the boat fast without any shocks. This was the conclusion I arrived at from this experience.

To drive the hydrofoil boat while watching the waves in front of it so as to minimize the vertical acceleration is exactly similar to the control method in an airplane to minimize its motion due to gust. I thought that if this control method was developed to make it automatic, then the fully submerged hydrofoil boat would become a commercial success.

The three-strut hydrofoil runs with its stern on the water at slow speeds. But if the area or angle of attack of the aft foil is increased, the stern would remain foilborne even at slow speeds, and the shock over the boat length could be minimized. So I was certain that automatic height control could bring about good soft riding characteristics even in smaller passenger boats.

3) Difficulty in watching the waves in front of the boat

In 1980, Yamaha's marine business expanded to include sailboats, fishing boats with outboard motors, large fishing boats and one-off boats for government use, some of which were more than 30 m in length. In those days, I had a feeling that the time had come to start studies and research on a fully submerged hydrofoil passenger boat.

I occasionally had the opportunity to ride the PT-20 type surface piercing hydrofoil passenger boat (Fig - 2) from Toba to Irako. I was embarrassed by its hard ride. The craft jolted so hard even on the leeward side of the island that I found it difficult to stand. After passing the Kamijima Island, the boat was tossed up and down by waves of nearly two meters height. The foilborne condition could not be maintained in such conditions, and the boat changed over to the hullborne condition, in which the entire hull touched the surface of the water, which means that the boat did not run as a hydrofoil boat but ran as an ordinary displacement boat.

When the hydrofoil boat was hullborne (not foilborne) in rough waves, it was like a submarine. At slow speeds, the boat plunged into every incoming wave and passengers could see nothing but waves. When the boat hit a wave, its bow rose very slowly because of the

Fig - 5

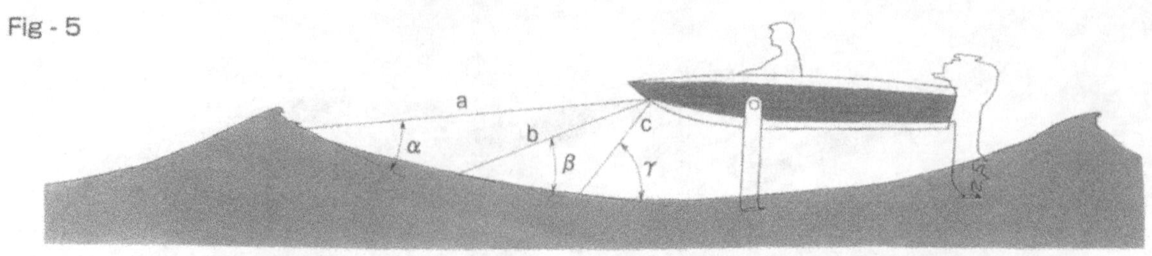

To emit beam and to measure the distance and angle of the water surface with the direction of the beam (Wave height 1 m, wave length 8 m)

hydrofoil drag. This experience served to confirm the poor performance of this type of hydrofoil boat in high waves. My feeling at that time was that a hydrofoil boat should be able to remain foilborne up to a wave height of at least ten percent of the boat length.

I also believed that if we used fully submerged hydrofoils with tall struts, we could have soft riding even in high wave conditions, and was convinced that an advanced control system and a strong structure would enable us to implement a good submerged hydrofoil design in the future.

I wrote a technical paper on the practical control method of hydrofoil boat enabling comfortable riding in waves at that time on the assumption that wave conditions could be measured, based on the ideas of my senior in the Aeronautical Department.

The gist of the paper was as below.

Like microwaves, three electric beams are emitted to the wave surface in front of the hydrofoil boat (Fig - 5). The distance and the angle to the wave surface are measured to analyse the waveform and the speed. If these data are input to the automatic height controller, the boat will be able to cruise safely and comfortably.

Fig - 6 General arrangement of YZ886

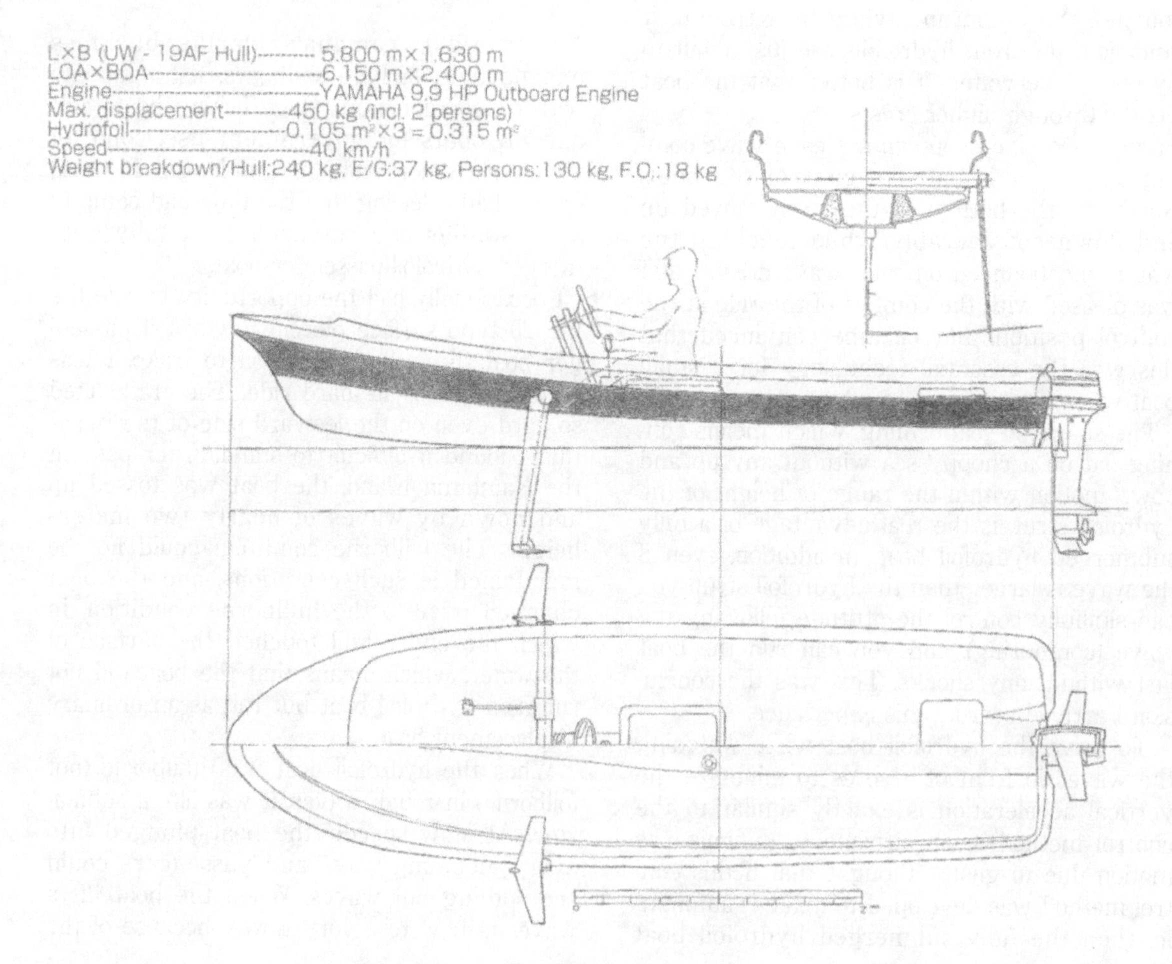

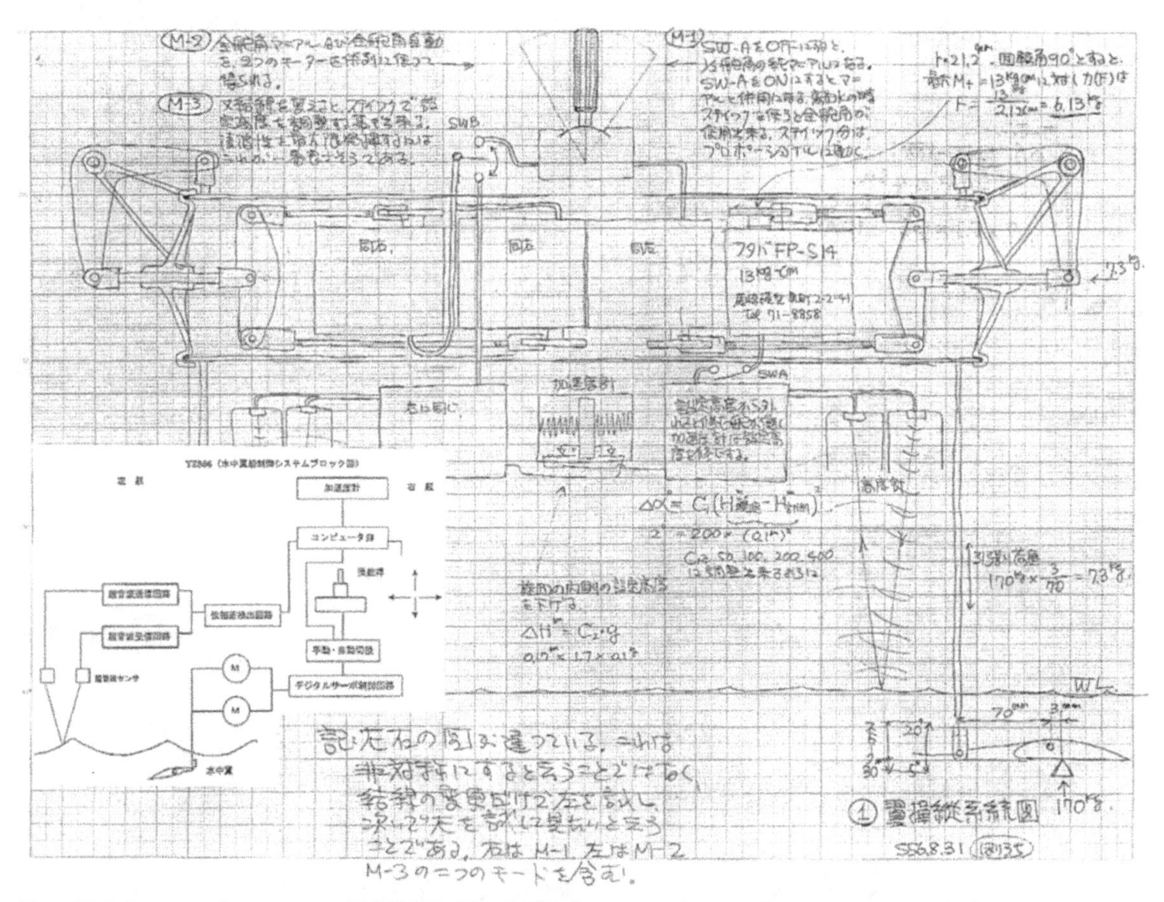

Fig - 7 Foil control system of YZ886 (Tentative)

This paper was written based on the experiences I gained after driving a three-strut hydrofoil boat in wave heights equivalent to 25 % of the boat length, and when I had struggled to drive a 18 foot power boat in 4 m waves as fast as possible during the Osaka to Tokyo 1000 km power boat race.

In 1981, a year after publishing the paper, I decided to start with the plans of a hydrofoil boat with automatic height controller. For the first time, I had to build an experimental boat to test the front wave sensor.

I was fond of the boat shown in Fig - 1, which was capable of banking 45 degrees. We called this type of boat a single strut hydrofoil boat because when you see the boat running toward you straight ahead, you can see only one strut, although it has two struts actually. This boat maintains transverse stability similar to a bicycle or a motorcycle. If the boat is small, you can control it with your driving sense similar to a bicycle or a motorcycle. But once the boat size increases, as in a passenger craft, driving is no longer dependent on the driving sense, like that of a motorcycle. Passenger craft should have a hydrofoil outboard on each side and transverse stability should be maintained by controlling ailerons, as in an airplane. Based on this concept, I used the three-strut type as an extension of the concept of the boat in Fig - 3.

The main purpose of this research was not the development of the boat itself, but to experiment with the height controller. As a test boat, I chose a 19 foot outboard type production fishing boat (UW - 19AF) that was on hand. The engine for this boat was a standard 9.9 HP long shaft outboard motor. Normally the shaft length of an outboard engine is for a transom height of either 12 inches (S : 305mm) or 15 inches (L : 381mm), but this 9.9 HP model fortunately had a super ultra long shaft for the 710 mm transom that had been specially made for fishing boats. Its shaft length was still not adequate for the hydrofoil strut, but we had no choice.

The development code number was YZ886 (Fig - 6).

The boat had an electric servomotor system for controlling the hydrofoils. The servo-

Fig - 8 Foilborn of YZ886

motor could change the angle of attack of the hydrofoil. In this way, when the sensor system capable of measuring wave height and shape was completed, both systems could be connected easily. Also, we planned and designed a height control system capable of both automatic and manual controls (Fig - 7).

For the manual control system, we adopted a joystick that had been used in model airplanes almost without modifications. But since the length of the joystick was too small for a beginner to control, we increased the length by adding a small bar so that the same feel as that of a control stick in an airplane was obtained.

Shozo Okuma worked out the detailed design of the hydrofoil and Norihiko Yoshikawa designed the control and sensor systems. Both of them thoroughly understood and shared my dream. As I was rather busy and only met them occasionally, the two of them actually completed the boat.

Adjustments to the hydrofoil boat were delicate and difficult; the boat could run long distances foilborne by manual control. I thought the boat was ready for testing (Fig - 8), but the sensor system did not work as we had intended.

It was possible to measure the foilborne height by emitting ultrasonic waves straight downward but it was difficult to sense in front of the boat, which was necessary.

Yoshikawa had studied microwaves and with the assistance of Yukio Kono, a senior mechanical engineer, he analyzed how each wave, either ultrasonic wave or microwave, was emitted and reflected back from the inclined water surface. It became clear that neither of the waves were reflected back adequately in disturbed water for use in our height control system. The reflected waves were buried in noise.

I was not very familiar with electricity and wave theory, and I might have been a bit too optimistic. Also, our facilities and our budget were insufficient. I could not understand why it was not possible to sense a wave only a few meters ahead, whereas in the air we had radar that was capable of sensing another airplane flying at several tens of km away. It was also very difficult to implement the necessary technology within the scope of our simple facilities.

I tried to network with Hamamatsu Photonics Co., a leader in optics research, and tried to get assistance from the Research and Development Corporation of Japan, but this did not work well. In addition, Yamaha's financial situation had worsened, and expenses were cut down. We lost the battle with Honda Motor Co. (called the HY battle, that is, competition in motorcycle development and marketing between HONDA Motor and YAMAHA Motor). In 1983, I resigned my post as director.

Because of the poor financial situation of our company, I gave up trying to push this project forward. The project could be started again with some budget when the technological level of each component rose. I considered my options and abandoned this project.

4) Order from Federation of Prefectural Association for Motor Boat Racing (FPA)

After the HY battle, our executives changed. The Horiuchi Laboratory was started in line with the wishes of the new president, Eguchi. We had not had such a department before for developing new products for the company. It was the first time that a laboratory was named after an individual in Yamaha Motor.

In May of the same year, we were consulted for developing an unusual boat for the FPA through our Tokyo branch office. They wanted to place an order for a dreamboat. Since details were not clear, I visited their planning department to understand their requirement better. They wanted to exhibit a full - scale dreamboat during a large - scale memorial race at the

Fig - 9 Rough sketch of the dreamboat for Kyootei

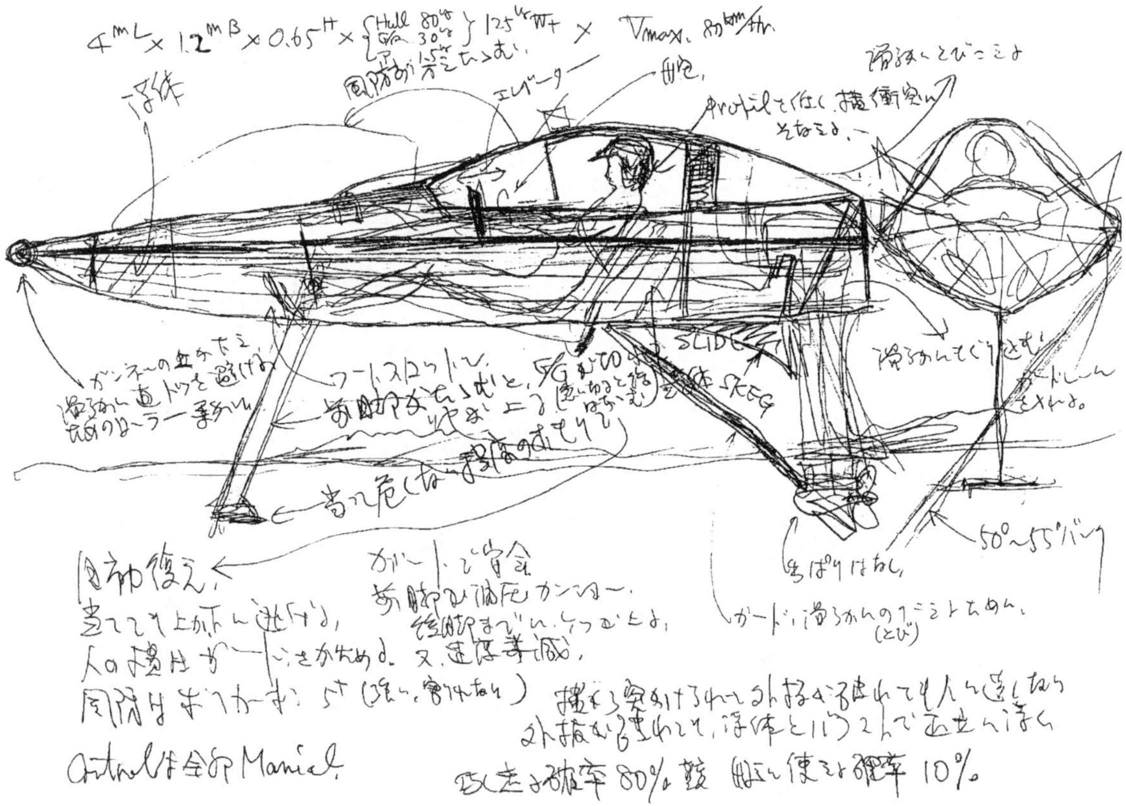

Fig 10 FLYING SILVER Dream takes shape

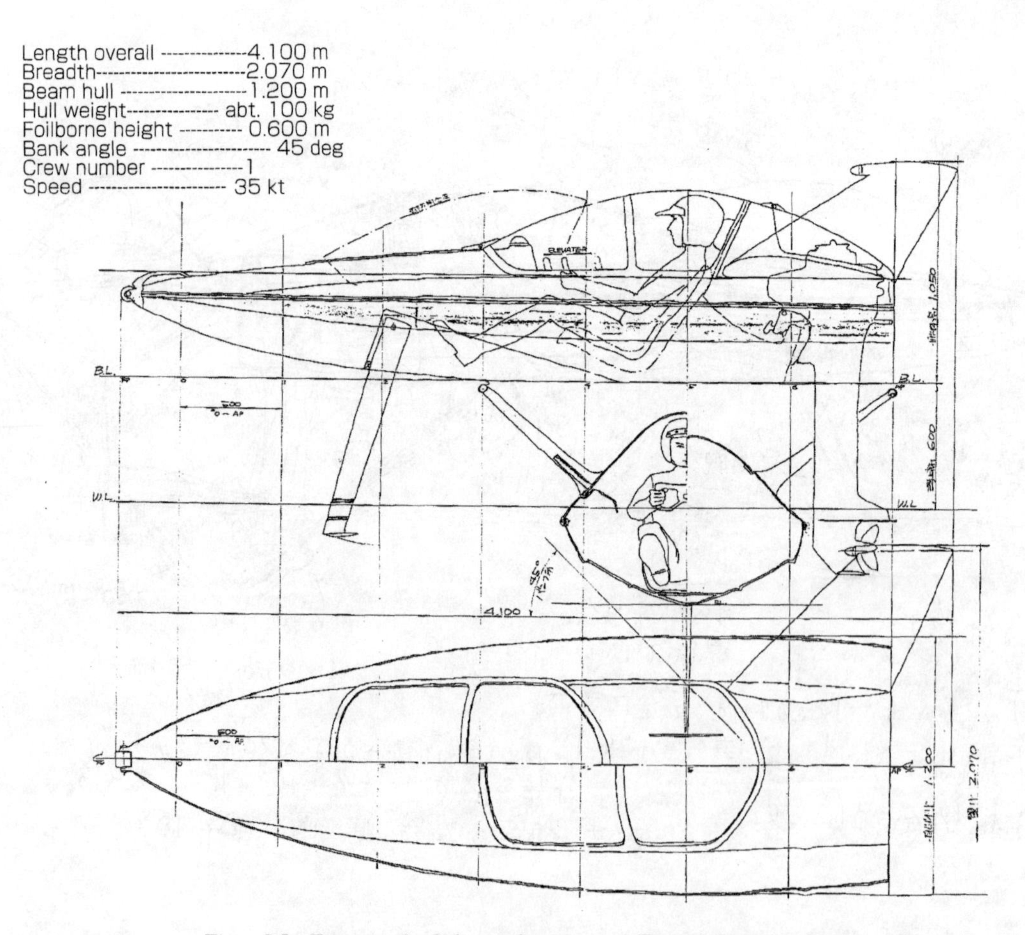

Fig - 11 Proposal of dreamboat Drawing submitted

Heiwajima racecourse in March 1985 to attract people. I remember that they did not want the dreamboat for racing at that time.

After I met them, I presumed that their intention was to realize a fanciful dream that would attract the interests of many people. But a boat builder like Yamaha cannot design a papier - mâché of an absurd dreamboat, meant only as a decorative piece without having any functions.

I was familiar with the design of racing boats. When I was working for Yokohama Yacht Works, I was involved in the development and production of racing boats since the Kyotei[1] race started, and I had some knowledge of these boats. My image of the Kyotei racing boat combined with a hydrofoil boat looked good. Fig - 9 shows the rough sketch that I made during my return trip by train, my heart thumping with joy. This rough sketch included 90% of the functions of the required boat.

Since the order from the client was a dreamboat, I drew a perspective sketch of the dreamboat with suitable accommodation for the Kyotei race (Fig-10). I presented its general arrangement drawing (Fig-11) to the client. Referring to both the three-dimensional sketch and the drawing, I will explain the intention of the design here.

One of the issues in the Kyotei race is the high probability of the leading boat at the first turn-buoy to maintain the top position up to the goal. This is because the wake of the leading boat disturbs the second and third boats trying to overtake the first boat. Since the wake of the hydrofoil boat is negligible, such an effect does not occur.

Another issue is that any differences between the boats or engines lead to irregular race results. For hydrofoil boats, control and adjustment techniques are more important for winning races, that is, the driver's skill and technique have more weight on the results of the race, so this issue of differences will also be minimized.

Near the first-turn buoy in the race, there is a likelihood of collisions between boats and serious injuries because every boat rushes to get to this buoy immediately after the start. An accident is likely to occur when the leading boat finishes the first turn and the following boats smash into the side of the first boat. To prevent such accidents and to maintain safety, I designed the hydrofoil boat with a strong structure that protects the driver, adequate stability with a range of stability of 180 degrees, and with a guarantee that the driver's head always remains above the water.

I also installed a rubber roller on the boat's sharp bow so any impact would be softened during a collision between this boat and other boats. Bow and stern hydrofoil struts were also designed to swing and fold back when the safety ropes attaching the struts to the bow and the stern break during an impact. The shape of the hull section was nearly rhombic enabling the other boat's bow to move up or down during a collision. It looked like a Stealth bomber rather than a dream hydrofoil boat.

The hydrofoil system used was a single strut type that I was familiar with. For this project, an automatic height controller to control foilborne height automatically was not necessary. Only those drivers, who had mastered the technique of maintaining a large foilborne height so that drag of the strut is reduced, could achieve higher speeds.

For propulsion, I used the Yamato outboard motor used normally in the Kyotei race, with extended spacer inserted between engine block and lower unit, which was used as the stern hydrofoil strut.

I was confident of this plan and presented it to the client with the necessary documents, and it was very well recieved. Some people in the FPA had taken a ride on the single strut hydrofoil boat that I had designed 30 years ago. It is evident that they became interested when a similar boat that they had experienced in the past appeared, which could be used in Kyotei races.

The initial plan of just exhibiting a papier-mâché boat was extended to exhibiting one boat and offering one boat for trials at the exhibition. However, the order was not finalized, and I thought that it would be necessary to start building the boat in October for the trial in January the following year so that we were confident of presenting a good product.

However, the order still did not come through. I waited impatiently, thinking that they must have had their own internal problems. Finally, we started building the boat without receiving the order.

In December, we received news that they had cancelled their plans.

It appeared that the Chairman of FPA, Mr. Ryouichi Sasagawa told his staff to build six boats and race them, but the investment was too high for his federation (FPA) to place the order at that time.

1) Kyotei : Special hydroplane race organized by FPA.This is gamble race similar to horse races.

5) YAMAHA boat OR51

I was very disappointed and felt personally responsible for the cancellation of the order. Director Tanaka, who was in charge of the Horiuchi Laboratory, encouraged me saying "since this is an interesting project, let us make it a company project and let us go forward and complete it." His words encouraged me to start afresh. The decision was made then to display the boat and make a video presentation of the trial of this boat at the forthcoming Tokyo boat show to be held at the end of February. We started work aggressively on a hastily prepared schedule after 2 to 3 days of disappointment after the order was canceled.

This boat was named OR51 and Masato Suzuki was in charge of this project. He was a graduate of the Aeronautical Department of Nippon University. He was the first person assigned to the Horiuchi Laboratory after being employed at YAMAHA. He had been enthusiastic about human powered airplanes and gliders since his university days, and had won the Birdman Rally, held every year on

Fig -12a First prize at the Birdman Rally in 1998
(distance flown - 23.7 km, flying time - 54 min. 33 sec.)

Fig -12b World record run
(18.67 knots, 10.409 sec./100 m)

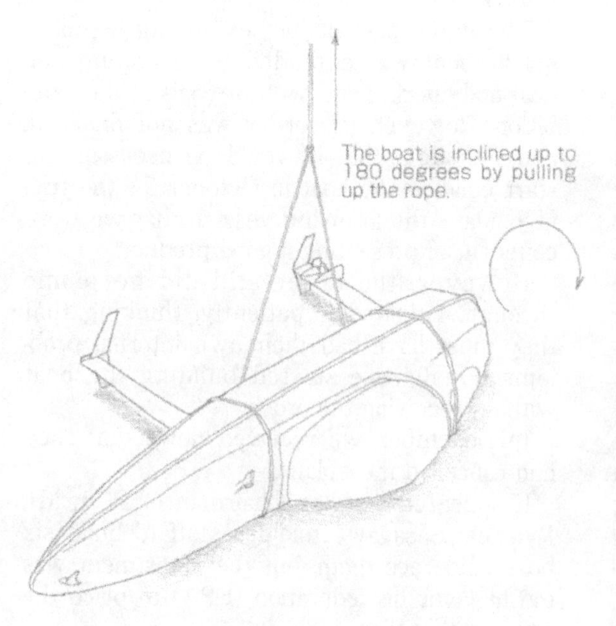

Fig - 13 Righting test of OR51

Fig - 14 Horiken - Color
White deck, gray hull, and three gradated colored lines with red, orange, and yellow aft of the windshield
The photograph by Mr. Seo was so good that I used it on the cover page of my book A LOCUS OF A BOAT DESIGNER (1) with his permission

Lake Biwa. His team's human powered airplane took first prize for flying across Lake Biwa. His experience in controlling airplanes and his enthusiasm on flying products made him the ideal person for this project (see Fig - 12a).

The next member of the Horiuchi Laboratory was Fumitaka Yokoyama. He was in charge of mechanical parts. He was a graduate of the Kobe Mercantile Marine University. After he joined YAMAHA, he worked for several years in the diesel engine development and technical administration departments. Beside his usual job, he participated in a car race called Mileage Marathon, in which the distance driven using one liter of petrol is the criterion for winning the race. He had personally designed and built the car body and also the suspension, engine, and gear train. He was an aggressive worker and an enthusiastic person, and even today, he is busy designing human powered boats. His record was approved as a world record in 2002 (see Fig - 12b and Chapter 5 of this book). His best record is under 10 seconds on a 100 m course, which is 50% faster than the record for the eight oars rowing shell. It must be the fastest human powered boat in the world.

Thanks to these two engineers and the technicians of the prototype building shop, the OR51 was completed successfully. Before the trial run, we had to test and confirm the safety characteristics of the boat. This included the self-righting test from the capsized position. The test method was as follows: The boat with one person (me) on the seat wearing a seatbelt is on the water alongside the pier, and a crane pulls up the rope tied to the base of the hydrofoil struts and looped once around the hull. The boat heels gradually until it assumes an upside down position, at which point, the boat should immediately right itself to its original state (see Fig - 13).

At the signal, the boat was inclined gradually until just before reaching the upside down position when the crane was stopped, which surprised me. Water penetrated the joint between the windshield shell and the deck and collected at the top of the windshield shell, as the boat was in an upside down position. The water level first wetted my head, increased to reach my forehead, and just when I was scared at the thought that it would reach my mouth, the crane pulled up the rope. At the 180 degree inclination, the boat righted itself rapidly. It was a scary experience, although only for a few seconds. After this experience, I am certain that I will not be scared even if the boat capsizes.

When we started the trial run of the OR51, I found that it was rather difficult for the boat to take off. Its engine specially made for the Kyotei race, seemed to have smaller torque at low speeds, which made it difficult for the boat to become foilborne due to poor accelerating thrust.

In addition, the engine installed in a separate compartment, was made especially for racing, and had neither clutch nor a starter-motor. It was difficult to control the boat and another boat must always follow it for assistance in an emergency.

I looked for another suitable engine, and found the YAMAHA 15 PS outboard engine with a super long shaft, which had been developed for fishing. We replaced the previous engine by this engine and got good results. As the project was no longer for the FPA, we thought replacing the engine was a natural thing to do.

Soon after the New Year, I tore my Achilles tendon while playing badminton with colleagues near my house and was admitted to the hospital. I prayed on the hospital bed for Suzuki and Yokoyama to do their job well and complete the boat as scheduled.

With the focus on the Tokyo International Boat Show, I was concerned about the coloring of the boat with which I was not satisfied. I asked Keiji Nakagawa, an industrial designer, for several color plans and studied them while in the hospital. I liked very much one of those plans that had three-gradated colored lines based on a two-tone color of white and gray. The boat was repainted with these new colors for the Tokyo International Boat Show. This color plan was refined and to our liking, and we called it the Horiken-Color (Horiuchi Laboratory Color). Since, then it has been used consistently on airplanes and boats (see Fig - 14).

This boat was received very well at the Tokyo Boat Show. The video presentation attracted many visitors. The show was just 40 days after I had torn my Achilles tendon, and I was just out of the plaster cast. Since I desperately wished to see the OR51, I visited the Tokyo Boat Show at Harumi with the help of crutches and a wheelchair. The Horiken-Color was bright and looked beautiful. I was satisfied with the display of the boat. Both Suzuki and Yokoyama were very busy replying to visitors' queries.

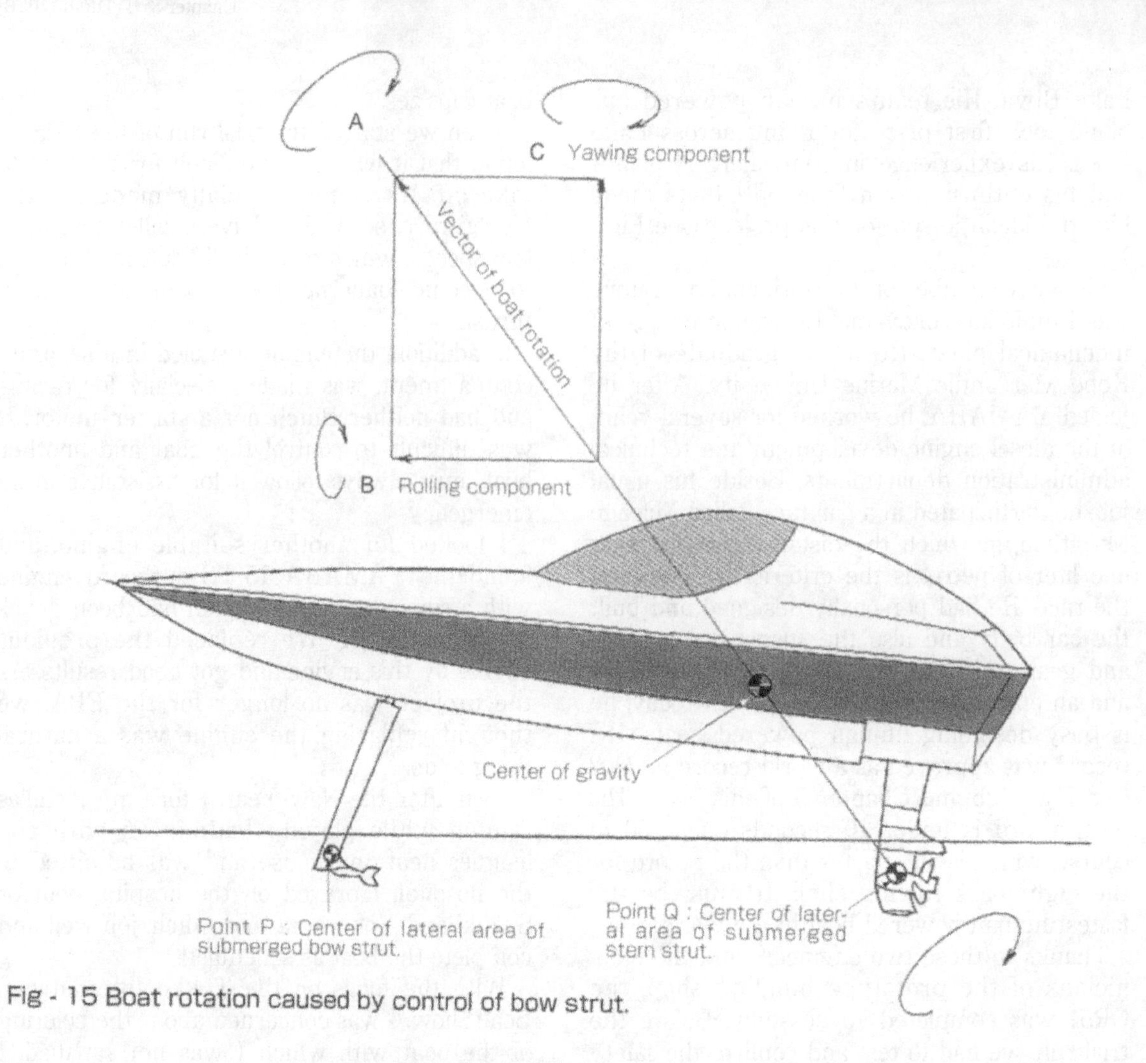

Fig - 15 Boat rotation caused by control of bow strut.

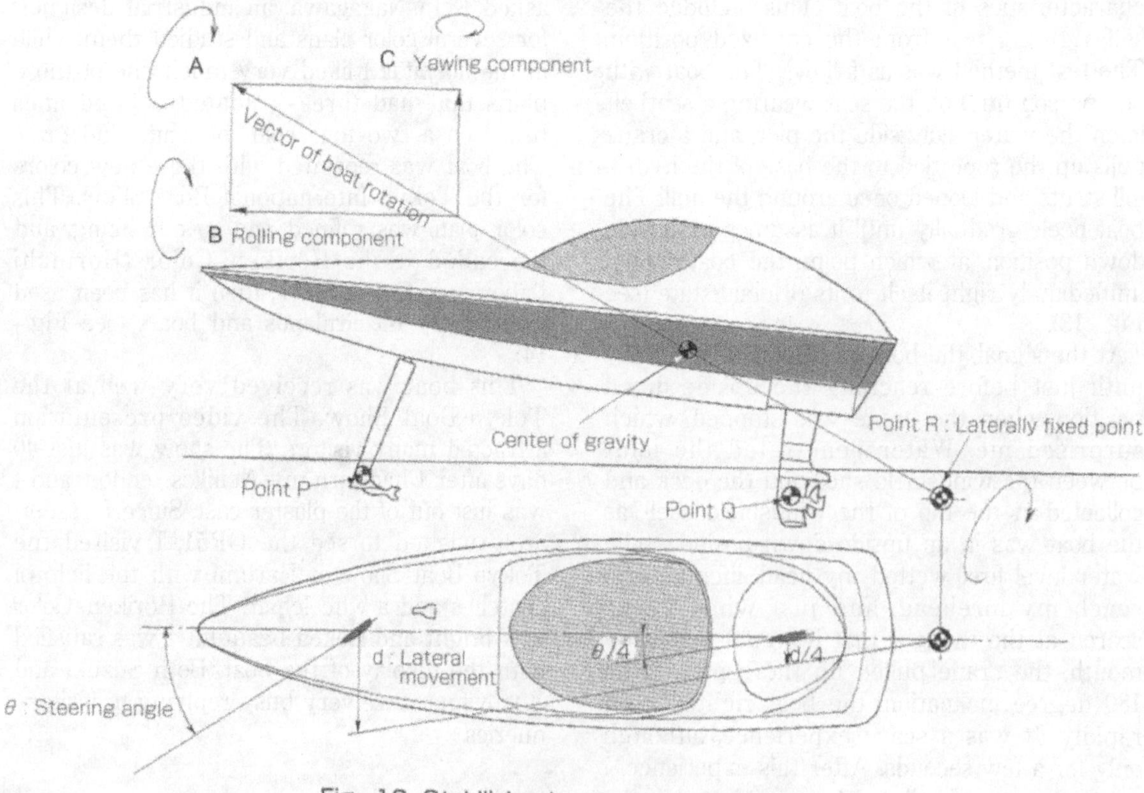

Fig - 16 Stabilizing by two - strut control

6) Improvements to OR51 (Part 1 Two - strut control)

We were able to finish the OR51 just in time for the Tokyo Boat Show. But we were not satisfied with its stability when foilborne. Although Suzuki managed to control the boat, it always yawed[2] while running, which indicated the difficulty of controlling it.

The single strut hydrofoil that I designed previously had adequate stability, but why not the OR51? I pondered deeply and found that it was because of the longitudinal position of the boat's center of gravity. Please refer to Fig - 15.

During the foilborne condition, when the steering wheel is turned, the bow strut rotates to one side (according to the direction of the steering wheel) like a bicycle's front wheel and a lateral force acts at point P on the submerged bow strut. This is the force that controls lateral stability and steering. On the other hand, the center of gravity of the boat and part of the submerged stern strut (point Q) can withstand lateral movement by the lateral force on the bow strut, which, as a result, rotates the entire boat around its axis in line with the center of gravity and point Q. The rotation is represented by vector A on that axis. The vector A can be divided into two vectors, the horizontal B vector and the vertical C vector. Vector B may be understood as a rolling component and vector C as the yawing component. In other words, you can now understand that by rotating the bow strut, both rolling and yawing motions occur in the OR51.

This is similar to the case of a bicycle or a motorcycle. You can consider the vector passing through the point at which the rear wheel touches the ground in case of bicycle or motorcycle the same as the vector passing through the point Q on the submerged stern strut in case of the hydrofoil boat. If the position of center of gravity is located more toward the aft, the axis of rotation tends to become upright, which causes the yawing factor to increase and the rolling factor to decrease. This means that it is not very effective to bring the heeled

Fig - 17 General arrangement of OR51

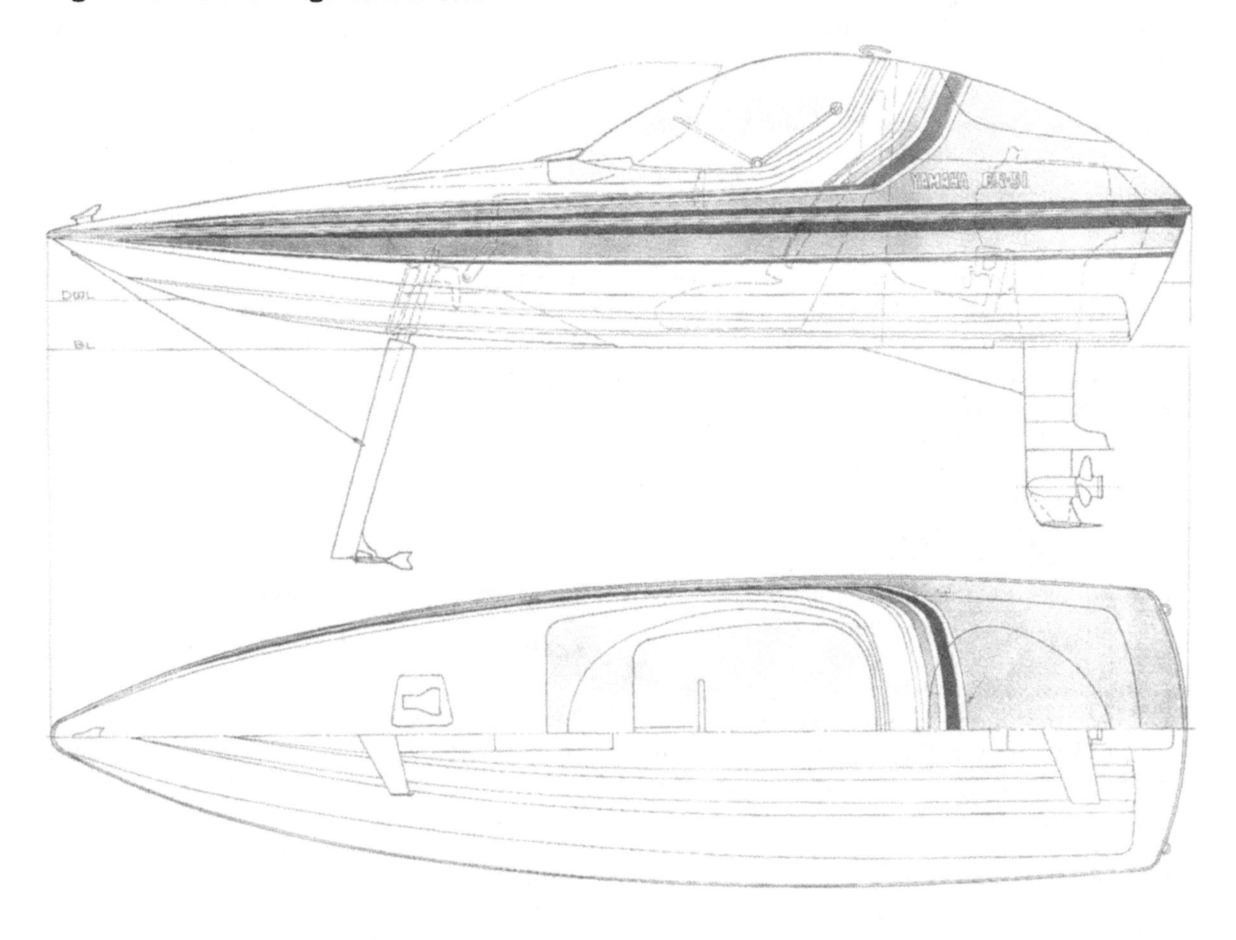

2) Yaw : Lateral swing of the bow

Fig - 18 Height sensor
To sense the depth of bow strut by emitting ultrasonic wave upward from the front edge of bow strut and measuring the time for the wave to be reflected from the water surface.

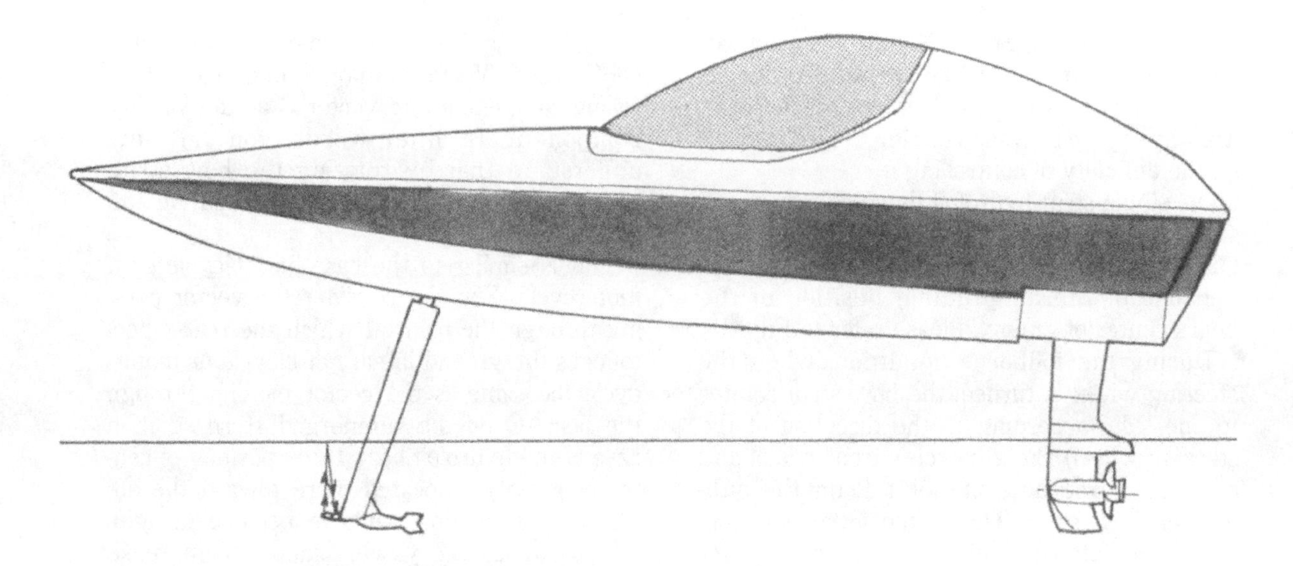

OR51 to the upright position by turning the strut by steering it. Because of this, the pilot tends to oversteer, and naturally, the side swing of the bow due to yawing motion becomes more noticeable.

Let us consider a bicycle. If the center of gravity moves aft because of a heavy weight or a person seated on the aft carrier, the steering handle becomes unsteady. A similar problem occurred on the OR51. To improve this, the center of gravity must be moved forward. But the layout of the OR51 did not permit moving the center of gravity forward. I hated installing a ballast weight forward because boat weight would increase, and the boat was already heavy compared to its engine power.

The idea of using a two-strut control system occurred to me. Around that time, the four-wheel steering system for cars was just making its appearance. Yamaha was also studying a two-wheel steering system for motorcycles. The two-strut control system was the marine version of these ideas.

Until then, only the bow strut of the OR51 was used for steering the boat while the stern strut had been fixed. If the stern strut could be steered to one-fourth of the bow steering angle on the same side as the bow strut (see Fig - 16), a lateral force would act on the submerged stern strut similar to the bow strut, and the point R would remain fixed laterally and move to a position further aft of the stern strut.

If the line passing through this point (point R) and the center of gravity becomes the new axis along which the entire boat tends to rotate, then this new vector axis becomes more horizontal than that of the single-strut control system and the rolling motion due to steering will increase. We tried to validate this concept. Since our outboard motor had the basic steering mechanism, only cables had to be routed without any major modification.

The result was as expected and the stability improved dramatically. If the position of center of gravity is correct, a two-strut steering system is not necessary. I was happy that we could confirm our theory on stability for both a two-wheeler and a two-strut hydrofoil boat (Fig - 17).

7) Improvements to the OR51 (Part 2 Height control)

We had previously confirmed by tests three years ago that the height sensor using ultrasonic waves had worked well. But it tended to be affected by noise and did not always give consistent results. Since we were considering commercialization of the hydrofoil product after OR51, we had to study the height sensor control in the OR51 in further depth, taking it as the prototype boat.

At that time, Yokoyama was in charge of this topic assisted by Masao Sugiyama, of the Electrical Engineering Department.

We were aware from previous experiments that emitting ultrasonic waves in the air was not feasible under conditions where these waves are exposed directly to both waves and spray. Considerable noise is generated, which causes performance to suffer.

We approached the research on the height sensor this time from another direction. We used a receiver/emitter from a fish finder that was already available in the market, and developed a system made from the parts of the fish finder such that ultrasonic waves were emitted upward from the front edge of the bow strut and the time after reflection from the water surface was measured. This helped us know the submergence of the hydrofoil (see Fig - 18).

This device gave us good results because the sound velocity was higher and we could consult with an expert on the fish finder. Thus, the OR51 was fitted with a height sensor enabling constant foilborne height to be automatically maintained. In addition, the boat had a manual control system, giving the OR51 two control systems. In special conditions such as take off, cornering, and contouring (to control the hydrofoil boat to follow the wave contour), manual control was necessary.

Experimental data showed that under steady running conditions, no manual adjustments were required and the foilborne height was maintained at a constant level. Additionally, it became possible to simulate the motion of the boat by height controller in parallel, and the foundation for operating a fully submerged hydrofoil was gradually firmed up.

This boat was not large enough to be effectively used as an experimental boat. Its cockpit was very narrow and we could not reach for instruments under the front deck. Yokoyama had to work or adjust instruments under the deck with his head and body upside down, and he often became very disgusted.

YZ886 was an experimental boat to test systems for sensing waves in front of the boat, while the OR51 was studied with the intention of popularizing its design especially for Kyotei races. Henceforth, we intend to develop our own plans for salable hydrofoil boats.

HYDROFOIL BOATS OU90 & OU96

These two boats were developed as production models based on the successful OR51. I present here a very simple saddle type hydrofoil boat of low cost and easy to popularize, and a high-speed hydrofoil waterbus with the minimum required specifications of a passenger craft. We were optimistic in developing an electronic stabilizing system and succeeded in developing a mechanical gyrostabilizer. All our hydrofoil boats were of the two-strut type. These two boats might have gone into production had the timing been suitable, but both faded away because of different reasons.

1. Saddle type hydrofoil boat OU90

1) Next steps for OR51

As explained in a previous chapter, I injured my Achilles' tendon in January 1985 and was admitted to a hospital. During my stay in the hospital, I had adequate time to consider the next steps in the development of the OR51. The current OR51, without modifications, would be difficult to commercialize because of its high cost. The new product should be a simpler boat but with the good features of the OR51.

While lying in bed at the hospital, I made a sketch of one of the solutions (Fig - 1). This saddle type hydrofoil boat was similar to the one I had designed 30 years ago. This time I adopted a surfboard-like hull shape to increase productivity and to cut down costs. In addition, I used a simpler and more compact hydrofoil system than the OR51, and used the lower part of the outboard motor as the stern strut with a hydrofoil under its skeg.

Although we did consider jet propulsion, we decided to use an outboard motor first. To build a prototype of this model seemed rather easy, and the OR51 had raised many expectations at the Tokyo boat show both from within our company and from outside. We wished to build a simple prototype by ourselves in the Horiuchi Laboratory and to evaluate its performance.

2) Prototype completed smoothly

This boat was named OU90 and Fumitaka Yokoyama was in charge of this project from the basic design stage, through building, and up to the trials of the boat. He also studied the electric attitude control system using the OR51. In the course of these studies, he finished all drawings of the OU90 within two months after the Tokyo boat show.

We entrusted the industrial design of the OU90 to YAC, a Yamaha subsidiary, with Keikichi Iwadate being nominated for performing the industrial design of this boat. He had written several articles on boat restoration in the BOAT CLUB magazine, a sister publication of KAZI. The restored boat was a wooden boat as old as his age. He was an industrial designer, very familiar with boats.

This industrial design of a boat would be a first for him. As he had worked on the industrial design of the Yamaha scooter at YAC

Fig - 1 Original plan of saddle type hydrofoil boat

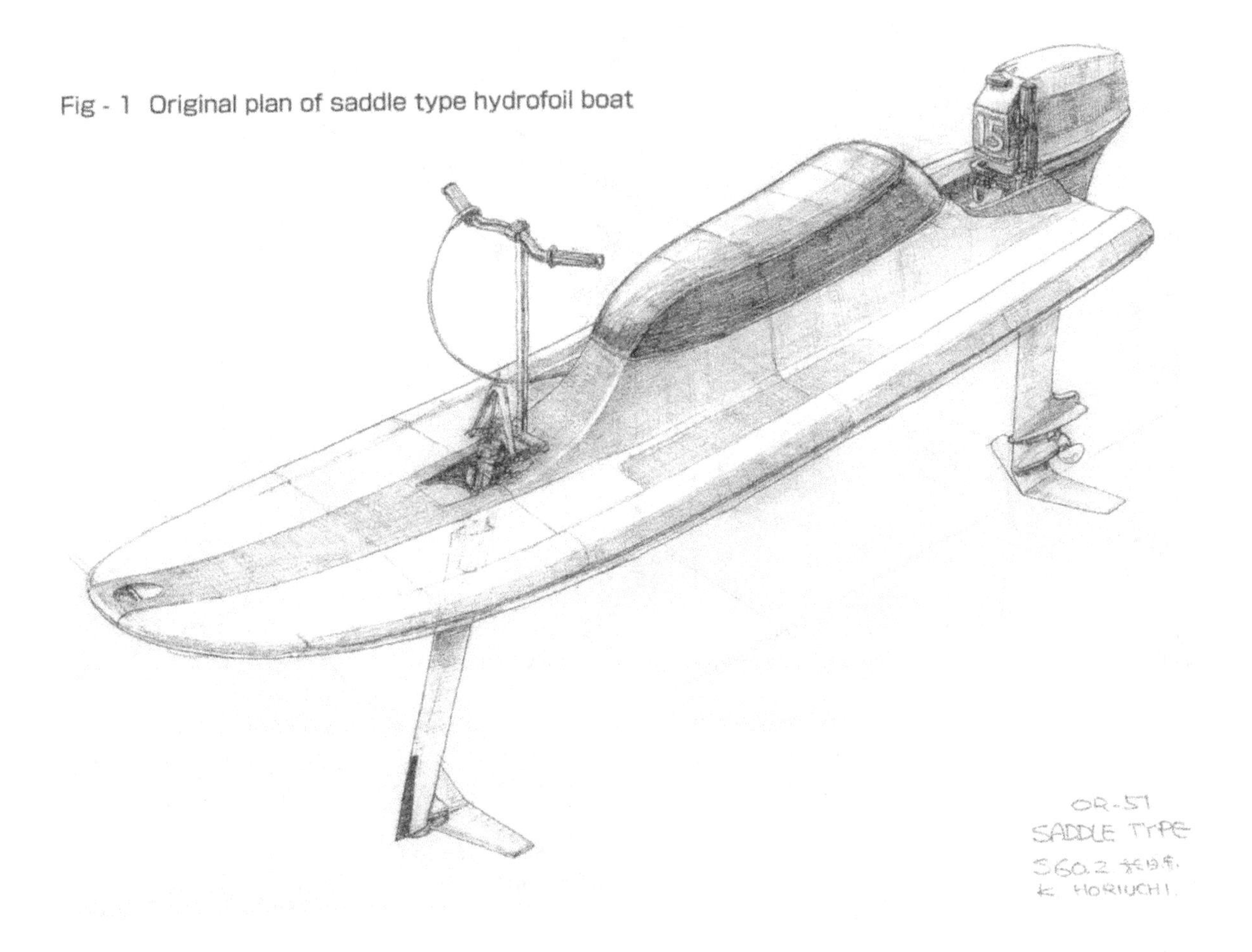

Fig - 2 FRP construction plan

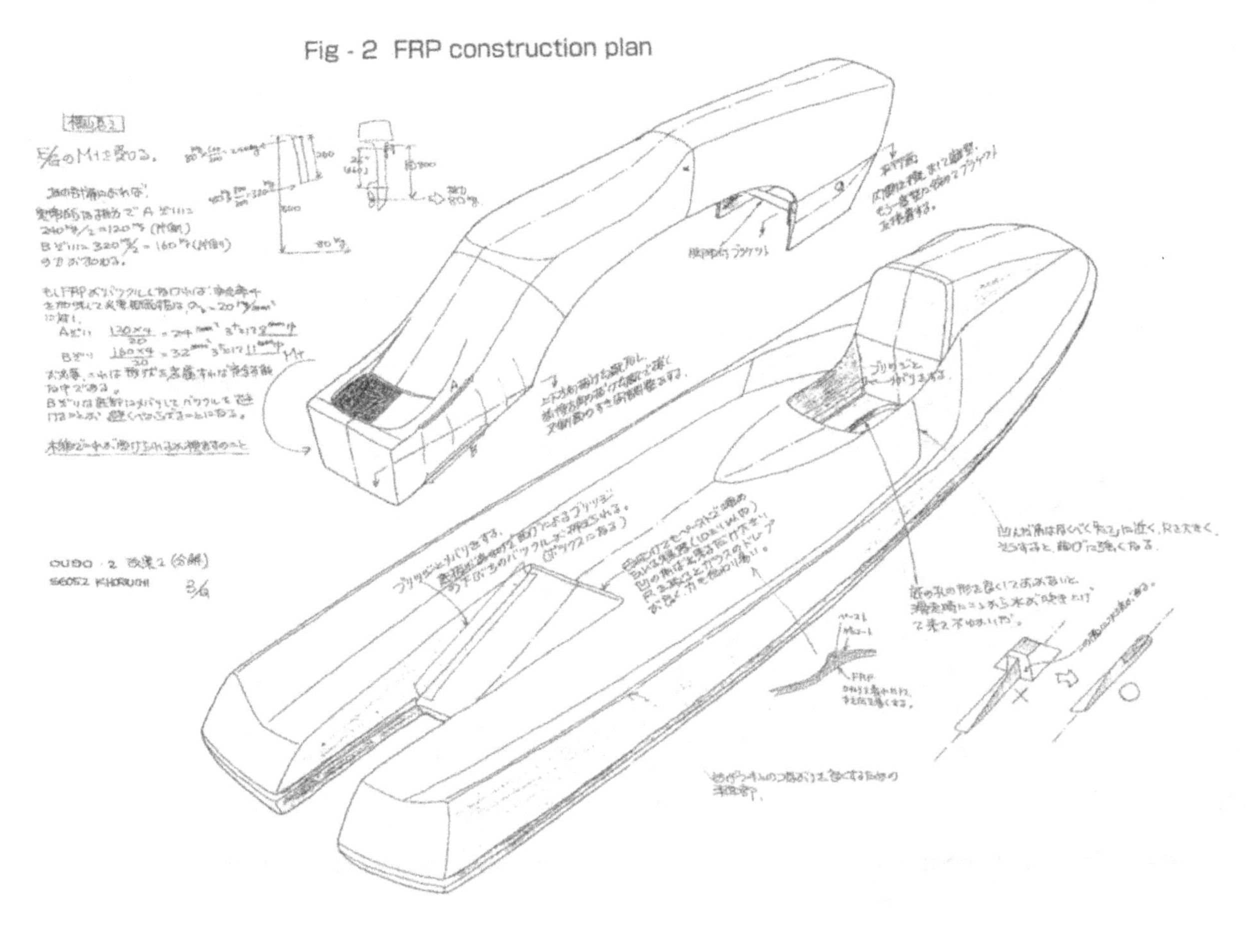

Fig - 3 Smooth maneuvering of OU90

Corporation, many aspects of this project were familiar to him, and he was an aggressive worker.

The control system of this boat was very similar to that of a scooter, such as the riding style, steering bar, two struts, and throttle grip, except for the lever and the pedal for braking.

Since this was just one prototype we were building, we made the hull by shaping large urethane form blocks and covering it with FRP. Neither Iwadate nor Yokoyama had any experience in the production design of FRP boats, so they discussed various matters with each other and with us, including ideas for construction and style on the perspective drawings of this boat (see Fig - 2).

I remember we did the design work in March and April, and the first trials probably

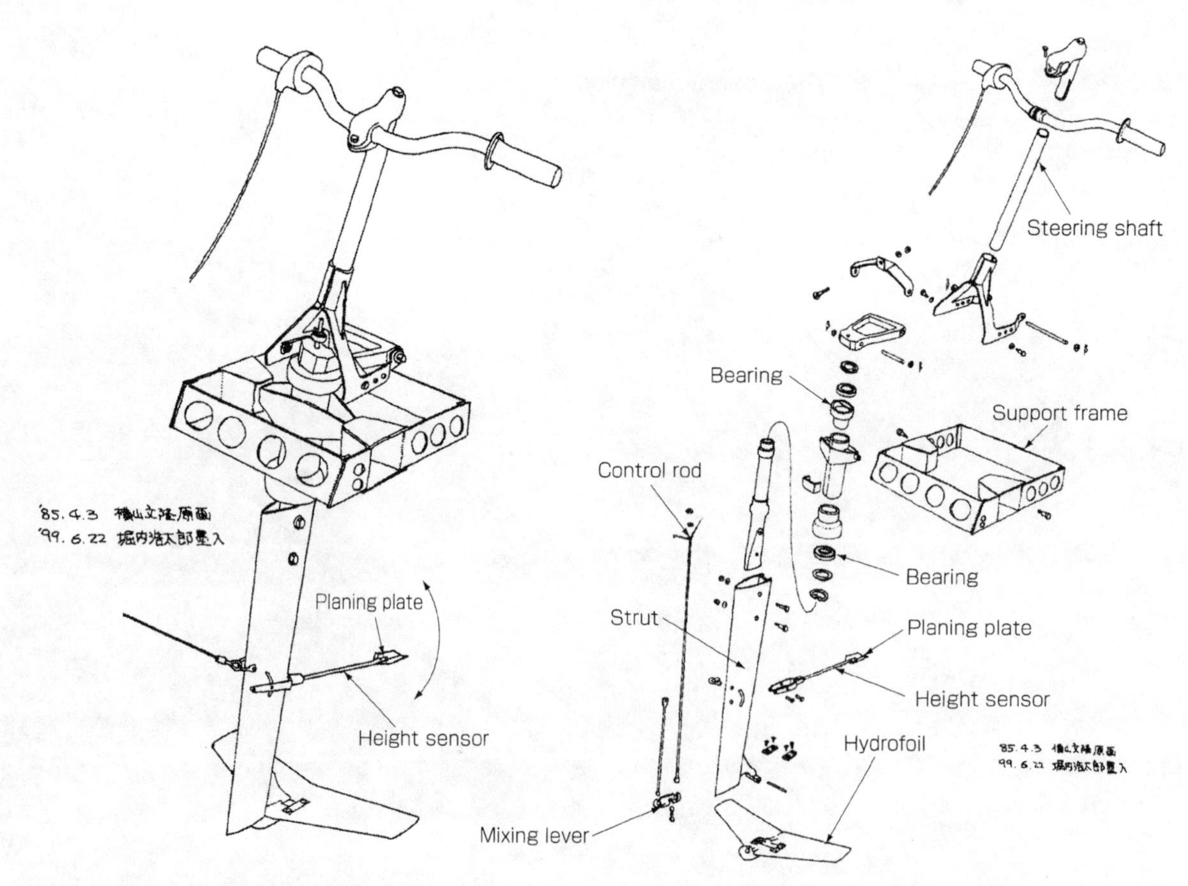

Fig - 4 Bow strut construction

Fig - 5 Exploded view of bow strut

began in June. The boat was constructed to a lighter weight compared to the OR51, yet was fitted with the same 15 hp outboard motor, and it ran very lightly right from the beginning (Fig - 3)

3) Mechanical height control

With the OU90, we could attain sharp turns with 45 degrees banking and become foilborne quickly. The boat was reasonably small and had such easy handling characteristics that Yokoyama planned to use this boat to study and gather various design data for development of passenger craft in the future.

There was one problem in the single strut hydrofoil boat that I had built 30 years ago. When the boat speed increased, the control of the angle of attack of the hydrofoil became too sensitive for the boat for manual control to be used. As a result, the boat moved up and down considerably and could not reach the maximum speed that should be attained for the horsepower provided.

To resolve this problem, we developed an automatic height controller using ultrasonic waves for the OR51. However, this system was meant only for passenger boats and was rather complicated for use in small boats, so we decided to develop a height controller using a mechanical sensor for the OU90 (Fig - 4, Fig - 5).

The system of the mechanical height controller is as described below.

Fig - 6 Maximum speed of OU90 was 57 km/hr

Fig - 7 Smooth foilborne condition of OU90 with two crew on board

The boat runs dragging a short rod (sensor). The sensor is pin-jointed at nearly the center of the bow strut and has a small planing plate at the other end. This planing plate moves on and follows the water surface. When the boat's altitude drops and it approaches the water surface, the planing plate rises upward relatively and this movement increases the angle of attack of the bow hydrofoil, which restores the boat to its previous height.

A mixing lever is provided between links (Fig - 5) that enables both automatic and manual controls. The system that we developed was quite successful. When the boat was running straight, there was no need for any control, and this was significant progress.

The boat's running performance was so good that Yokoyama could investigate the effects of hydrofoils of different shapes and sizes on speed, maneuverability, and stability. In October, we attained the maximum speed of 57 km/hr in the OU90 (Fig - 6).

Generally, it is not easy to obtain such a speed with only a 15 hp engine, except in race boats. The boat I designed while at Yokohama Yachts

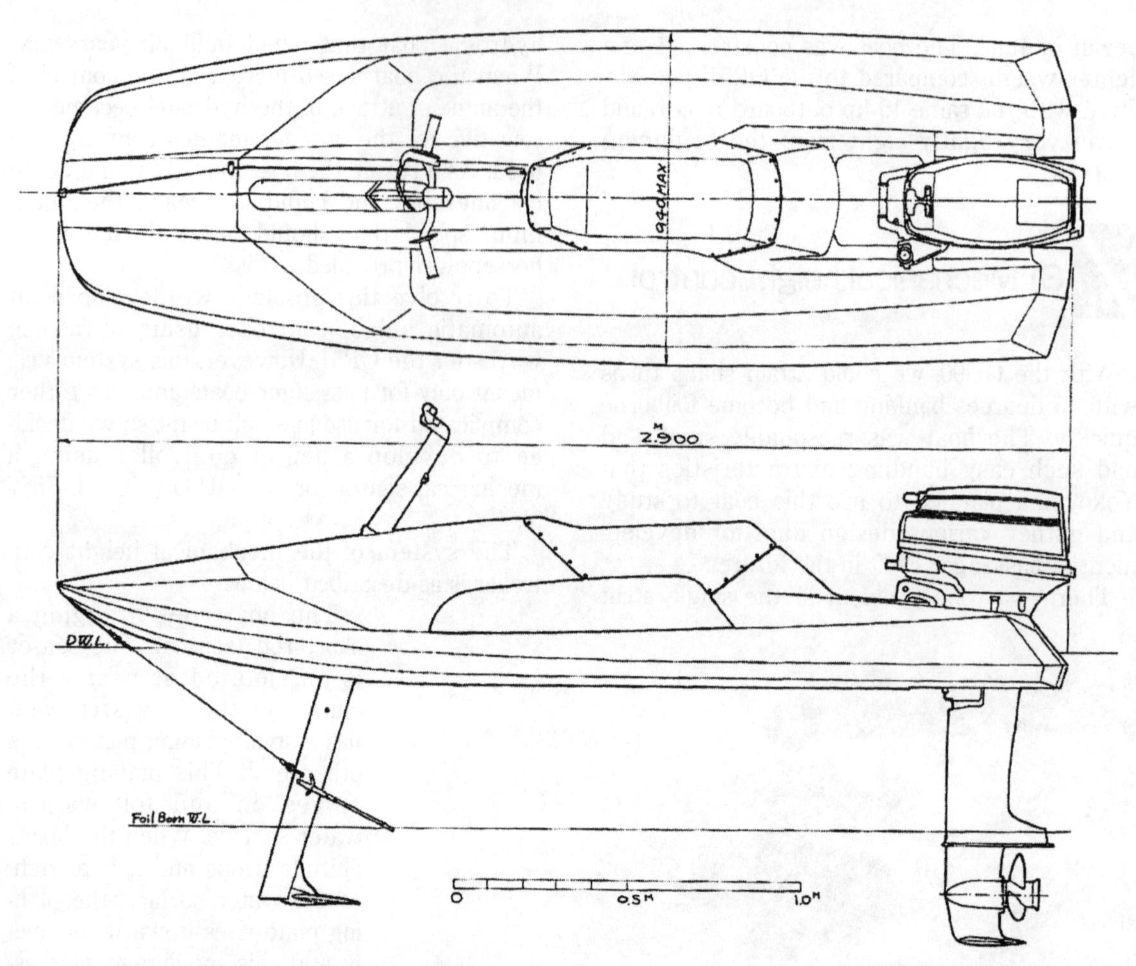

Fig - 8 General arrangement of the OU90

many years ago could not attain more than 35 km/hr because of control problems. Compared to this boat, the speed of the OU90 was 60% greater. The speed was largely dependent on the sensor system. The mechanical control system was very suitable for these kinds of small boats because of its light weight and simple construction.

4) Knowledge acquired from the OU90

Yokoyama conducted the running trial with two crew on board the OU90. Due to the light hull weight, the boat became foilborne smoothly with two crew without any problems (Fig - 7). Looking at the boat on its trial run from the shore, I was delighted with the impression it gave – it looked as if the crew were flying on a magic carpet.

Yokoyama studied the influence of the size and shape of hydrofoils on the OU90 performance. We confirmed that a larger hydrofoil area enabled the boat take off easily above the water (become foilborne) at low speeds. On the other hand, lift was proportional to the square of speed, so the area would be excessive at high speeds, and would result in increased drag.

Conversely, if the area of hydrofoil decreased, the boat would need a higher speed to take off, and the maximum speed would not increase, the range of possible foilborne speeds would become smaller, and this would make the hydrofoil inconvenient for actual use.

Accordingly, deciding a suitable area of the hydrofoil became one of the most important topics related to take off speed, maximum speed, maneuverability, and a proper balance of these items.

During the trial, Yokoyama reduced the hydrofoil area gradually and changed the wing loading (displacement/total foil area) sequentially.

With the increase in the weight/area ratio from 14 kg/dm^2 to 18 kg/dm^2 and to 22 kg/dm^2, the take off speed increased from 20 km/hr to 30 km/hr and to $(30+\alpha)$ km/hr respectively, and the time necessary to become foilborne increased proportionally. The maneuverability became poor because of the smaller lift margin. The maximum speed was attained at a weight /foil area ratio of 18 kg/dm^2, which was at the middle of the three alternatives.

A series of tests helped us understand how hydrofoils should be designed for these kinds of boats.

The drag of the struts was larger than I had estimated; therefore, when the foilborne height became unsteady, it was difficult to increase the boat speed due to the larger drag when the foilborne height decreased. The aspect ratio (ratio of breadth to length of the hydrofoil) would affect not only the boat speed significantly, but also the stability. Likewise, we could acquire a wide range of practical knowledge on hydrofoils through these tests. We proceeded to carry out many tests by exchanging different parts quickly and without incurring high costs, by using the compact boat and by working unhurriedly toward the completion.

Accumulating knowledge step by step through research as mentioned above, enabled Yokoyama and Tsuide Yanagihara to create splendid records at the Dream Boat Contest held in 1991 and the next two years (both belonged to the Horiuchi laboratory; Yokoyama won the first prize once and Yanagihara twice). These two persons are still competing with each other today in the speed race for human powered boats. I still remember the good memories of those days when they studied and acquired basic technical knowledge on hydrofoils.

5) Capsizing considerations

If the OU90 capsizes, its outboard engine will be submerged causing seawater to penetrate the engine and naturally, we will have to repair it. To prevent such an occurrence, for the first time, I considered using a jet propulsion system. This system can use the same power unit as the outboard motor. The centrifugal water pump operated by this engine can take in water from the lower end of the aft strut and pump out the water aftward. The power head can be fixed in the boat and shielded by a cover to prevent water from entering the boat in the event of a capsizing.

The jet propulsion system was not realized, however, because we could not focus our energies on a project not planned to go into the production stage; thus, the outboard motor continued to be used.

Actually, the OU90 did not capsize at all. Yokoyama's control technique was probably excellent or the boat probably had good stability because of its struts (legs supporting the hydrofoil) that served as large ballast. So, the boat never capsized throughout the long test period, but the crewmembers on board have fallen overboard several times. This means that the engine has never been submerged at all.

The electrical equipment of a modern outboard motor is very durable in sea water and can be restored to its original condition without imposing a large burden on the user even if the boat capsizes. Thus, it will be better to use an outboard motor than a jet propulsion system, which has a rather complicated power train because of its watertight construction. The outboard motor is more economical and users can avail of good maintenance services.

6) Concerns about product liability

The OU90 was completed to the anticipated performance and was just on the stage of reaching the production stage. On one of those days during this period, I met Shoji Muraki, the vice president of Yamaha Motor USA, who was in Japan at that time. I demonstrated the OU90 to him and asked him for his comments. His reply caught me off guard.

His concerns were as follows:

If this boat was sold as a water sports product, the struts of the hydrofoil might injure people swimming in the water. In those days in the USA, several lawsuits had been filed for accidents involving the propeller of outboard motors. The hydrofoils and struts of the OU90 might be liable to Product Liability (PL) suits. Muraki was worried about such events.

After becoming aware of this problem, I was depressed and did not feel like pressing forward with the production of this boat. However, I comforted myself with the thought that we had studied and acquired considerable knowledge on fundamental performance theories for hydrofoil, and that the OU90 had worked very well as a test boat for the mechanical height controller.

2. High speed passenger craft OU96

1) Chairman's order

In January 1989, Mr. Kawakami, the Chairman of Yamaha Motor Co., Ltd., summoned me to his office. He told me that his boat Toyotama-maru (see Fig - 1) was being overtaken by locally-made boats in Ishigaki Island and asked me why we could not build faster boats. At that time, Toyotama-maru had a maximum speed of 20 knots, while a locally-made boat named Sabani had a maximum speed of nearly 30 knots

In those days, the advancements in performance of diesel engine were rapid and newer boats with such engines had faster speeds than old boats. However, Chairman Kawakami did not accept this and ordered me to build the fastest boat that we could build. I thought of a fast passenger craft with a speed of more than 30 knots. At the same time, the boat should be able to earn and pay for itself through commercial services.

One of our targets for product development of hydrofoil boat was the development of a passenger craft. Jet Foil (Fig - 2) was already operating on the Sado sea route in those days. The craft had a good reputation in commercial service because of its soft riding characteristics and capability of cruising in 3 m high waves. However, the diffusion of the Jet Foil did not appear easy because it was expensive, costing nearly 1.5 billion yen, and its gas turbine engines were so delicate that they had to be flashed after every cruise.

The Chairman's request was the same as the approach mentioned above. I thought that we were encouraged by him on this issue. However, the Chairman's boat was used for

Fig - 1 Toyotama - maru

Fig - 2 Jet Foil

Fig - 3 Perspective view of the OR96 by Nakagawa

cruising the coral reef near Ishigaki Island, and a hydrofoil boat was unsuitable for this shallow sea area. A jet boat would be better there. Therefore, I did not want to offer a hydrofoil boat to the Chairman.

I concluded that the Chairman's intention was to develop a model of a fast passenger craft in the series of a Yamaha production boat, and that we had to build another boat for our Chairman that could safely cruise shallow waters at high speed.

2) Boat we would like to build

The boat we would like to build was smaller than the Jet Foil and capable of being fitted with normal engines, which meant a normal boat with soft riding performance and high percentage of days in operation while in service. This was our intention, but the budget allotted to the Horiuchi Laboratory, which had a small number of staff, seemed too large for this project.

Nevertheless, I thought the project would be worthwhile, as it could lead to significant progress of these products in the future. We cut down expenses for the prototype and spread the total expenses over several seasons to make it a long-term project, and thus the conditions for starting the project were satisfied.

Initially, we aimed for a taxi boat that could be used in sheltered waters like the Seto Inland Sea. The boat would be light in weight, of low cost, have a capacity of 12 passengers, and be driven by one person who would manage all relevant tasks, such as berthing and de - berthing, selling tickets, handling baggage, and so on, similar to a public transport bus handled by one person. If such a boat is to have a speed of 40 knots, and silky - soft riding in waves of 1 m height, I was certain that it would be widely used, and it would naturally pave the way to the application of the same system to larger boats.

After finalizing the general idea, I asked Keiji Nakagawa to sketch an image of this boat named, the OU96 (Fig - 3). Nakagawa was a talented industrial designer, who had been performing design work since Yamaha had started its boat business and had many excellent products to his credit. Earlier, he had designed the color scheme for the OR51, and I was so fond of it that we decided to use it as the standard color scheme of the Horiuchi Laboratory (Horiken-Color).

This time too, I was happy with the sketch of the OU96, and we finalized the image of the OU96.

Since the 1950s, I had been dreaming of building a hydrofoil passenger craft, but after some time I had given up the idea because I thought that even if a large high performance hydrofoil passenger craft was developed, only around 100 units or so would be adequate for operations in the domestic market and no more. Today, I can say that even 100 units was too high an estimate, but on the other hand, I cannot imagine why I thought of restricting them only to the domestic market.

3) Attitude control and progress of materials

Since then forty years have elapsed, and my thinking has changed. An ordinary boat fitted with an easily available engine and simple hydrofoils having high speed and soft riding characteristics is likely to have a bright and prosperous future. I thought a boat like the OU96 would be in high demand all over the world.

Electronic components such as ultrasonic height sensor system, servomotors and so on, will become smaller and have a better performance, and will become cheaper. These components can be easily adopted in any small boat without worrying about cost and weight.

With the advancements in materials such as advanced fibers like carbon fiber, Kevlar, and so on, lightweight, high strength and highly rigid materials are now easily available. On this aspect, the advantages to smaller boats are more numerous.

Materials with high strength and rigidity are indispensable for hydrofoils as the boat size increases. Therefore, hydrofoil craft with weight greater than 500 tons are difficult to realize.

4) From three struts to one strut

To make the structure simple, it is desirable to use a single-strut hydrofoil (actually it has two struts) that we had studied earlier. A simple system, short struts and no structure outboard of the boat offer major advantages considering the work to be done by a single person operating the boat.

As mentioned previously, the driver has to ensure transverse stability when foilborne on a single strut hydrofoil, similar to driving two wheelers. Therefore, I believed for a long time that the single-strut hydrofoil would not be suitable for large passenger crafts. That was the reason that I adopted three struts for the YZ886 (see Chapter 8).

By this time, however, we had learnt considerably through experiments on the OR51 and could develop an attitude control sensor using gyro and accelerometer.

Yoshihiro Nagami of the Boat Department estimated attitude control by computer simulation. We gained confidence in the simulation by referring to and collating with the actual test data of the OR51.

Koji Sakuma, who was Nagami's junior, joined in the planning work of the passenger craft. Both Nagami and Sakuma were excellent engineers and graduates of the Naval Architecture Department of Osaka University and were very good at analysis. I planned for the simulation of the OR96 to be done mainly for longitudinal stability. Simulation for lateral stability was to be done manually at the beginning of the trial, similar to what was done previously. Later, we planned to gradually switch over to the automatic method.

After watching the progress of the simulation, I invited people who had experience in developing attitude control to the meeting and received their advice, including Akira Hasegawa, an expert in the theory of dynamic stability of motor cycles, and Masao Sugita, who was in charge of attitude control for helicopters, besides Nagami and Sakuma. Through these meetings, we confirmed the possibility of controlling the dynamic stability of the single-strut hydrofoil.

In 1987, the following year, Yanagihara of the Horiuchi Laboratory worked out a simple system with which a two-wheel vehicle could run automatically without loss of dynamic stability, and the tests were successful. A two-wheeled vehicle with small wheels could be driven stably at slow speed with hands free and without loss of dynamic stability using a gyro. Some

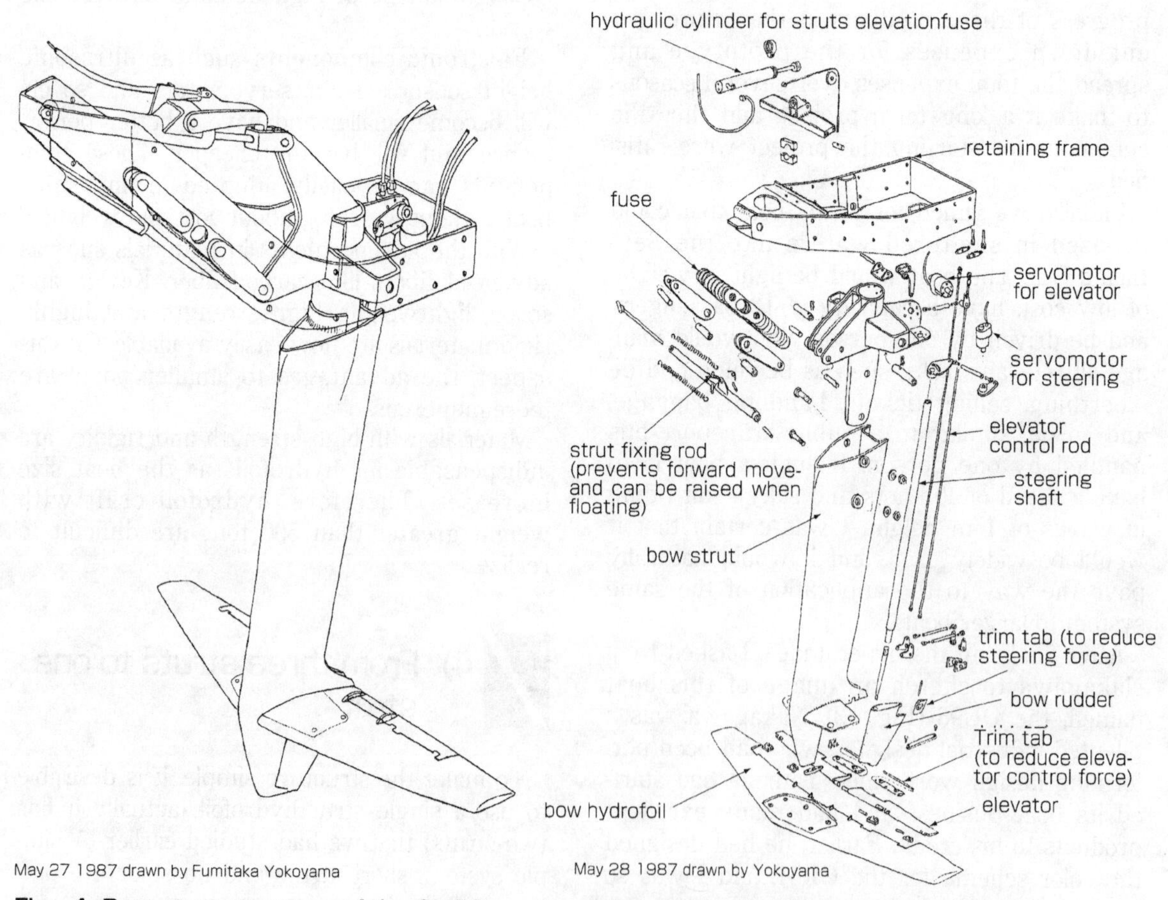

Fig - 4 Bow strut structure of the OU96

Fig - 5 Exploded view of bow strut

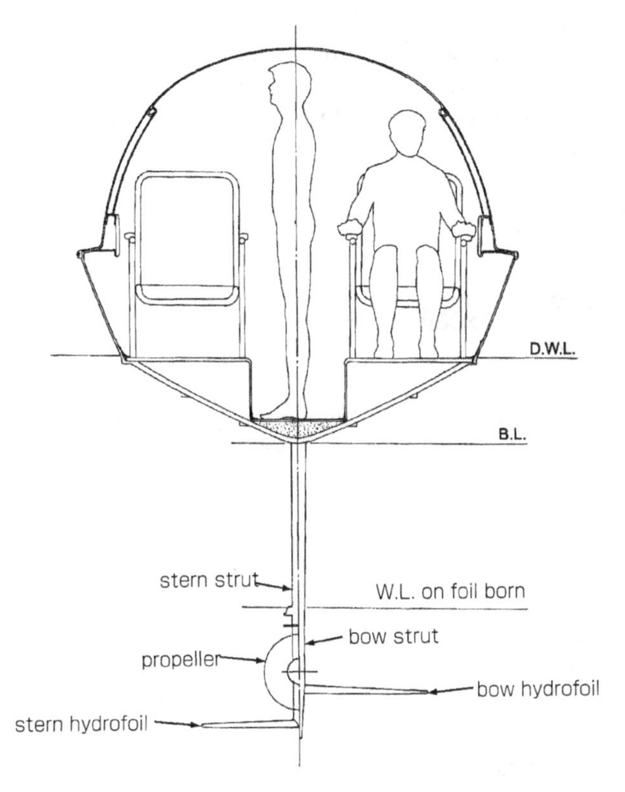

Fig - 6 Transverse section of OU96

day I will write an article about this invention (see Chapter 12). This system will be useful for the OR96 in the future.

As mentioned above, the conditions of dynamic stability of single strut hydrofoil by both electric and simple mechanical devices were satisfied.

5) Collision with floating objects

Next, we had to consider collision with floating objects. I have heard that if the size of the hydrofoil is as large as that in the Jet Foil, it can easily cut even a floating log. However, our boat was small and some damage from floating objects may occur.

In case of a collision, we considered an arrangement by which both bow and stern struts would fold down during a strong impact. When a collision occurs, the two struts fold back successively absorbing energy to reduce boat speed, and as a result, shock is reduced to some extent (Fig - 4, 5).

The boat loses lift and comes down on the water, but if the resistance of the folded struts

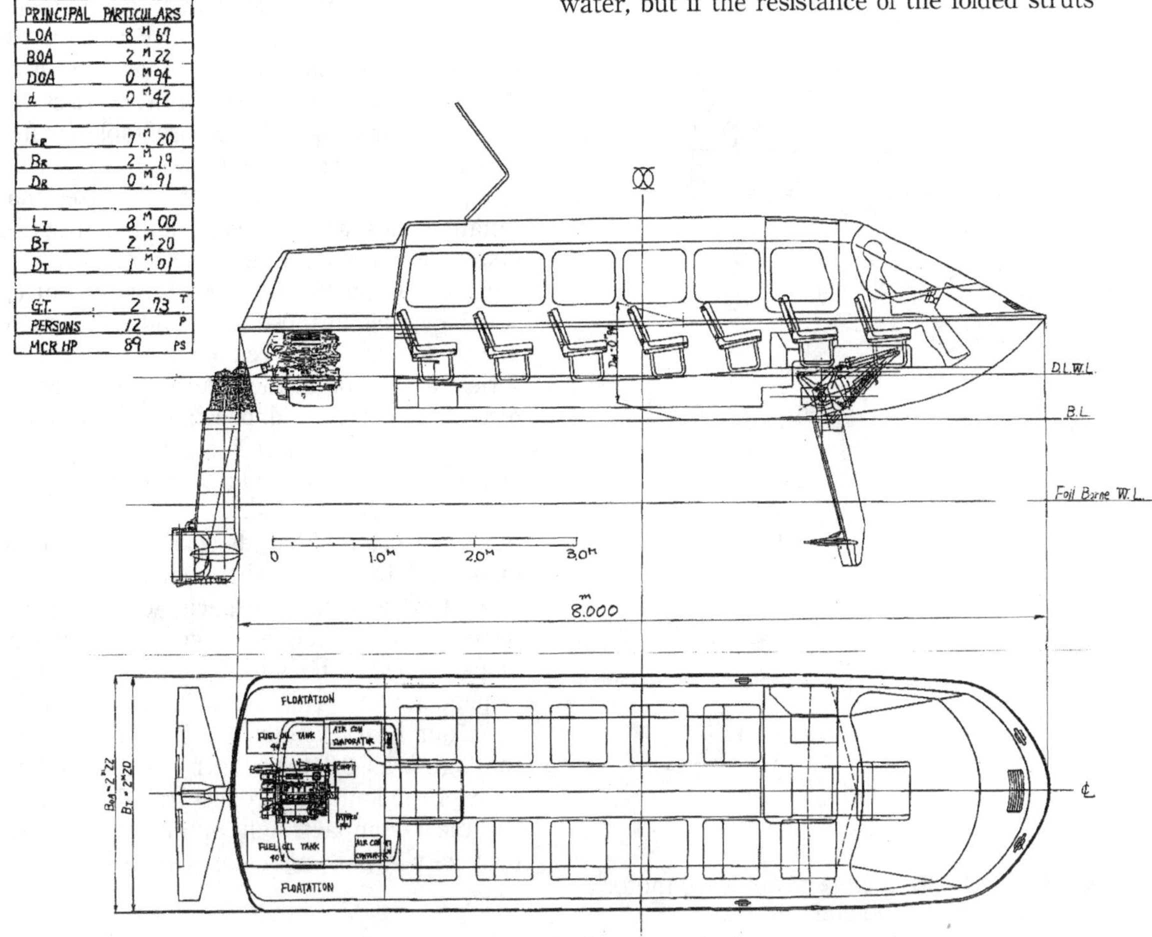

PRINCIPAL PARTICULARS	
LOA	8^{M}67
BOA	2^{M}22
DOA	0^{M}94
d	0^{M}42
L_R	7^{M}20
B_R	2^{M}19
D_R	0^{M}91
L_T	8^{M}00
B_T	2^{M}20
D_T	1^{M}01
G.T.	2.73 T
PERSONS	12 P
MCR HP	89 PS

Fig - 7 General arrangement of the OU96 (inboard diesel version)

is minimal, the boat will continue planing for a while before coming to a dead stop. If the folded struts can be moved back to their original position, the boat can cruise again in the foilborne state.

Here again the single strut has an advantage over three struts. When either the left or the right strut in a three-strut boat hits an object, the three-strut boat tends to capsize but a single-strut boat does not because external force works only in the plane of symmetry and no capsizing force occurs. Accordingly, the boat will come down to the hullborne condition without heeling. However, if the bow strut twists laterally after hitting some object and a capsizing moment is generated, the boat may capsize. Therefore, it is important to ensure that this does not happen in the structural design.

You can sense that the OU96 will soon return to upright position even if capsizes (Fig - 6), by referring to the sketches and drawings of the boat. This is because the hull cross section is nearly circular and both bow and stern struts with hydrofoils work like large ballast keels. However, to ensure safety during capsizing, the passengers have to fasten their seat belts and rely on them. Although such an accident could occur, we wanted to adopt all possible mesures to prevent it in the OU96.

Safety is one of the most important aspects for a passenger craft. We planned to perform collision tests against floating logs at various speeds with the intention of identifying every possible hazard. We believed that capsizing analyses could be clarified through such tests.

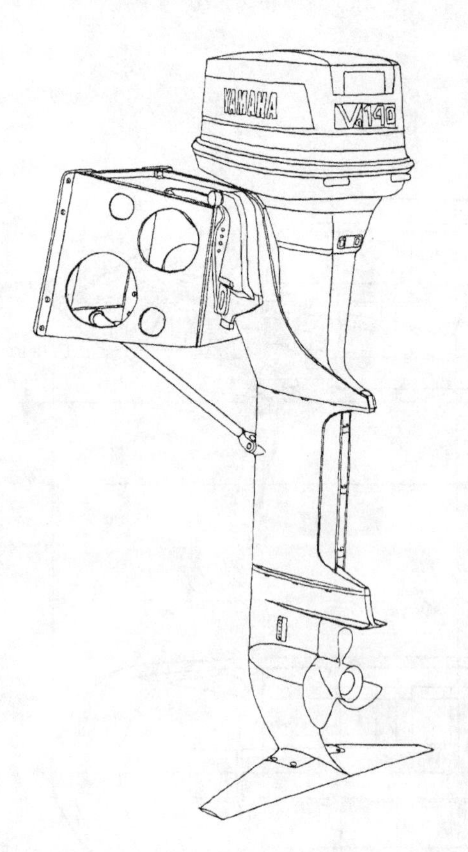

Fig - 8 For OU96, the leg of outboard motor was extended and stiffened

6) Boat size

The Japan Craft Inspection Organization (JCI) cannot inspect this boat, although the OU96 falls in the category of small craft because it is a hydrofoil boat. The OU96 has to be inspected by a branch office of the Ministry of Transport on behalf of the Japanese Government. The branch office deals mainly with inspections of large ships. That means that we have to prepare equipment suitable mainly for large ships, to install approved (by Japanese Government) engines, and to apply FRP construction requirements applicable to large ships. These were excessively harsh requirements for a small craft aiming for low costs, lightweight construction, and high performance. Considering only FRP construction, the weight increases to more than 100 kg and the position of the center of gravity rises. The mission of this boat is to be profitable for ferry service. To realize the goal, we have to select an appropriate boat size that minimizes maintenance expenses, repairs and inspections under the given circumstances (Fig - 7).

Inspection by the Japanese Government (JG) was inevitable in those days, and I thought we had to do our best to reduce expenses through series production in the future. I expected that hydrofoil boats would become reliable and be regarded as small craft like normal motorboats in the future. This would depend on how well our boat was completed.

Normal boats of length below 12 meters are inspected by JCI. If the boat is less than 5 gross tons and has a capacity of less than 12 passengers, the required accommodation and inspection can be simple. That is why the inspection of a normal pleasure boat is simple.

As our boat's length was 8 m and its gross tonnage was 3.3 tons, both quantities being clear of the limits, we concentrated on building the most compact, lightweight and low cost boat with a passenger capacity below 12 persons.

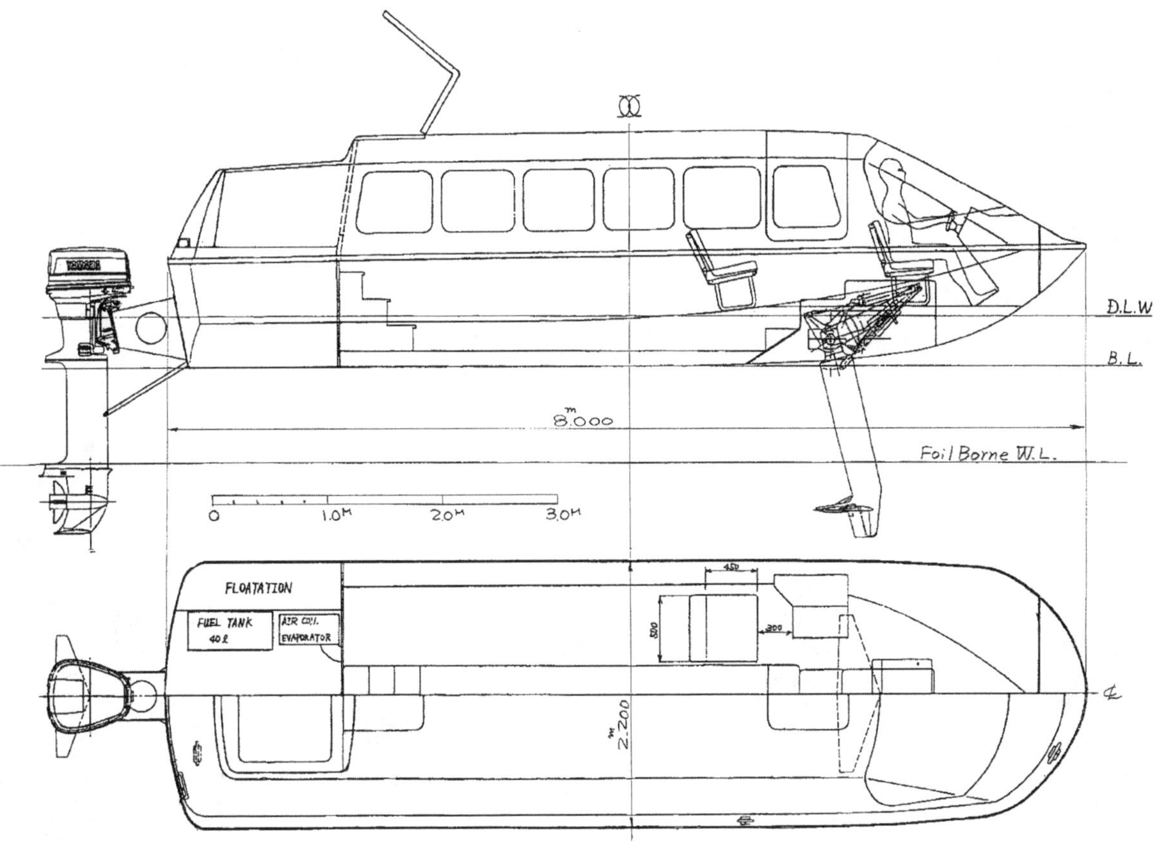

January 17 1987 drawn by Fumitaka Yokoyama

Fig - 9 General arrangement of the OU96 (outboard motor version)

7) Engine

During the development stage, we faced a problem. We found we could not receive boat inspection because the 89 HP Yamaha diesel engine that we intended to install had not been certified by JG. We could apply for certification of the engine for future use, but it would take half a year and cost us 300 million yen. It was unthinkable to waste half a year and to spend so much money.

Soon Yokoyama began to compare the certified Yanmar diesel engine and the Yamaha outboard motor for introduction in our boat. Initially, we thought a diesel engine would be suitable for passenger craft, but later found that the outboard motor had advantages such as lighter weight, more horsepower, and it already had a strut.

The weight of a 90 HP outboard motor is 200 kg lighter than a diesel engine with the same horsepower, which was very welcome. Moreover, we could freely choose either a 140 HP or 200 HP outboard motor, giving us alternatives to test up to high speeds using these motors. On the other hand, the boat with a diesel engine would attain only 30 knots, and it might not be easy to get the boat foilborne.

In those days, diesel engines were rapidly becoming lighter in weight and higher in power. I think there may have been better diesel engines when our passenger craft was completed.

In January 1987, we decided to use a 140 HP outboard motor in our boat and Yokoyama began to redesign the boat. The outboard motor strut had neither adequate length nor adequate strength, so Yokoyama designed the strut extension and its stiffener (Fig - 8) carefully. The hydrofoil span (foil breadth) decreased to nearly half and its foil weight also decreased from 94 kg to 30 kg. I estimated the maximum speed of our boat (Fig - 9) to increase from 33 knots to 46 knots, exceeding that of Jet Foil.

We thought its fuel consumption would deteriorate, but there were no uncertainties because we intended to finish tests up to high speeds with that engine, and using the test data, we intended to select a suitable engine when designing the production passenger craft.

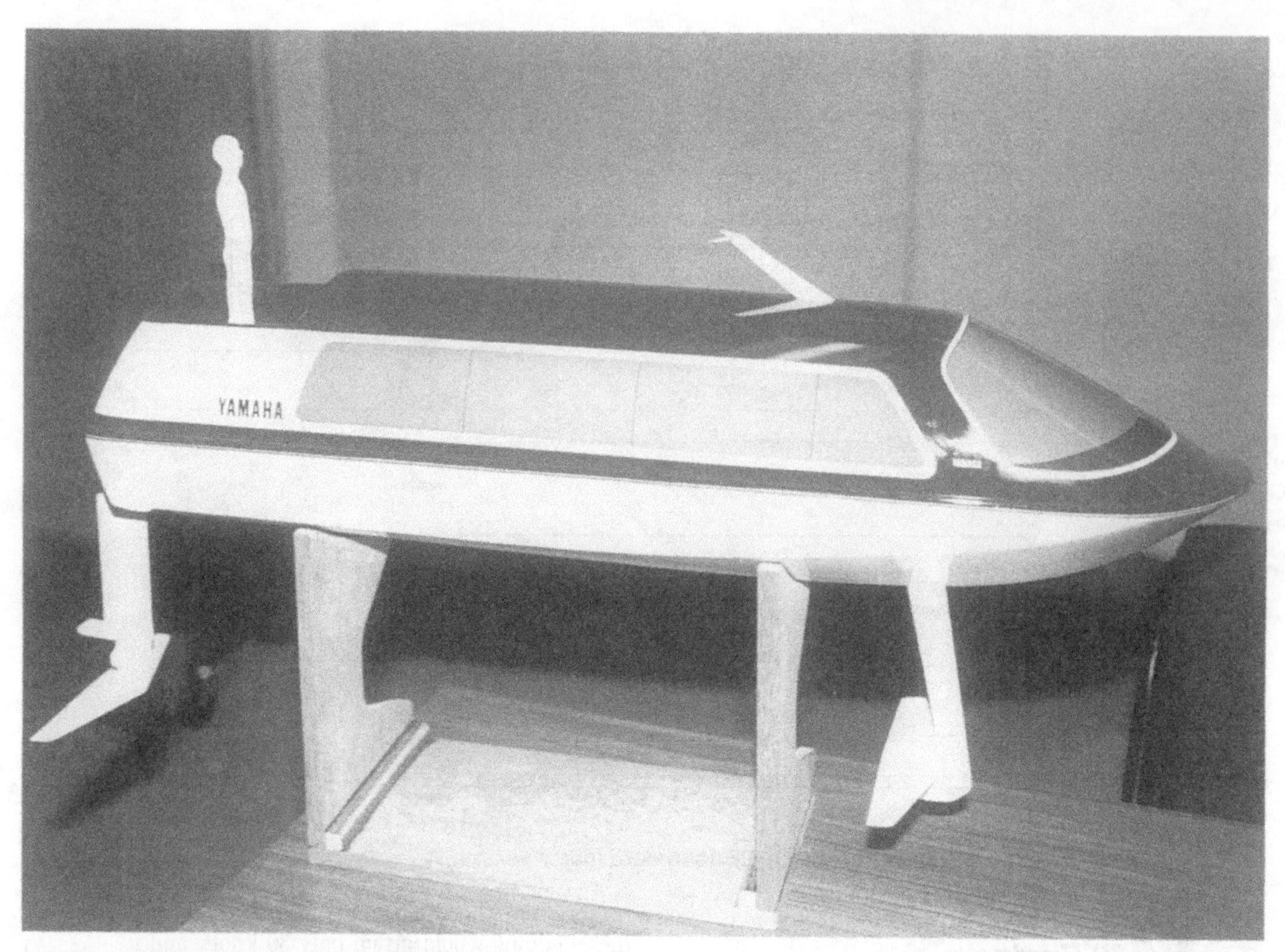

Fig - 10 Scale model of the OU96

8) Hull

If you plan a hull that is simple, has low cost, is lightweight, and has a capacity of 12 passengers, it will be similar to that of an airplane. I reworked the design to minimize the circular section of hull, providing a sectional arrangement containing comfortable seats on both sides and a central passage. In this way, the diameter of the hull section and its perimeter decrease, which results in reduced surface area which in turn has the effect on reducing the weight and stiffening the structure (Fig - 6).

As a result of our design efforts, the hull had a constant circular section and began to look similar to a modern business jet airplane in shape and volume. In the longitudinal sectional arrangement, the passageway is lowered by one step, and a single seat is provided on either side. The cross sectional shape is almost a circle and it is constant throughout the passenger cabin. The shape is ideal also from the point of productivity considering interior furnishings and so on.

To strengthen the area of the hull-deck connection to withstand side forces from the pier, a side deck of some width was provided, although we did not anticipate to use it as a passageway.

To confirm the image of the full boat, we made a scale model, which looked good, and even Nakagawa was satisfied with it (Fig - 10).

9) Building

At the end of 1986, we started building the boat at the Gamagori boatyard. Just in time, we discovered a problem related to engine inspection, and we decided to replace the diesel inboard engine with outboard motor at the beginning of the following year, and started re - designing the boat.

However, we did not change the main hull. We made the structure simple, maintained the same position of the CG, and retained the engine room as it had been so that we could install a suitable engine in the future. Therefore, the earlier arrangement remained unchanged, and a new outboard motor bracket was fitted outboard of the stern.

As we intended to build only one boat at that time, it was built using the temporary female mold method. The female mould was built as described below.

Within the outer frame, which has same shape as the outer contour of the hull, thin plywood sheets were fitted and nailed to complete the female mold. This boat had complex curved faces at both bow and stern ends with areas that were difficult to finish. However, practically the entire length of the hull had a constant cylindrical section, which enabled us to build the female mould easily.

Consequently, expenses for the prototype were not very high, and the boat was much larger than normal boats that we had designed in the Horiuchi Laboratory

Because of the female mold building method, the hull did not look very nice with the natural FRP color (without gelcoat), but when the hull was placed on a higher cradle so as to accommodate struts and hydrofoils, it had the dynamic feel of business jet airplane. I wanted the boat to run exactly like an airplane.

10) Suspension of the project

After building the hull, manufacture of struts, hydrofoils, and control system continued. Unlike struts and other parts in small craft that could be made easily with steel plates in stock, our new boat had a weight of 10 tons and required stronger and more rigid materials. Moreover, we had to use anti-corrosive materials, which could be used in boats in service. Accordingly, we had to procure materials that were of high quality and difficult to process, causing Yokoyama to struggle with various problems when manufacturing parts.

On one of those days, I ordered the project to be stopped. I do not remember what I had felt and thought when the project was stopped. At the time of this writing, I asked Yokoyama and Hattori, who were actively building the prototype, about the situation at that time. They said I had not clearly explained the reason for stopping the project, but since it was many years ago, they also had only vague memories. I felt very sorry for them, but I still cannot remember the details as to why I had ordered the project to be stopped. We can only guess from various circumstances in those days.

As far as I was concerned, the project was my final target, but it was probably stopped on account of money matters, although I do not remember the details.

Originally, the expenses for building this prototype were more than adequate for the budget estimated by the Horiuchi Laboratory. I was going to distribute the expenses and the work over several seasons to prevent incurring large expenses at a time. However, as the work progressed, the expenses tended to concentrate, and I probably thought of shifting the project to a later date to prevent this concentration of expenses. The price of struts and hydrofoils also increased more than we expected because of the difficulties in manufacturing them, and the total expenses for building a prototype tended to increase. These were probably some of the reasons, but anyhow I had to explain clearly these reasons to persons who were working on this project.

The postponement of the project was inevitable considering the circumstances in those days, but now I deeply regret missing the opportunity to re-start the project. I became busy in other jobs following the suspension of the project, and in the meantime, expenses were curtailed because of the severe economic recession. Under such circumstances, I was appointed Director of Marine Operations, and I was unable to proceed with the project again. Later, during the long recession, I resigned from the Board of Directors and the Horiuchi Laboratory closed down.

I heard that the hull and parts of the OU96 on which Yokoyama had worked, had been kept until 1997. I had not clearly explained to Yokoyama the reasons for stopping the project and also had not communicated to him my intention of what to do with the boat. I suppose these were the reasons the project could not be re - started. I have also lost the chance to convey to him my intentions and ask him to take over my dream.

Today, I am a bit sad and feel very sorry for all concerned persons who worked on this project.

10

HYDROFOIL BOAT OU32

The plan for the OU32 started with our intention to display it at the boat show. It was also a hydrofoil boat intended for commercialization. The design of the boat was based on the OR51, and considerable progress was made by modifying it to a two - seater arrangement, using jet propulsion, and installing a height sensor. Its maneuverability and soft riding characteristics were almost perfect, and the boat had the performance that I had been dreaming of since many years. A movie of the boat run was exhibited at the boat show, and it was received favorably even in Australia and the USA.

1) Start of the project

In the previous chapter, I wrote that I could not remember the reasons for stopping the OU96 project. Likewise, I cannot remember the situation at the start of the OU32 project. I asked Yokoyama, who was in charge of this project, but he also does not seem to recollect the details.

According to the records, the project started in September 1987, which was four months after the suspension of the OU96 in early summer. Since the development schedule was submitted to Yoshiaki Murakoshi, General Manager of the Planning Department in Marine Operations, the Planning Department would draw up the plan. The original idea probably came from Mr. Arata, the Director of Marine Operations at that time, who was determined to carry it through. This project had a very tight schedule, with the completed boat scheduled for exhibition at the boat show in 1998.

In those days, Yamaha used to exhibit a dreamboat annually at the boat show. Perhaps, following the same custom, I presume that a dream hydrofoil boat, more advanced than the OR51 and having two seats, would be exhibited at the boat show.

2) Tandem two seater

The OR51 was designed as a special racing boat, so naturally it had a single seat. However, with its size and with only a single seat, the boat was not of much practical use. In case of light airplanes too, a single-seater is meant only for a special purpose such as races or circus flights; many gliders have two seats.

A two-seater tends to be more useful than a single-seater for recreation, training, or other practical uses. You can use a two-seat vehicle with a family member or with your friend, and if an extra fuel oil tank is installed instead of the second seat, the cruising distance can be extended considerably.

For two seats in both airplanes and boats, it is important to choose the ideal seating arrangement: tandem (longitudinally in line) or side-by-side. In general, a tandem seating arrangement has a smaller frontal area resulting in less drag, a simple structure, lighter weight, and higher performance, but communication between the seated persons is inconvenient.

With the side-by-side arrangement, the two persons can speak to each other conveniently and feel comfortable. Therefore, to decide the seating arrangement, one has to choose between performance and comfort.

In practice, we did not have any choice in the seating arrangement for the OU32. To obtain the same maneuverability as the OU51, a 45 degree inside bank was essential, and a side - by - side arrangement was not desirable because the large cabin width might reduce the banking angle significantly. As the basic performance was more important, we adopted the tandem seating arrangement without hesitation.

3) Engine

The OR51 was fitted with a 15 HP outboard motor because only this engine had a long shaft. This engine had been developed with an ultra long shaft for use on fishing boats in South East Asia and Africa.

An outboard motor is normally fitted to a transom of height 38 cm or 50 cm, depending on the shaft length. On the other hand, only an engine that fits a transom height of 70 cm is suitable for a hydrofoil boat. This engine had lesser horsepower than was planned for the OU32, intended for carrying a crew of two, but no other outboard motor satisfying the required shaft length was available.

Just at that time, Yamaha introduced the Marine Jet in the market and its 32 HP engine with a jet pump unit appeared suitable for use as the propulsion system for the OU32.

I thought if we utilized the propulsion system of the Marine Jet, we would be able to improve the water jet intake, the jet pump would take up water through the long strut and discharge it from the aft, and the length of strut could be freely selected. Additionally, larger horsepower engines appeared very attractive to us because they were likely to be developed successively.

There could be some energy loss in bringing up water taken in from the intake, but this system had already been used in the Jet Foil and its overall efficiency had been satisfactory. Of course, the shape of the intake and the long strut up to the entrance of the jet pump had to be designed properly. If the design was not appropriate, efficiency would degrade to an extent about which we had no idea.

As only this point was new in the design for us, we concentrated our energy on it. The project started in the middle of October and the boat show was scheduled in early February. The time available was only three and a half months. During this time, we had to perform trials, make improvements, finish the boat beautifully for the boat show, and prepare an attractive video movie of the boat. This was an absolute must. In addition, we hoped to receive good feedback on the project at the boat show. Thus, the schedule of the project was very tight.

Looking at the schedule chart, only two months were available from the start of the project to the launching, which meant that we had to place orders within a few days after the start of project for cast parts such as hydrofoils, control system, and the aft strut for the jet pump. Nevertheless, we enjoyed the work since it would put together all our previous studies on hydrofoil boats and we were happy to do this work. Our days were full of busy and lively activities.

4) Potential for commercialization

Since the project started as described above, the only requirements were making a two-seater and a good impression at the boat show; we were free to prepare the rest of the specifications. We wanted not only to exhibit this boat at the boat show, but also to sell some units after the show. We also wanted as many people as possible to enjoy this boat.

Considering the characteristics of this boat for commercialization, I can visualize it as a commuter boat with attractive controls. It can cruise at high speed in 50 cm wave heights, maintain smooth and soft riding characteristics, and can also be used as a small-scale commuter boat on lakes in the USA, given that it is small and has only two seats.

Additionally, the splendid maneuverability of this boat is sure to give pleasure to its driver. Because it has two seats, it is convenient for training and since its range of stability is 180 degrees, it can be used for training safely even if it capsizes due to poor handling. Moreover, there are no crash hazards such as in an airplane.

In case of the OU90, we were concerned about injuring swimmers, but this time we did not worry because we thought this boat would not be used in waters frequented by swimmers.

We had been worried about the inadequate stability of the OR51 because of the center of gravity being too much aft. Therefore, this time we designed the boat positioning the center of gravity forward at an appropriate position to ensure yaw stability when foilborne, in which case a two-strut control system (Chapter 8) would not be necessary. Also, we used a compact height sensor (for foilborne height stabilizing system), which had been successfully developed for the OU90, to reduce the cost, to enable us to concentrate on designing a simple structure, and to ensure stability when the boat was foilborne.

As this boat would be the last hydrofoil boat to be built by the Horiuchi Laboratory, many of the technologies accumulated in previously developed boats were used in the design of this boat. The original plan was implemented in a very short time.

5) Hull

The development schedule was so tight that we made use of the molds of the OR51. First, fiberglass was laid on the female hull mold of OR51. After curing, an FRP hull with the same shape as the OR51 was taken out of the female mold. The FRP hull was extended to make a male mold for the OU32. As for the deck, the original male mold was extended and used for a crew of two. The shape around the cockpit was modified to fit a large windshield for two persons. Thus, the deck male mold was completed with the modified area restricted to a minimum.

The hull was extended 500 mm in the aft, so as to make room for two persons and to arrange the engine and jet pump longitudinally. We designed the hull bottom with a step such that the bottom surface extending toward the aft became 50 mm higher than the original bottom (Fig-1). By introducing a step, we intended to keep the boat adequately trimmed, thereby reducing drag and also minimizing the splash resistance of the upper portion of the aft strut, which was difficult to shape cleanly.

The aft strut had a mechanism that folded the aft strut during collision with an object. Therefore, several hinges and other fittings were exposed, which could result in large splash resistance. The step was designed to prevent this splash resistance.

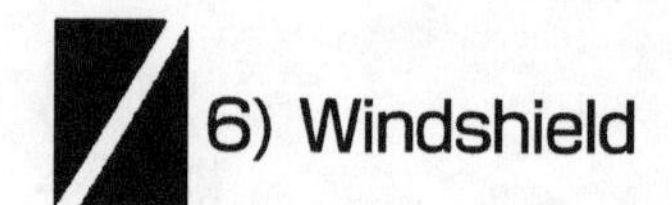

6) Windshield

The windshield of the OR51 was a one-piece acrylic plate from forward to aft, made by the vacuum method, but I was not satisfied with its shape. When looking forward through the windshield from the driver's seat, we had often been troubled by its distortion. This time we designed the forward windshield with a fixed-frame structure, and a thin acrylic plate elastically bent and fixed on this frame (Fig-2). This system minimized the distortion of the windshield because of the smaller deviation in its thickness.

The aft windshield was made by the vacuum method and was openable. It also had a solid frame around it. Moreover, we planned to use inflatable tube packing, which was used in airplanes, to prevent water from entering the cabin through the windshield edge when capsized.

When viewed from the side, the connection between the fixed and the openable windshield had a knuckle on the centerline of the wind-

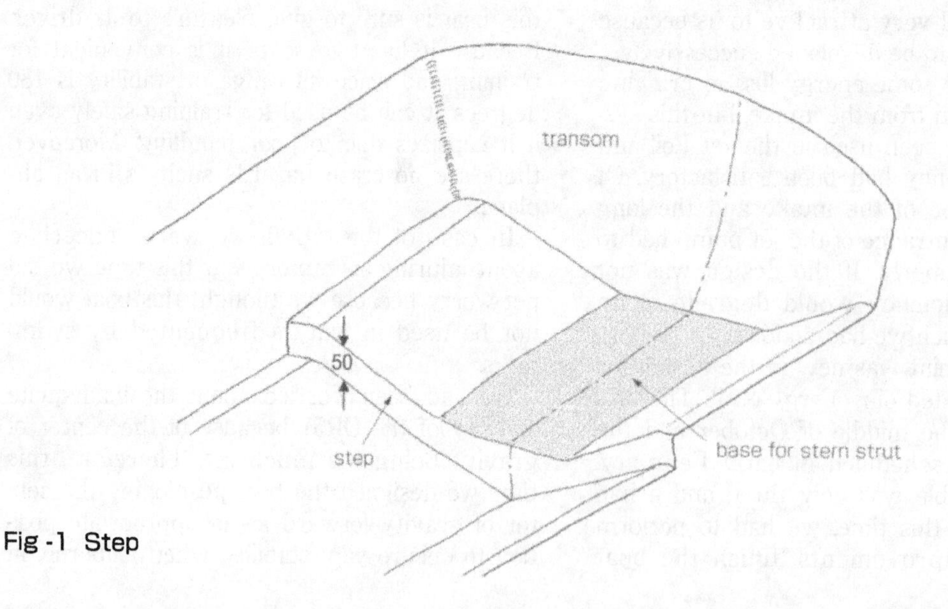

Fig-1 Step

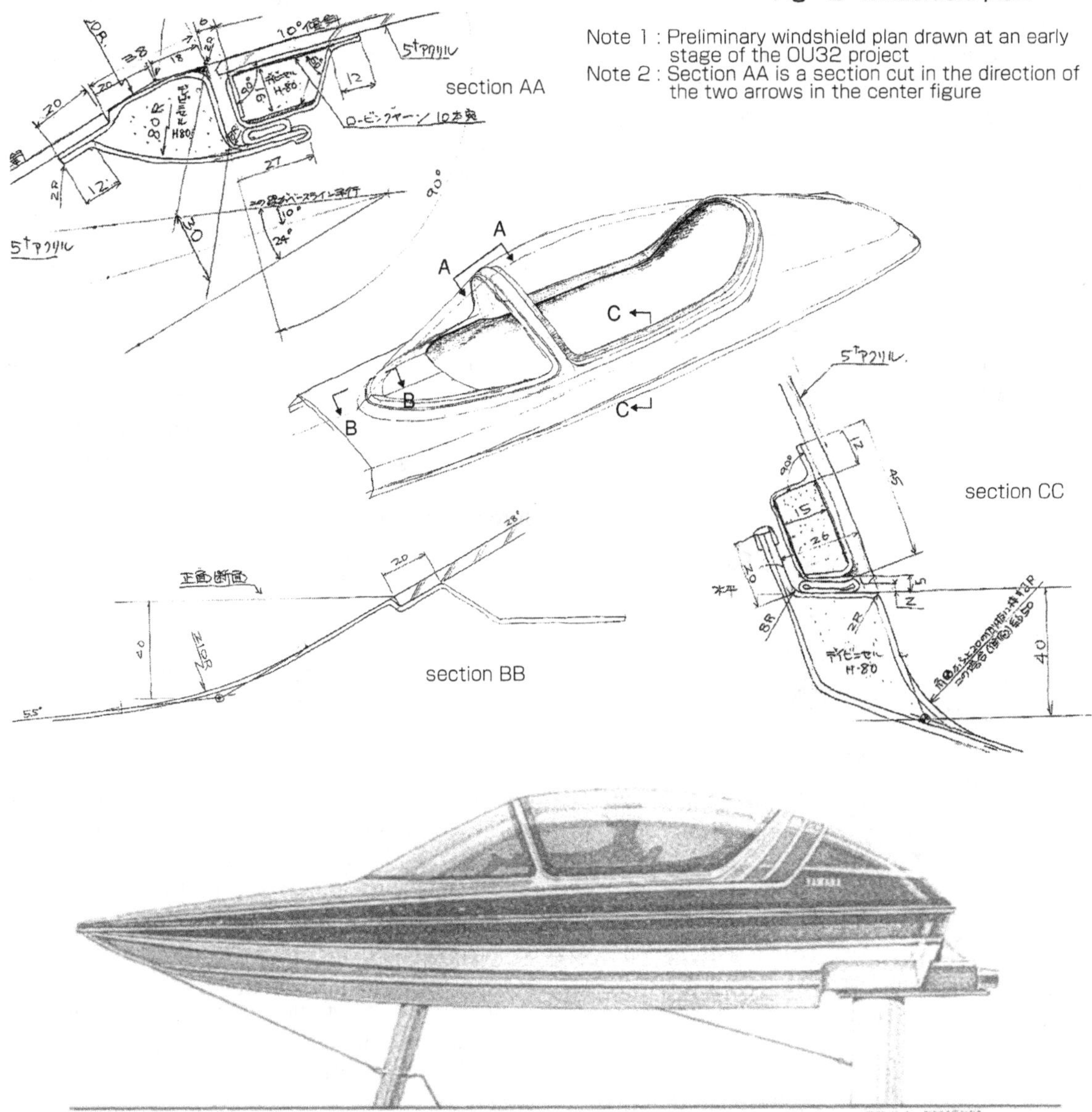

Fig - 2 Windshield plan

Note 1 : Preliminary windshield plan drawn at an early stage of the OU32 project
Note 2 : Section AA is a section cut in the direction of the two arrows in the center figure

Fig - 3 Sketch drawn by Nakagawa

shield structure, but this was not a major problem. Automobiles also have the same kind of knuckle when viewed from the side.

Thanks to this design and manufacturing method, we could minimize distortion, reduce the cost, and make a vacuum formed openable windshield with a firm structure easily.

Thus, the boat became longer with good overall proportions and an improved windshield shape. I was quite satisfied with its beautiful and balanced shape.

I asked Nakagawa to draw a sketch showing its appearance (Fig-3). The Horiken colors were bright and on the drawing, they suited the boat perfectly. It did not look as though it was built by modifying an existing mold. I thought that the collective experiences gained through designs of many hydrofoil boats previously were effectively utilized in this boat. I also thought that the outline of the boat presented a good impression of what was probably the best hydrofoil boat developed by the Horiuchi Laboratory.

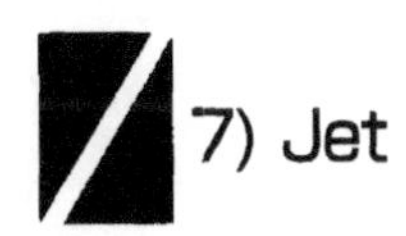

7) Jet

When we started the OU32 project, we had practically no technical experience of water jets. Therefore, our primary concern when using the engine and jet pump of the Marine Jet was how smoothly water from the intake would be transferred to the jet pump through the long stern strut.

In those days, the patents of Jet Foil were open to the public. So as a starting point, we referred to other examples of the strut through which water was taken in and felt reassured (Fig-4). However, it was not easy to decide the actual dimensions because of the differences in scale and type of pump. We were not confident of our plan, but we had no time to start the basic study from the beginning.

Fig - 4 Jet Foil patent

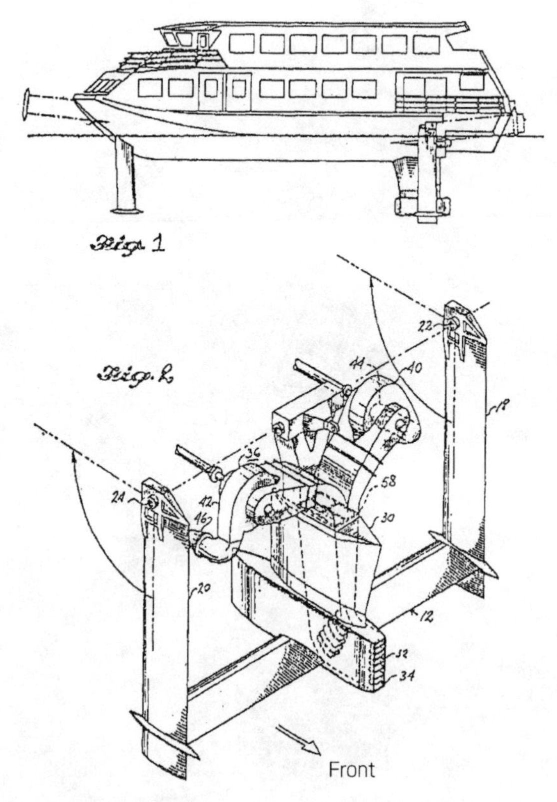

Reference : Japanese patent publication (s51-7907)

Intake : 33. 34
Nozzle : 44. 46
Pump : 40.42
Hydrofoil (stern) : 12
Strut : 19. 20

Under such circumstances, Yokoyama suggested making a full-scale model of the strut and allowing water to flow through it in order to identify problems and find solutions. This was a good suggestion and was implemented as described below.
Half of the side of the route from the intake hole to the jet pump, guide vanes, and other parts, were made of wood. Clear acrylic plate was fixed on the model, its position corresponding exactly with the centerline, so that the internal water flow was visible (Fig-5).

Quantitative measurements were not performed; we could observe stagnation and separation of the water flow through this acrylic plate, which helped us to make effective improvements. To my regret, this test apparatus had a different intake shape from the actual one. If the intake is placed in water flowing at the same speed as the boat, we could simulate the actual conditions. But since this was difficult to implement, we connected a fire hose directly to the intake. Therefore, we did not know whether the shape of the intake was appropriate or not. We had no means to resolve this except by performing actual boat trials.

Here, the size of the intake is a very important factor. When the boat is running, water should ideally enter the intake at the same speed as the boat speed to arrive at the jet pump smoothly.

If the intake size is too large, a large amount

Fig - 5 Experimental apparatus for water flow through stern strut

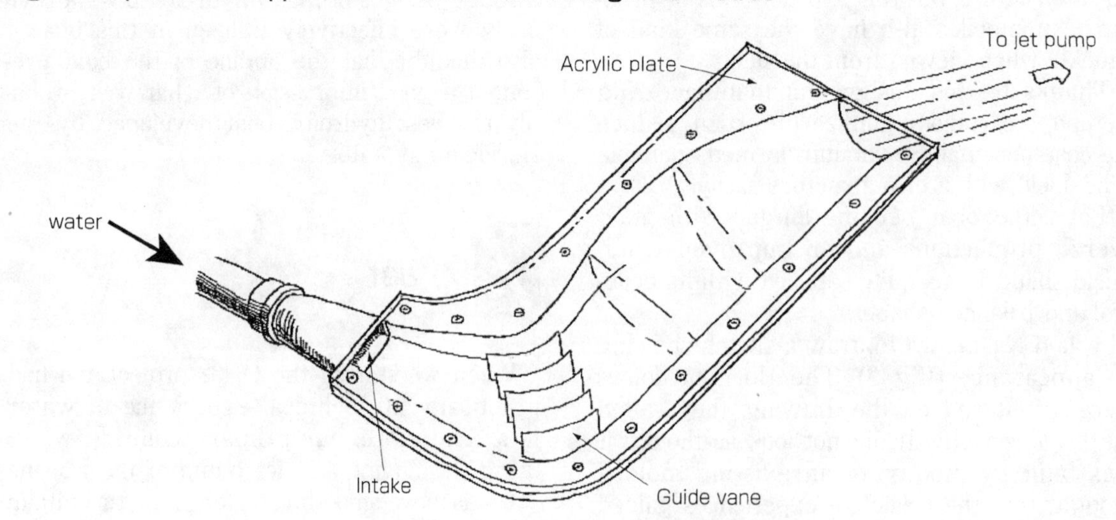

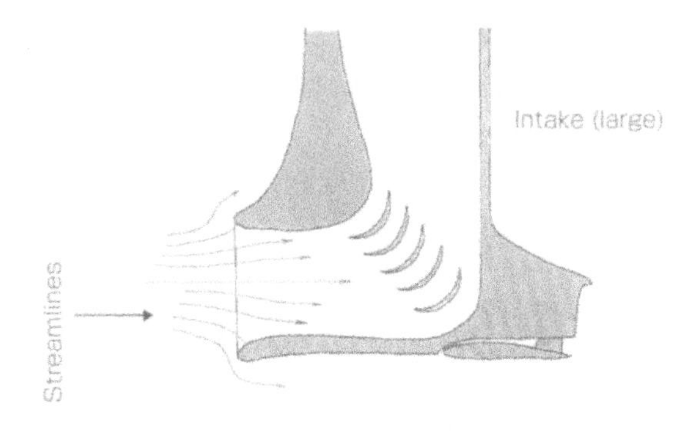

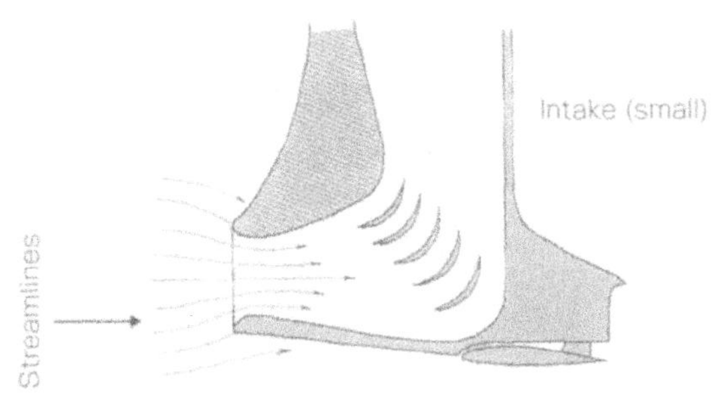

Fig - 6 Size of intake

If the intake area times the boat speed is greater than the amount of water that flows into the jet pump, excess water flows outside the intake. If it is smaller, insufficient water is supplied to the jet pump, which results in smaller thrust.

of water tends to enter the intake but flows outside because the route to the jet pump is not large enough (Fig-6). This disturbs the flow lines and the intake creates a large resistance. On the other hand, if the intake size is too small, the amount of water taken in is insufficient and the jet pump is less effective.

There is an optimum intake size for each boat speed. However, since the boat speed varies, the intake size is not always optimum. Therefore, it was rather difficult to decide an intake size that would not cause problems during normal use of the boat.

We decided to start with a tentative size, carry out trials, and then modify the size if necessary, according to the trial results. This intake size, about which we were not sure, gave unsatisfactory results. The boat could not get foilborne during the first trial. We enlarged the intake area so that the boat became foilborne more easily because the thrust at low speed increased. Trial-and-error method is necessary when there is no previous experience.

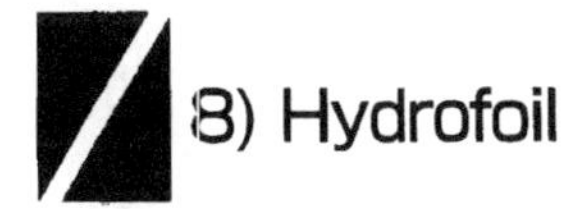

8) Hydrofoil

We had no particular problem regarding the hydrofoil as we had extensive experience in this area. Hydrofoils with smaller surface area have been used but we used a hydrofoil with a slightly larger area firstly because we had to prepare a video film showing the high maneuverability of the boat during the first trial immediately after launching (Fig-7).

The mechanical height sensor that had been used for the OU90 was used again with a mixing lever to lighten the load on the driver, making full use of our experiences with the OR51 and the OU90.

The steering system was also modeled on that of the OR51, as it was known to be reliable (Fig-8).

We concentrated on using our combined technical experience to build the boat with low cost, simple structure, good looks and high performance, since we dreamed of this boat entering the production stage.

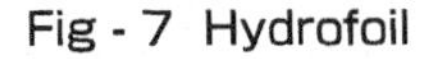

Fig - 7 Hydrofoil

Note : This is the preliminary design. Later the bow foil was modified to NACA2R(2)12 profile with 700 mm span. The stern foil was also modified to the Gottingen593 profile.

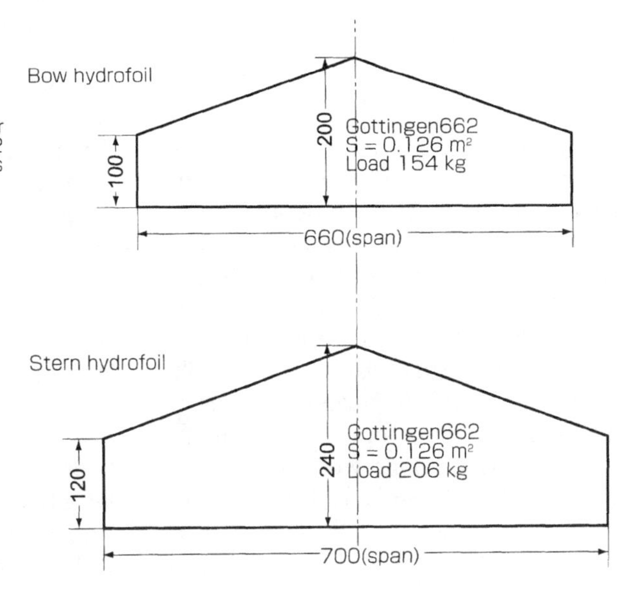

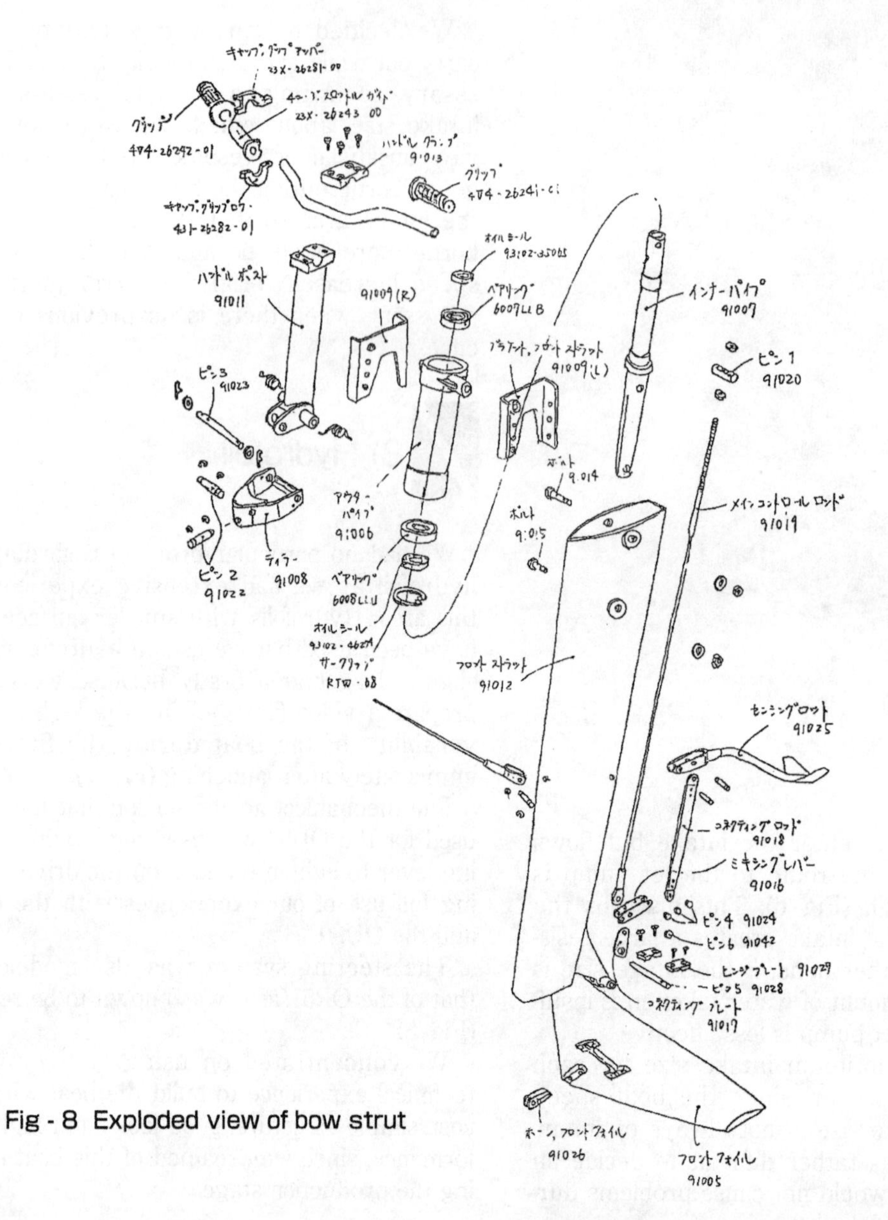

Fig - 8 Exploded view of bow strut

9) Schedule

In the middle of December, the Design Administration Department circulated a project notification to the concerned departments, informing them of the importance and urgency of the project. The schedule was very tight from the beginning. On top of that, an additional boat for exhibition at the International Leisure Exposition in Australia was to be built in the beginning of the following year, as a result of which the members of the departments in Marine Operations were tensed up. Additionally, the OU32 also had to be finished, packed, and made ready for dispatch to the exposition by February 22.

All of us at the Horiuchi Laboratory made all - out efforts to implement this project smoothly and support Yokoyama. We also kept the concerned departments, including the Planning Department and the Design Administration Department, informed about the progress, the time table, and allocation of work, and so on. Thus, we took care to be through.

Another issue was the government inspection of the boat. Every hydrofoil boat had to be inspected by the authority (JG) according to the prevailing regulations. As the inspection system was originally designed for large ships such as the Jet Foil, a large number of drawings and documents were to be submitted. Our boat was not a commercial product, so our intention was to receive approval for temporary operation of the boat for a limited period. But without satisfying the requirement of submitting the large number of drawings and documents, the project could not go forward smoothly because the government officials had to study and understand the new boat (Fig-9) before giving approval.

The Shimizu branch office of the Chubu Maritime Bureau had earlier approved temporary operation of the OR51. So, we assumed that it would be the same for the OU32, but this time the Shimizu branch office asked the Chubu Maritime Bureau and the Ministry Of Transport (M.O.T.) as to how they should treat our application. For a while, the Chief Inspector was inclined to take the decision to not allow operating the boat before the boat show. Wherever they went, the government officials criticized them and asked why they had not come for advice earlier? Shisyaku and Yokoyama, who were responsible for submitting the application, had a hard time, but they has no answer. Nevertheless, with the help of the Shimizu office and the efforts of some concerned persons, we finally managed to arrive at an arrangement for inspecting stability of the boat on January 18, in the presence of the inspector.

10) Launch

On January 17, 1988, we launched the OU32 and performed the 180 degree stability test. The results were excellent and good photographs of the test were taken. These are shown in Fig-10. During the test, Matsuse, who was one of the members of the testing section, was inside the cabin to check the seals for water leakage. He observed slight leakage of water from the windshield edge packing, but the amount was very little.

As mentioned above, the stability test was performed without any problems. During the first run, however, the boat did not rise to the foilborne condition due to the low thrust. The low thrust at slow speed was due to the insufficient area of water intake, so we cut off the mouth of the intake to widen it (Fig-11).

On the18th, the next day, the boat easily became foilborne in the presence of the inspector, thanks to the increased thrust after improvements were made. When the inspector was inspecting the boat in operation, an unex-

Fig - 9 General arrangement of OU32

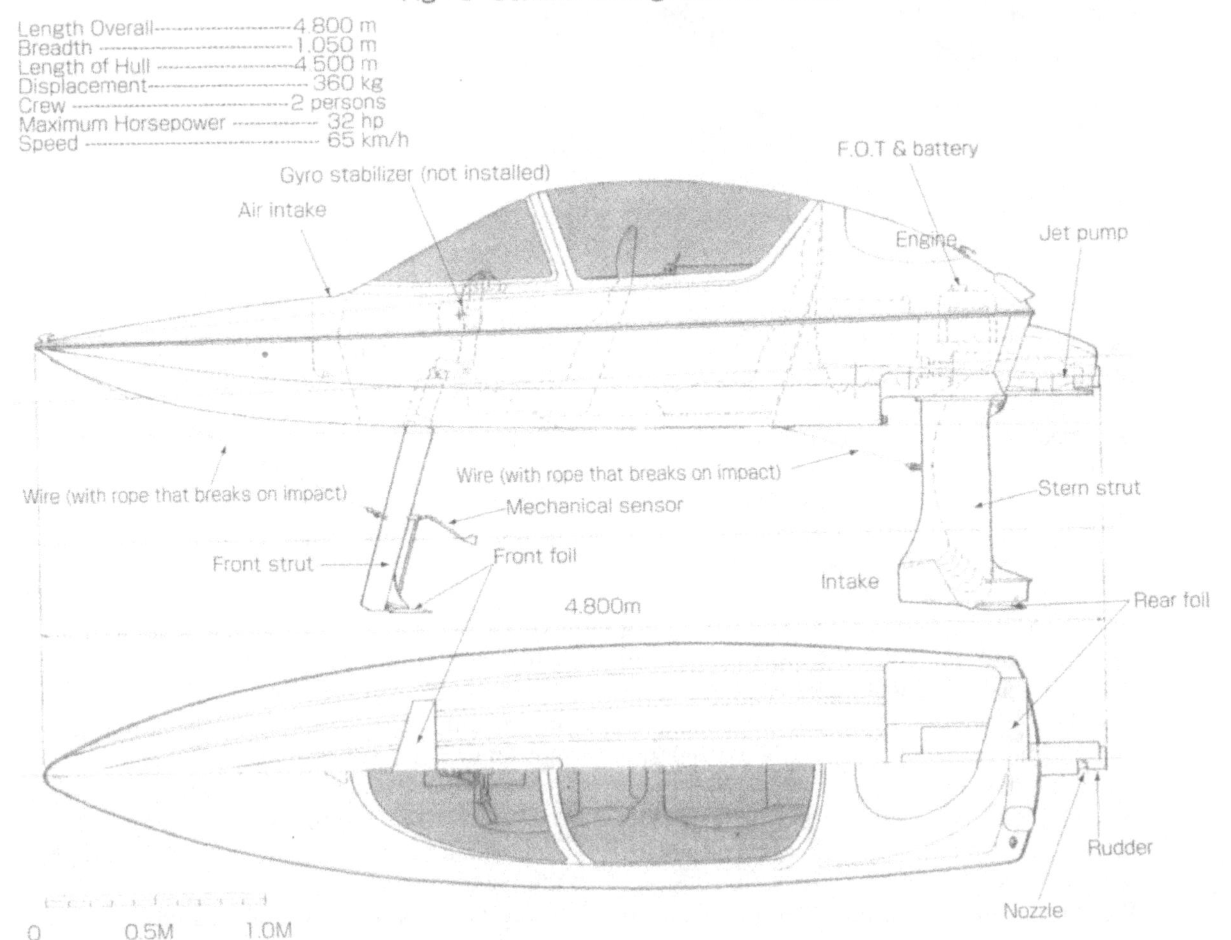

The boat is pulled almost to the upside down position by ropes.

Once beyond the critical point, the boat tends to return to the upright position quickly. Note that the ropes are loose.

Boat returns to upright position quickly.

Fig - 10 Stability test

pected accident occurred. Just as the OU32 began a turn, its bow suddenly plunged into the water. I could not tell what had happened. After bringing the boat back to land at slow speed, we found that the bow strut had broken transversely in the middle.

I did not remember what excuse I made to the inspector. According to Yokoyama, we had asked the inspector to wait until the reinforcement was completed, and the trial run was performed again for inspection.

To our relief, we received special approval for temporary operation of the boat from January 23 to February 5. Perhaps both the Ministry of Transport and the Chubu Maritime Bureau gave us special treatment considering our tight schedule.

I presume that the authorities judged the boat as not dangerous even if it met with some problem because it had a stability of 180 degrees and did not lead to a serious hazard even when its bow strut broke.

We recalculated the strength of the damaged strut and found it insufficient. We must have been off guard because of our success with the structures of the OR51 and the OR90. Looking back, I think that we had not performed thorough checks since the design limits were exceeded because the speed and the weight had increased.

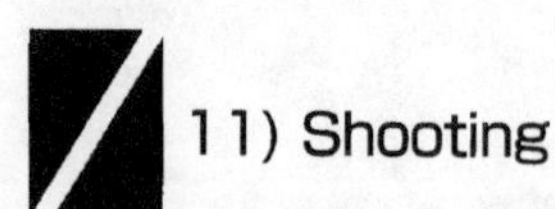

11) Shooting

On January 30, the video movie was shot. During the preparations for one week starting January 23, several trial runs and minor improvements seemed to have been made. I do not know the details exactly because I was not present, but we did not have any major problems. The most important result was that Yokoyama became familiar with the controls of the boat.

The video movie of OU32 was taken from an

escort motor cruiser. Yokoyama drove the OU32 perfectly, and its performance exceeded our expectations.

In the morning and afternoon, we ran the boat and we shot the video. In the morning session, we had almost no waves, so the boat exhibited flat running and tight 45 degree banking turns one after another, which was its strong point. I looked proudly at the boat turning tightly, exposing the entire hull bottom with very little wake (Fig-12). I was onboard the accompanying boat with the video camera operator, pressing the shutter of my own 35 mm camera to get some excellent shots.

The OU32 cut the wake of the motor cruiser left and right. The wake was rather high, at about 50 to 60 cm. But the OU32 heaved very little with a bow heave of only about 10 cm, whereas the motorboats would jump high out of the water and slam down on the water.

The heaving motion of the bow was due to the height sensor, which sensed the wave height and controlled the elevator to change the attitude. If we cut off the sensor and Yokoyama controlled the boat manually, we could eliminate heaving. The boat ran splendidly.

In the afternoon session, the wind increased. Normally, Matsumigaura Bay in Lake Hamana is calm. But this time we had white waves, a wind speed above 10 m/s, and a wave height of around 40 to 50 cm. In these conditions, the OU32 ran very smoothly, as is clear from the photo of the model waving her hand looking as if she was enjoying the ride greatly (Fig-13). She did not seem to be scared of the rough sea. I was reminded again of the splendid soft ride given by hydrofoil boats.

The shooting of the video movie was very successful and we eagerly awaited its completion. After the fourth day, on February 3, we had a chance to watch the video movie. It was excellent. It presented good shots indicating the boat's performance, which I had previously observed. The background music had been added and its was nicely arranged. I was thoroughly satisfied with it, more than I can express by writing. I wish all the readers of these articles could watch it. When I compile these articles into a book, I hope to attach a DVD. Members of both the Horiuchi Laboratory and of Marine Operations worked

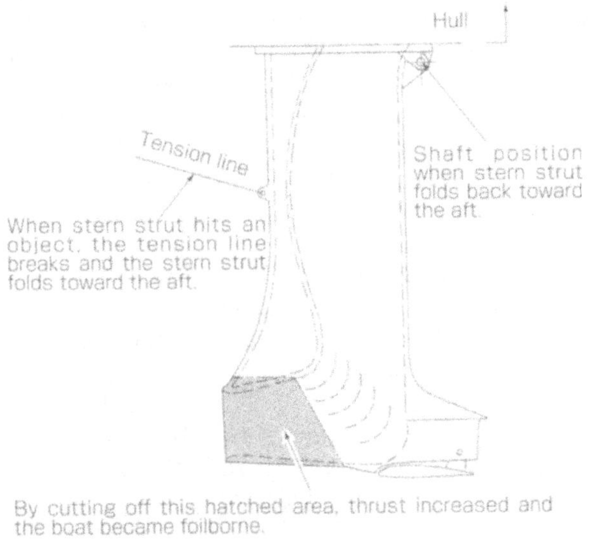

Hull

Tension line

Shaft position when stern strut folds back toward the aft.

When stern strut hits an object, the tension line breaks and the stern strut folds toward the aft.

By cutting off this hatched area, thrust increased and the boat became foilborne.

Fig-11 Cut out part of intake area

Fig-12A Tight turns

Fig-12B Tight turns

Fig-12C Tight turns

Fig-13 Running through waves

on this project to be on schedule for shooting the video movie. Finally, we were rewarded with the excellent video movie. I would like to express my gratitude to Akoh Co., who produced it.

Other still shots that I took included many fine scenes which were used effectively later. The day the video movie was shot was a very fruitful one.

We were convinced that the video fully demonstrated the excellent manoeuvrability and attractive features of the OU32. We had 6 days remaining before the boat show and we concentrated on the final touch-up and decoration of the boat.

12) Tokyo boat show

The Tokyo boat show was at the Harumi exhibition center from February 9 to 14. Yamaha exhibited the OU32 in the main booth with brilliant decorations suitable for a dream-boat.

The exhibition of the OU32 enabled us to present advancements realized after refining the OR51 and the slender design of the boat, with which we were fully satisfied. Visitors to the boat show were also interested in the boat and stood riveted by the video performance asking questions one after another.

I was pleased to see that Mr Arata, Director of Marine Operations, and Mr. Murakoshi, General Manager of the Planning Department were very satisfied with the boat. The development schedule had been very tight but we were fully satisfied with the results, thanks to all the people who had worked on this project.

13) Presentations to the media

After the Tokyo boat show, the OU32 caught the attention of people from various fields and was also featured in various media.

The Chunichi Shinbun (newspaper) planned a special colored issue on the theme, Wing Beats for the 21st Century, in memory of the seventh anniversary of the establishment of their Tokai Head Office. The OU32 and Yamaha's marine business prospects were highlighted in the April 28 issue of this newspaper.

In addition, foilborne photos of the OU32, its drawings, and detailed explanations were introduced in the new column called Forum, in the August issue of the Nikkei Mechanical Magazine.

Both the Ministry of Transport and the Ministry of International Trade and Industry requested us to exhibit the OU32 and the OU90 at the World Expo 88 scheduled to be held in Brisbane, Australia. For this purpose, an additional OU32, different from the one exhibited at the Tokyo boat show, was packed, and dispatched on February 20.

The World Expo 88 took place in the center of Brisbane city and was a large-scale exhibition attended by 8 million visitors. Firstly, Yokoyama visited Brisbane to supervise the assembly and exhibition of the boat. After that, I visited the opening ceremony to observe the reaction of visitors and to suggest improvements at the exhibition (Fig-14). The OU32 and its video show also attracted many Australians who loved boating.

It was probably the result of this exhibition that ATN TV Station of Sydney, Australia, made a proposal to collect more information on the OU32 for their news. This TV station had its own original science and technology program called Beyond 2000 broadcast for one hour during prime time every week. The program is a popular and authoritative scientific program in Australia.

The ATN in cooperation with Fox Network Co., of the USA planned to produce a US edition of Beyond 2000, and wished to gather data for this program. As we were marketing motorcycles and outboard motors in the USA on a large scale, we accepted the proposal and a video movie was completed.

Later the movie was broadcast in the USA and I heard that many inquiries related to the OU32 were received at the branches of Yamaha Motor. Since we had not given them detailed information about the boat, they might have found the questions difficult to answer. Regretfully, we also were unable to follow up on the responses to the boat.

14) Later

As the OU32 was featured in various media, our purpose of building the OU32 was fully satisfied. It probably contributed to enhancing the image of Yamaha Motor Co., in no small way.

From the beginning of the project, we had hoped that a limited number of the boats would be built and sold as a commercial production model and not only for boat shows. However, this was merely the designers' wish and not the company's policy. Therefore, the boat was treated as a dreamboat. Since we did not expect customers to buy the boat, we did not make preparations for marketing the OU32.

On the other hand, the quality and finish of the OU32 were adequate only for the boat show and for video movies. If it were to be made into a production model for the market, many improvements would have been necessary. According to Yokoyama's assessment, the improvements were required to over 30 items.

In those days, the Horiuchi Laboratory had many projects under development and its members were always busy. I asked Yokoyama to prepare drawings for the production model, including drawings for the necessary improvements, when he had the time and was not very busy.

Yokoyama completed all the drawings for the production model by working at the end of 1988, in 1989 and in the spring of 1991.

Fig - 14 Exhibition at The World Expo 88 in Brisbane

I was the Director of Marine Operations in those days, when the economic recession began and reduced our sales volumes. A large investment was necessary to bring the OU32 into production, and it gradually became difficult to envisage this situation especially when profits were dropping.

Under such a situation, the commercialization of OU32 was frozen at a time when all the drawings had been completed.

Looking back, the OU32 appeared in various media but we did not perform any market surveys. We never came across a customer who insisted on owning that boat, even though it had been presented as a dreamboat. Considering the situation those days, it might have been difficult to meet the conditions of commercialization and to sell the OU32 in the market.

Now I am grateful for having had the chance to develop the OU32, which was the pinnacle of the hydrofoil boats of the Horiuchi Laboratory. A beautiful boat and a video movie that shows the boat's splendid operation are all that remain. These offer the greatest pleasure to a person who worked on the development of the boat.

I would again like to express my gratitude for being given such an opportunity.

Note : I heard that one OU32 would be remade and kept on permanent display in the Communications Plaza (Exhibition hall for Yamaha products) of Yamaha Motor Co., when its construction is completed. You may get the chance to see the OU32 there.

11

HIGH-PERFORMANCE LIGHT PLANE OR15

Our chairman, Genichi Kawakami used to urge us to enter the aircraft business. But we hesitated because product liability suits were increasing in the USA at that time. Also, at that time, I visited the OSHKOSH Aviation Event with Mr. Ohmori, technical consultant of Yamaha Motor, and was very impressed with the event. Around the same time, Yamaha Motor started development of an aircraft engine. We started to formulate a plan for development of a two-seater, high-performance light plane, and to aim for breaking world records. Our wind tunnel test results were excellent and we were delighted. Suddenly, we had some unhappy news. I believe however, the appeal of the base plan will remain for a long time to come.

1) Order of Chairman

Since 1980, Mr. Kawakami, the Chairman of Yamaha Motor had urged Directors (I was one at that time) repeatedly to enter the aircraft business.

Chairman Kawakami took over the business of music instruments from his father and later diversified into many new businesses successfully, including motorcycles, archery, skiing equipment, boats, outboard motors, and various leisure facilities. His success in these fields is well known.

We faced a major problem in the aircraft business. Our studies showed that we could not find a solution for product liability[1] issues, and we were hesitant in starting the project.

In 1981, we bought two ultra light planes and flew them in the International Hang Glider Championship at Beppu. We approached the aircraft industry in such a timid manner (Fig - 1).

The development of remote control helicopter for agricultural use was under way, (Fig - 2, Chapter 6) aiming for business in this field, but both these projects were not in line with the Chairman's intentions.

The development of the OR15 began under such circumstances.

Fig - 1 Yamaha's ultra light plane and concerned personnel

From left, the author, Mr. Hasegawa, later president of Yamaha Motor, and other concerned personnel

1) Product liability : Liability of any or all parties along the chain of manufacture of any product for damage caused by that product.

Principal Particulars
Fuselage length············2.555 m
Breadth·····················0.640 m
Height······················1.000 m
Main rotor diameter·······3.000 m
Tail rotor diameter········0.550 m
Aircraft weight············44 kg
Take-off weight···········67 kg
Fuel oil capacity···········4 liters
Payload······················20 kg
Flight duration·············30 min
Engine: water-cooled···12 ps

Fig - 2 First flight of the remote control helicopter R - 50

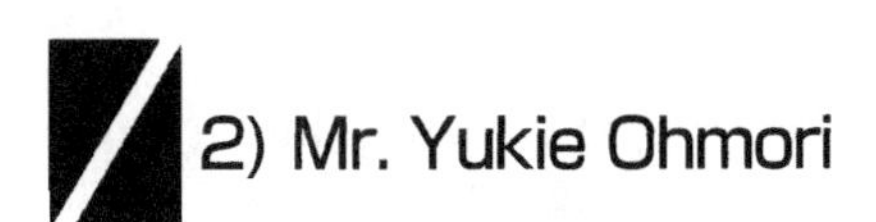

2) Mr. Yukie Ohmori

From January 1986, Mr. Yukie Ohmori, with whom I was acquainted with for long time, started coming to Yamaha once a month as a technical consultant.

In 1952, the N-52 was the first Japanese aircraft to be built after the war at the Okamura Company in Yokohama (Fig - 3). Mr. Ohmori was a member of its design team and so was I.

Mr. Ohmori had worked at Nippon Hikouki Co. during the war designing Navy aircraft, and was a respected senior. Since I was then a fledgling designer with an experience of only one year in boat design, I joined the practical aircraft design team at the Okamura Company for the first time and I studied aircraft design under Mr. Ohmori.

After N-52 at the Okamura Company, I worked on the development of LBS-1 (Fig - 4), a soarer[2] with two side-by-side seats ordered by Tokyo University. As I was involved in the work from the planning stage, I learnt a lot and understood the flow from the start of design to the completion of the plane.

The activities at the Okamura Company lasted only two years, after which Mr. Ohmori moved to the Technical Research Center of the Self-Defense Agency and was the Head of the Third Research Department for eight years. This department mainly handled aircraft and missiles. He then served as Senior General Manager of that Center for seven years. After retiring from that position, he came to Yamaha as a consultant. Thus, he was an active aircraft designer with extensive experience.

Fig - 3 N - 52

Principal Particulars
Wing span····················8.600 m
Length Overall··············6.000 m
Wing area····················12.0 m²
Engine: Continental·········65 Ps
Aircraft weight··············300 kg
Total weight··················500 kg
Max.speed····················180 km/h
Cruising speed···············160 km/h
Climbing ratio···············160 m/min.
Cruising range···············500 km
Take off distance············180 m

2) Soarer : High performance glider

Fig - 4 LSB-1 (side by side two-seat soarer) and concerned persons
Mr. Ohmori is second from right. Note the V-shaped tail wing of the aircraft.

Principal Particulars

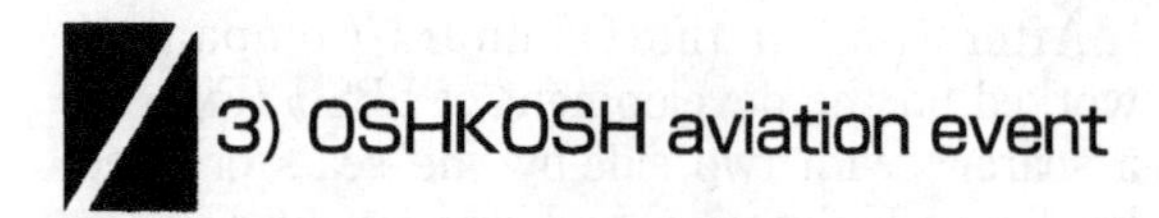

Wing span	13.500 m	Total weight	320 kg (two crew)
Length Overall	6.900 m	Max. gliding ratio	nearly 20 (two crew)
Wing area	17.2 m²	Min. descending ratio	1.1 m/sec (two crew)
Aircraft weight	180 kg	Stall speed	52 km/h (two crew)

3) OSHKOSH aviation event

At that time, I was in charge of both R&D[3] bases, that is, R&D California and R&D Minnesota in the USA, and I visited these centers every other month.

One day R&D California suggested that I visit the biggest aviation event in the world named, the EAA FLY-INN[4] to be held in August at Oshkosh, Wisconsin.

Since I had wished to see this event someday, I decided to go with Mr. Ohmori and Mr. John Gill, a respected naval architect who had supervised engineers in Yamaha for several years and was one of my good friends.

I had heard about this famous OSHKOSH aviation event, but I was astonished after actually seeing it. It was originally an exhibition for home-built aircrafts, with over twelve thousand personal aircraft owners coming to Oshkosh from all over the USA to participate in the event (Fig - 5).

About fifteen thousand aircraft owners parked their aircraft and stayed on the field, pitching their tents at the allocated places for a week during the event. I was astonished to see the huge, extensive area where a great number of aircraft could park and moreover, to know that nearly 800-thousand visitors gathered here.

During the event, the visitors looked around the home-built aircraft, saw acrobatic flights, air circuses, Harrier[5] performance during vertical take off and landing, visited many tents where materials, parts, new products, and demonstrations of home built aircraft were exhibited or performed, and made plans for restoring or replacing their aircraft. They were very excited to be at the exhibition. For these aircraft enthusiasts in the USA, this is a once in a year event that has no substitute.

While we went around looking surprised, Mr. Ohmori's reaction seemed to be different.

Mr. Ohmori had been designing and building aircraft at the Nihon Hikouki Company, Okamura Company, and Self-Defense Agency for a long time. Here people looked like they enjoyed building aircraft by themselves out of their own interest, without any responsibility or for the sake of business; therefore, under these circumstances, they could easily take up the challenge of designing new layouts or new mechanisms without hesitation.

The world of home-built aircraft is a mix of good and bad aircraft. There will be poor products. We saw many aircraft that looked like high performance aircraft, which were assembled with kits and drawings, and most of them were designed by Burt Rutan[6]. There were also several very interesting products. Mr. Ohmori was very impressed after seeing such freely-designed aircraft.

It is natural that there will be large differences when building aircraft by two methods: first, building aircraft as a part of one's work, and the second, building it of out one's desire and interest. In the event, people seemed to

3) R&D : Research and development
4) EAA : FLY-INN: Experimental Aircraft Association FLY-INN, renamed in1999 as EAA AIR VENTURE
5) Harrier : Jet propelled fighter developed in UK that has the ability for vertical takeoff/landing
6) BURT RUTAN : Famous American aircraft designer, well known for his designs such as the home-build Quickie model and the development of the round-the-world nonstop airplane Voyager

Fig - 5 Visitors' aircraft parked at the OSHKOSH aviation event
Source: Nigel Moll, EAA OSHKOSH - the world's biggest aviation event (photo by Osprey Publishing Limited)

have built the aircraft out of their own interest.

Mr. Ohmori seemed to be regretting that he had almost finished his career as an aircraft designer without knowing the method of designing and building such kinds of aircraft. These people build aircraft in relaxed manner, and their attitude will be different from ours. The quality of their airplanes will also be different from that of ours.

As fuselages were made of FRP, their shapes could be freely designed. They used aluminum in the past, and aluminum plates on the fuselage surface overlapped each other and were fitled rivets. The fuselages and wings were shaped from thin aluminum flat plates by just bending the resulting discontinuity. Modern aircraft, however, have smooth and beautifully shaped three-dimensional curved surfaces.

Air flows very smoothly along modern fuselages. I think the performance of small aircraft was better than expected because of the refined shape. Many kinds of ACM[7] were incorporated in aircraft construction such as sandwich or unidirectional fibers in parts where aluminum had been predominantly used earlier. Compilation of such advanced techniques would enable us to build high performance aircraft. I felt that Mr. Ohmori wished to build a dream aircraft once again.

After the OSHKOSH aviation event, we went to the R&D Minnesota Center, visited the transport exhibition held at Vancouver, and then returned Japan. For those five days, I traveled all the time with Mr. Ohmori only, and I understood how he felt.

Mr. Ohmori began to feel more strongly of the need to build superior aircraft that would set world records through which he could transfer good aircraft design techniques to the younger generation.

Mr. Ohmori was 70 years old at that time and he felt it was his immediate task to pass on his knowledge and techniques of general aircraft design to young people.

4) Masaaki Kusunoki

Nearly at the end of 1985, Masaaki Kusunoki came to Yamaha and joined the R&D department. He had always been building hang gliders, had participated in the Birdman Rally held at Lake Biwa, and had got good results. He had worked as a designer for an ultra light plane manufacturer before joining Yamaha.

There are many who enjoy flying an ultra

7) ACM : Advanced Composite Material. Lightweight and strong composite materials like carbon fiber, Kevlar, sandwich structures, uni-directional fiber, and so on.

light plane or a hang glider and also build these by themselves. But I did not know any person who did this better other than Kusunoki, who had studied aircraft engineering techniques correctly and designed it systematically.

His articles that appeared in magazines were exceptional. His simple explanations on theoretical concepts in his articles convinced me of the confidence in his own ability.

We had ten or so aircraft enthusiasts in Yamaha Motor including the Suzuki brothers, named Masato and Hiroto, who had all been participating in the Birdman Rally for a long time. Kusunoki was known to them as a tenderhearted senior.

The person who invited Kusunoki to Yamaha was Mr. Oguma, general manager of a dept. that become Sky dept. later. His job was to promote aircraft business, and he expected Kusunoki to become the head of engineering of that section.

Kusunoki was a talented engineer having extensive knowledge of small aircraft and was very enthusiastic about building them, although these could not be sensed from his small stature and outwardly calm appearance. I was very impressed by his personality and ability that I came to know through mutual contacts.

5) Start of the project

After I returned along with Mr. Ohmori, I consulted Director Yamashita, who was also the Divisional Manager of the Automobile Engine Department because new types of engine were always being studied and developed in his department.

Fortunately, they were also studying the development of engines for small aircraft, and moreover, studying the German motor glider market to explore the possibility of developing the glider itself.

I was happy to note the possibility of using our company's engine and planned to have Kusunoki design excellent aircraft under Mr. Ohmori. Mr. Oguma was also happy with this arrangement.

Kusunoki, who was invited to Yamaha, finally got his opportunity to design aircraft. He would learn the basics of aircraft design related to aircraft before-and-after the Second World War from Mr. Ohmori. For Kusunoki himself, this was a golden opportunity. This idea was decided and went forward smoothly. Kusunoki began designing aircraft under Mr. Ohmori's guidance, who came to Yamaha once a month.

Since I was busy with more than ten projects of the Horiuchi laboratory and also held the post of Assistant Manager in Marine Operations, I was not fully involved in this project but only in decision making, and I worked as a co-ordinator between Mr. Ohmori, Kusunoki, and the company.

At that time, there was a plan to move Kusunoki to the Horiuchi Laboratory; therefore, drawings and reference reports arrived at my desk. Occasionally, I gave advice on products and on refining the industrial design of the products.

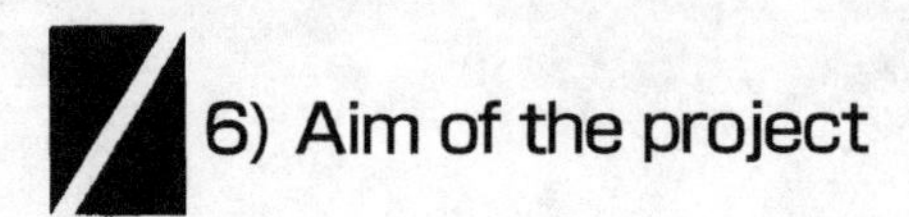

6) Aim of the project

Mr Ohmori, Kusunoki, and I held discussions and the aim of the project became focussed. To develop a conventional aircraft was meaningless. We wanted to develop a superior aircraft, the design of which would at least be equivalent to the Burt Rutan design. Our aim was to design an aircraft with a performance that would break various world records. Accordingly, Kusunoki began to classify and study the official records of FAI (Federation Aircraft International) in those days.

We were aiming for a small high performance aircraft. FAI classified aircraft by weight and there were two categories of small aircraft. One was aircraft of weight less than 300 kg and the other was of weight between 300 kg and 500 kg. We proceeded with our plan bearing in mind the records in these two categories.

At the meeting on September 6, 1986, just one month after our visit to the OSHKOSH aviation event, the fundamental development policy was decided as follows:

1. The product should be developed to the level of a light plane, not an ultra light plane.
2. The product should adopt advanced technologies befitting a Yamaha product.
3. The product should have a performance superior to that of conventional aircraft.
4. Initially, its engine should be a water-cooled two stroke engine based on that of the outboard motor, but in the future, a new four stroke engine that would be developed by

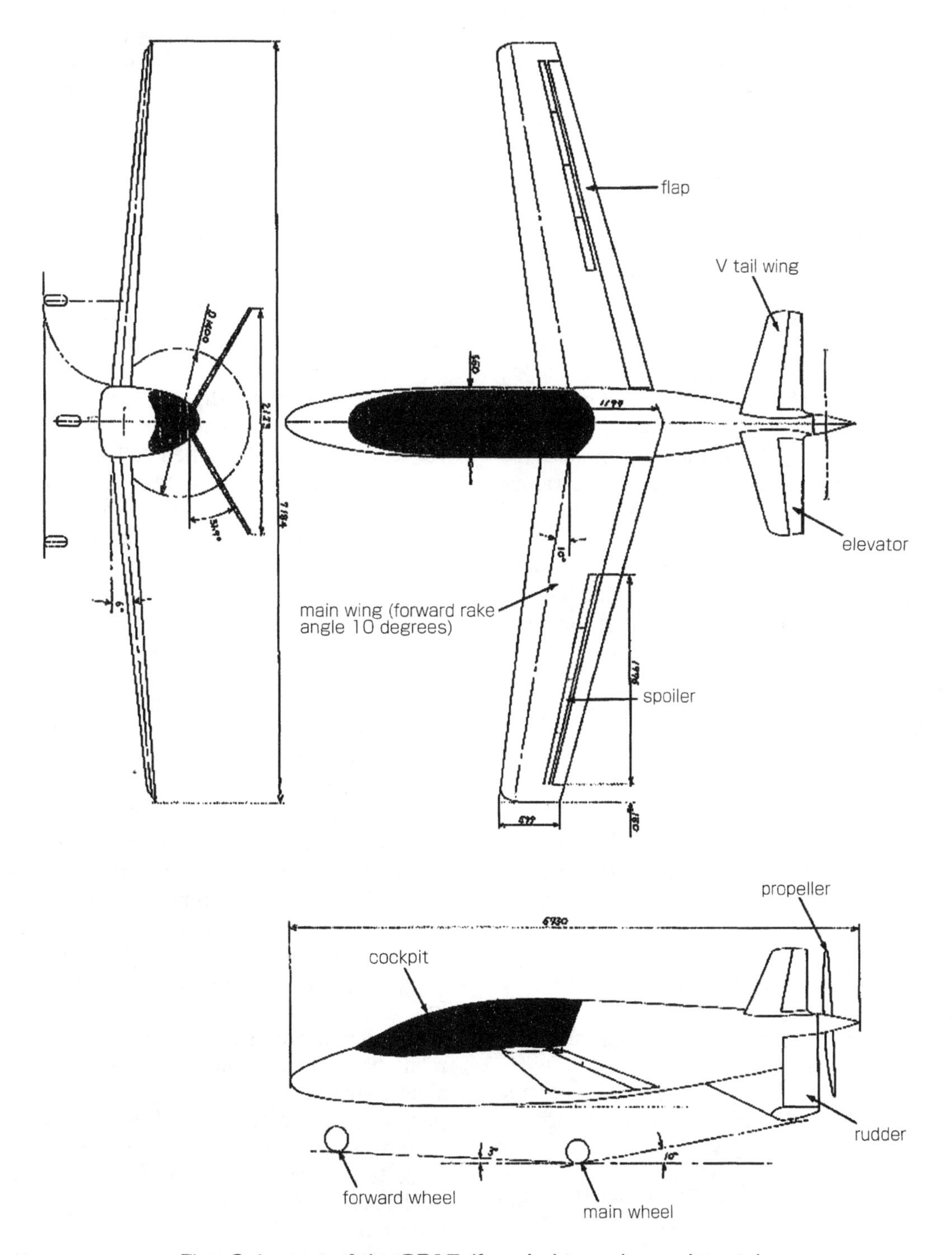

Fig - 6 Layout of the OR15 (for wind tunnel experiments)

our company should be used. The aircraft could also be used as a test bed for the new engine.

5. The fuselage should be made of ACM (Advanced Composite Materials) and have a beautiful curved surface with a shape that has the least drag.
6. The product should have an arrangement for two seats considering marketability.
7. The American market should be targeted otherwise there would be no business. Naturally high performance, design and style should suit the US market.
8. To attain high performance with lower horsepower engine, mechanisms such as retractable landing gear and constant speed propeller should be incorporated.
9. Focus should not be on the ability to perform acrobatics. The aim should be for a structure of category N with two crew and

category U with one crew (these categories indicate strength levels. N is for passenger aircraft and U is for normal flights including some acrobatic flights).

10. The rudder balance and landing gear arrangement should be suitable for pilots who are beginners (landing gear arrangement of two forward wheels and a tail wheel may result in failed landings for beginners; the arrangement should be a three-wheel type arrangement having one wheel forward and two wheels at the aft).
11. A side-by-side seating arrangement would be preferable from practical aspects. On the other hand, tandem arrangement is better considering aerodynamic behaviour, strong structure, and performance. If possible, the design should proceed with tandem arrangement, introducing ideas such as higher positioned aft seats.
12. More economical high lift device should be adopted (this would enable easy landing for beginners because of lower landing speed).

7) Aiming for high performance

To avoid the disadvantage of placing the fuselage in the aft stream of the propeller wake, and to make the fuselage attractive, we used a pusher-propeller type arrangement (propeller positioned aft). Naturally, the engine was located aft of the crew.

Mr Ohmori's insight at that time was that we should also prepare the layout considering small turbo-jet engines that may appear in the near future (Fig - 6).

The craft had forward raked[8] wing, or the so-called swept forward wing. Mr. Ohmori had emphasized since many years that the angle of attack of normal wings of many aircraft was designed so as to decrease toward the wing tip, that is, the so-called twist down of tip to prevent wing tip stall. This was one of the reasons the performance of the aircraft deteriorated.

On the other hand, if the swept forward wing is provided, wing tip stall is not likely to occur and stall starts from the wing root in this case. That is why the swept forward wing prevents the hazard[9] of wing tip stall. Twist down is not necessary for the swept forward wing, so a higher performance may be expected.

For the swept forward wing arrangement, the main spar must be of bent shape at the centre of the fuselage and this weakens the strength of the wing. Also, this type of wing is likely to tear off due to torsion by gust of wind because of its shape. To prevent such accidents, high rigidity is required.

On this point, ACM are highly reliable. All wing surfaces are covered with ACM. If glass fiber or carbon fiber is arranged diagonally on the surface, the torsional rigidity of the wing will increase dramatically. The disadvantage of the bent main spar at the center is also compensated by the outer ACM cover. ACM are very useful materials for producing curved surfaces and strong structures.

Next, we adopted the V-shaped tail wing. Normally, the tail consists of horizontal and vertical wings. A V-shaped tail wing has no vertical wing. In this case, the horizontal wing has nearly 30-degree dihedral angle and works as vertical wing at the same time, therefore the drag is smaller but a little dynamic stability is lost.

The American four-seater light plane named Bonanza was offering options of conventional tail wing or V-shaped tail wing. In Fig - 5, you can see Bonanza in the second row at the second position from the right hand side. Also, LBS-1 in Fig - 4 has a V-shaped tail wing.

In our case, we adopted the V-shaped tail wing because of our experience with LBS-1 and its good performance. However, we had to design the structure like a vertical tail wing under the aft fuselage to prevent the propeller located aft from touching the ground. By using this structure that would work as a vertical wing, we would get adequate directional stability.

As mentioned earlier, tandem seat arrangement was adopted and the entire layout was almost finalized. Based on the layout, we began performance calculations. For four cases of weight less than 300 kg and six cases of weights between 300 and 500 kg, we calculated performance by altering the wing area, the aspect ratio[10], the engine horsepower, and the number of crew to find the ideal combination that would break past records significantly.

In the end, we found it difficult to use a two-seat arrangement of less than 300 kg weight and could not find the ideal combination for good performance in this weight category. So we focussed on the category of weight between 300 and 500 kg.

The total weight was decided and layout was finalized. The wind tunnel test was scheduled to start in May 1987, the following year. A 1:5 scale model was chosen and we planned to

8) Forward rake : Opposite of aft rake, wing tip is more forward than the wing center (see Fig-6). Such a wing is called a swept forward wing.
9) Hazard of wing tip stall : If one side wing tip stalls, the lift of that side wing tip is lost very quickly, resulting in a quick roll which is dangerous. If stall starts from the wing root, there will be no roll and less chance of an accident.
10) Aspect ratio : Ratio of wingspan and wing breadth. A long and narrow wing has high performance, but it is difficult to obtain good strength. Aspect ratio = b/c, where b is wingspan and c is mean wing breadth.

confirm the high performance of the craft that we were aiming for by model tests.

We especially wished to confirm through this wind tunnel test, the performance of forward raked wing with no twist down and the effect of V-shaped tail wing, for both of which we had insufficient past data. Additionally, we also wished to confirm through wind tunnel tests that the attractive fuselage had lesser drag.

8) Result of wind tunnel tests

Thanks to the Divisional Manager, Mr. Oguma, several persons were able to assist Kusunoki to perform wind tunnel tests of the

Fig - 7 Lift-Drag ratio of the OR15
(Case 16, two-seat standard type)

Fig - 8 Final layout of the OR15

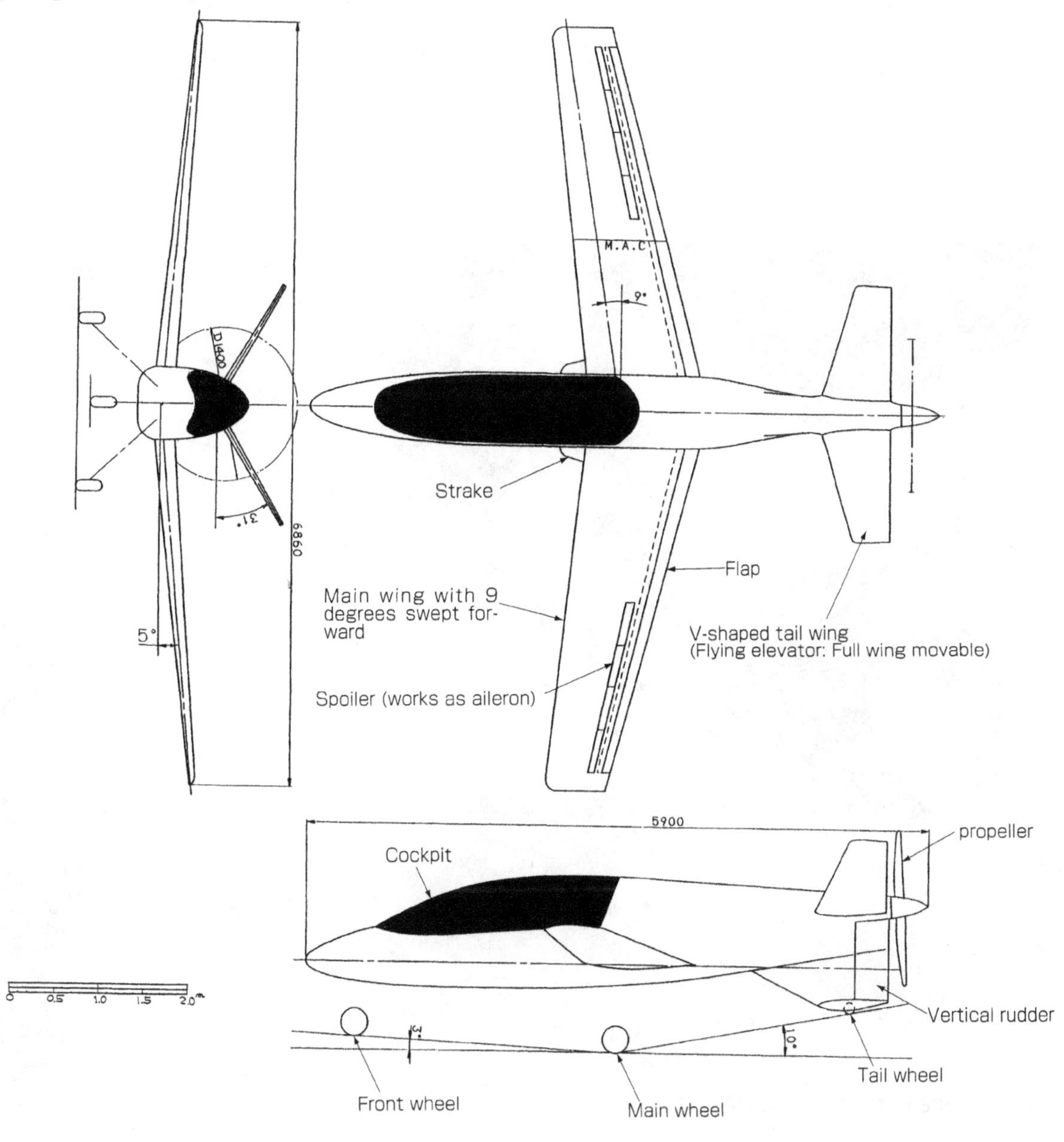

Fig - 9 Principal particulars of the OR15 (CASE16 and CASE17)*

Length overall	(L)	5,900 m
Main wing span	(b)	6,860 m
Main wing area	(S)	5,882 m²
Main wing aspect ratio	(AR)	8.0
Total weight	(W)	500 kg
Weight per wing area	(W / S)	85 kg/m²
Horsepower	(hp)	90 hp
Weight per horsepower	(W / hp)	5,556 kg/hp
C_L max.	(Flap 0 degree)	1.25
	(Flap 30 degrees)	1.80
C_D min.		0.018
Fuel weight	(Wf)	53 kg (CASE 16)
		130 kg (CASE 17)
Fuel consumption	(f)	0.28 kg/hp · h

*CASE16 is for two crew and fuel with a total weight of 500kg
CASE17 is for one crew and 130kg of fuel with a total weight of 500kg

Fig-10 Performance data table of the OR15

CASE16

Stall speed (flap up)	118.75 km/h
Stall speed (flap down)	98.96 km/h
Max.flat speed (100 % power)	327.81 km/h
Max.cruising speed (75 % power)	299.44 km/h
Max. cruising range at speed of 170 km	1791 km
Cruising range at max. cruising speed	947 km
Possible max. speed on 100 km course	327.81 km/h
Possible max. speed on 500 km course	327.81 km/h
Possible max. speed on 1000 km course	251.46 km/h
Possible max. speed on 2000 km course	(Impossible)

(Above data are calculated with remaining fuel for 200 km)

CASE17

One crew removed from CASE 16 and fuel is increased so that total weight becomes 500kg. This is the status for attaining record.

Max.cruising range at speed of 170km	4813 km
Possible max, speed on 1000km course	327.81 km/h
Possible max, speed on 2000km course	310.99 km/h

(Above data are calculated with remaining fuel for 200 km)

Fig - 11 Scale model of the OR15

OR15 in August 1987.

The lines drawing was scheduled to be completed in the third week of June, the scale model in the third week of July and the tests in August. A 1:5-scale model with controllable wings was to be used, and the main wing was to be interchangeable with one having a 10-degree swept forward shape and the other having a standard shape to enable us to compare the performance of each.

From the wind tunnel test, we could evaluate the high performance of the OR15 that we had expected. The Lift-Drag ratio (Total Lift L by Total Drag D of aircraft = L/D) that affects the performance of aircraft to a large extent, was excellent for our OR15 to attain a maximum of 22.68, which was close to that of the Soarer (Fig - 7).

The value 22.68 means that you can glide forward by about 23 m for a drop of 1 m in attitude, which is a very good performance for a light plane. This performance was probably because of the small sectional area of the fuselage due to the tandem seat arrangement and because of the highly streamlined shape that Kusunoki designed.

On the other hand, we found several mistakes in design. When the flap was lowered, the elevator did not work satisfactorily. We had to re-design and enlarge the tail wing area to solve this problem.

Since the main wing had such a swept forward amount, the stall at the wing root was likely to occur easily, resulting in a low lift coefficient of the entire aircraft. This problem was resolved by placing strakes on the wing root (Fig - 8). The strakes were effective in preventing separation of airflow around the wing root.

If we reduce the swept forward angle to nearly half, better performance might be obtained. However, since we had not prepared such a wing, we had to make plans for another test, which we did not conduct. The swept forward angle was decided as 9 degrees in the final design, 1 degree less than in the preliminary design.

We made some modifications such as changing the swept forward angle from 10 to 9 degrees and added strakes. The basic performance of the OR15 was still excellent and we were delighted.

In November, Kusunoki re-calculated the performance based on the data of wind tunnel tests to find large improvements. Fig - 8 shows the final layout at this stage and Fig - 9 shows the principal particulars. Fig - 10 shows the table of performance data.

In parallel to wind tunnel tests, we made the scale model to confirm the general impression and to decide the color scheme (Fig - 11). The scale model painted to the so-called Horiuchi-ken colors looked beautiful and gave the impression of high performance. We were delighted and very proud of it.

9) Forecast to break world records

According to past records of the Federation Aircraft International (FAI), the maximum speed in the same class was 420 km/h, attained by an aircraft named Owl Racer, which had been built mainly for breaking speed records.

As the OR15 was meant for normal use, we gave up the idea of trying to break that record.

Calculations however, showed that its average speed when flying 1000 km on a circular course would be 327.81 km/h, which was 9.4 % higher than the past record of 299.63 km/h. Also, the average speed when flying 2000 km on a similar course would be 310.99 km/h, which was 36 % higher than the past record of 228.26 km/h (Fig - 10).

Moreover, the maximum cruising distance (circular course) of the OR15 was 4,813 km, which is 32% greater than the past record of 3,641.70 km, according to our calculations. These calculations were based on a take-off weight of 500 kg that included one pilot and fuel oil filled so that the total weight became 500 kg.

The performance above is the one expected on a circular course. Records for straight courses would be affected largely by factors other than the performance of aircraft, such as geographical features or weather conditions, and these would be difficult to consider in Japan. Therefore, we eliminated studies on straight course. The estimated performance was noteworthy for a multi-purpose aircraft and our predictions showed that we had a good chance to attain the target we had set initially.

We were very delighted. The actual aircraft was not completed, but since the calculations were conservative, the margins considered would cover several problems that might occur.

It was my task to decide when and how to start building the aircraft and how it should be done, considering the development of engine also. Kusunoki was in the process of completing the third report of wind tunnel tests following the first and second reports.

10) Unexpected event

A series of calculations was completed in November 1987. Kusunoki was fully engaged in the completion of his work throughout that year. It was December 21, I was thinking that the year about to end as a delightful one because of the successful progress of the OR15 project. Just then, my wife heard news on the radio about a hang glider accident near the Mt. Fuji hillside and the name mentioned was Kusunoki. I was shocked to listen to the news that followed and it became clear to me that our Kusunoki had met with an accident. He had been frequently testing and improving his hang glider, which he had built himself, and during one such test, he had met with an accident. Considering his careful nature, it was difficult for me to believe the news. Since the hang glider was being tried out as a prototype, there was possibly an element of danger. I was very confused and shocked; I did not know what to do and kept hoping that the accident was not fatal.

According to reports, the left side of his head had taken a strong blow, there was no external wound, but he was unconscious and both thighbones were broken, with the left thighbone severely damaged. He had difficulty in breathing and had a high temperature.

On 22nd the next day, I visited him in hospital but he was still unconscious, had high temperature, and was in a bad condition.

On 23rd early in the morning of the following day, Kusunoki passed away.

I visited the place where he met with the accident. I saw a huge Japanese cedar forest adjoining the flat area. When landing from the flat area side, it seems his hang glider had collided with the trunk of the cedar tree. What kind of misjudgment did Kusunoki, such a careful person, make? Was the accident because of gust or loss of control? I could not believe it even after looking over the scene.

I considered several points for the future after returning to work. We had lost Kusunoki, who had accumulated all technical data almost by himself. Without him, it would be difficult to continue with the OR15 project because a suitable engineer to take over Kusonoki's job would not be found soon. From the beginning,

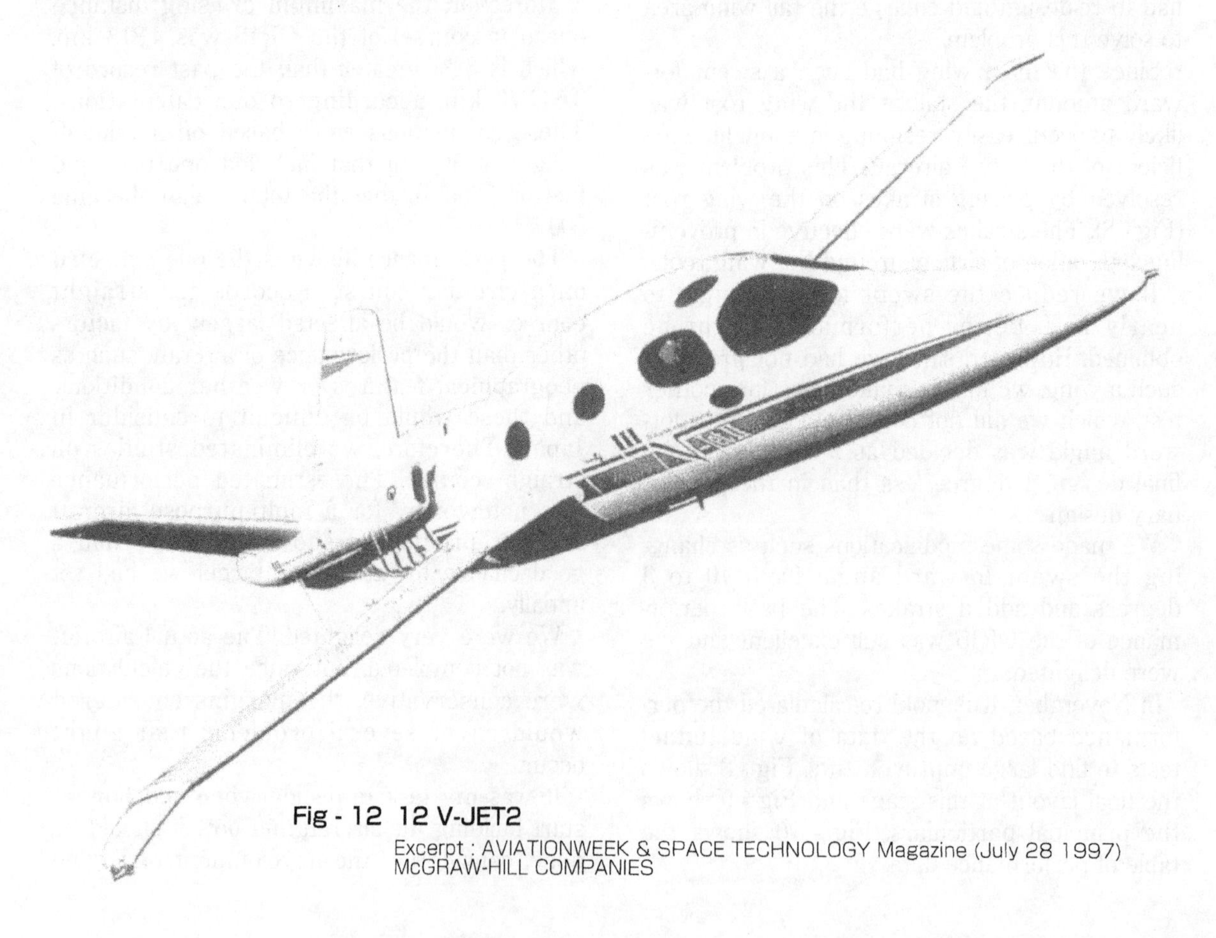

Fig - 12 12 V-JET2

Excerpt : AVIATIONWEEK & SPACE TECHNOLOGY Magazine (July 28 1997) McGRAW-HILL COMPANIES the project had been managed almost fully by Kusunoki himself. The project was suspended subsequently.

At about the same time, I had a talk with Director Yamashita to report the accident. He said the development of the new engine also faced difficulties. Although, the engine itself had some good prospects, the prospects for procuring element components were not good because manufactures had refused to supply the same.

At that time in the USA, aircraft engine was considered a very risky product because PL suits were growing rapidly. We could somehow understand such a situation because we had been hesitating to go ahead with the development of aircraft because of our concern for PL suits.

The developments of both airframe and engine could not go forward, and so the development of the OR15 was suspended.

11) Later

Ten years after the suspension of the OR15 project, probably at the end of 1997, I received a phone call from Mr. Ohmori. He said that Burt Retan had developed new aircraft similar to the OR15, but a larger one accommodating 5 to 6 passengers.

According to the article I received, this was part of NASA activities. In recent years, the US general aviation appears to be in stagnation because only the Lycoming and Continental engines developed 70 to 80 years ago, are available for small, general-purpose aircraft. That could be the reason why development of airframes has shown no progress.
The turbo-jet engine is available for business aircraft but it is too expensive and can be afforded only by companies or very rich people. Therefore, the number of production units would be small.

NASA came out with a plan to develop a turbo-fan engine with small fuel consumption, good durability, small size, light weight and economical for general-purpose aircraft with the aim of stimulating the aviation field. NASA called on Williams International, a manufacturer of small jet engines, to develop this engine as one of the members of the GAP (General Aviation Propulsion) team in NASA.

They decided to use 40% of the budget for the demonstration of the new engine. The new aircraft V-JET2 was scheduled to be built as a test bench for the new engine to be developed (Fig - 12).

It was reported that V-JET2 was completed at the Williams International factory in 1996 but the design and construction of the airframe were made under the supervision of Burt Retan at the Scaled Composite Company owned by Burt Retan and located in the Mohave Desert.

The aircraft was basically developed for testing the new engine and for demonstration; it was not intended for mass production. However, it can be considered that much of the airframe incorporated Rutan's dreams. The fuselage was made fully from ACM (Advanced Composite Materials), and had a new look and streamlined shape. It was powered by two ready-made jet engines each capable of developing 550-pound thrust initially, but according to the article, it was going to be powered by new turbo-fan engines FJX-2 capable of developing 700-pound thrust by 2000 and would attain a 3000-km flying distance at a speed of 500 km/h.

The layout of the aircraft was very similar to that of the OR15 considering its swept forward wing, aft engine, V-shaped tail wing, three wheels, and small lower vertical tail wing. The impression obtained by a mere glance would be different because of the outline of the window and low profile on ground when fitted with jet engines. But the concept of its design was doubtless the same as that of the OR15. If the OR15 had been fitted with jet engines, both craft would look and become more similar.

My heart was filled with emotion when I learned that Burt Retan had developed an aircraft similar to the idea Mr. Ohmori had dreamed of 10 years earlier. The OR15 project, which we had started with the aim of surpassing and overtaking Burt Retan, had failed, and we felt sad; I think, Mr. Ohmori was also sad.

The article reported that besides the twin-engined V-JET2 with a capacity of 6 passengers, the GAP (General Aviation Propulsion) Team was planning to develop another new aircraft having four seats and a single engine. The new one will be announced in 2000. It would be very interesting to see whether this design would be more similar to the OR15.

We did not get the chance to restart the OR15 project. Nowadays, I think of the lovable OR15 with its advanced design with nostalgia and fondly remember Kusunoki's face of bygone days.

12

LEAN MACHINE OR49

A long time ago, Honda made a scooter named Juno. Even after driving it for about 45 kilometers in the rain remained the driver's clothes remained dry. Since then, I have considered developing a smaller and comfortable vehicle. After the Horiuchi Laboratory was set up, I wanted to develop this kind of vehicle for the US market and we started the development work. We developed a gyrostabilizer, which was an advanced device for the said vehicle. The performance of the gyrostabilizer was so good that we decided to use it in hydrofoil boats also. Such vehicles offer hints on how to resolve traffic issues and reduce

1) Juno

For three years from 1957, I used to commute to work on a scooter named Juno made by Honda. The name Juno was also given to a second-generation model, quite different from the first-generation Juno that I used.

The model I used had a large body with broad wheels and was fitted with a 200-cc four-stroke engine. If memory serves me right, its weight was 195 kg according to the owner's manual, but it felt heavier.

Juno was a very ambitious product and had many new features, which overturned the established concepts of scooters such as Rabbit (made by Fuji Heavy Industries) or Pigeon (made by Mitsubishi Heavy Industries).

Juno had cantilever[1] front and rear wheel shafts, an exceptionally large windshield, large volume not seen before in similar vehicles, and an FRP body, which was introduced just for mass production (Fig - 1).

The FRP body was made by the Matched

Fig - 1 Juno

Fig - 2 Slit above small windshield of Juno

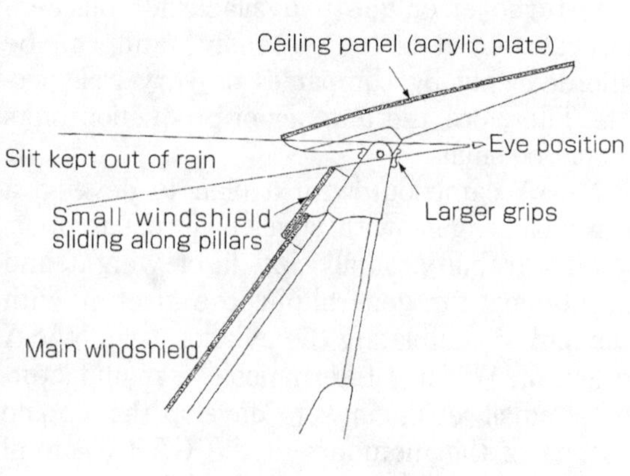

1) Wheel shaft supported only on one side by fork. Wheel replacement is easier in this system.

Metal Die (MMD) method using male and female metal molds. This method enabled a deep hollow shape to be created resulting in a more attractive look than obtained by pressed metal plate. I liked its finish.

The scooter encountered many serious issues and Honda had a hard time coping with them. It was reported at one time that as a result these issues, the company faced a serious crisis.

My scooter was an already improved one and I had no problems as such, but its frame seemed to be inadequately stiffened, and its aft body twisted on every cushion stroke. I thought this was a part of the cushion stroke and did not mind, but its running attitude looked strange when viewed from the back.

On the other hand, the points that attracted me to this scooter far outnumbered its defects. Firstly, its gentle and soft riding characteristics were excellent. In addition, the protection of driver against wind, rain, cold, and its ability to keep the driver comfortable were splendid. Its windshield, which I felt was too large looking at it the first time, was designed carefully with many good ideas that became more convenient after each use. Although difficult to believe, while riding the scooter I did not feel cold, almost did not sense the rain, and I experienced soft riding, all of which gave me a very comfortable feeling.

One day during a downpour, I tested and studied the scooter. I drove the scooter from my house in Kamakura to a factory in Tsurumi, Yokohama City, a distance of 45 km via Inamuragasaki, Shichiriga-hama, Enoshima, Fujisawa, and the National Route 1.

During the drive, the visibility was good and I drove easily in the rain without wiper. I was then wearing a long US-Army weather coat. When I arrived at my office and took off my coat, only my elbow was wet, and I was dry within.

How did this happen? The design ideas were good. Firstly, the windshield covering the outer steering grips fully protected the hands from rain and wind. Secondly, the field of vision was excellent. There was a small windshield provided in front, which could slide vertically and could be adjusted to suit the driver's seated height so that the driver could look ahead directly through the slit between the small windshield and the ceiling, without facing the wind and the rain (Fig - 2).

Additionally, a ceiling panel was mounted with large knobs in front of the windshield. If necessary, the driver removed this ceiling panel and fixed it on the upper end of the windshield pillars so that it protected his head from the rain and could serve as a ceiling above the head, for instance, while waiting at a signal. The front end of the ceiling was positioned slightly forward of the small windshield mentioned earlier. By adjusting the height of the small windshield, the slit between the small windshield and the ceiling could prevent rain from coming in. Through the slit, the driver could look ahead easily without getting water on his face and head, and had good visibility. It was remarkable that the slit height could be adjusted and the slit itself also could be made wider or narrower by the driver.

As the ceiling panel was not as large as an umbrella, the driver's shoulders became wet when waiting at a signal in the rain; I was not aware of this probably because there were not many signals at that time. The windshield was so good that I presumed that Mr. Soichiro Honda had himself developed it with all his heart in it and must have been fully satisfied with its finish.

Thinking about the production stoppage of such an excellent model because of other problems and the inability to attain its target, I could imagine Mr. Honda's bitter disappointment. I also regretted that Honda and Yamaha did not have a scooter like Juno during the last 40 years or so.

Juno triggered my belief that even a two-wheeled vehicle can be an all weather type vehicle, and I hope to experience the same comforts as in Juno in another vehicle some day in the future.

I thought however, that the development of an all weather type motorcycle like Juno was a job for motorcycle engineers. I therefore wished to design and develop a fully-covered type two-wheeled vehicle. This article gives a report on the development of such a fully covered type two-wheeled vehicle.

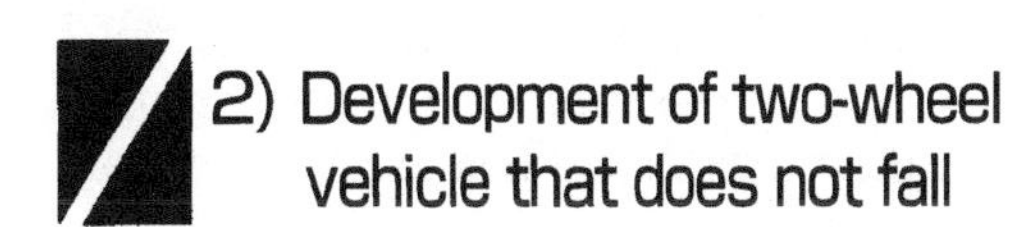

2) Development of two-wheel vehicle that does not fall

From the end of the 1970's to the beginning of the 1980's, the motorcycle had become a nearly technically-complete vehicle. The new target of motorcycles was to develop one that does not fall.

Several such studies were being carried out in Yamaha Motor too. Honda Motor began selling a three-wheeled scooter named Stream from the beginning of 1982.

Fig - 3 Stream

Since I had some work related to other small vehicles at that time, I was very interested in the direction these vehicles would take and had already started my own studies.

Firstly, I borrowed Stream and experienced driving it (Fig - 3). This scooter is called a tilting three-wheeler[2] and the same type of vehicle is now being used by pizza delivery services having a roof. We can see it everywhere. On this scooter, the two rear wheels and engine unit do not tilt but the front body tilts inside when negotiating a curve similar to an ordinary two-wheeled vehicle. You can drive Stream with almost the same feel as a two-wheeled vehicle. At a very slow speed as when the vehicle is about to stop, the tilting mechanism is locked, the vehicle stands on three wheels, and you need not stretch out your leg to support it. I thought its completion level was very high, but when driving on a bumpy road, the rear part had a tendency to sway unlike the comfortable feel of a two-wheeled vehicle. Another point I disliked was the discontinuity when changing from the tilt-free condition suddenly to the tilt-locked condition.

The rear sway motion can be decreased by lowering the tilting axle as far as possible but I could not find any solution for the discontinuity between the tilt-free condition and tilt-lock condition.

When the speed of a two-wheeled vehicle decreases, its steering bar falters and it can be steadied only by a large steering effort. Since the lateral speed of the front wheel at the point of contact with ground is decided by the product of the steering angle and vehicle speed, if the speed decreases, the steering angle must be large.

In Stream, the tilting motion at such a slow speed was locked, as mentioned above. I thought that it would be preferable to retain the same feel as during high-speed motion until the vehicle speed reduces and it comes to a stop. Unlike Stream with the characteristic discontinuity when moving suddenly between the high-speed phase and the slow speed phase, I would have liked to maintain continuity of the feeling.

It would however, require a powered auxiliary moment to return the vehicle to the upright position with a small steering effort. Basically, such an additional system would not be necessary for small vehicles but when the vehicle becomes larger and more sophistication is required, a powered system would become necessary. Based on these considerations, I examined electric, hydraulic, and other systems.

The two rear wheels of a three-wheeled system always touch the ground, while in an auxiliary-side-wheel system, the side-wheel must touch the ground when required. As this seemed rather difficult to develop, I did not take the decision immediately.

In the meantime, I designed the open three-wheeler, my dream of a three-wheeled vehicle (Fig - 4, Fig - 5). This is a vehicle in which the front body tilts. To protect the driver from wind and rain, I used a wide windshield, having small windshield in front, similar to that in Juno but with a larger ceiling. The seat height was set as low as possible to minimize air drag and to get on and off easily. Adequate space for a business bag beside the seat was also provided.

For lateral stability, I used a powered stability system mentioned earlier, which enables the vehicle to remain erect only by steering con-

2) Tilting three-wheeler is a three-wheeled vehicle in which the two rear wheels and engine unit do not tilt but the front body tilts. If the tilting axle is close to the ground, the sway of the rear body decreases. But if it is practically difficult to lower the axle, complex mechanisms may be used to lower the axle.

trol until its speed reduces and it gradually comes to a stop. When the parking brake is activated, the tilt system is locked for the first time and the brakes are applied on the wheels. My intention was to develop this vehicle as an alternative to a car.

If the windshield is designed well with this layout, one can easily drive in rain while traveling to work wearing only a raincoat.

The Horiuchi Laboratory (Horiuchi Lab) was set up in January 1984. It was here I began considering a lean machine, a topic suitable for the Horiuchi lab. Fig - 6 shows the drawing prepared at that time.

Regarding open three-wheeler (Fig - 4, Fig - 5), Stream already existed and we were capable of developing a similar type of vehicle. It was preferable to aim for a more advanced, fully-covered type vehicle as a project of the Horiuchi Lab.

Fig - 4 Open three-wheeler

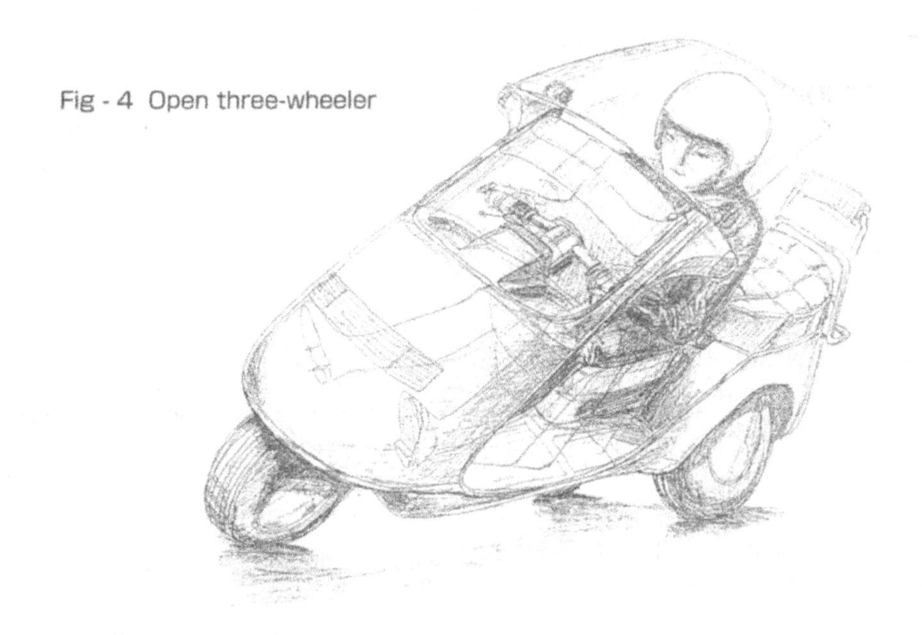

Fig - 5 Three orthographic views of open three-wheeler
High seat position for passenger

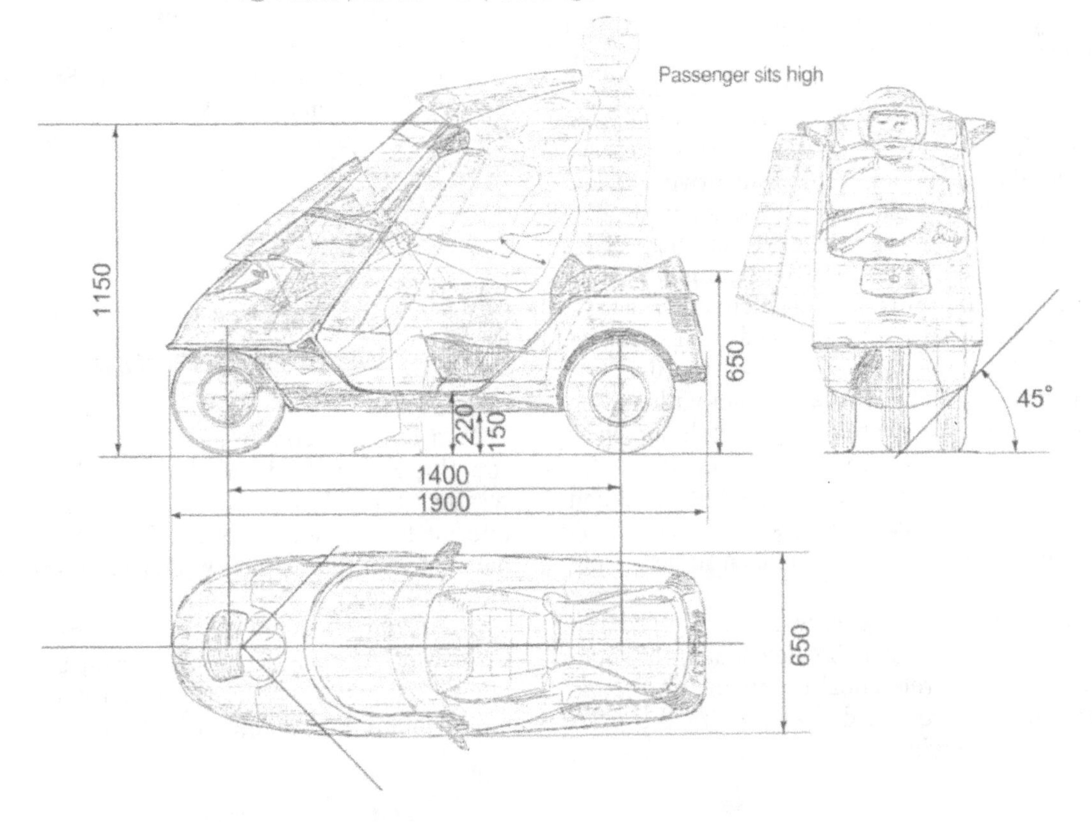

In this new concept, we would need to conduct trials for some unknown techniques, so we planned to develop and to test as small a model as possible.

From the beginning, we thought that the profile should be low enough to reduce the air drag and to prevent the effect of side wind and the center of profile area should be as far aft as possible. Additionally, we thought of adopting a gyrostabilizer to retain stability at slow speeds.

3) Start of R & D Center and proposal from Stevens

The R&D Center started in July 1985 and I was put in charge of the center. In the USA, R&D California and R&D Minnesota started, both of which were under the R&D Center (Japan). Both the American R&D centers were responsible for finding and developing new products suitable for their respective areas in the USA. I visited both the R&D centers in the USA every two months and had discussions with the researchers and offered advice and assistance.

At that time, R&D California had a close connection with the scientists of the Jet Propulsion Laboratory (JPL), and had a meeting with these excellent scientists related to product development. In that meeting, I received a proposal from James Stevens, a member of the JPL.

Stevens needed one and half hours to travel to work through dense traffic, although the driving time would be only thirty minutes had there been no traffic jams. His proposal was that Yamaha should develop a small vehicle, preferably a safe and comfortable two-wheeled vehicle which could be driven quickly through the dense traffic, and which would help eliminate such dense traffic. He strongly wished that Yamaha would develop such a vehicle.

Stevens was a very talented man. Even at that time, he suggested that we use the Internet and e-mail. He was a man with a fountain of ideas. We were annoyed with ourselves for nothaving responded to his ideas. He favored Japan and was very fond of Sushi and pickled Chinese cabbage. As he looked happy and enjoyed meeting us, we liked to hear from him.

His suggestion was basically very correct. If such a vehicle could be developed, it would probably get preferential treatment traffic regulations. I have heard that now in the USA only cars carrying more than two persons can drive on the lane near the centerline, which is always vacant. Perhaps new cars will be given such favorable treatment and the time to drive to work will reduce. If everybody uses such cars, naturally traffic jams will be eliminated.

Until then, I had been interested and had been studying systematically, the all weather type two-wheeled vehicle and the stable two-wheeled vehicle that did not fall. According to his suggestion, there was a possibility of a huge market for this kind of vehicle. I was encouraged and felt strongly about continuing the study in this direction.

In addition to the open three-wheeler (Fig - 4, Fig - 5) and the minimum model (Fig - 6), that is, the all weather type model for one person, I designed a large two-wheeled vehicle for the US market, a tandem two seater (Fig - 7). We considered the spread of products through these designs and the direction for starting development first (Fig - 8).

The design of the tandem two seater was a practically usable model, which was exactly in line with Steven's idea. The reason for using two seats was that the vehicle's weight and cost did not differ appreciably compared to those of a single seater, moreover it could be used as a family vehicle and carry more luggage. Additionally, I presumed that a two-seat arrangement would result in a finer aerodynamic shape and look more attractive. I had not studied side wheels yet.

Comparing the three designs (Fig - 8), I thought that the open three-wheeler would be suitable as the theme of the new product at the Motorcycle Department, like the Stream of Honda. The tandem two seater was similar to Steven's idea but it had a large body that made it an unlikely project theme for the Horiuchi Lab, which was a small group that mainly studied new technologies.

4) Planning of single seat all weather type vehicle

As a result of comprehensive studies, we at the Horiuchi Lab, decided to develop an all weather type single seater, which was in the intermediate position. Tsuide Yanagihara was nominated as the person in charge of this project.

He was a graduate of the Tokyo Metropolitan College of Aeronautical Engineering and was known for his excellence in design work, having adequate knowledge of both airplanes and

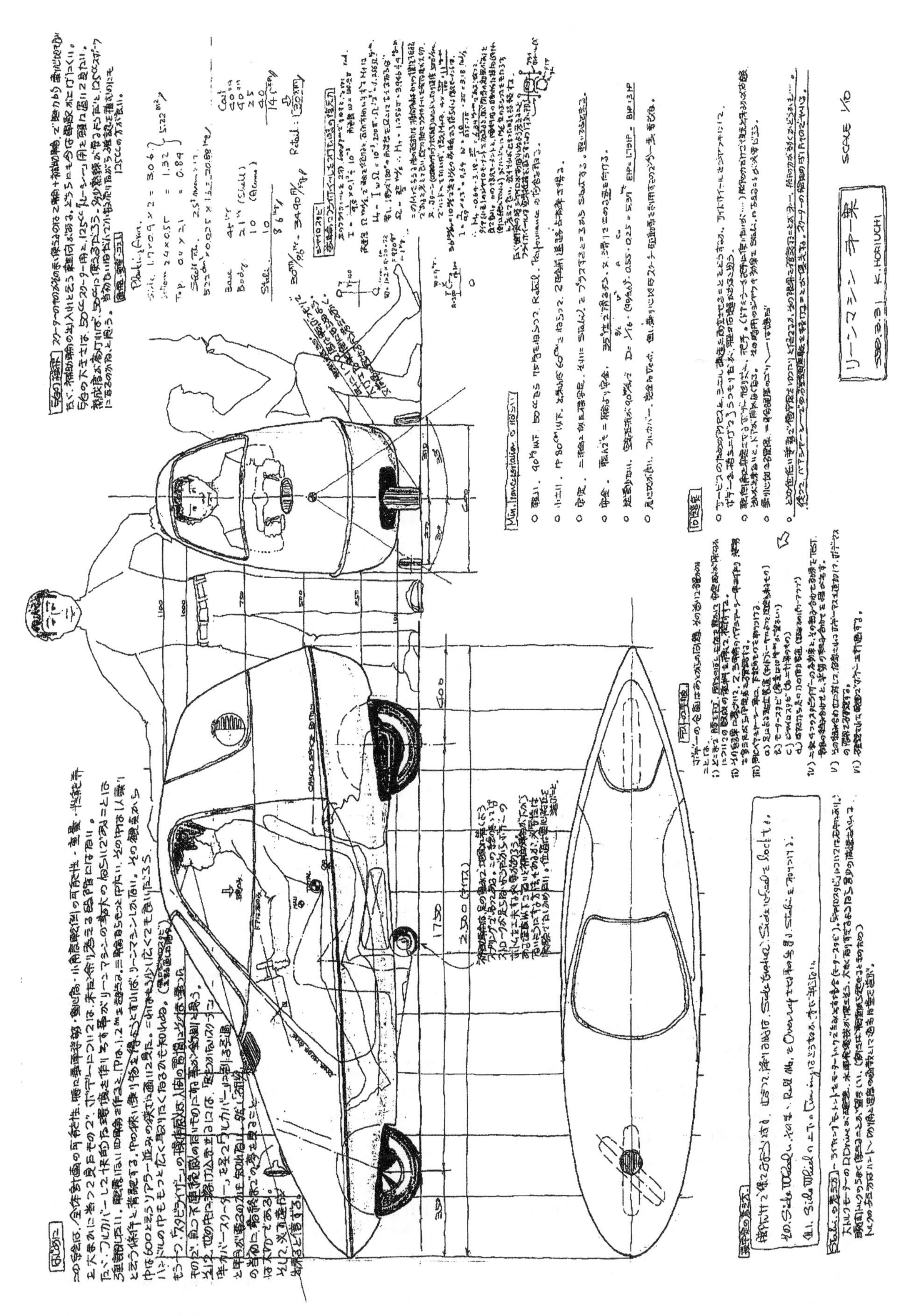

Fig - 6 Lean machine - first proposal

Fig - 7 Lean machine tandem two seater

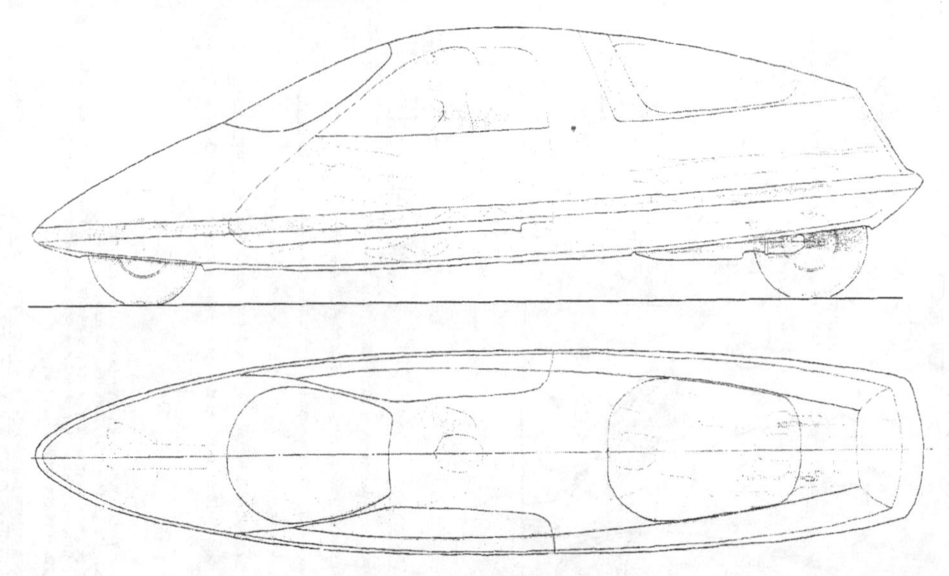

Fig - 8 Comparison of three types of lean machines

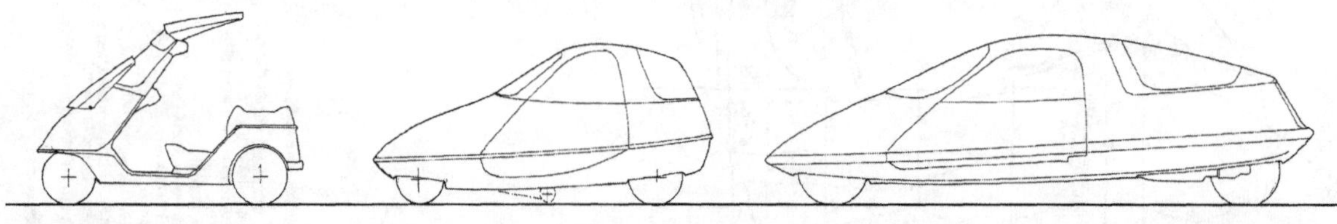

Open three-wheeler Lean machine - first proposal Lean machine - tandem two seater

boats. I have always admired his ability for analysis and design of light structures. He mainly designed the Cogito, which was adjudged the fastest boat in the recent human powered boat race.

Yanagihara and I decided the following course based on the first proposal of the lean machine (Figures 6-8): -

① Aim for minimum weight transportation; weight should be less than 90 kg. If engine of 50 cc is used, the weight should be less than 75 kg. Balance between cost and performance should be proper.
② Breadth is to be less than 80 cm, and if possible, less than 60 cm so as to minimize occupant area.
③ Dynamic stability at high speed is to be of the same level as a normal two-wheeled vehicle. During slow speed to stop, vehicle to be kept erect by auxiliary side wheels. About four stability systems should be designed, and finally a suitable one chosen.
④ Vehicle to be safer than the normal two-wheeled vehicle by limiting the angle of inclination and considering the non-slip system after a fall. Arrangement of safety air bag should be considered.
⑤ Fuel consumption should be small. Air drag would be 5 kg at 90-km/hr speeds, which would need 2 HP, and the vehicle should give 150 km per liter of fuel.
⑥ It should be comfortable to ride in, not too warm or too cold, and adequately ventilated, although fully covered and with no air-conditioning system. Since propulsion and suspension systems of scooter are used, riding comfort should be considered for paved roads; riding might not be excellent, but this is unavoidable.
⑦ To minimize sway by side wind, profile area is to be minimum and the center of the profile is to be positioned as aft as possible.

When setting of the basic proposal, our concerns were the three points below:

Firstly, lateral stability was inadequate because the riding position was low and the diameter of front wheel was small. As a measure, we decided to use a gyrostabilizer system.

The second concern was the side wind. Even

cars are pushed to the side of the road by the side wind blowing on bridges. This trend is particularly apparent in tall wagon-type vehicles. We were concerned that our vehicle would be pushed downwind because it was extremely light in weight although it wasn't tall. We thought this problem should be studied by wind tunnel tests.

Thirdly, we were concerned that it would fall due to a strong wind when parked. To prevent this, the stand width must be increased; but on the other hand, body width was to be made narrow, and we disliked side wheels protruding outside the body. Finally, we decided that it should withstand a wind speed of 20 m/s; in case stronger winds were predicted, users would be advised to park the vehicle in a suitable parking place.

The stability system has to be developed first then the planning of the body follows, in general, when developing such kinds of products. But for this project, the plan started from a body with the smallest shape, the lightest weight, and a seat for one person, and then the stability system was to be tested and developed on its prototype.

For designing this vehicle, it is essential to harmonize a lightweight structure as used in aircraft and aerodynamic considerations skillfully, otherwise the design may fail. Therefore, I thought that we should take care of the considerations above and test them practically. If the development succeeds, it can serve as a good reference for designers to develop new vehicles like this in the future. In this way, I decided to start our development work. The project number was decided as OR49.

5) Gyrostabilizer

I mentioned gyrostabilizer earlier, but here I would like to explain what a gyrostabilizer is. As shown in Fig - 9, in the past there were some two-wheeled prototype vehicles installed with a huge flywheel to forcibly keep them erect using the uprighting ability of the gyro. In such a system, the flywheel was too heavy and stiff to roll, and a banked turn was not possible. This means that the vehicle could be driven only at slow speeds, and we cannot enjoy the nimble maneuvering of the two-wheeled

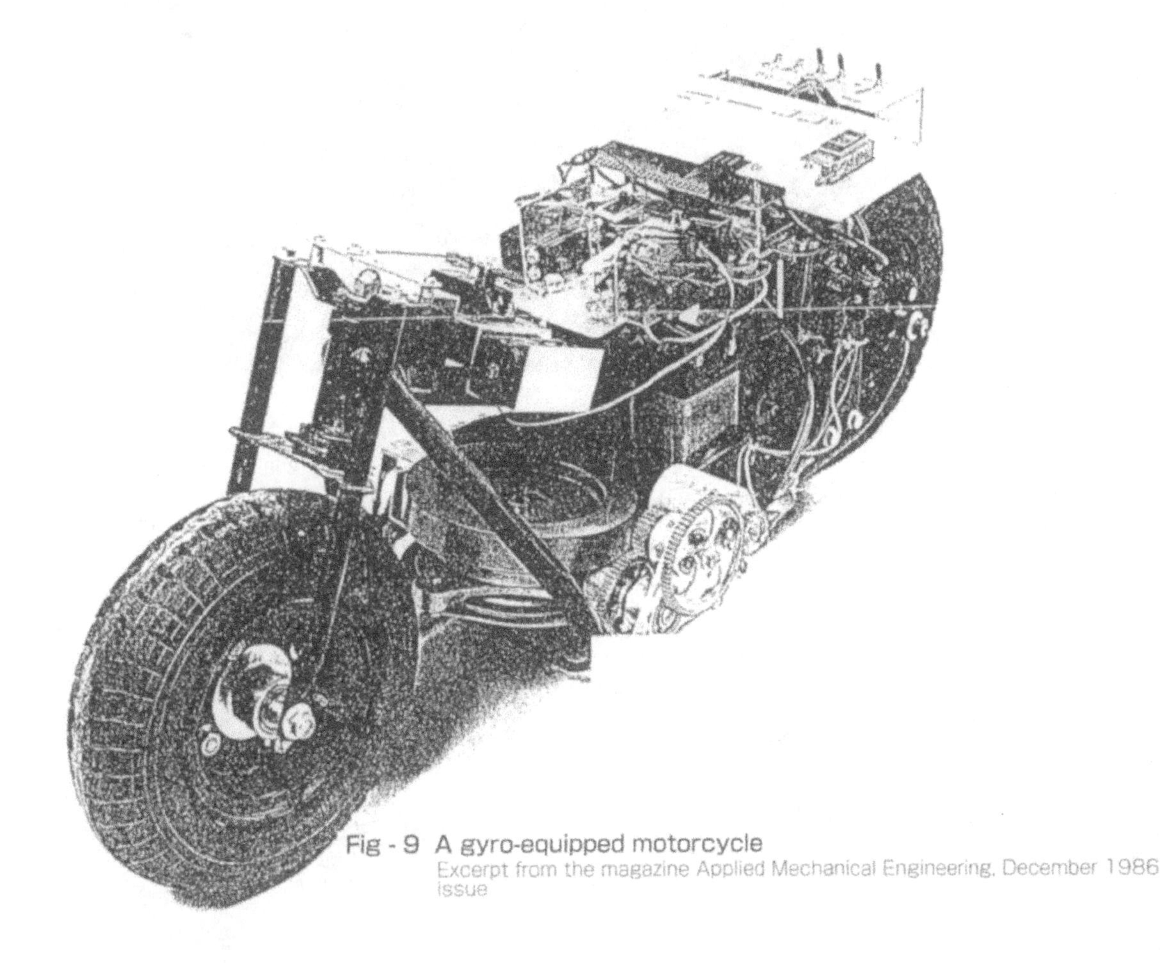

Fig - 9 A gyro-equipped motorcycle
Excerpt from the magazine Applied Mechanical Engineering, December 1986 issue

vehicle. So we considered a complete different use of the gyro.

As is well known, we can drive a two-wheeled vehicle stably using the gyroscopic effect of its front wheel. The theory is as follows: If you incline its body to the right side when the front wheel is spinning fast, the front wheel turns right due to the gyro effect. Its body turns to the right, which generates a centrifugal force to roll the body to the left side. Thus, the body inclined to the right side, returns to the upright position. If the front wheel is heavier and larger, the gyro effect increases. That is why a large motorcycle has that quiet feel of solid stability.

You may have driven a bicycle hands free. In this situation, if you sway your waist to the left side when driving a bicycle straight, the saddle moves left causing the bicycle to incline left. The front wheel turns left due to the gyro effect. This is to say, that you can control the front wheel by swaying your waist to the left or the right, enabling you to ride a bicycle hands free.

The scooter has a light and small front wheel and a smaller gyro effect, and its steering handle moves in an unstable manner. The reason is that the gyro effect of the front wheel of the scooter is one-fourth of that of a 250-cc class motorcycle at the same speed.

We intended to use the front wheel of the scooter for the OR49. The seat height of the OR49 was only 35 cm from ground level, so we needed to improve its lateral stability. For this purpose, we adopted a gyrostabilizer with new ideas. We chose a method in which the gyrostabilizer wasn't fixed on the body but which assisted in the gyro-effect of the front wheel.

The gyro effect of the front wheel was increased by an additional rotating flywheel powered by an electric motor, which was con-

Fig - 10 Gyrostabilizer

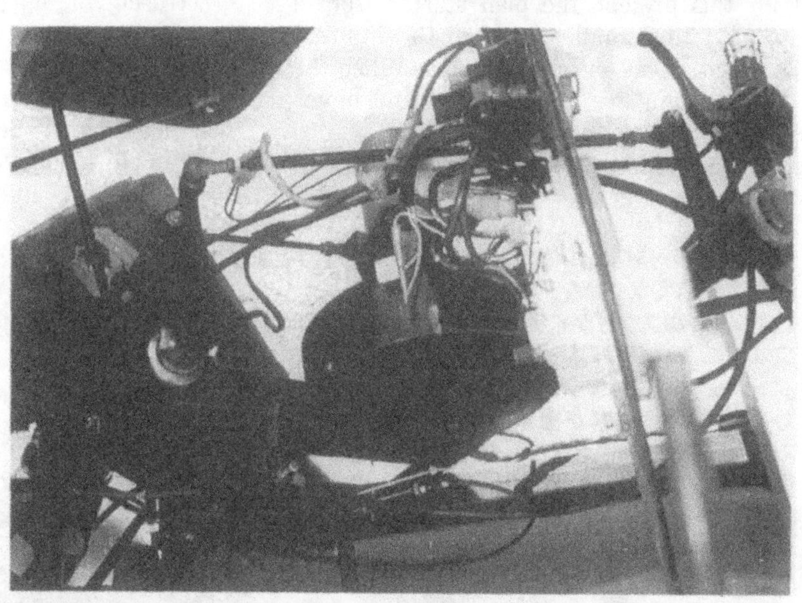

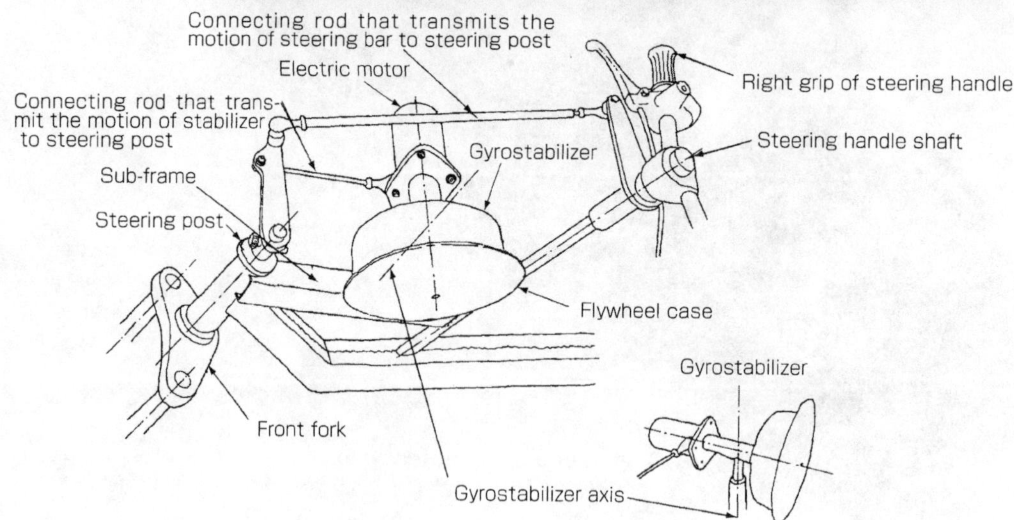

nected to the front fork (Fig - 10).

A flywheel with a diameter of 20 cm and weight of 5 kg spinning at 3200 rpm generates the same gyro effect as that of the front wheel of a 250-cc motorcycle running at 30-km/h speed. This effect continues even at a speed of less than 10 km/h, its stability improves and exceeds that of a normal motorcycle. This resulted in remarkable stability, as we came to know later.

When we started, our image of the system for our gyrostabilizer was as follows: the higher the revolution of the gyro, such as ten or twenty thousand revolutions per minute, the more compact the system would become; but higher revolutions need a more powerful motor because of the large air drag. The flywheel should therefore spin in a shielded chamber which is almost fully evacuated so that a compact and smaller-powered gyrostabilizer could be used. However, we did not follow up to that level since our priority was to finish the body.

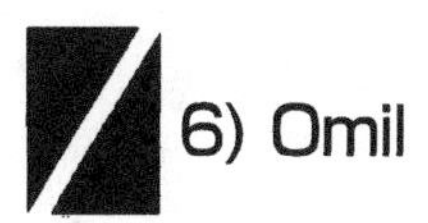

6) Omil

At the beginning of 1987, we had some talks with Pelaves Co. in a small village named Binteltour, 25 km from Zurich in Switzerland

Mr. Arnold Wagner, the President of Pelaves Co., was eager to develop and sell an all weather type two-wheeled vehicle using engine and parts around the wheel from a 1000-cc BMW motorcycle. He named the vehicle Omil short for OEKOMOBILE.

The concept of Omil was similar to our OR49 and the developed product was almost complete. We considered taking up Omil as one of the products of Yamaha Motor and negotiated to buy one Omil. Mr. Wagner was originally an aircraft engineer and fond of motorcycles. He had ten motorcycles, five of which were made by Yamaha. This was the start of our relationship with him.

The number of products was inadequate

Fig - 11 Photo and three orthographic views of Omil
Excerpt from EINSPUR-ZEITUNG, Feb. 1990 issue

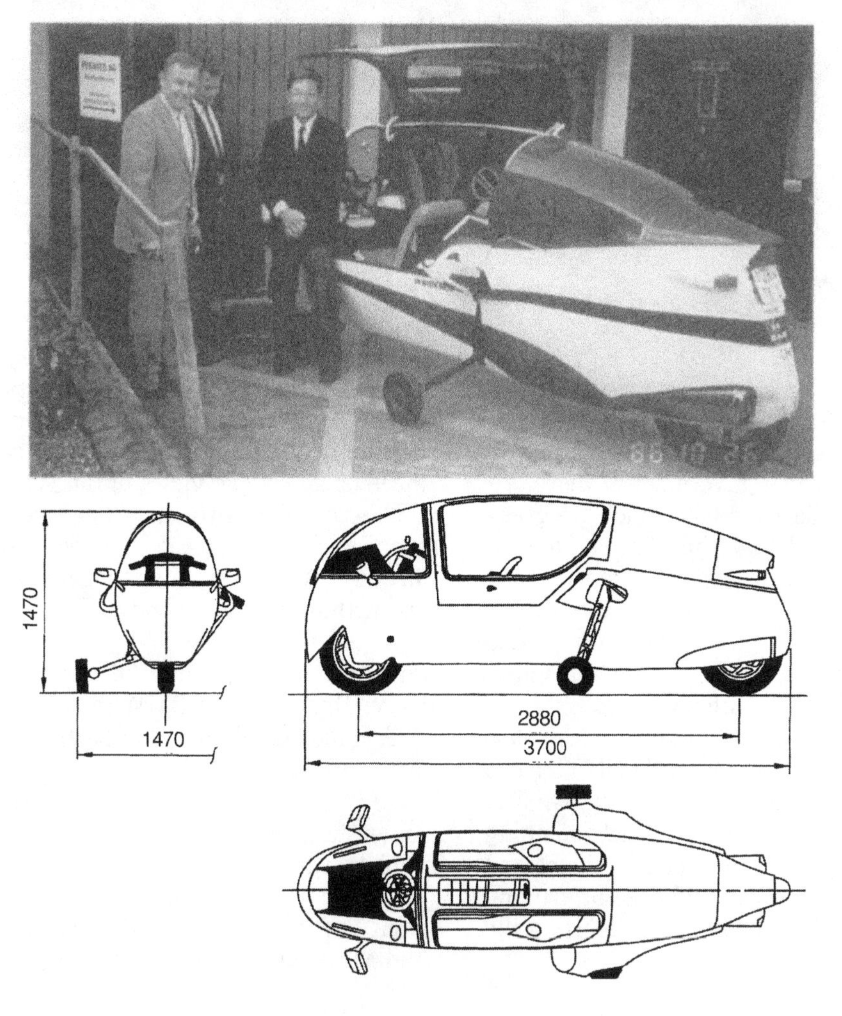

Fig - 12 Omil in a bank
Excerpt from EINSPUR-ZEITUNG, Sep. 1988 issue

Fig - 14 Bare chassis (prototype without body) of the OR49

Fig - 13 Omil driving in line
Excerpt from EINSPUR-ZEITUNG, Feb. 1990 issue

because very few people were involved in its trial production. They were also very busy trying to get the type approval certification from the Swiss and German governments, therefore, the negotiations did not make much progress. In the meantime, I had the chance to visit Europe and I stopped over at his factory.

Mr. Wagner was pleased to guide me around his factory and gave me an opportunity to ride the Omil. On this occasion, I came to understand the overall image of his vehicle.

I would like to describe here the main features of Mr. Wagner's vehicle, Omil. You can see a photograph and a sketch of the Omil in Fig - 11. The photo shows the vehicle standing with its windshield in the open position, and small auxiliary wheels (side wheels) protruding from the center of the body and supporting the vehicle.

Mr. Wagner is on the left, Mr. Bean, a test driver from Yamaha is in the middle, and the author is on the right. Also, you can see the three orthographic views of the vehicle.

The vehicle has two seats in tandem and the shape of its windshield is similar to that of an aircraft. When running, the auxiliary wheels are withdrawn in the body (Fig. 12) and it was reported that the vehicle could bank inside at an angle of more than 50 degrees, which is larger than that of a BMW motorcycle.

A K100 BMW engine of 90 HP was installed in the vehicle, capable of deriving a maximum speed of a remarkable 260 km/hr. I could understand Mr. Wagner's statement that the vehicle was a sports car like a Ferrari.

During the trial run, Mr. Wagner was in the front and I was in the rear seat. Mr. Wakano and Mr. Bean, who were accompanying me, followed us in a Mercedes observing its running behavior.

It was unexpectedly quiet in the cabin but the auxiliary wheels had to be taken up and down frequently because we drove in the town initially. When speed reduced, Mr. Wagner pressed a switch on the handle bar to put out the auxiliary wheels. Every time the switch was pressed, the auxiliary wheels went up and down in 0.5 seconds with a slam. I was astonished with the sound and its shock for the first time, but got used to it later.

Although I had the feeling of riding almost like a car, I was scared initially when the vehicle banked but became used to it.
After passing through the town and entering the highway, he promptly accelerated the vehicle to 140 km/hr. The feeling of acceleration was similar to that of a motorcycle. I was concerned a little that the vehicle might sway slightly because of the inevitable disturbed airflow when following a large vehicle such as a trailer, but there was no harm. I was surprised that the vehicle was influenced by the rear airflow even when the large moving object was 1 km ahead.

According to Mr. Wagner's explanation of the side wind effect, they tested the vehicle at the BMW wind tunnel and found that its rolling moment was slightly greater than that of a motorcycle while its course deviation was less than half that of a motorcycle. I thought this was not a problem.

We drove at more than 200 km/hr on a course where few cars were running and I did not hear any wind noise, and thought that it ran with good stability. I thought the vehicle was a very good product.

Next, we got off the highway and climbed a mountain. The course was a wet winding road where snow still left remained on the shoulders of the road. I enjoyed the climb with the vehicle banking left and right just like a motorcycle. Since the vehicle had good acceleration characteristics and a narrow body, it overtook cars in an instant, an action that excited and pleased me.

The movement of Omil's auxiliary wheels was very interesting in the context of our OR49. Omil had a simple system in which its auxiliary wheels moved in and out when the driver pressed a switch. If the driver forgot to press the switch, the vehicle will fall. I was impressed however, because the simple system worked well and was useful. I thought we might have elaborated too much for our system. On the other hand, it was good to see that even this simple system worked.

Later, Omil was given type approval in Switzerland and is now being produced although the number is not large. Subsequently, it is to be type-approved in Germany also.

Several Christmas cards that Mr. Wagner sent me every year showed more than ten Omil vehicles driven in line (Fig - 13) and also indicate that they enjoyed driving and meetings. The vehicles were also making many users happy. Mr. Wagner must have succeeded in attracting users who enjoyed Omil's high performance, and in creating a select group of users. I sincerely wish him success in the growth of Omil and in the growth of users in the future.

In 1988, Omil was introduced in detail in the US magazine Cycle World and was also introduced in a Japanese magazine in the same year, which mentioned that a Japanese importer for the vehicle had been decided. Later I called the importer and came to know that they were having problems in getting certification from the Japanese Ministry of Transport. So far, I have not heard of Omil in the market; they must have given up the idea, which is sad.

7) Prototype of the OR49

In 1988, the year after we experienced the ride on Omil, we started to build the OR49, a one-seater all weather type lean machine. After the basic research, Yanagihara started practical design work from April. I was too busy in other projects and did not find time to work on this design, but I discussed it with him. The bare chassis (prototype without body) of the OR49 was completed in the middle of May.

On May 16, we brought the prototype to the roof of our building and the members of the Horiuchi Lab conducted trial runs. It had stable handling and appeared to have good maneuverability, although its seat height was only 35 cm above the ground level (Fig - 14).

On May 26, we tried the prototype installed with a gyrostabilizer. As we expected its stability was excellent, and we were able to drive it hands free. Its acceleration could be controlled by the foot pedal so we could drive it hands free, circling on the roof as long as we liked.

Yanagihara tried driving it hands free in a figure of eight shape. As the width of the build-

Fig - 15 Photo of OR49 - type 1

Fig - 16 OR49-type 1 driving hands free

ing was 22 m, he was able to do the figure of eight shape hands free within this width. The length of the figure of eight shape was less than 20 m, and it could still do the eight-shape even when the length became smaller and smaller. This kind of performance cannot be not be attained in a bicycle and of course, in a motorcycle. The stability at slow speed of the vehicle was excellent. We admired the wonderful effect of the gyrostabilizer.

I thought that if this system were to be installed actually, the low seat height and the small front wheel would not be a problem. As

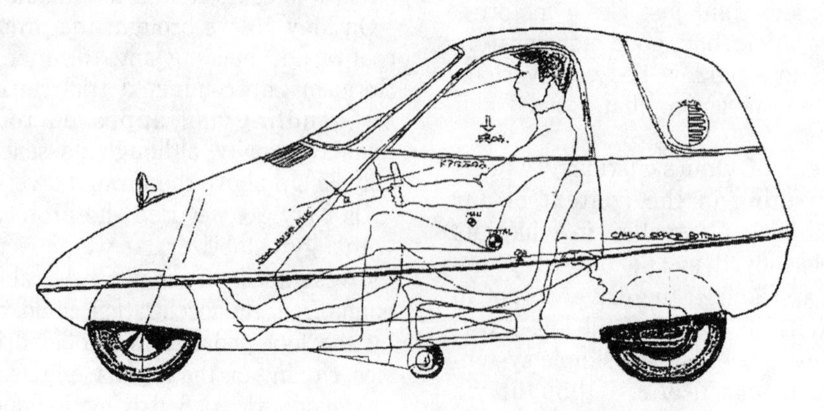

Fig - 6 Lean machine - first proposal (re-printed)

the vehicle was very stable until walking speed, we could easily prevent it from falling by either extending the auxiliary wheels or have the driver extend a foot when the vehicle came to a stop.

At this point, we became convinced of the success of this project because we were fully satisfied with the output of the bare chassis prototype, and Yanagihara soon started the design and proto-building of the complete model, based on the first proposal of the lean machine (see Fig - 6).

The body was to be made of FRP and windshield was to be made of acrylic panel. Both parts needed moulds. Also, as these parts had many openings, many metal parts would be necessary to open, shut, keep, and fix them.

The OR49 type 1 body was completed in September (Fig - 15). Until then, we had made many attempts with auxiliary wheels to prevent the vehicle from falling when it came to a stop. You can see one such structure (see Fig - 6). The system was as follows: When the driver bends his knee to pull in his heel, the auxiliary wheels touch the ground because of the weight of the leg. The driver then extends his feet to support the body when the auxiliary wheels are locked after pulling the parking brake.

Unfortunately, when running through many tight curves, the auxiliary wheels with small width often generated noise and wore out because of sideslip, since the auxiliary wheels were located too far in the front. In some rare cases, the vehicle fell because the auxiliary wheels were not able to support the body when the vehicle banked excessively.

Fig -17 Force-measuring instrument

Although we knew that it would be better to land the auxiliary wheels on the ground far from the body as in Omil, we wanted to prevent extending the wheels outside the body width considering Japanese traffic conditions. Looking back, I think now that I stuck to that idea too rigidly. It would have been better to adopt the system used in Omil.

Finally, we cut off nearly one-third the area under the side doors where the driver could extend his feet and support the body. It was a halfhearted solution.

The stability of the vehicle was excellent after the gyrostabilizer was fitted. Similar to the bare chassis model, the OR49 - type 1 model could be driven hands free in a figure of eight shape easily (Fig - 16). Therefore, it was easy for the driver to extend feet to support the body only at the last moment.

After its trials under various conditions, the direction of our development was summarized by the following two points:

1 We wish to proceed with the full fairing (fully covered) type. For this purpose, we have to design smaller doors from where the driver's feet can extend when necessary.
2 As we are concerned about side wind, we want the profile area to be smaller and want to move the center of the profile area more toward the rear.

To confirm 2 aerodynamically, wind tunnel test was essential. So I thought it would be natural to decide the outer shape through wind tunnel tests.

8) Wind tunnel test

We had no large wind tunnel facility in Yamaha those days. When designing motorcycles, the designers rented and used the wind tunnel facility of the Japan Automobile Research Institute (JARI), in Yatabe, Ibaragi Prefecture. But Yatabe was far from our office and we incurred considerable expenses. JARI's facility was too large for our tests that were required to be performed repeatedly after modifications and measurements.

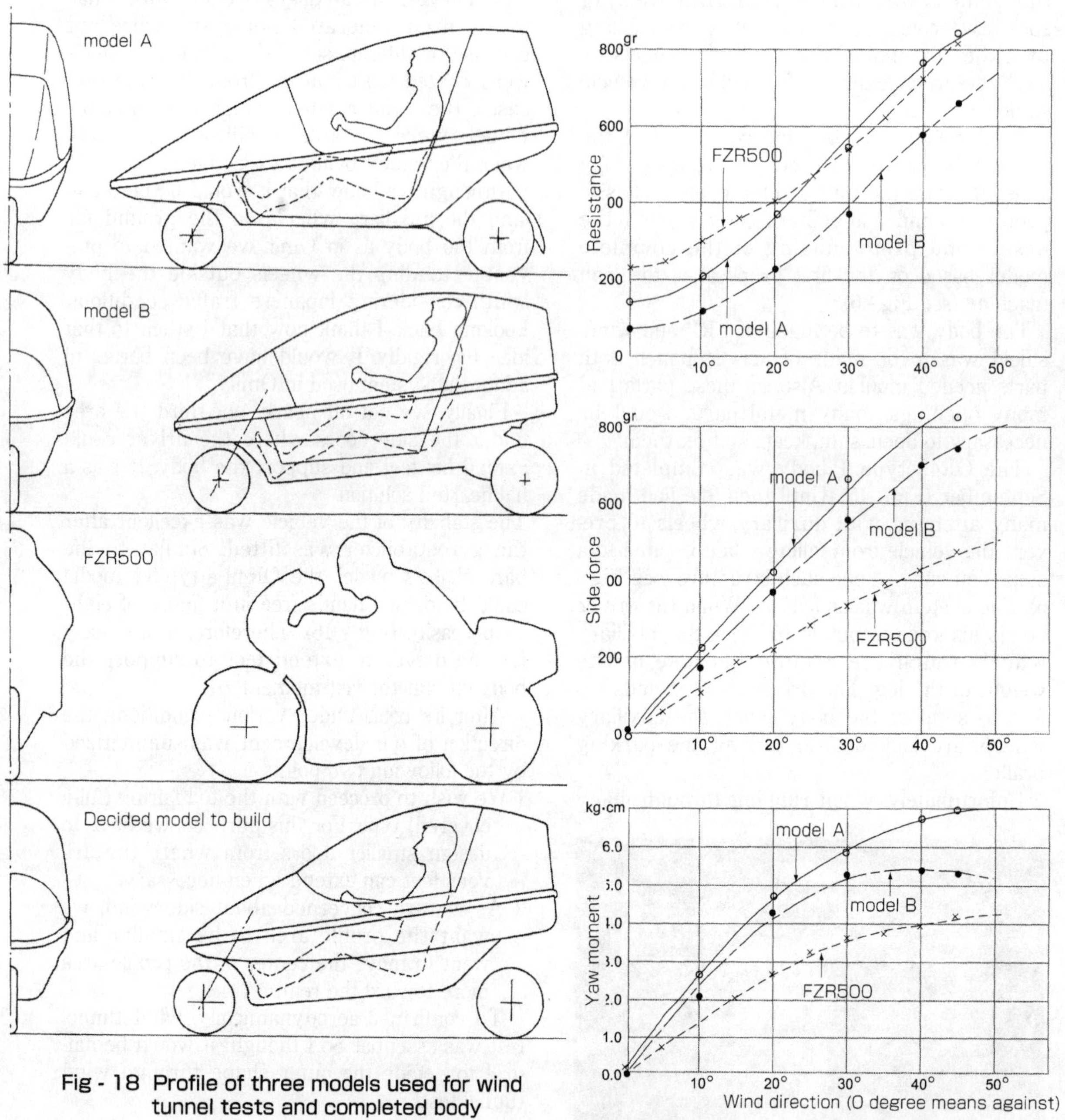

Fig - 18 Profile of three models used for wind tunnel tests and completed body

Fig - 20 Results of wind tunnel tests

Model	Front area	Cx
Model A	276.6 cm²	0.41
Model B	250.6 cm²	0.21
FZR500	116.0 cm²	0.60

Fig - 19 Front area and drag coefficient (Cx) of each model

On the other hand, we had already had a simple wind tunnel facility near our boat factory. I had requested Shigemitsu Aoki to make this facility for finding the center of wind pressure of the boat, on which side wind acts and for finding a measure to prevent splashes that engulf the boat cockpit aft of the superstructure. This wind tunnel had been used also for observing airflow around helmets for motorcycles. We decided to use this simple wind tunnel facility for our project

This simple wind tunnel had an airflow section of one-meter square and its wind speed was 15 m/s produced by a 10-hp motor and hand-made wooden propeller. It was a very small-scale facility with all equipment installed in a canvas tent. Its original force-measuring instrument was very large and less accurate than the new force-measuring instrument (Fig - 17) that Yanagihara had designed and made by himself for our purpose, which could measure three components (drag, side force, and yaw moment).

We prepared three one-fifth scale models for this wind tunnel, namely Model A, Model B, and the plastic model of a motorcycle (a model of Yamaha FZR500) on the market (Fig - 18).

The results of measurement can be seen in Fig - 19. Although the area around the wheel covers was not adequately streamlined, the drag coefficient of Model B was extremely small (Fig - 19 and Fig - 20, upper graph).

On the other hand, the side force was smaller in the motorcycle model (Fig - 20 middle graph).

Also, the yaw moment (around the center of gravity) was smaller in the motorcycle model (Fig - 20 bottom graph). The yaw moment is presumed to be nearly proportional to the course deviation (the amount the motorcycle is moved downwind). The wheelbases of Model A and Model B, however, were designed to be 24% longer than the motorcycle model, which adds to the resistance against yaw moment, and thus the actual difference in deviation among Model A, Model B and the motorcycle model becomes very small.

It was reported that rolling moment of Omil was slightly larger than that of a motorcycle and its course deviation was less than half that of a motorcycle. Since Omil had a wheelbase of 2.88 m, about twice that of motorcycle, I could accept their data.

Model B had a smaller drag coefficient and yaw moment compared to Model A. So we decided to go ahead based on Model B. As the vehicle would be used at moderate speeds and with smaller influence of side wind, the aerodynamic characteristics of Model B were satisfactory for our purpose.

To further improve the characteristics, Yanagihara consulted Mr. Ohmori, our consultant (an authority on aircraft engineering; see Chapter 11) and tried to add steps outside the body, modify the body section, add a spoiler (an obstacle plate), and a small upward inclination at the end of the body to compare the changes in side force and yaw moment.
Although each of his attempts did have the effect desired, they were not adequate, and unfortunately, they did not bring about large improvements.

Comparing Model A and Model B showed that the original shape practically governed the side force and the yaw moment. If we had continuously carried out tests while modifying the original body shape, we probably might have got better results.

But it was more important for us to confirm the feasibility of the prototype as a product. We decided to build the vehicle with the modified design based on Model B after shortening its nose (Fig - 18 bottom).

9) Building of type 2

As mentioned above, we decided on a practical measure for the influence of side wind, but we found it difficult to arrive at a reasonable solution for the question of full fairing. Finally, our decision was as described below (Fig - 21).

When the driver is driving the vehicle while sitting in a relaxed manner on the seat, his knee should be pressing slightly against the footboard. We concluded that this light pressure on the footboard could be used to open or close the small doors enabling the driver to put out/withdraw feet as necessary. That is, when the driver is sitting in a relaxed manner on the seat, the pressure exerted on the pedal keeps the small doors closed. When the foot is withdrawn to prepare for a stop, the door can be opened with the force exerted by a gas spring. (Fig - 22).

Yanagihara designed a smart four-pivoted link mechanism for the small door so that it lifted up while opening outward. This is to ensure an adequate opening size and also to ensure that the door does not touch the ground when open and when the vehicle leans at corners in the running condition.

Fig - 21 Front view of the OR49 type - 2 (Yanagihara is in the cabin)

Fig - 22 OR49 type - 2
— its small doors are open

Fig - 23 OR49 type-2
— its front windshield is open

Assuming the existence of the small doors mentioned above and considering the method of opening the main door for the driver to get on or get off, a construction by which the entire windshield unit swings up on hinges located forward of the body was found to be the best solution. The small doors accessed by the driver's feet also open together with the windshield unit (Fig - 23).

Getting on or getting off the vehicle was easy with this structure, but on the other hand, we were afraid that the vehicle would fall in a strong wind when its main windshield was kept raised while parked. However, we did not think this was a serious problem because the probability that the driver leaves the vehicle with the windshield kept open is small.

In this way, we finished type-2, which had a completely different body compared to type-1.

Yanagihara brought the vehicle to the Yamaha test course for trials. Thanks to the extremely small air drag, its acceleration was excellent and it reached a speed of over 100 km/h almost instantly although its power was just the same as that of an 80 cc scooter engine.

The gyro effect gave excellent lateral stability, good handling characteristics, and a comfortable ride. We also did not feel the effect of the side wind. The gyro would automatically cancel the rolling moment due to the side wind. When this stage was reached, I judged the OR49 as complete.

10) Marketability

This project started with the proposal made by Stevens. However, the OR49 was too small for the US market. If anything, the Omil or a two seater lean machine (Fig - 7) would suit his proposal better. On the other hand, I intended to develop as small a vehicle as possible, considering the restrictions of the Horiuchi Lab. Problems can be easily identified in small vehicles. Thus, if problems are resolved in the small vehicle, larger vehicles can be developed with comparative ease. In case of boats also, larger boats can be developed without major problems if we have studied a type ship (reference boat) thoroughly. However, a major problem may occur in a lightweight, small boat such as a rowing dinghy, the characteristics of which are largely influenced by crew weight, even if a small mistake is made. I thought the situation was the same for our project too.

I did not think that a market existed immediately for such small vehicle. But I feel that someday in the future this type of vehicle will be in demand in the market.

Nowadays, use of cars and consumption of energy are concentrated in the Western Countries and in Japan. However, the economic growth in China, India, and other Asian countries is rapid, and people in these countries will soon have their own cars. Since the population of these countries is much greater than that of the Western Countries and Japan, it is evident that consumption of energy, discharge of CO2, global warming and balance of these on a global scale will become issues, assuming that cars become popular in these countries.

It will become difficult for only the Western Countries and Japan to continue to grow and develop as before; we must take the lead in making the move to a vehicle that has a fuel consumption of one-tenth or even one-hundredth that of existing vehicles.

I think that considerable efforts have been made in improving engines and developing fuel cells, but these developments are not fast enough. I think the time has come for vehicles to necessarily have very small fuel consumption.

Cars used for commuting to work carry only one person. To transport one person, we use a car 20 to 30 times heavier than the person's weight. This is a basic problem, and if people in the world want to enjoy a "car society" in the future, they have to ensure that the weight of the car is reduced significantly.

I am sure that a vehicle like the OR49 will be in demand some time in the future.

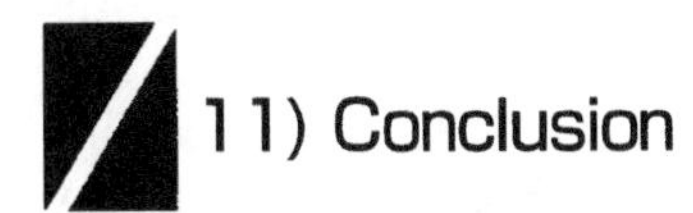

11) Conclusion

The weight of the OR49 when complete was only 85 kg, thanks to the elaborate work of Yanagihara. This is approximately the same weight as a person's weight. Therefore, with a new engine and with a very low fuel consumption, the vehicle will probably ease the environmental burden on the earth by 1/20 to 1/50. This was the target for the vehicle body, and I believe we have almost achieved this target.

A project like the OR49 may be launched probably ten years or twenty years later, we don't know. I believe firmly that this stage will surely arrive some time in the future.

13

THE CREST RUNNER STORY

I could not find practical ideas for President Kawakami's request for a boat that runs while piercing wave crests. Before long, I came to know that a large boat with the same concept had been developed in Australia. We also predicted good prospects for such boats, but smaller in size, by confirming performance through model tests, and also the advantages by building a prototype boat. However, unexpected difficulties related to commercialization of the boat awaited us.

1) President Kawakami

Yamaha Motor Co., Ltd., was established in 1955 and was the newest of over ten motorcycle manufacturers. At first, Yamaha Motor started business with only one production model, namely the YA-1, a 125-cc motorcycle with a two-stroke engine (Fig - 1). Several months later, Yamaha Motor won the Mt. Fuji mountain climbing race and the Asama Heights race, two of the most famous motorcycle races in Japan at that time, defeating Honda Motor, the top motorcycle manufacturer at that time.

As a result of winning these races, Yamaha Motor's sales increased. People in the company thought that the success achieved after starting the company was entirely due to the challenges taken up by President Kawakami at that time. He himself directed developments and also participated in endurance tests. His desire to take up challenges is well known and many of his episodes are still discussed in the company. In those days, I was not with Yamaha, but I was very impressed by its overwhelming victories.

Five years later in 1960, President Kawakami again took up the challenge and started the boating business. At that time, I joined Yamaha as a member of the design staff. Before the Yamaha boat presentation (product launch) in April 1960, President Kawakami was

Fig - 1 YAMAHA YA - 1

Excerpt from a Yamaha story in Exciting Bike, edited by Yanagawa Research, published by Bike Connection.

on board the new boat himself to travel from Lake Hamana to Tokyo, a distance of nearly 300 km, and to be welcomed by a military band at the Takeshiba Pier. However, because of rough seas, he had to land at Shimoda, and subsequently, he went to Lake Ashinoko by land where the presentation was scheduled. He went on board boats by himself at the slightest opportunity.

Boats were often used to study other new businesses. President Kawakami cruised to Toba and Ago Bay frequently, and as a result of these cruises, the Toba International Hotel, and Nemuno Sato, a large resort were set up. Naturally, we had to go out to the rough Sea of Enshu by boat whether we liked it or not, and we had some scary encounters, but the rough sea was also interesting. By studying boat behavior and experiencing high waves, we felt

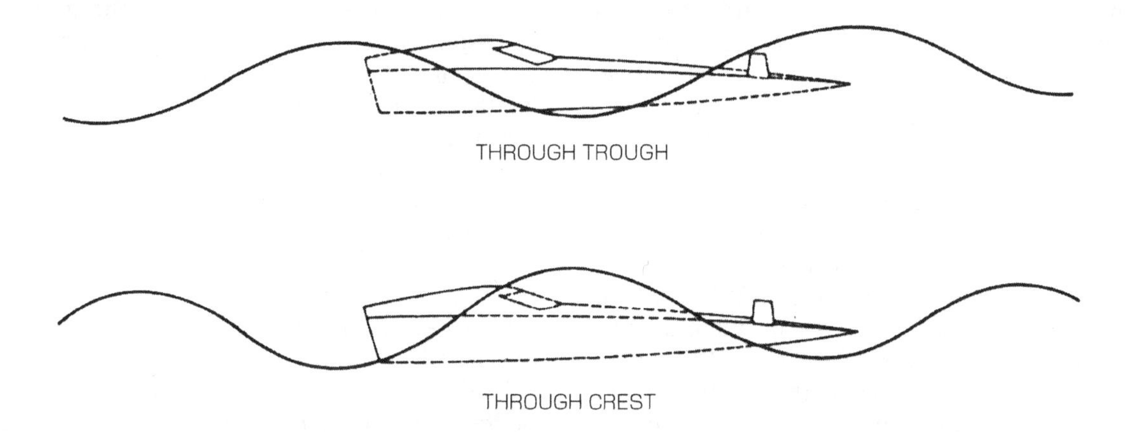

Fig - 2 Wave-piercing boat sketched by Levi
Excerpt from the book Dhows to Delta, Page 220, author: RENATO LEVI

fear for the first time, and our bodies could sense which wave was dangerous and which one was safe. Gradually, we began to enjoy cruising in rough seas.

Since we were accustomed to rough seas, we won comprehensively (all three days) in the 1000-km offshore races between Tokyo and Osaka in both 1961 and 1962. These victories gave confidence to Yamaha boats and the company's market share increased within a short period so that Yamaha became the top boat manufacturer in Japan.

President Kawakami himself participated in the development of new products in both motorcycles and boats, and he pushed the business forward. With the victories in races, confidence in the company's products increased, and business picked up. That was Kawakami's method as well as Yamaha's method of doing business.

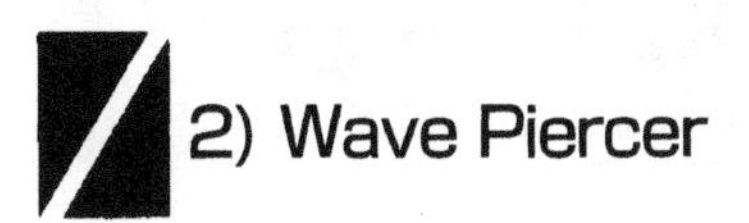

2) Wave Piercer

As mentioned above, President Kawakami had in-depth knowledge of boats after riding them and his suggestions were reasonable. Our work progressed smoothly as he understood boats and accepted our proposals of new hull forms. However, there was one item ordered by President Kawakami for which we had not yet arrived at a solution. He was not comfortable riding boats in rough seas and he asked me if it were possible to develop a boat that could run through wave crests without impact. I studied this question hard but it was not easy to come up with an answer. To realize such a boat, we have to build a very slender hull like an arrow with a fully sealed structure keeping out the water, like a submarine. Could it be a pleasure boat? I did not think so.

In the meantime, I found the same concept in a book written by a famous Italian naval architect, Renato Levi (Fig - 2). It was a submarine type boat, just as I had been thinking. But was it necessary for a pleasure boat to run in such a style and by such a method?

Levi drew the sketch visualizing a racing boat, capable of a speed of 130 km/hr. As he was a naval architect designing racing boats as well as a racer, it was a reasonable dream idea. We were fully occupied with the designs of many popular pleasure boats, and we could not afford to dream of such a design that was so far displaced from our normal designs.

Around 1980, we came to know of a British magazine named High Speed Surface Craft (hereafter called HSC), which contained articles on high speed craft, catamaran, hydrofoil boats, hovercraft, and so on. We read this magazine with great interest. Since nearly 1985, we saw some articles and advertisements for an unusual craft called a wave piercer (craft running through wave crests) in this magazine. According to the Incat Design Company that designed the craft, the wave piercer named Spirit of Victoria had a length of 28 m, and a hull that could pierce the crests of waves (Fig - 3).

I, as well as Levi, had thought of a full boat that ran through the crests. On the other hand, the idea of the wave piercer was quite different. As you can see from Fig - 3, the craft has two slender sponsons (side floats) that pierce the wave crests, and a high cabin is located on top of the sponsons. No watertight cabin is needed for this layout. I do not know why we did not think of this layout although our president had made the same suggestion earlier. It was not necessary for the cabin to run through the wave crests.

For this reason, I was very much concerned about the wave piercer. Some time later, I found an article on the accident to the Spirit of Victoria in the HSC magazine. According to the article, when the Spirit of Victoria was cruising at night through waves of 3-m height, it slid down along a following wave, its bow ran through the crest, and the shock due to deceleration caused several seats to be torn off, injuring several persons. The injuries were non-fatal but again looking at the photo of the craft carefully, we thought that such an accident was bound to happen. When sliding down along the front slope of the wave, the narrow sponsons would run through the forward crests, and the blunt face of the cabin would hit the wave strongly. This is a scenario that can be thought of naturally.

Since I admired the concept of the wave piercer, I felt very sorry for the craft and its designer after the accident.

I saw an article again about a new wave piercer, probably around the end of 1986. The craft looked smart and attractive, unlike the Spirit of Victoria.

Unlike craft built earlier, the new wave piercer had a fine conventional bow, by which its sponsons that tended to submerge, could be pulled up soon. This could be noted at a glance. I felt a strong desire to study the details of the craft.

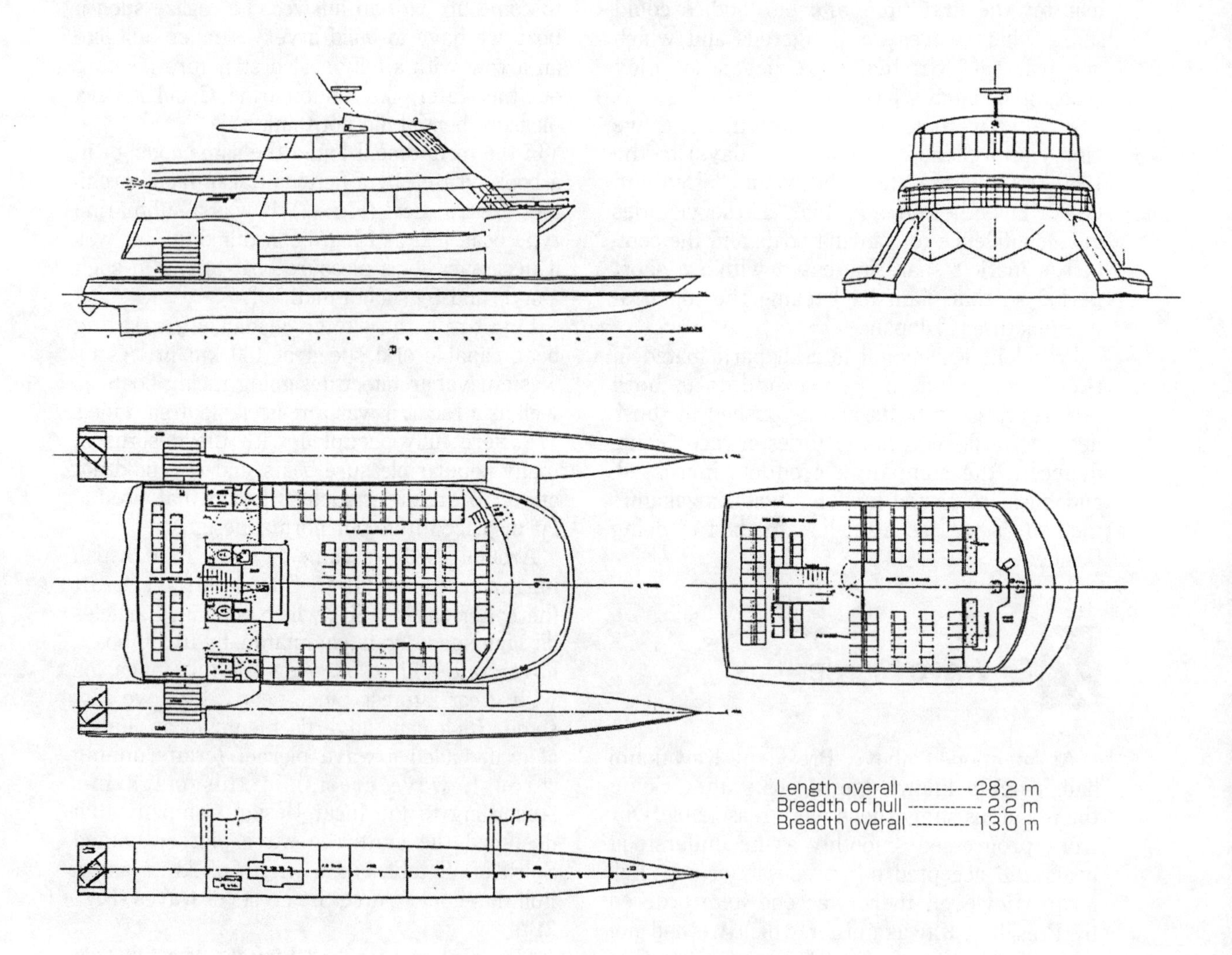

Fig - 3 Spirit of Victoria

Fig - 4 Located Tassie Devil 2001

Located Tassie Devil 2001 running at speed with full load of observers on board

Fig - 5a Tassie Devil 2001 arriving in a port

Fig - 5b Tassie Devil 2001 arriving in port

Fig - 6 Meeting with Mr. Hercus

From left, Mr. Phillip C. Hercus, Kazuyuki Higashijima (Yamaha), Mrs. Taeko Imura (interpreter), Mr. Geott Eldrid (Yamaha Australia).

3) To Freemantle

At the Directors' lunch meeting in January 1987, President Eguchi requested me to go and see the America's Cup Race. I hesitated initially, but finally accepted the offer. To see and study the race I visited Fremantle, a harbor town near Perth in southwestern Australia.

I stayed in Fremantle from February 2 to 8 observing and studying the America's Cup Race. There were many observation boats at the race full of people, and cruising in 2 to 3-m waves.

On one such day, I located the wave piercer. The craft, with the name TASSIE DEVIL 2001 painted in large letters, was running at high speed carrying a full load of people. It looked wonderful to me (Fig - 4).

For the America's Cup race, a tough battle between the Australian Kookaburra and the American Stars and Stripe continued on the rough seas, and finally Denis Connor's Stars and Stripes triumphed and they took back the America's Cup with them.

I wanted to take a ride on the wave piercer and looked for an appropriate contact, but I could not get in touch with the contact before the race ended. Later, when I went for a stroll along the harbor after the race, I found TASSIE DEVIL 2001 arriving at the port. I traced the craft to its pier and finally studied it closely from the outside (Fig - 5). I soon asked for permission to visit the craft and got it the next day, as far as I remember.

On the day of the visit, I met Mr. Philip C. Hercus, the designer of the craft on board and I conversed with him (Fig - 6). He said that the craft had been launched at the end of last year and had soon sailed to Fremantle. He also said happily that during the America's Cup Race, the craft was put into service for observing the race, had taken many observers on board, and had been received very well by passengers because of the comfortable ride.

According to him, the unconventional shape of the craft attracted observers onboard and since the heaving motion was minimal (Fig - 7), observers could even watch the race from the bow deck. This high acclaim from the passengers immediately after the launching of the craft probably made the designer very happy. Mr. Hercus' explanations helped me understand the advantages of Tassie Devil 2001 very well. It was a very useful meeting.

I heard that there had been many inquiries

① Typical deep vee (30 degree) planing craft
Length Water Line = 30 m
② Round bilge patrol boat
Length Water Line = 30 m
③ SRN4MK3 Hover craft
Loading 400 persons and 50 cars
④ Incat conventional catamaran
200 persons
⑤ Surface piercing hydrofoil
200 persons
⑥ Wave Pierver 2001
200 persons

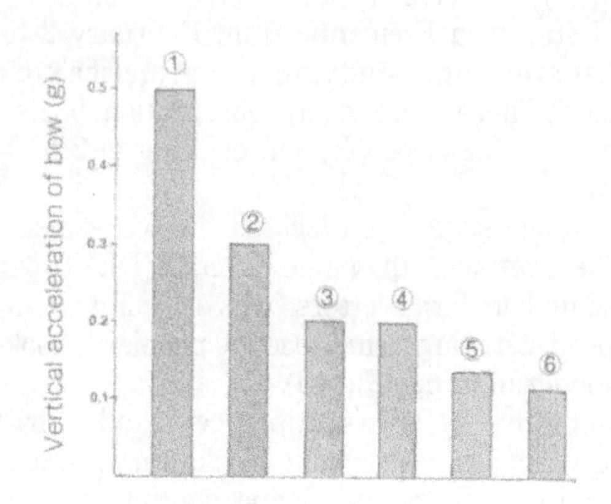

Source : CATAMARAN UPDATE Sept. 1988, author: Phillip C. Hercus

Fig - 7 Very little bow motion

for the craft from transport companies and shipbuilders all over the world. The demonstration at Fremantle was very effective. I heard also that the craft was to be chartered at Hamilton Island for three months after a few modifications. It was a very exciting experience for me to be at the place where the wave piercer made its glamorous debut.

4) Study of the wave piercer and model tests

I understood the concept of the layout of the wave piercer. Mr. Hercus separated the cabin from the sponsons, which were connected to each other by struts so that only the sponsons pierced the waves, as opposed to the idea of the complete hull including cabin piercing waves, which Levi and I had, and his idea was excellent.

My concerns were: What was the resistance of the craft and what horsepower did it need? Was there any advantage in using this layout in the small boats that we were building?

Generally, a catamaran has a large wetted surface and tends to be heavy compared to the conventional mono hull. The outer surface area of the boat is approximately proportional to boat weight and the building cost. On the other hand, the volume of the hull is directly related to the number of passengers carried, number and area of cabins, size, and payload. Therefore, a catamaran is rather expensive and the wave piercer is even more expensive. Mr. Hercus emphasized that his boat attracted more passengers because of its appearance, but this was a difficult point to consider for our case.

Another concern was performance such as speed. A catamaran has larger wetted surface compared to a mono-hull; therefore, the frictional drag of the catamaran tends to be larger. For example, when we compare a catamaran and a mono-hull both with a semicircular sec-

Fig - 8 Wetted surface area of the catamaran is 1.4 times that of the mono hull

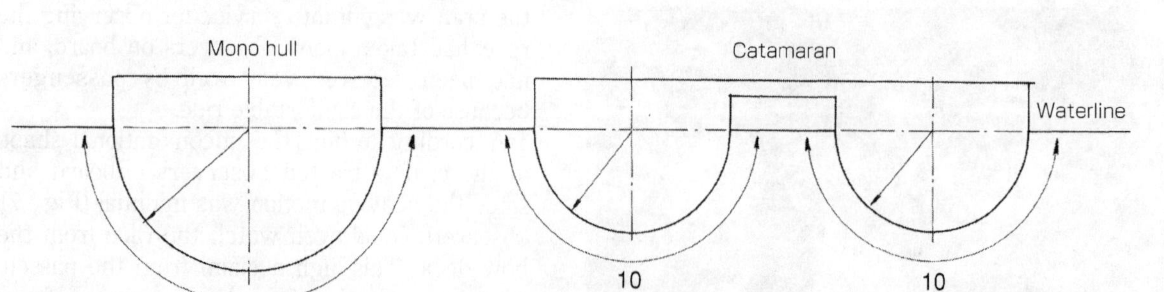

Sections below waterline of mono hull and catamaran are assumed to be semicircular (minimum girth length, minimum wetted surface area). If both crafts are to have the same length and the same displacement, the sectional area below waterline of each craft must be the same. If the radius is set so that the above condition is satisfied, then the ratio of girth length below waterline becomes (10+10)/14 ≒ 1.4

tion under the waterline and with the same waterline length, the wetted surface of the catamaran becomes 1.4 times that of the mono-hull (Fig - 8). On the other hand, the catamaran has less wave-making drag because of its slender hull. The total drag depends on the balance between friction and wave-making drag.

What makes it complicated is the size difference. For example, if the wave piercer has a 100-ft (about 30 m) length on the waterline and is capable of a 30-knots speed, its speed-length ratio (ratio of speed to length) is calculated by the following formula:

Speed-length ratio of wave piercer 2001 = $30\text{-kt}/\sqrt{100\text{-ft}} = 3.0$

And if our craft has a 25-ft (about 8 m) length on the waterline and is capable of 30-kt speed, then:

Speed-length ratio of the 25-ft craft = $30\text{-kt}/\sqrt{25\text{-ft}} = 6.0$

If the speed-length ratio of both boats is to be made the same, the shapes of the bow wave and stern wave become similar. If the hull form of the 100-ft craft is ideal, then a similar hull form is ideal for the 25-ft craft. Accordingly, we can verify the performance of a large craft by testing a small-scale model with the same speed-length ratio.

However, if speed-length ratios are largely different such as 3.0 and 6.0, the ideal hull form for each case is also different. It is reasonable to consider that said craft is in the planing mode if its speed-length ratio is over 6.0.

We even considered buying the design of the Tassie Devil 2001 for building and selling it in the domestic market. But according to Mr. Hercus, another Japanese company was already negotiating to import the craft; moreover, we did not have the capacity to build such a large aluminum craft at Yamaha. It was natural to consider application of the technique to small craft.

Fumitaka Yokoyama was in charge of this project. He had also been in charge of hydrofoil craft. He made eight different models for towing tank tests and investigated their drag. Initially, we were concerned that the drag would be larger than that of the mono-hull, but by modifying the aft bottom shape of the round bilge hull, the hull could plane and coming up on the water, its drag decreased considerably to become much smaller than that of a mono-hull planing craft at moderate speeds (Fig - 9).

After obtaining good results for drag performance, we needed to verify the general performance parameters. For small craft especially, capability of tight turns is essential. If this is

Fig - 9 Resistance of new hull form

This figure compares resistance of a new hull form and a conventional planing craft of 8m-hull length and 4-ton displacement. As the new hull form is slender, it has less drag at speeds in the intermediate range (20 to 30 knots).

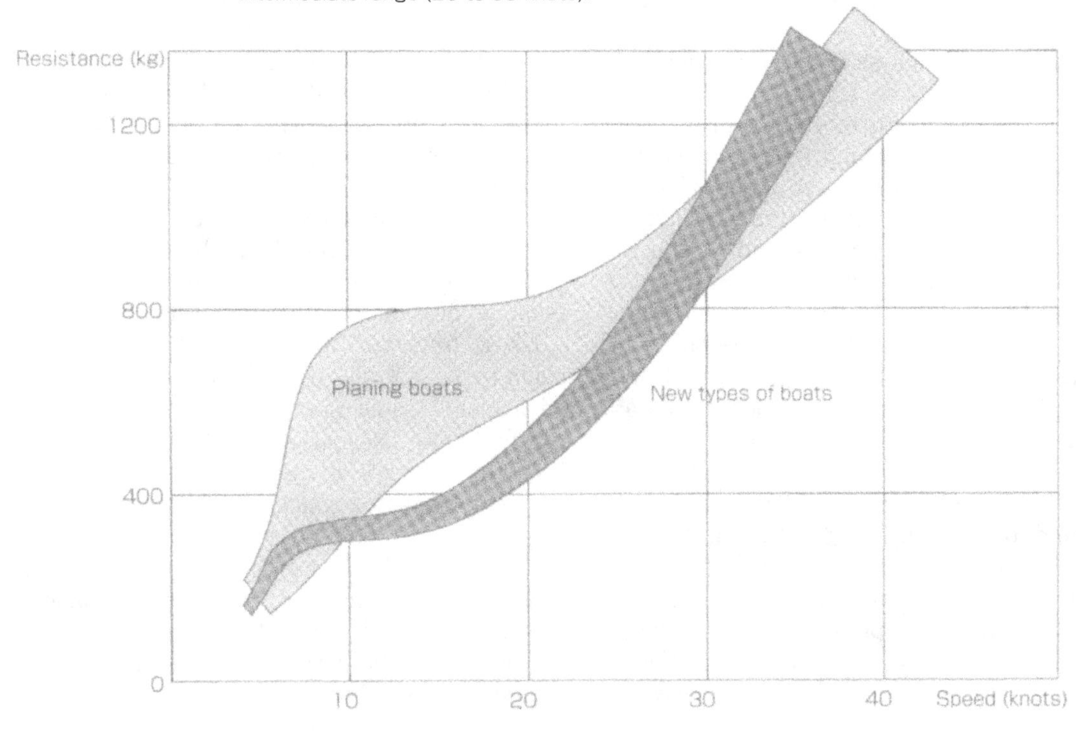

inadequate, the craft may not be acceptable as a pleasure boat. To verify this capability, we need to test the actual craft with people onboard and run it in the sea.

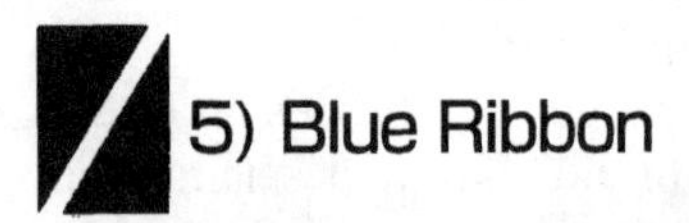

5) Blue Ribbon

Since the Titanic, the trans-Atlantic cruise has been the ultimate dream of European people. When airplanes were not commercialized, the fastest way to travel between Europe and America was by ship. Technically, a ship with greater waterline length has a greater speed; therefore, a large luxurious cruise ship possesses attributes such as most luxurious, largest, and above all fastest speed.

The Blue Ribbon was awarded to a ship that realized the glamorous dream of crossing the Atlantic Ocean within the shortest time. It has become a title identifying the best ship. Newer and larger ships have taken over the Blue Ribbon from large ships. Ultra luxurious cruise ships of 70,000 to 80,000 tons, such as Queen Mary and Queen Elizabeth II of pre-World War times, have also competed for the Blue Ribbon on their maiden voyages.

After the World War, airplanes were used to cross the Atlantic Ocean and the Blue Ribbon seemed to have faded away. Evidently, the competition for the Blue Ribbon among ultra luxurious cruise ships no longer exists, but the title of Blue Ribbon given to the most prestigious ship, will always remain in the memory of people who love ships.

Recently, some kinds of small craft, such as the offshore racer, have been competing for the fastest trans-Atlantic Ocean crossing. These are high-speed planing type small boats of length from 20 to 70 m. As the total fuel weight is generally heavier than the craft itself, its speed at the start is far less than its speed at arrival, which is maximum.

I also read in a magazine that a wave piercer was trying to get the Blue Ribbon. As far as I remember, a large wave piercer of 51 m length, for which negotiations to operate as a high speed ferry boat across the Dover Channel were in progress, was going to take up the challenge of the Blue Ribbon on its way to England, but the challenge did not materialize because the negotiations failed. Yokoyama became confident of the performance of his hull form of the wave piercer during towing tank tests and he even performed calculations of the new hull form of a 25-m wave piercer aiming for the Blue Ribbon.

The output of the calculations was as follows: The craft of 40 tons carrying fuel of 70 tons would start at a speed of 32 knots and arrive at the goal at a speed of 47.5 knots. It would cruise 3000 nautical miles in 77 hours at an average speed of 39 knots.

We even planned to modify the craft after the record run to a high-speed passenger craft carrying 150 persons, which was capable of higher working speeds like the Jetfoil, but the plan did not materialize.

6) At Hamilton Island

The World Expo 88 was held at Brisbane, Australia, in May 1988. We exhibited the hydrofoil boat OU32 and the hydrofoil boat OU90 made by the Horiuchi Lab at the Exposition, at the request of both the Ministry of Transport and the Ministry of International Trade and Industry. I visited Australia to check how our boats were exhibited and to attend the opening ceremony

On my return trip, I visited Hamilton Island, a resort located about 150 km southeast of Brisbane to take a ride on the wave piercer. The Tassie Devil 2001, which had been used as an observation boat during the America's Cup Race, was being chartered and used at Hamilton Island. In February of that year, a new craft 2000 had been introduced, and was being used. I wished to board this craft and study it.

The 2000 sailed for nearly one hour between Hamilton Island and Shute Harbor on the mainland. I wanted board the craft when the sea was rough, but the wave height was only 30 cm during both trips to and from Shute Harbor, and the bows of the sponsons were never submerged. Unfortunately, the trial ride was not very useful for my purpose and I made no new discoveries related to hull shape.

I was however, astonished that the craft could move laterally when it arrived or departed a pier. I could not understand the theory the first time because the craft had neither a bow thruster nor a stern thruster, but after thinking logically, I concluded that it was possible.

The craft had water jet units in each hull. If you steer the starboard jet to the starboard side and reverse it (Fig - 10), and likewise steer the port jet to the port side and set it to give a forward thrust, then the resultant of the two thrusts is a lateral thrust at the intersection of the two lines of thrust. The position of the resultant force shifts longitudinally with the rudder angle. Thus, you can adjust the heading of the craft and move it laterally. This kind of control is not possible in a normal craft. The wave piercer had two jet propulsion units and each water jet could be steered in opposite directions. In addition, the wave piercer had a very wide hull and each water jet was located a large distance apart transversely. The craft can be controlled conveniently because of these two characteristics.

I learned this convenient control method and found it to be very good. The same method can be applied to boats equipped with stern drives and outboard motors. When we built a prototype of the wave piercer and when I built my own boat Hateruma, I used this control method and controlled the boat with ease. However, Hateruma was a boat with a smaller breadth and fitted with outboards, so it was not very easy to use the same control method, and I reluctantly added a bow thruster to the boat to facilitate control.

This control method is very convenient for a large boat operated by one person. Persons who have not controlled a large boat single-handedly do not seem to be interested in this control method, probably because they think of it as a special case. When wind is blowing from the pier, berthing a light boat becomes very difficult, and this control method becomes very convenient. If the boat is to be moored for a short while, you can move the boat to the pier let it stand beside it without using ropes. I think this method is a necessary feature in a pleasure boat; unfortunately, most people are not interested in learning it.

Fig - 10 Catamaran moves laterally

Note : If twin water jets, twin outboard motors, or twin stern drives are installed in catamaran as above, the craft can be moved lateral to the pier.

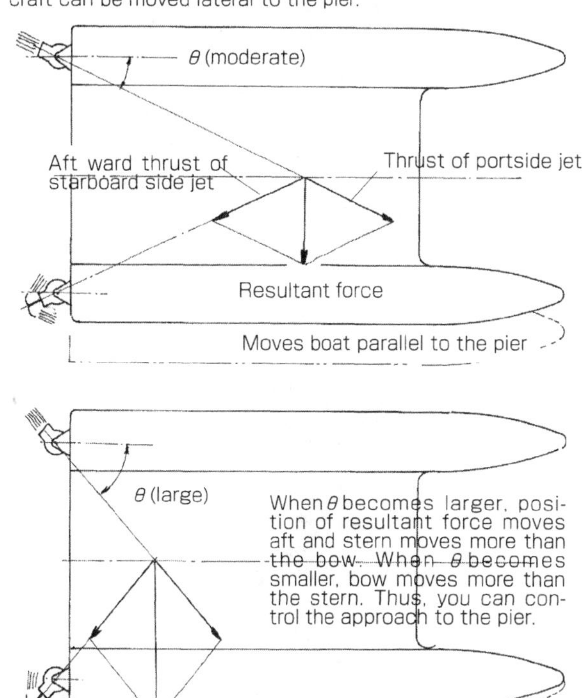

Fig - 11 5-m simple prototype boat

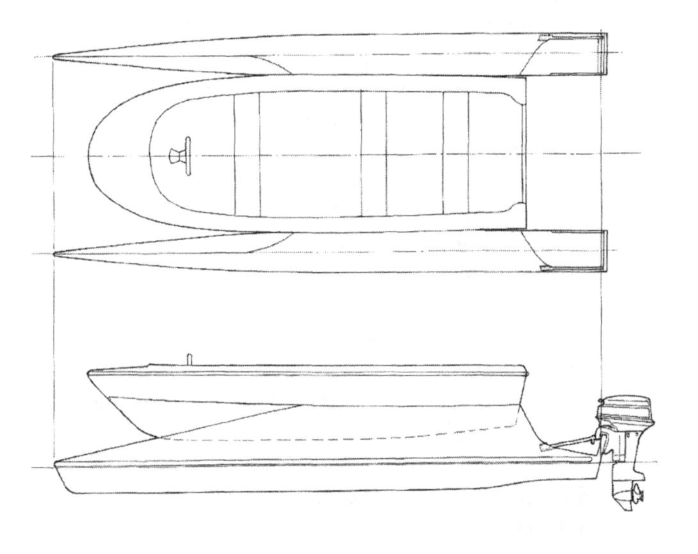

7) Actual boat tests and the OR10

To construct a test boat for maximum effect with minimum expenses, Yokoyama used an existing boat (F - 14) as the main hull and he prepared two sponsons taken off the same mould. He installed them on the main hull by a simple truss frame[1]. The result was a simple prototype boat (Fig - 11). The length of the main hull was 4 m, that of the sponsons was 5 m, and a 15-hp outboard motor was installed at the end of each sponson.

The test results were very good. We did not feel the shock usually experienced in normal

1) Very effective framework constructed by bar elements. Example: A bridge girder, the Eiffel Tower, a racing car body frame.

Fig - 12 Test ride on the simple prototype boat

Fig - 13 Three photos of the simple prototype boat

planing craft and the test boat landed on the water softly after a jump, and we were pleased. Many persons including those in other sections in our company took the test ride on May 20, 1988. Their reaction was very good, and many enjoyed the wave piercing action of the boat at sea (Fig - 12, 13).

There were some faults. Splash scooped up by the bow of the sponson came over the cockpit some times. Another fault was that the boat did not bank inside but banked outside when turning, unlike boats in our experience. But the test riders thought that these faults could be rectified before production, and they confirmed the performance of the new hull shape. Thus, the development of the wave piercer came closer to realization.

After the prototype comes the production design stage. What kind of boat would be suitable as the wave piercer? Yokoyama, industrial designer Nakagawa, and I drew various layouts. We planned four sizes from 16 feet to 35 feet. The drawings consisted of runabout, fishing boat, and cruiser versions.

I cannot remember the method by which we decided the design for development, but finally, we decided to build a 5-m mini cruiser nearly the same size as the simple prototype boat. The development number was decided as OR10, and construction of the prototype started at the beginning of 1988. I thought that a smaller boat would be suitable for the Horiuchi Lab because it consisted of a small group of persons.

It was not so easy to plan a small boat as a cruiser, because the floor level of the boat would be higher than the waterline unlike a normal boat in which the floor level was nearer to the hull bottom. In a normal catamaran, some part inside the sponson may be used, but this was not possible in this wave piercer, since its profile would become higher. However, an excessively high profile for only a 16-feet boat would be unnatural, and it would look ugly.

On the other hand, if the floor level were too low, waves would hit the bottom of the main hull making the ride uncomfortable. As the boat has a small waterplane area, it would immerse easily with many persons onboard and the bottom of main hull would tend to

Fig - 14 Plan and profile of the OR10

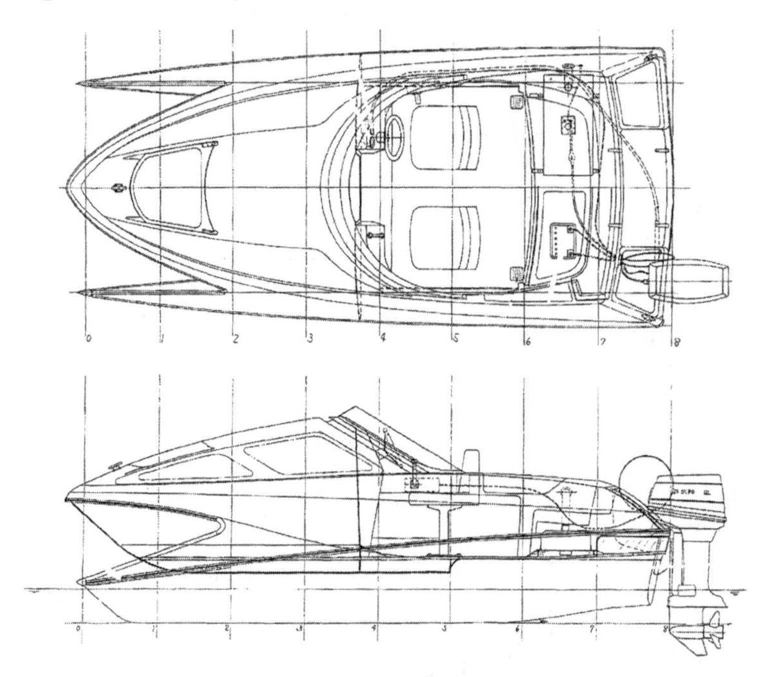

Length Overall------------------4.95 m
Breadth---------------------------2.18 m
Displacement (light condition)
---------600 kg
Engine------------------------- 50 hp
single outboard motor
Speed--------------------- ab. 25 knots

Fig - 15 Color design of the OR10

Fig - 16 The OR10 with Hateruma

become lower. After considering these points, we decided upon a rather low floor level.

For this boat, Yokoyama designed an ambitious hull shape with two unsymmetrical sponsons so as to have smaller resistance and a natural inside banking ability during a turn (Fig - 14).

In the meantime, Nakagawa came up with a suitable style and coloring for the boat that made the boat look chubby and attractive.

In April 1989, the OR10 was completed and launched (Fig - 15). Its chubby style and colors were novel, and I felt it was one of Nakagawa's best efforts.

After the test ride, I had the overall feeling that it was similar to a simple prototype boat. Comfort deteriorated as its weight increased and its floor level became lower. However, compared to normal planing boats, it offered a different taste during the ride.

We tried running the boat for a distance of 160 km from Lake Hamana to the Bay of Ago, escorted by my boat, Hateruma, which had been launched at the end of the previous year (Fig - 16). The OR10 was a small 16-feet boat. I thought riding it through one to two meter waves must be difficult for us, but in practice, I did not feel it so much because of its soft riding characteristics.

It had one characteristic that I was concerned about. Several times during cruising when sliding down a steep wave and when the forward part hit a steep wave wall, the boat would not clear the waves smoothly. This situation seemed to be similar to the situation when the Spirit of Victoria met with an accident. The OR10 never became submerged under waves, but would decelerate suddenly, and a large splash would occur in the forward region. This was not good.

If the bottom of the main hull extended more to the front or was positioned at a higher level, this situation might have improved. But the OR10 had already been completed, and we could not do anything.

Yokoyama made minor modifications to the OR10 and corrected faults such as splashing and directional instability when striking forward waves. But by adopting that hull shape, the surface area, weight, and the cost would increase. The question that remained was: is the product attractive enough to cover such problems? The cost of the hull accounts for a large share of the total cost and accommodation accounts for a very small part of the total cost of the 16 feet boat. Therefore, an increase in cost of the hull affects the selling price directly. Finally, no decision was taken to start production of the boat, and tests continued.

8) Negotiation with Mr.Hercus

At that time, we were negotiating with Mr. Hercus for the production of wave piercer type boats, including the OR10. As mentioned earlier, the speed length ratio of the Tassie Devil 2001 and our small boat were very different and therefore, we might say that they were very different hull shapes. Mr. Hercus' craft was evidently a displacement type hull while ours was a planing type hull. Mr. Hercus understood and recognized this difference well.

While the hull shapes were different, the combination of wave-piercing sponsons and cabin mounted high was evidently Mr. Hercus' idea; we wished to conclude an agreement with him for granting us the right to use part of his patent and design.

The negotiations continued from July 1987, and the draft of the agreement was almost fixed when he visited Japan. Mr. Hercus probably thought that a small wave piercer would not have adequate performance, so he waited for a proposal from Yamaha for FRP boats under 20 m. The negotiations were mainly for the production and marketing of both aluminum and FRP crafts of length greater than 20 m, in which Tassie Devil 2001 was included. For these larger craft, each item was decided in detail.

9) Crest Runner OU93

It was decided within the company that our high-speed wave piercing boat should not be called a wave piercer, as this name was not appropriate, therefore, our boat was named Crest Runner. Although this name did not indicate its inherent characteristics like wave piercer did, we could not find any other appropriate name.

Since we struggled to decide upon a reasonable cost for the OR10, we studied a large cruiser as the next product development topic. For a large cruiser, the percentage of accommodation increases, and the effect of increase in cost of the hull decreases relatively. If this large boat were designed to have splendid, comfortable riding characteristics, overall a

Fig - 17 Side view of the OU93

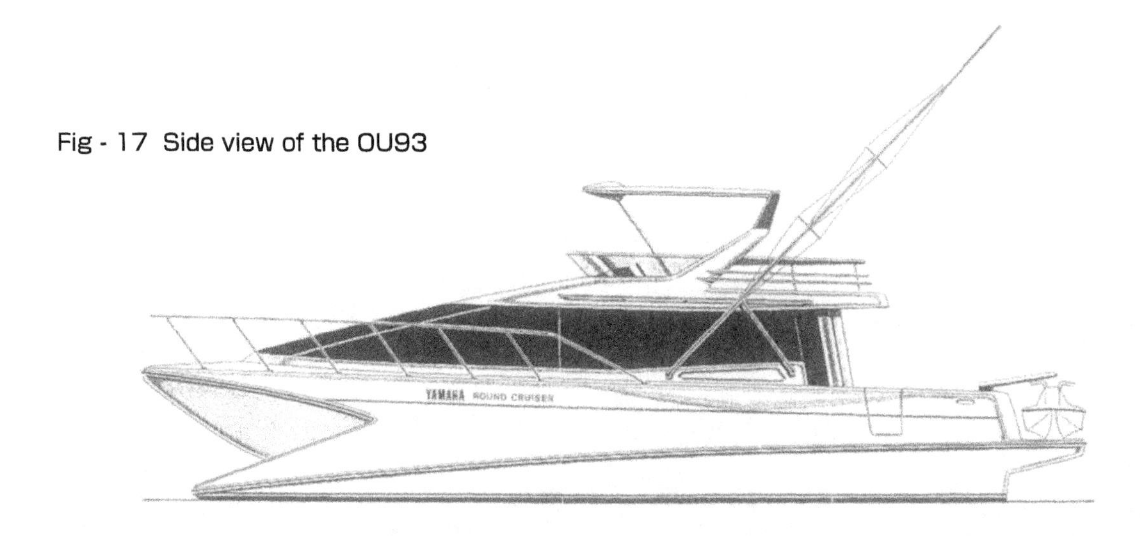

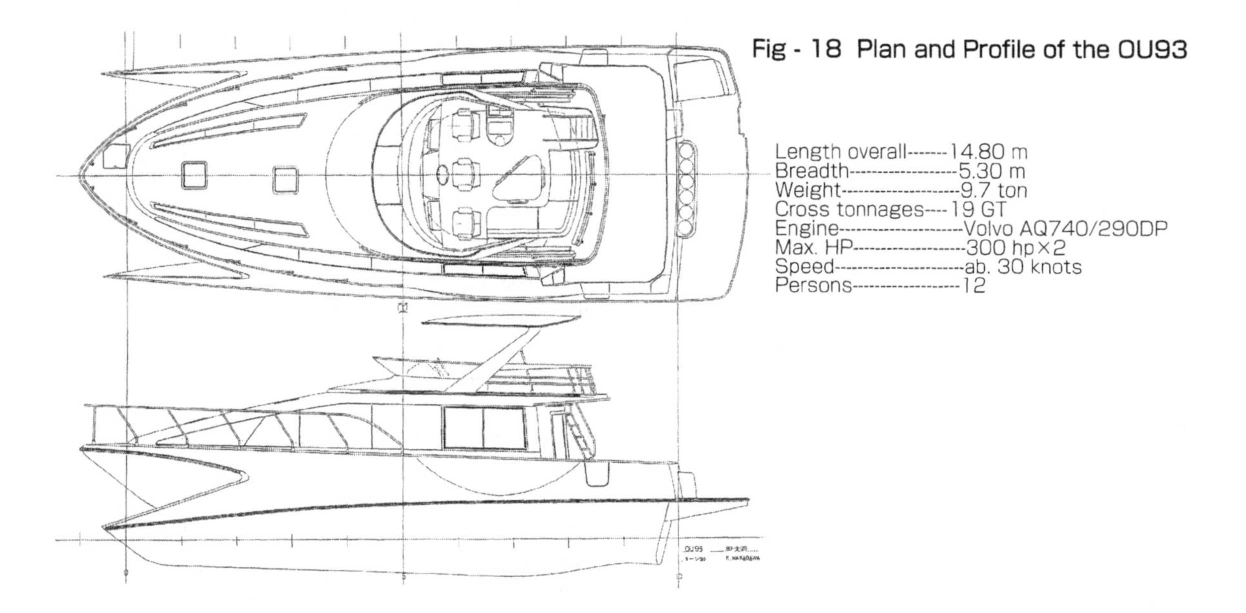

Fig - 18 Plan and Profile of the OU93

high cost performance can be anticipated. With these considerations, we decided a large cruiser of about 13 m in length and 20 gross tons (Fig - 17, 18) as our development target.

In 1988, the economy was at its peak and we had some reserve budget. In those days, Yamaha Motor used to exhibit a dreamboat at the annual boat show. The company decided to build a prototype of a large Crest Runner in cooperation with Marine Operations and the Horiuchi Laboratory and exhibit it at the boat show.

In June 1989, the project number was decided as OU93 and it was scheduled to be completed at the end of the year, tested in January and exhibited at the Tokyo Boat Show in February, as the boat show model.

Since Yokoyama alone would not be able to complete the design by the required time, Kazuhiko Satouchi and Mitsuharu Tazura from Marine Operations joined the team. These two were experts in the design of accommodation and structure. Nakagawa was responsible for the styling design. The development proceeded smoothly, thanks to the efforts of this excellent design team.

In this way, the OU93 was completed. The team of four designers worked to bring out an excellent finish to the boat, which was highly acclaimed at the boat show. But I was a little concerned and wondered whether the boat had adequate performance to justify the cost increase due to the new hull shape.

Later, we accompanied the cruising fleet to

Fig - 19 The OU93 in waves (A)

confirm the performance of the OU93 to Mera in the Izu Peninsula (Fig - 19, 20), and we also took the OU93 on a separate cruise to the Yura harbor near Osaka Bay to preview the Cruiser Rally organized by the Japan Boating Industries Association in 1989. Both times, I boarded the OU93 or looked the boat from outside to confirm its performance. I enjoyed the cruise, as the boat's performance was good. But my concern related to the balance between cost and performance of the boat remained. I could not say with confidence that its performance justified the cost increase.

10) Later

In 1990, the economic bubble burst. The sales of production boats decreased sharply and it became difficult for us to take up all kinds of work. Like many other projects, product development of the Crest Runner was not discussed anymore, and development work stopped after the exhibition at the boat show.

In the mid 90's, Yokoyama started studies on an improved OU93, but the opportunity to realize the plan never arose.

Looking back, I must express my remorse. We knew all the faults of the OR10, yet we developed the hull shape of OU93 by basically extending OR10 and making a few small improvements.

Now I feel that this was unfortunate, but do not remember why we did so. Being too busy, I might have probably given instructions to proceed without major modifications. Anyhow, the OU93 was built inheriting all the merits and the faults of the OR10.

I found from past documents some statements made by Mr. Hercus repeatedly emphasizing tank tests to be conducted before the completion of Tassie Devil 2001 after Spirit of Victoria. I also found that adequate tests were conducted especially on the behavior in waves and the motion in waves improved noticeably because of some minor modifications (Fig - 21).

We must have missed this point. Particularly, we missed the point that we had to continue improving the hull shape after studying its motion in following waves.

It is really unfortunate because we found the simple prototype boat to have very good riding characteristics. Mr. Hercus completed the wave piercer making full use of his experience after studying the accident. We probably relaxed and did not work as hard after our initial success.

Considering the above, I feel responsible for closing the doors to the bright future of the Crest Runner. However, it is no use crying over spilt milk. I hope that some day in the future people will admire the wonderful performance of a more-meticulously planned Crest Runner.

Fig - 20 The OU93 in waves (B)

Fig - 21 Tassie Devil 2001

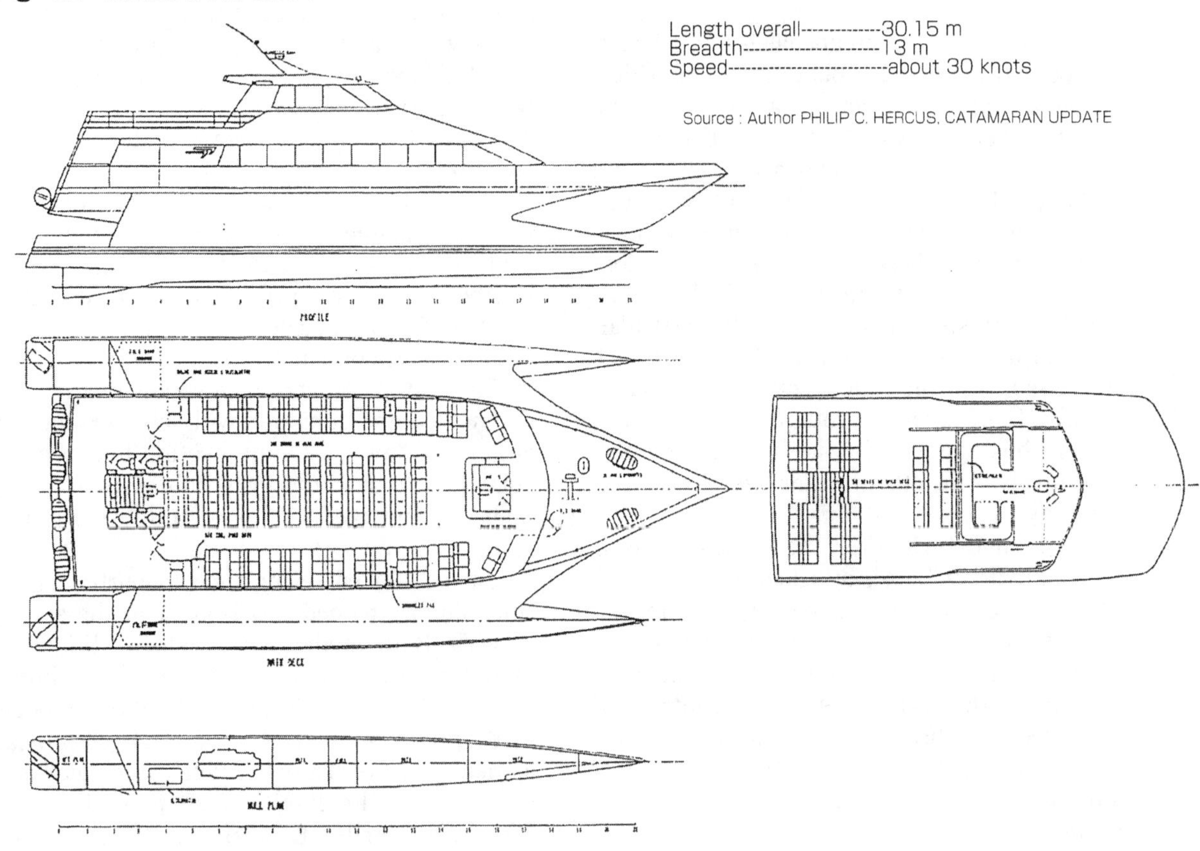

14

PASSPORT-17 & FISHERMAN-22

The oil crisis of 1973 gave the boating industry a deathblow and brought about the cancellation of the national boat show.

At that time, we at Yamaha, developed boats named Passport-17 and Fisherman-22, which satisfied the users' demands. Compared to similar- sized boats, our boats had half the weight, one-third the horsepower, and were less than half their prices. The concept was a great success and the recession in the boating business was eliminated within a short time. The rational design method that we adopted tformed the keynote of our designs. The boats also became the starting point of a large flow of entirely new types of motorboats and recreational fishing boats in the market.

1) Oil crisis

Popularization of recreational boating began in 1960 and increased rapidly from 1965 (Fig - 1). Yamaha accounted for a share of 70 to 80% of the market, and was the top manufacturer.

As boats became more luxurious, most of the boats over 17 feet were installed with stern drives and they had specifications that were more or less similar to those of some of the most luxurious US boats. Even 15-feet boats were installed with stern drive engines. In particular, the shipments of such luxurious boats increased and the sales volume increased rapidly. Good profits brought about by the boating business caught the attention of persons involved in other businesses within our company.

In October 1973, after the boating season ended, the first oil crisis attacked us. This crisis affected even products necessary for normal living such as cars; sales volumes decreased sharply and boats were much more affected. Motorboats that consumed large quantities of petrol took the brunt of the criticism, and we ourselves began to question whether it was right to sell the luxurious boats mentioned above.

All persons concerned with the boating business were adversely affected, and the national boat show scheduled in spring 1974 had to be canceled. The Japanese economy, which had achieved a miraculous, rapid growth for 30 years stalled and the economic growth experienced a downturn the likes of which had never been experienced after the Second World War. The year 1974 ended dismally under these circumstances.

In view of this situation, we made plans to radically change the trend of development of Yamaha boats . Until then, our main product had been had been large, heavy, and luxurious motorboats. We decided three new points for product development.

The first point was to develop motorboats that were light, equipped with low horsepower engine, of low cost, but useful. The second point was to shift the focus to development of sailboats, and the third was to develop motor sailers.

Yamaha started selling the Passport-14 and a small dinghy named Sea Hopper in the spring of 1975. Both models sold well. Additionally, motor sailers MS-21 and MS-24 were introduced in the market, and both these models were received well. Also, a racing sailboat named Wing of Yamaha won the single-hand race across the Pacific Ocean held at the Okinawa Marine Exposition. This victory

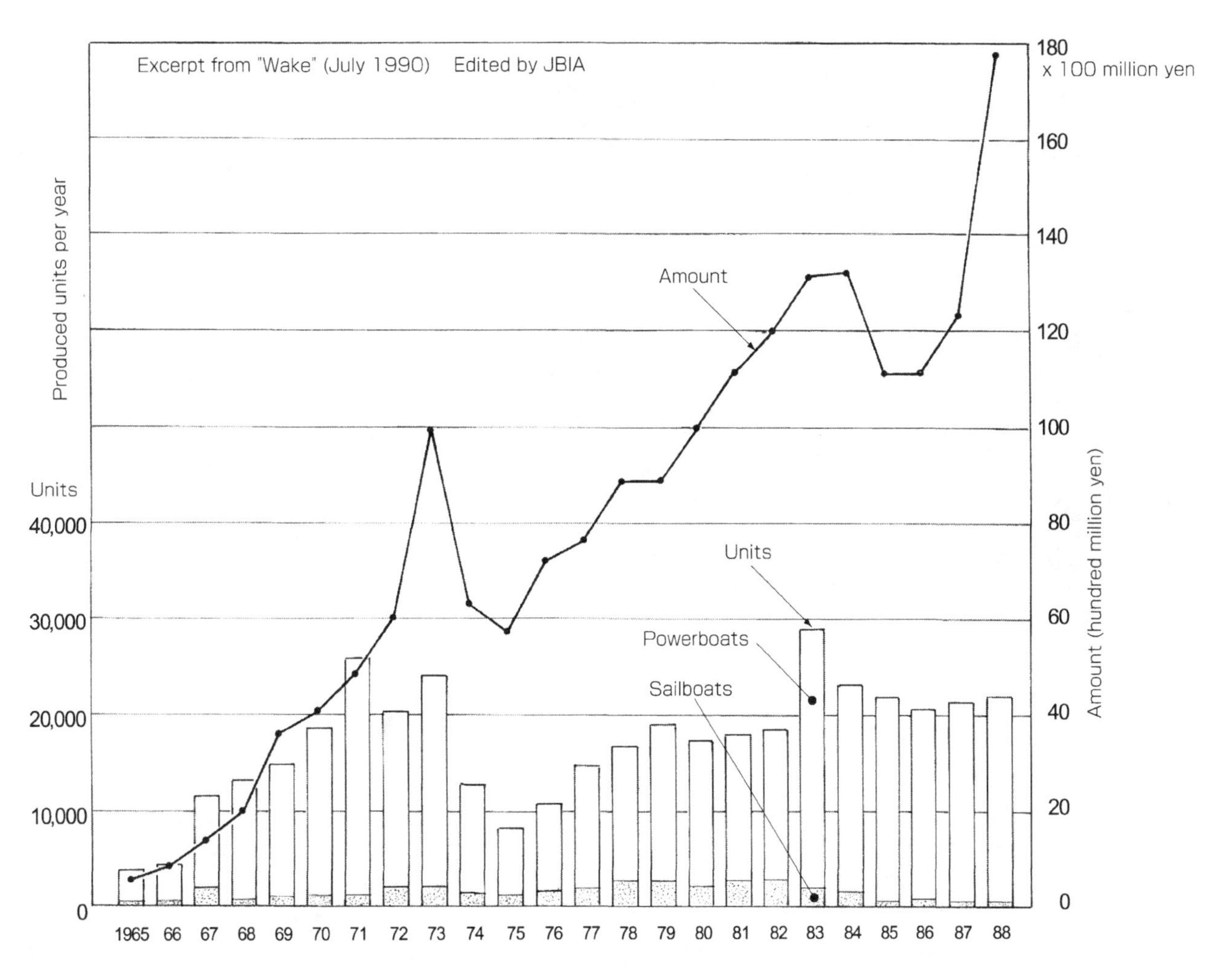

Fig - 1 Annual production units and shipment value

demonstrated Yamaha's keen enthusiasm and its focus on the development of sailboats.

Such changes were successful in individual products, but these products did not account for a major share of sales of Yamaha boats. We needed to develop a larger motorboat that better suited the needs of a next-generation product.

2) Hiflex hull form

We developed a boat named Hiflex-14 in 1961 when Yamaha Motor started selling boats, and this boat was well received. It had a round bilge (outer part of the bottom shell) and a deep keel, and Hiflex (name given by Yamaha's President) hull form (Fig - 2). The bottom part of the hull surface was almost a curved surface, therefore the bottom shell possessed form rigidity[1] and needed no stiffeners. This enabled the boat structure to be made simple and light in weight.

As shown Fig - 3, the part under the chine is almost a curved surface, which gives the boat strength. The hollow portion of the keel was covered by a plywood of small width, to form a strong backbone of the boat. Because of this bottom shape, the upper surface of the hull bottom could be used as floor and did not need floorboards. Also, the seat bases were installed laterally and they fixed the relative positions of keel, chine, and hull bottom. The seat bases served as of strong frames (Fig - 4). As each part of the outer shell and structure had dual functions, a light and simple structure that did not need any stiffener at the bottom or any floorboard, was realized

Moreover, the boat had such excellent maneuverability and good ability to make tight turns that it surprised our President, Kawakami, who named it Hiflex (high flexibility). Careful selection of keel size, round bilge, and width and height of the chine resulted in

1) Form rigidity : A characteristic of shape that gives rigidity and strength on account of the 3-dimensional form

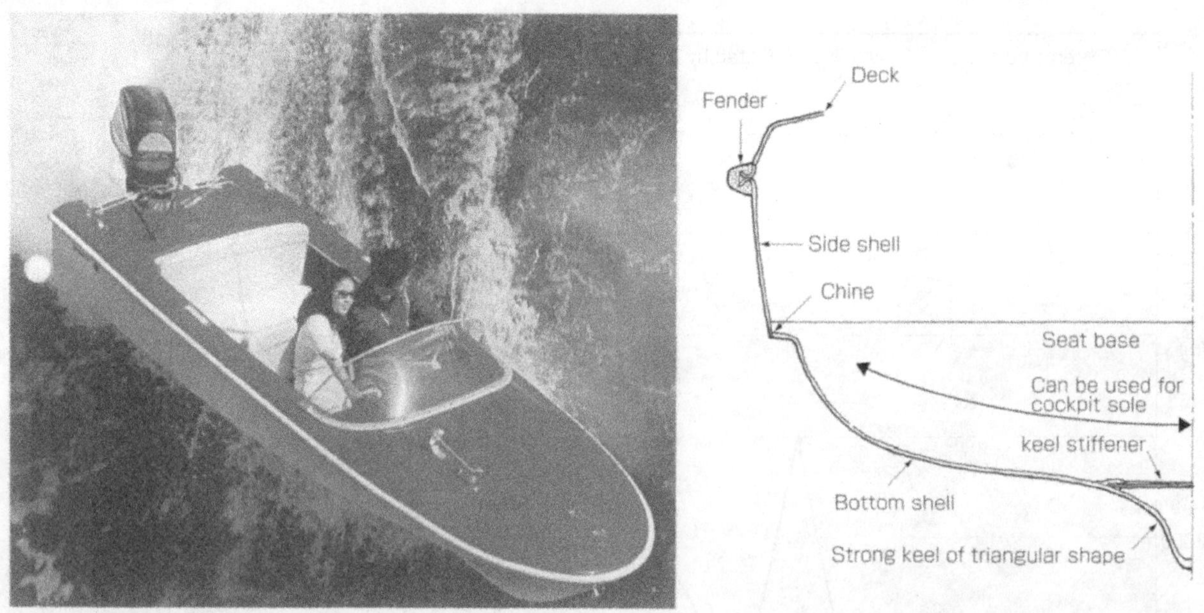

Fig - 2 Hiflex (H - 14)

Fig - 3 Hiflex-14 Structural section

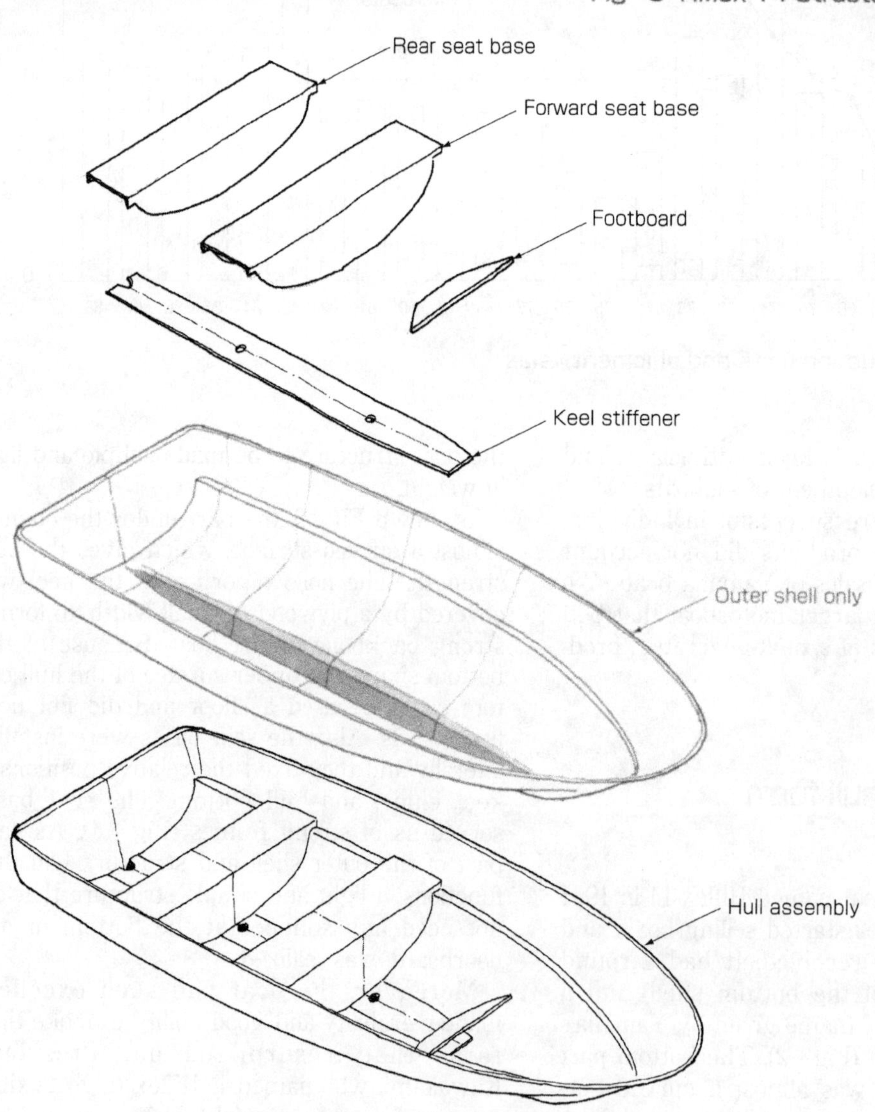

Fig - 4 Hiflex-14 Structural perspective drawing

Fig - 5 Passport -14 (P -14)
Length overall : 4.21 m, max. Breadth : 1.68 m, hull weight : 170 kg, horsepower : 25 PS,
max. horsepower : 40 PS

Fig - 6 Passport -14 NEW (P -14)
Length overall : 4.21m, max. breadth : 1.67 m, hull weight : 178 kg, horsepower : 25 PS,
max. horsepower : 40 PS

superb maneuverability.

As mentioned earlier, from the spring of 1975, we started selling Passport-14, a successor to the Hiflex-14, the production of which had been terminated more than ten more years ago, to re-start the flow of Hiflex. Nanao Murai, in charge of engineering, and Keiji Nakagawa, in charge of the styling, planned the new layout to suit the new policy placing high value on practical use, light weight and low cost, and we announced a new model (Fig - 5).

Four years later, they developed the new Passport-14 with the same hull form but with a highly ambitious deck style. The windshield frame of the boat was made of FRP in one piece within the deck mold. Normally, the windshield frame is a time-consuming and expensive item among the small quantity of outfitting items in these kinds of boats. This windshield construction was simple, yet strong, and resulted in a good-looking style (Fig - 6). The name Passport includes the wish of the designer to make beginners to boating feel comfortable when taking a ride on the boat. Therefore, a hull that was light in weight, less expensive, and with good performance in the low horsepower range was required.

In view of these circumstances, we recalled the Hiflex hull form that had not been used for more than ten years. Again, we made 100% rational use of the Hiflex-type hull form in the Passport series, and in this way, we came out of the difficult period of about 10 years after the oil crisis.

3) Passport -17

During the development of Passport-14, I found that we could apply the hull form and structure of Hiflex-style to larger boats.

Until then, the shape of the hull with a length of more than 16 feet had to be deep-vee. However, from this time onward, there was no

Fig - 7 Stripe - 17 Sport Cruiser (S - 17 SCR)

Length overall : 5.24 m, max. Breadth : 2.14 m, hull weight : 550 kg, boat weight with engine : About 870 kg, horsepower : 120 - 140 PS, max. horsepower : 170 PS

need to adhere to this limit strictly because the hull weight reduced.

At that time, I studied the possibility of designing a boat with the same length as boats in the past (Fig - 7), but with half the weight and one-third the required horsepower.

If the boat size is the same, yet has one-third the horsepower, it means that if the boat weight decreases, the hump[2] resistance decreases significantly. Thus, I expected one-third of its horsepower was needed for planing. Getting on the plane easily is an important factor deciding the required horsepower.

Until then, a normal 17-feet boat was fitted with a stern drive engine. In this propulsion system, a car engine is installed inboard and connected to an outboard propulsion unit like an outboard motor. The weight of the stern drive is naturally much heavier than that of an outboard motor. Therefore, the total boat weight can be made significantly lighter by merely replacing the stern drive by an outboard motor. The target of reducing the horsepower to one-third was easily attained because of the reasons mentioned above.

In those days, the maximum horsepower of the Yamaha outboard motor was 55 PS. We had to install a stern drive if an engine of more than 55 PS was required. Additionally, all stern drives were imported, which meant that the cost would increase. These factors helped us to go ahead with the new project because of the expectation that the boat to be developed by us would be less expensive.

After drawing a sketch of a 17-foot boat fully reflecting the features of Hiflex, I could see the practical feasibility of a boat that was not luxurious but had adequate volume, good cockpit space, half the weight, and satisfactory performance with one-third the engine horsepower.

A 17-foot boat of such simple specifications already existed abroad, especially in European countries. Since Yamaha boats had become too luxurious, the simple 17-foot boat of lightweight construction and economical cost probably stood out and attracted attention.

One day in the summer of 1975, I approached Shoji Muraki, Manager of the Sales Department with the drawing to explain the three aims mentioned earlier - namely that the new boat would have the same length as the carlier 17-foot boat, but would have half the weight and one-third the horsepower.

Muraki, a national water skiing champion in his university days, and had in-depth knowledge of boating, and he understood the customer's mind very well.

I did not say anything about the cost but he must have guessed that the boat's price would be half and the engine price would be one-third, especially the price difference between a stern drive unit and an outboard motor would be large, resulting in a large difference in the total end price. He agreed to the project immediately.

The agreement to the project by the manager of the Sales Department encouraged me because Muraki knew a lot about boats, and I returned to the boat factory at Arai feeling satisfied.

2) A peak in the resistance just before a boat starts planing. At that time, the boat's bow rises and its resistance increases significantly.

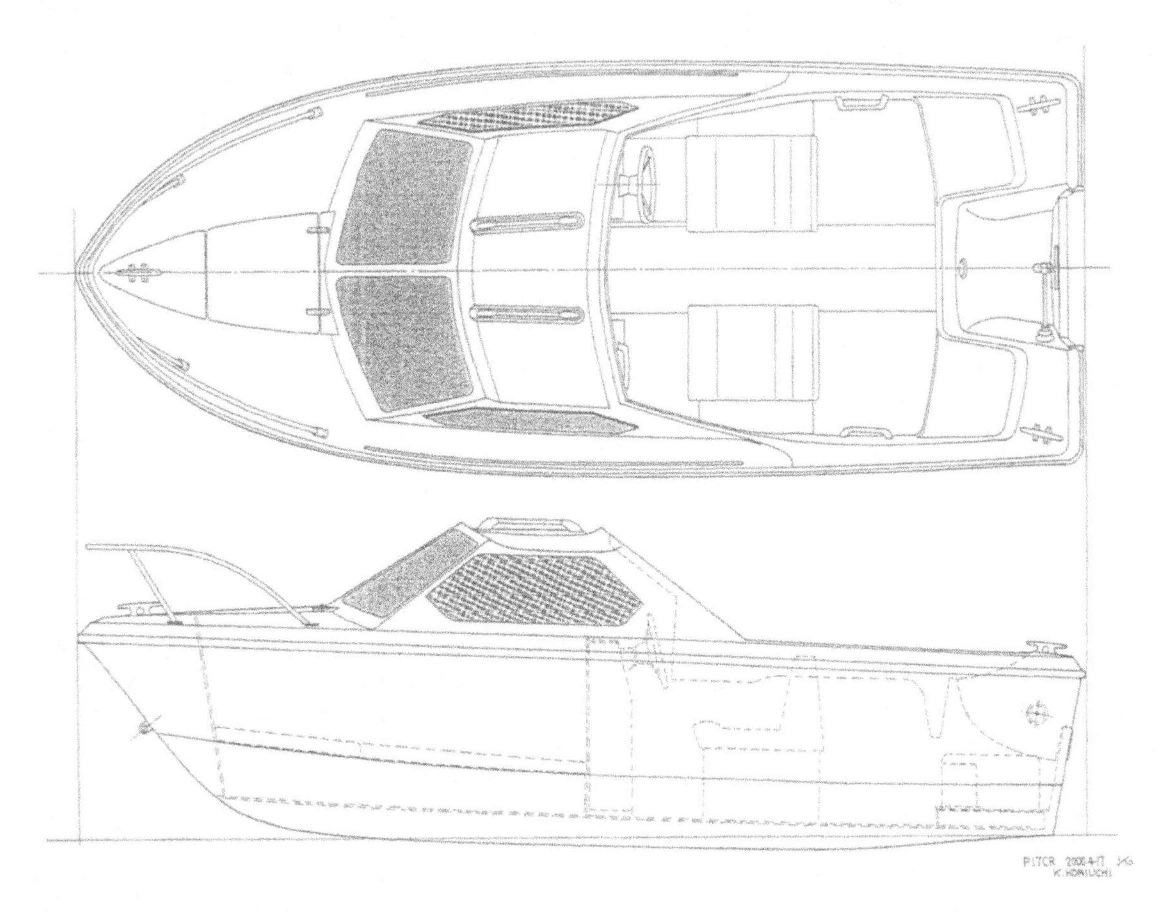

Fig - 8 P -17 CR general arrangement

Fig - 9 Passport -17 (P-17CR)

Length overall : 4.96 m, max. beam : 1.99 m, hull weight : 350 kg, boat weight with engine : 430 kg, horsepower : 55 PS, persons : 7

Fig - 10 Passport -19 (P -19 CR)

Length overall : 5.88 m, max. beam : 2.36 m, hull weight : 715 kg, horsepower : 85 -115 PS, max. horsepower : 140 PS, persons : 7

Unfortunately, I do not remember the days after the agreement to the beginning of the development work. I only remember that I had a meeting soon with Murai, who was in charge of this project. As the design team of Murai and Nakagawa had worked on Passport-14 developed with the same concept, they accepted and understood my feelings on the project.

A simple structure with no floorboard, no bottom stiffener, no windshield frame, and which utilized the advantages of the Hiflex hull form to the maximum extent was designed, and its attractive shape appeared as seen in the figure (Fig - 8).

The round bottom hull continued to the deep keel, the hollow part of which was covered with a cover of small width. Seat bases that served as frames were placed on top of the cover. This construction was similar to that of Passport-14.

This time we made plans for both cruiser type and open type simultaneously, and we concentrated on the cruiser type boat. Naturally, the cruiser type had cabin and roof, which were made in one piece with the deck. Front window made of glass and acrylic side windows were later directly fitted on the outside of the cabin made of FRP. The glass and acrylic windows formed part of the structure that contributed to strength and rigidity together with the FRP cabin.

Naturally, some rigidity and strength were required around the windows. Nakagawa designed the windows with a reasonable shape using his experience in the design of Passport-14. The result was that we attained the targets we had set and completed a very attractive boat that each one of us wanted to own ourselves (Fig - 9). (Nakagawa became the owner of this boat later.)

From October 1976, we began to sell Passport-17 and it sold like hot cakes. The Boat Division that had been ailing since the oil crisis became lively again. Muraki's business prediction had hit the mark beautifully.

Yamaha introduced Passport-19 to the market in 1978 aiming for similar success (Fig - 10). The same pair of Murai and Nakagawa designed this boat also. We did not have the courage to apply the simple Hiflex structure on this large heavy boat planned to be installed with a stern drive unit, but we adopted the deep-vee hull form, which became the successor to the Passport series of boats.

In this way, the mainstream of Japanese motorboats after the oil crisis centered around Passport-17 and Passport-19. Especially, Passport-17 took the valuable first step in rejuvenating the Japanese motorboat industry and restarted the large flow of motorboats to the market after the oil crisis.

Again look at Fig - 1. Note that the newly developed boats were Passport-14, Passport-17, Passport-19, Sea Hopper, and motor sailers, which were introduced to the market in rapid succession, and which led to recovery of the sales volume that had been dropping because of the oil crisis.

Boating business in Japan has been always unstable and eventful. During my stay in

Fig - 11 Bottom of F - 22 (Fisherman - 22)

Yamaha for 36 years, the really best times in boating business were about three times and each period continued only for four to five years. Besides, the best times were generally followed by very difficult times.

Accordingly, new products that throw off the hard times need to be developed promptly, so as to keep the business going. In the future, I hope people engaged in the boating business will refer to the stories of the P-17 and the F-22, and that these stories serve to assist them from escaping the hard times that are likely to come again.

4) Development of F - 22 (Fisherman 22)

Looking at the recovery of motorboat sales because of the P-17, the P-19 followed by the P-14, both Oshita and Hasegawa thought that the situation of fishing boats could also be changed. Oshita was an industrial designer as well as a fishing enthusiast, and Hasegawa was a skilled engineer engaged mainly in designing open fishing boats and motorboats.

In those days, Yamaha was supplying open fishing boats to pleasure fishermen, but the open fishing boats were built only for professional fishermen and other type of fishing boats intended for pleasure use were limited to only large boats powered by diesel engines. This is because the commodity tax those days for pleasure boats was 30% for boats of length greater than 8 m, 15% for boats of length greater than 5 m, and 10% for boats of length less than 5 m. As the price of engine was included in the boat price, this tax was rather high. On the other hand, no tax was imposed on open fishing boats built for professional use. Therefore, pleasure fishermen did not have to pay tax so far as they bought such kinds of

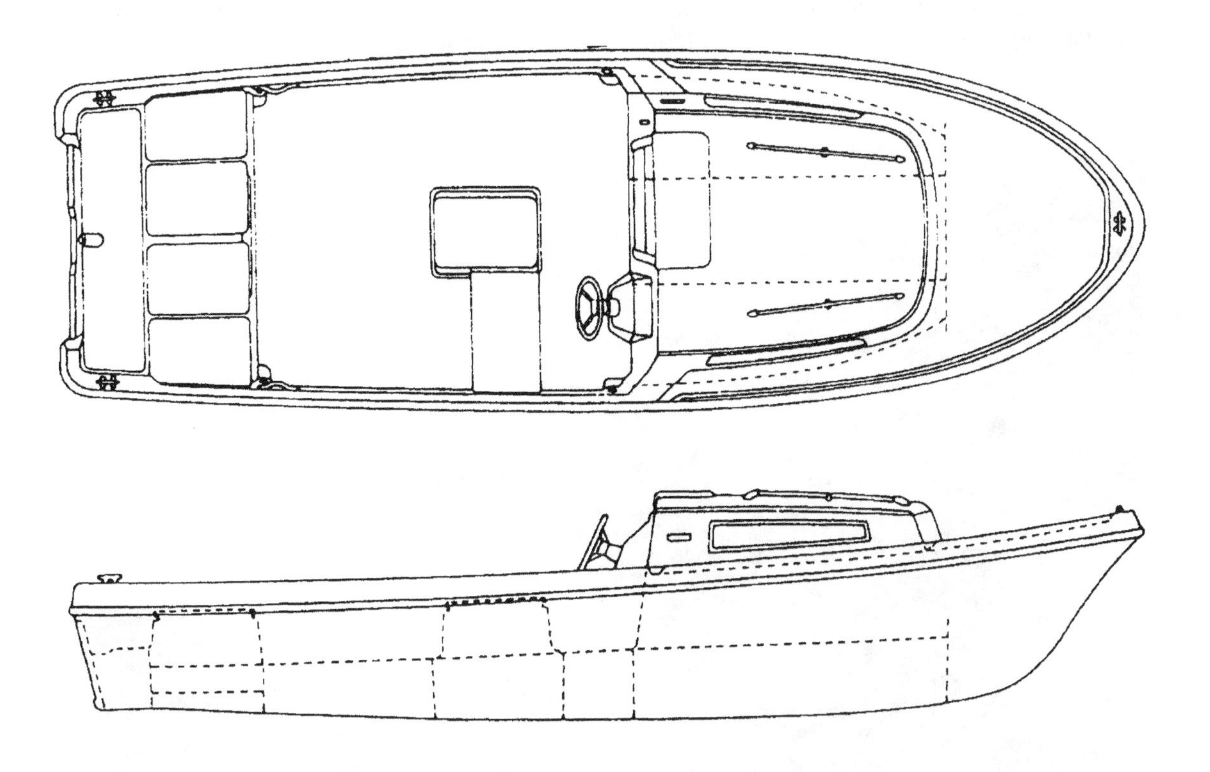

Fig - 12 General arrangement of F - 22
Length overall : 6.58 m, max. beam : 2.14 m, hull weight : 520 kg, engine : 55 PS or 2X25 PS

open fishing boats.

Pleasure fishing had already become a huge market. Looking at the circumstances, both Oshita and Hasegawa wanted to develop a fishing boat that was particularly designed for pleasure use. The time was probably ripe. Mr. Tadanori Arata, the divisional manager at that time, approved the development of the pleasure fishing boat and the project started.

Since I often heard Oshita mentioning that he wanted to develop such a boat, I expected him to make good progress in the development work. The two persons firstly designed a new hull shape, which was Hi-flex type based on the open fishing boat.

5) F - 22 topics

Open fishing boats had a tendency to slide laterally during high-speed turns, to drift because of wind, and to ride hard on waves. These deficiencies had to be corrected. Therefore, they used the Hiflex type round bilge with a deeper keel on the open fishing boat hull shape. This was a kind of hybrid hull shape, which was their solution for the pleasure fishing boat.

The hull had a round bilge like Hiflex but its keel was separated into two keels from midship to stern to facilitate landing (Fig - 11), which is the characteristic of an open fishing boat used by professional fishermen. The boat had a cabin that increased the profile area, which might cause drifting of the boat by wind. To prevent this wind drift, Hasegawa designed a deeper keel with a skeg.

They introduced several ideas in deck design. One of the ideas was a cabin where one could take a nap. Normally, cabins in our boats were luxurious but a cabin in a fishing boat that permits one to take a nap and allow storage of tackles and other fishing gear would be adequate. By minimizing the cabin space, the cockpit could be made larger. Its layout was re-arranged based on this concept. The boat had a layout that permitted two or three crew to take rest while waiting for the tide in the small cabin, and also had a long, wide, open deck to enjoy fishing, and an ideal cockpit depth (Fig - 12).

In those days, open fishing boats were increasingly being provided with a self-bailing deck. Since the boat was not flooded even when moored on the water without a boat cover, and since the exposed parts could be washed after fishing without worrying about water flowing into the bilges, the boat could be kept clean at all times. The self-bailing deck and cabin top were made using one mold. Other ideas included stepped transom height

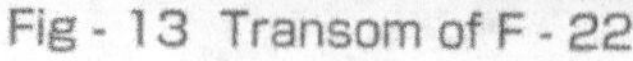
Fig - 13 Transom of F - 22

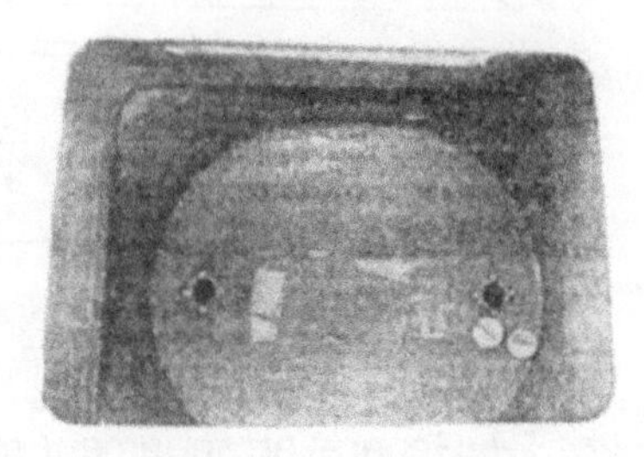

Fig - 14 Bait well of F - 22

Fig - 15 Profile of F - 22

Fig - 16 Overall view of F - 22

Fig - 17 F - 22 fitted with windshield

suitable for both single and twin outboard engine installations, which helped retain the correct propeller depth (Fig - 13), a circular bait well to keep fish fresh (Fig - 14), and so on. Thus, many ideas of both Oshita and Hasegawa were included.

I was very interested in the project but not familiar with fishing, so I kept telling Hasegawa that the weight of the boat should be 100 kg lighter than in the original plan.

As the weight of a simple boat like the F-22 is nearly proportional to its material cost and labor cost, I felt strongly about preventing any cost increase and the need to realize the initial target.

Hasegawa did not show me the detailed data for weight reduction but I remember that the boat weight was reduced by about 70 kg. By weight reduction and rationalization of structure, the price of F-22 initially became 950,000 yen. The price was lesser compared to the 1,600,000 yen for P-19CR or the 900,000 yen for P-17CR in spite of the larger length. The F-22 price attained was amazing.

6) Success of the F - 22

As described above, the F-22 was completed as an ideal pleasure fishing boat of lighter weight, greater length, and a small cabin. Since it had a rather flat bottom, the boat was not a tough runner in rough seas but was suitable

Fig - 18 F - 24

Length overall : 7.34 m, max. beam : 2.42 m, hull weight : 830 kg, engine : 85 PS or 2x55 PS, persons : 8

for fishing because of its stability. The boat got on the plane easily without the hump resistance being felt because of its length, and it had a comfortable ride characteristic better than we expected (Fig - 15, 16).

Its forward cabin could be used while waiting for tide and it could also be used for stowage of fishing gear, which specially made the customers happy because of its convenience, since the customer could go home after storing the fishing gear in the cabin and locking the door. In the first year, 300 units of this boat were sold. There was an article in one a boating magazine named "Kazi" describing the boat as "a hybrid of a motorboat and an open fishing boat". Thus, the boat was rather simple and blunt. There was no windshield and the cabin entrance was fitted with a closing arrangement, which was like a washboard with lid as in a sailing boat.

Subsequently, the F-22 model, which was very simple initially, was improved year after year reflecting users' demands, and these models were named F-22-2 and F-22Dx. The model of 1983 was fitted with a windshield (Fig - 17).

In 1981, F-24, an advanced version of the F-22, was introduced in the market (Fig - 18). This was the form based on which the modern Japanese pleasure fishing boat developed subsequently.

Yamaha found the key points of pleasure fishing boats through its development of the F-22 and F-24, then refined them and developed larger or smaller models, expanding the range of fishing boat series to meet the needs of pleasure fishermen.

We also increased our lineup of pleasure fishing boats by developing newer series of boats according to grade and application naming them as FR, UF, and SF series with letter F indicating fishing. In this way, we increased the types of Yamaha fishing boats significantly.

Our competitors followed in a similar manner after observing our success, and almost all Japanese boat builders became fishing boat builders with Yamaha taking the lead. I believe that the F-22 was the first to set the standards of the modern Japanese fishing boats, and should be acclaimed for initiating the vast flow of modern fishing boats in Japan.

7) Conclusion

Like the Passport-17 or F-22, new types of boats developed with novel concepts found new markets, increased their families, and became the main force in the market. This will remain a noteworthy event in the history of the Japanese boating industry.

When the sales of boats dropped steeply, and boating business became depressed because of the oil crisis, the development of new models such as the Passport-17 and the F-22 led to a recovery of the boating business, and these models became the main products that contributed to its rejuvenation.

The sense of growing crisis and tension had heightened considerably in those days. Therefore, when we found bright prospects for a development project, and when a new model was very well received in the market, our satisfaction and delight were equally high. Looking back at this series of events, each event was an exciting experience for us and we were proud of ourselves for achieving good results.

Finally, I would like to thank all those persons whose splendid efforts made such achievements possible.

15

DREAM OF A PADDLE WHEEL GENERATOR

Around 1980, we successfully developed a small ferryboat to cross the river and go upstream without power, at the request of ATCHA[1]. The paddle wheel used at that time could effectively extract power from water current. If this paddle wheel is placed in a rapid current, it can produce electricity and pump up water. I thought this was thus the most ideal technology for Nepal, and I decided to build and test one practically. This is the story of the development of the paddle wheel.

A boat or a paddle wheel to be used in the backwoods of Nepal should be sized so that it can be carried on the back of a person. I considered giving the paddle wheel adequate buoyancy so that it floated by itself and no boat would be necessary. The result was that although the size became smaller, its development became more difficult technically.

1) Development of ferryboat for Nepal

Mr. Jiro Kawakita, the Chairman of ATCHA, visited Yamaha Motor in the summer of 1979 to meet our managing director, Mr. Eguchi. At the meeting, it was decided that Yamaha would assist in the development of a ferryboat that could be used to cross rivers with rapid currents in Nepal.

Nepal is a long and narrow country in the East-West direction located near the Himalayas. The country has three large rivers which form the upper streams of the Ganges River and numerous branch rivers flowing from north to south. Except in the south, these rivers have rapid currents and flow through deep valleys, separating the transport and distribution of goods between the eastern and western parts of the country. To resolve the problem, Professor Mori and his assistant Ogawa of the Tokyo Institute of Technology were to perform model tests of a boat that could cross the rivers with rapid currents for transportation of goods and people. They requested Yamaha to build a full-scale boat for the tests. I immediately visited the Tokyo Institute of Technology for further details and to hear about their dream for the future.

Chairman Kawakita and Professor Mori created a group named the Jizai Laboratory, to engage in activities for producing articles with free ideas. This activity was one of the ATCHA activities.

The principles of the ferryboat shown in Fig - 1 are as follows:

One end of a rope with a length equal to approximately twice the river width is fixed on the bank of a river and the other end is fixed on the boat. The concept is to adjust the angle between the rope and the boat so that the direction of current with respect to the boat side can be changed. The boat is pushed by the current from the side to reach the opposite bank of the river; if this angle is reversed, the boat can return to the starting bank of the river. This idea was confirmed by model test, and I too was convinced that it would work. However, the subsequent part of the dream was to develop a boat without a source of power upstream and downstream of the river. We called this boat a paddle wheel climber or upstreamer. It was a very interesting dream, but every sketch they showed me indicated that the possibility of realizing it was dim.

My interest was aroused. If a boat could

1) Association for Technical Cooperation to the Himalayan Areas

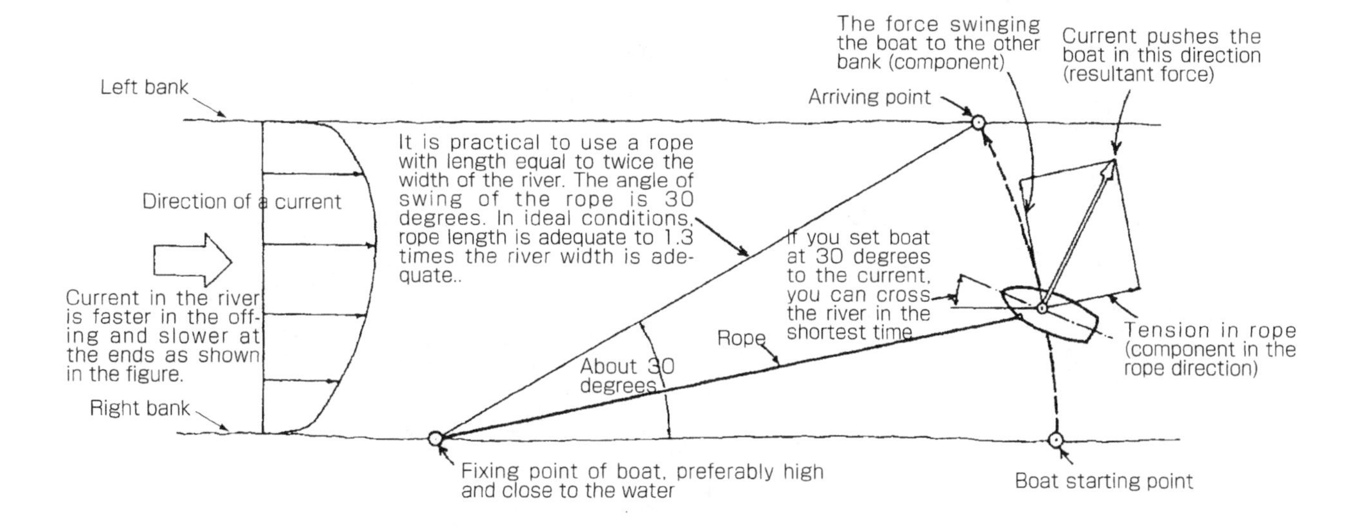

Three ways to incline boat with respect to the current

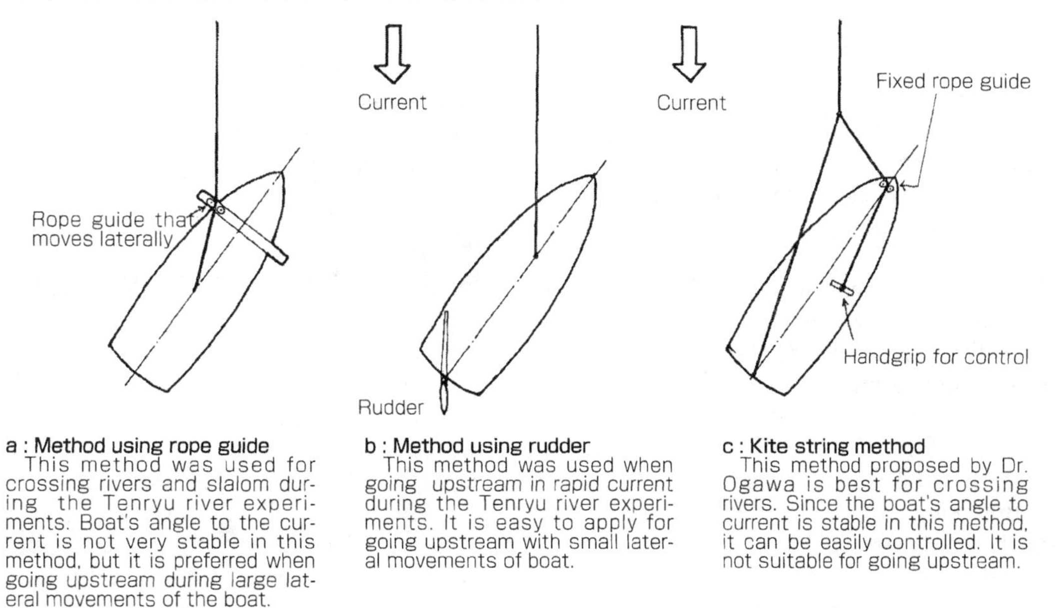

a : Method using rope guide
This method was used for crossing rivers and slalom during the Tenryu river experiments. Boat's angle to the current is not very stable in this method, but it is preferred when going upstream during large lateral movements of the boat.

b : Method using rudder
This method was used when going upstream in rapid current during the Tenryu river experiments. It is easy to apply for going upstream with small lateral movements of boat.

c : Kite string method
This method proposed by Dr. Ogawa is best for crossing rivers. Since the boat's angle to current is stable in this method, it can be easily controlled. It is not suitable for going upstream.

Fig - 1 Principles of ferryboat operation
If you turn the boat 30 degrees to the current, the boat can cross the river at the fastest speed.

cross a river freely and in addition, go upstream or downstream in a current, the boat has the freedom to move in any direction on the surface like a powered boat. It might be able to slalom like a water ski. We thought that the hull form and structure of a Yamaha outboard fishing boat could be used as the full-size test boat, so we decided to choose an appropriate one from the production models. Tsuide Yanagihara, the leader of the boat experiment group, made a device with sailboat hardware that could change the angle between the boat and the rope. He fitted it on the test boat, which was towed by a powerboat, and he confirmed that it worked as a ferryboat. The towing speed used was the same as the speed of the river current to simulate boat behavior on the river.

I could not come up with a practical layout of the upstreamer for more than a month or so, but it came to me suddenly while lying in bed one morning just two weeks before the official trial. The boat could be made to go upstream considering the theory that the river current rotates paddle wheels and winch-drum, both of which are connected through a shaft, and the winch-drum winds up the rope. It was a very simple system without using gears or other equipment. The rope, one end of which is fixed on the bank upstream, is wound for about

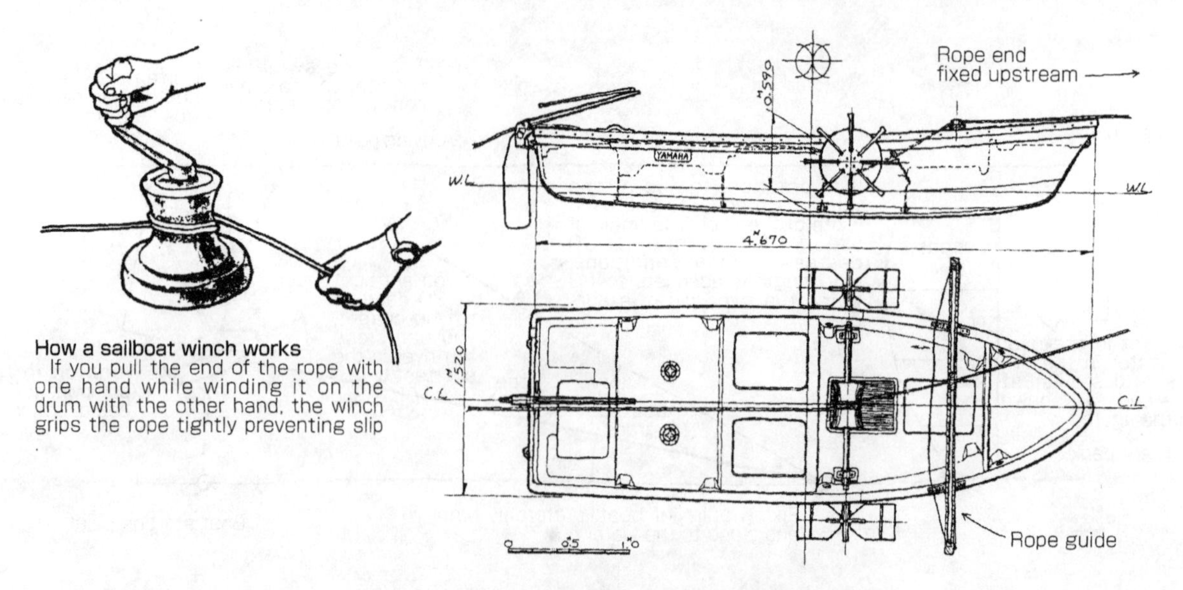

Fig - 2 Boat, paddle wheel, winding drum

three rounds around the drum and the other end is thrown on the water behind the boat. The current pulls the rope on the water so that it tightens on the drum and can wind up without slip just like a sailboat winch. This is the theory of a boat to go upstream on a river (Fig - 2).

On the other hand, when a person onboard takes up the rope behind the drum and loosens its winding on the drum, slip occurs between the rope and drum, and the boat stops moving upstream and starts moving downstream of the river. Soon by simple calculation, we worked out the optimum diameter of the drum and upstream speed. I quickly prepared drawings for making the prototype and sent a request to make one set of paddle wheels. After they were completed, Yanagihara installed the system on the intended fishing boat to complete the prototype boat (Fig - 2). We towed the boat to confirm its performance upstream and downstream and also confirmed that it moved freely sideways.

After working out the drawings of the paddle wheel, I did not find time to assist further. Yanagihara completed the prototype boat for official tests and finished the validation test in only two weeks. This was an admirably quick job.

In November 1979, the official trial was held at the Tenryu River attended by a large number of people from the Jizai Laboratory and the press. The performance of the boat as a ferry and upstreamer was excellent. We also successfully completed the slalom trial around floating buoys in a zigzag manner, encouraged by shouts of joy from the attendants on the bank. Going upstream with the current at a speed of more than two meters per second created a splash and it was very exciting.

2) Charm of the paddle wheel

I was satisfied for having thought of the paddle wheel system. Sturdy paddle wheels and winch drum were fitted on a single shaft supported on two bearings. The system was simple and strong. Professor Kawakita always emphasized the necessity of using appropriate technology. If we suddenly take advanced technology to Nepal, it would not take root. From this point of this view, the concept of the paddle wheel was ideal for this environment.

Additionally, we found after the trial run that when the boat encountered drifting garbage or timber, the paddle wheel would climb over them like wheels on land, whereas such items would become entangled with a propeller. If one imagines the measures required if such items are entangled with a propeller, the advantage of paddle wheel becomes evident. Moreover, I found an extended use for this paddle wheel system. If you fit an electric generator instead of winch drum, the boat can be converted to a generator boat; if you fit a pump, it can be converted to a pump boat.

Technologies required in Nepal beside the ferryboat were pumping systems that brought river water to higher locations, and small-scale

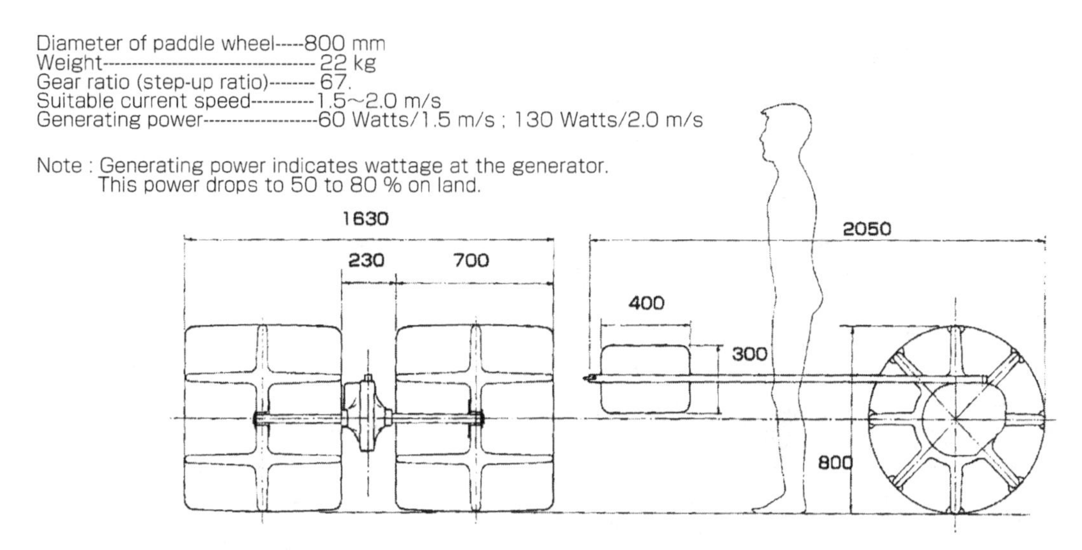

Fig - 3 Initial Plan designed by Yanagihara

electric power generating facilities. These technologies were already being implemented as part of the activities of ATCHA. Our paddle wheel generator system might contribute to small-scale power generation.

However, I kept this idea to myself. Unexpected difficulties were likely to be encountered when putting into practice new technologies in Nepal, so simple ideas on such topics might not be suitable. The prototype trials of the ferryboat were successful, but later it was reported that it must be possible for people to carry the boat on their back to the required location, and that the ferryboat should not be very small since it should be able to ferry cattle and horses. This suggested that a ready-made fishing boat was too heavy to carry and too small to use.

Finally, we continued to study selection of materials for building a large boat that could be carried manually to the intended location for ten years after the experiment at the Tenryu River. We were not in a hurry to bring the ideas on the generator system and the pump system to the table.

3) Horiuchi laboratory

Since then, I had been thinking about developing both generator boat and pump boat someday. In the meantime, I explored the idea of giving buoyancy to the paddle wheel blades and the generator case; in this case, we might not need a boat anymore. This project seemed to me to be rather important.

Imagine such a paddle wheel generator, connected to the end of a rope fixed on a river bank, flowing with the current, and the electricity generated coming to land through a cable fitted along the rope from a generator fixed between the right and left paddle wheels. If we can develop such a compact paddle wheel generator, it could be carried on a person's back, and there will be no need for a boat. Yanagihara had good knowledge about Nepal, but he was then working on developing a wind generator in a different department. I again consulted him. Fortunately, I got permission from his boss and I asked Yanagihara to make a plan for a paddle wheel generator without boat. His plans were made in detail and were perfect; they also included forecasts for commercializing the generator (Fig - 3).

I was, however, concerned about the large diameter of the paddle wheel, which was 80 cm, and the total weight of 22 kg. Since the plan was to use a motorcycle generator, the rotation of the generator had to be geared up to 70 times that of the paddle wheel. I was also worried that its structure would become complicated. However, the base of the total plan was fixed at this stage.

The Horiuchi laboratory started in January 1984. My thinking was that the laboratory should work on the development of new products difficult to implement within the conventional structure of the company and should foster talented persons suitable for those jobs. So, I gathered suitable projects for the laboratory and looked for appropriate talented personnel in the company.

The first person who joined the laboratory was Masato Suzuki. He joined the company

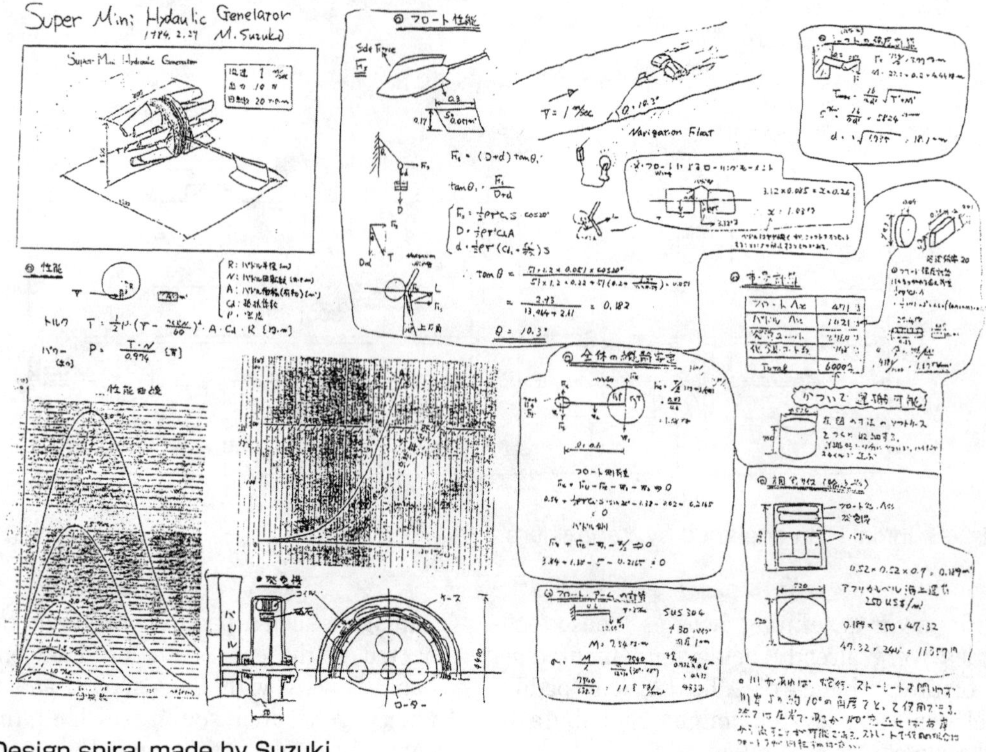

Fig - 4 Design spiral made by Suzuki

the previous year and worked in a department related to airplanes. Suzuki was a graduate of the Aeronautical Engineering Department, Faculty of Engineering, Nippon University. While in the university he participated in the Birdman Rally, and he had even operated a plane and won a race. He was an airplane addict. I had even recommended that the company employ him because he was introduced to me by Professor Akira Naito, the successor to Professor Hidemasa Kimura[2]. Suzuki had a twin named Hiroto who had joined Yamaha the previous year. Therefore, we called them by their personal names of Masato or Hiroto, and not by their family names.

I was worried that Masato might not be useful in some other section of the company because he joined the Horiuchi laboratory soon after his employment, and because the laboratory was not a conventional section. But I thought that after three to four years he could be moved to a conventional technical department to prepare him for a career in the company. Thus, Masato became the first member of the Horiuchi laboratory.

I looked for topics that would be appropriate for the Horiuchi laboratory. These were selected from the topics I was engaged in at the Head Office and while in the Boat Department. Finally, the number of topics came out to more than ten. I selected the paddle wheel generator and assigned it to Masato Suzuki.

The concept of the paddle wheel generator was as follows: it should be carried easily and once it was floated on the water, it should soon generate electricity adequate to light up a room, to listen to radio and possibly to watch television. That would require about 10 to 30 watts, which was adequate to change the current life style of Nepal radically. To realize these targets, the paddle wheel should be made smaller than half the size and lighter than the initially-planned paddle wheel, and if possible, without any gear to increase the rpm.

From May of that year, the actual work on hydrofoil OR51 (see chapter 8) started and Masato was assigned to work on both the paddle wheel generator and the hydrofoil. While the OR51 was an fixed-order project, the date of completion of the paddle wheel generator project was not fixed, and its work was likely to be intermittent and behind schedule.

At that time, Masato drew the complete plan of the paddle wheel generator on a sheet of paper (Fig - 4). This was the first step of the design spiral that I had recommended. Although it was not precise, it included all essential factors to be studied for the success of the project, including total checks from performance, size, weight, and structure and the balance of those items.

Masato then made the working model of the

2) Famous aeronautical engineer, previously Professor of Tokyo University.

paddle wheel without generator. He worked with Masayuki Hattori, who was employed in the Marine Department, and who was an expert in manufacturing prototypes. Masato measured the power performance (Fig - 5) by the model. We wanted to know its behavior and performance by placing the model in the river and actually performing measurements. Finally, the entire paddle wheel project from beginning to finish, was entrusted to Hattori.

The test was carried out near a camp area upstream of the Tenryu River not far from Hamamatsu City. The river width was small, and its depth and current speed were appropriate. To measure the power generated by the paddle wheel, we varied the frictional resistance on the shaft, and then measured the rpm from the outside from which finally we calculated the power, and evaluated its overall performance.

However, we were not ready with the carry-out system[3] in this model, which we had studied, as shown in the center of Fig - 4. To bring the paddle wheel to the offing[4] of the river where its current was maximum, Masato had to wear a wet suit and guide the paddle wheel in the fast current, utilizing the bend of the river, or he had to set a rope across the river and connect the paddle wheel system from the mid point of the rope and let it flow with the current. In this way, Masato worked hard and was keenly aware of the importance of the carry-out system.

During the trial, he encountered another problem and had a difficult time solving it. Originally, the paddle wheel was designed with buoyancy in the blades and generator case to eliminate the boat hull. Its waterline was adjusted to an appropriate level by the provided buoyancy when it floated on calm water without any current.

In practice, however, it submerged more than anticipated in the presence of a current. The reason was that firstly the forward blades were submerged because of the strong downward-acting water pressure and secondly, the forward water level rose because of the resistance of the paddle wheel, which made it sub-

Fig - 5 Working model (with torque meter)

3) Rope connected to paddle wheel generator is fixed on the river bank, but paddle wheel generator itself must be in the fast current in the offing of the river. It is important to send it to the offing of the river. We called this sending system the "carry out system".
4) Offing means nearly center of the river, far from the bank.

merge more. Because of these problems, the paddle wheel with only one-fourth of its diameter submerged in calm water, became submerged to the shaft level in the current and it rotated with great difficulty. Moreover, both directional stability and lateral stability deteriorated.

Naturally, unnecessary tension occurred in the towing rope and the efficiency of the paddle wheel degraded. This problem made us aware of the need for the carry-out system and gave us good experience at the stage of the functional model test.

4) Generator

Masato and I were not very familiar with electrical engineering and we often consulted Mr. Kitano, who was the technical general manager of Moriyama Kogyo Co. This company was a 100%-subsidiary company of Yamaha Motor. It developed and manufactured electric parts for both motorcycles and outboard motors. Mr. Kitano was previously general manager of the Electrical Department in Yamaha Motor. I knew he was always positive to new techniques in Yamaha. He understood the problem in the initial plan that used a motorcycle generator, and he designed for us a new generator that had no gear (directly connected to paddle wheel).

In general, it is not easy to obtain high output from a generator without gear because of the very low rpm and also because the number of poles tends to increase the size of the generator. Considering these characteristics and the portability, we planned on a new generator that had a paddle wheel of diameter 500 mm, two sizes smaller than the initial one and with a total weight of nearly 10 kg.

General Manager Kitano and Masato co-operated and made the generator. When we began the towing test of the generator with the paddle wheel using a motorboat, to our surprise we found that it did not rotate at all. The generator had 36 poles, which meant it had 36 small permanent magnets around the rotor disc, with wire-coiled steel cores outside each magnet facing each other (Fig - 6, 7). As the steel core and the permanent magnet exerted a pull on each other, the rotor did not start turning. When we forcibly turned it, the rotor rotated 10 degrees and stopped. To change the combination, we needed a large torque. The torque was called a cogging torque, a term that we heard for the first time.

Unless pressure acting on the paddle wheel overcomes the cogging torque, the paddle wheel does not turn. But once it starts turning, it continues to turn smoothly by inertia. At that time, we turned the paddle wheel manually to record data tentatively since we had no choice. This method was impractical, however, because one could not turn the paddle wheel manually in the offing (see note 3) in a current.

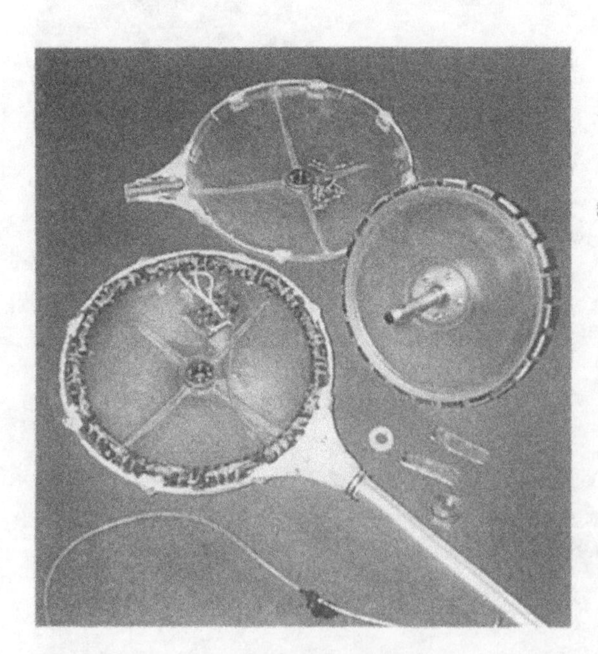

Fig - 6 Photo of disassembled generator

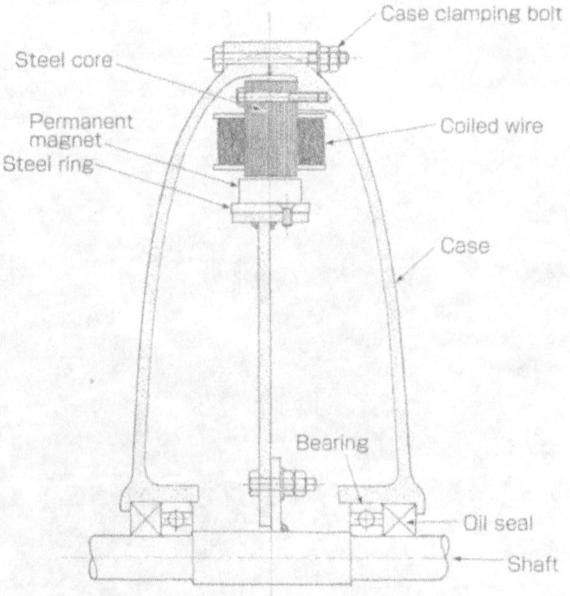

Fig - 7 Cross section of generator

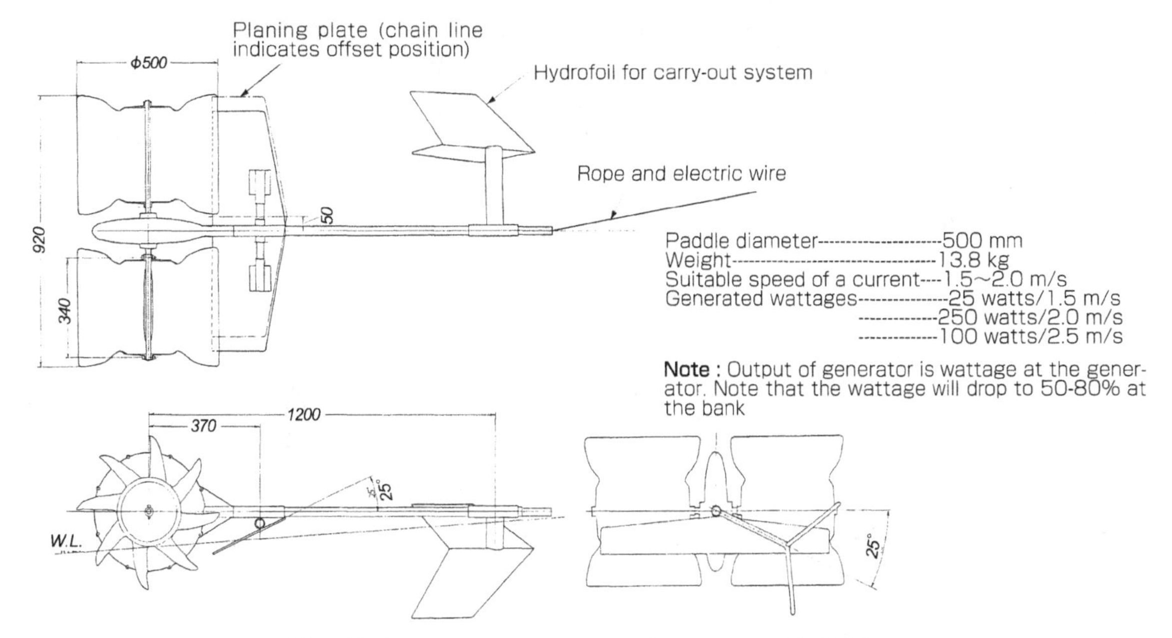

Fig - 8 General arrangement of prototype (OR48)

We had no idea how to solve this problem.

By using an ultra-strong permanent magnet named Neosium that was just developed at that time, the cogging torque increased. The price in those days was 700 yen per gram[5] of the new material. Although we used a new and expensive material, it caused a major problem later, which we did not expect, and that was unfortunate. We wished to get higher output from the directly connected generator, but we did not succeed.

Mr. Kitano helped us out of the crisis by modifying the shape and arrangement of permanent magnets. By arranging the magnets diagonally, the cogging torque was spread over the circumference and the peak toque decreased to one third. It could now be started in a current speed of less than 1.5 m/s and we somehow overcame the problem. At the critical current speed when it would not start, we jerked the rope attached to the paddle wheel in one swift motion so that it started turning.

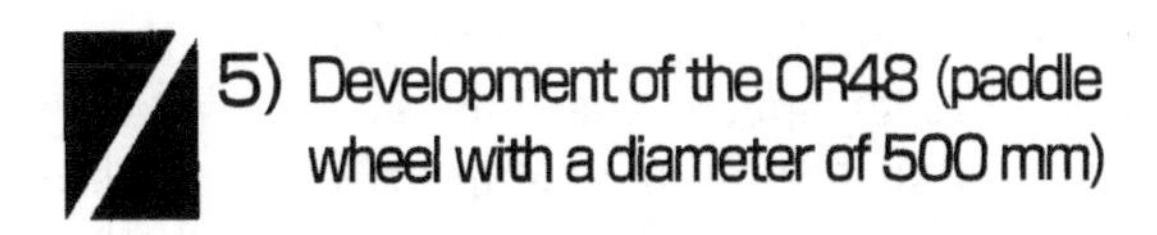

5) Development of the OR48 (paddle wheel with a diameter of 500 mm)

The development number was decided as OR48 in early 1985. Masato was developing both a new blade that had adequate buoyancy and a watertight case for new generator (Fig - 8) while simultaneously working to improve the generator.

The blade was molded with thick Styrofoam that had a complex curvature. Its casting mold was made of aluminum alloy, suitable for small-scale production. There were eight blades each on the left and right sides, and the total number of blades required for one generator was sixteen. If the blades were made one by one by hand, it would be a time consuming job. These blades were fixed to the shaft, eight on each side, by tightening cones that were fitted to the shaft (Fig - 9).

For the waterproof case of generator, we made a silicon female mold from the wooden plug, suitable for small-quantity mass production. The method for manufacturing the case was to pour liquid urethane in the mold and harden it to obtain solid urethane. By assembling these parts and a generator with small cogging torque, we intended to complete an improved generator. But Masato was very busy in producing the main parts and had no time at all to work on the carry-out system.

In those days, Isao Toyama of the Boat Department was responsible for the transfer of Yamaha production techniques for Yamaha professional outboard fishing boats fitted with Yamaha outboard motor to 25 developing countries of the world and to raise the level of fisheries in those countries.

His specialty was chemistry and I knew him since he joined our company. Since I had even arranged his marriage, we knew each other very well.

To teach lamination techniques for boats and molding techniques for small mass-produced FRP parts of lawn mower and others, Toyama

5) Price of normal permanent magnet made of ferrite was only one yen per gram.

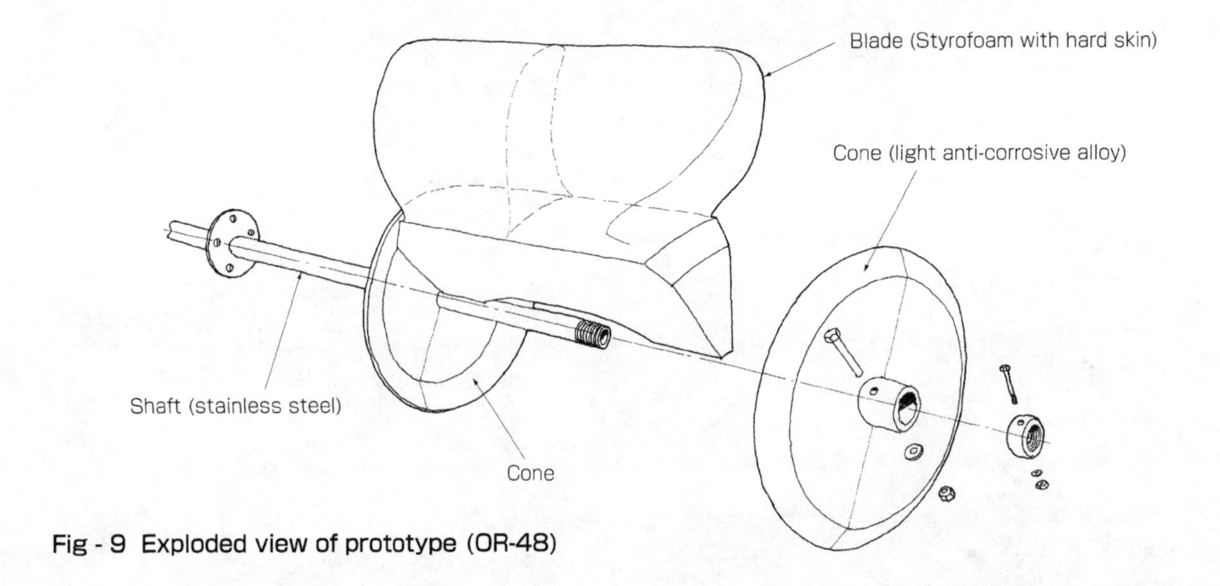

Fig - 9 Exploded view of prototype (OR-48)

often visited the Londonio Company, a Yamaha boat licensee located in Mederin city (second largest city in Colombia). Toyama was a technical adviser to the Londonio Company. It was natural that Toyama thought that the paddle wheel generator should be introduced to the Londonio Company. There are many big rivers in Colombia, building electric facilities was not easy, and electric supply was available only in cities. Rich farm owners of Spanish descent used electricity supplied by individual generators, but most indigenous people were poor and spent their life under lamps.

With 5 to 20 watts of electricity, they would lead a happy life at night. Londonio was already working to sell such a system combined with a solar battery (imported from Kyocera) and a charging device in Columbia. But the solar battery was too expensive and was useless at night. In addition, a charging system was necessary. These were problems that prevented popularization. On the other hand, a paddle wheel generator generates electricity even at night and needs lesser maintenance.

Considering these conditions, Toyama thought the paddle wheel generator would become popular and he contacted the Londonio Company. He approached both Londonio and the Horiuchi laboratory with a request to perform local tests of the paddle wheel generator in Colombia.

The negotiations that had begun in March 1985 were completed soon. The plan was to send the generator to Colombia in the middle of July for local tests. We were very astonished because our initial plan was to test it in July by ourselves. Therefore, Masato advanced our schedule and assembled the paddle wheel generator desperately to test it before sending it to Columbia. But the testing did not proceed smoothly. As I explained earlier, the increased buoyancy of the new paddle wheel blades should have solved the problem of the tendency of the paddle wheel to submerge. However, we found the paddle wheel still floated at a lower level than expected. The buoyancy was inadequate. This interfered with the work on the carry-out system and further degraded the stability of the generator. As an emergency measure, Masato inserted a planing plate just in front of the paddle wheel. The planing plate was a plate with an angle of attack(Fig - 8, 10) that produced lift and lowered the level of inflow of water, and made the rotation of the paddle wheel smooth so that its stability increased.

But the adjustment between the carry-out system and the planing plate was so delicate that Masato made it just before our departure. Finally, he got it into shape. I could not help him during the adjustments, but advised him after receiving his reports. We went downstream of the Akiha Dam on the Tenryu River. The current speed was only 1.5 m/s at the first trial at the Sasago camp area, but at the Tenryu River after the typhoon passed, the speed of current was presumed to be 2.5 m/s. Although it was a little dangerous, we performed the trials, confirmed that the generator could work at the current speed of 2.5 m/s or more. Finally, we could send the paddle wheel generator to Colombia.

Fig - 10 External view of prototype (OR-48)

6) In Colombia

On the morning of July 14, Toyama and I reached Bogota, the Colombian capital, with the packaged paddle wheel in the same flight. The Colombian Customs required three hours to inspect and clear the package, which appeared strange to them. But predicting such an event, two persons from Londonio were present to clear the package through Customs, after hard negotiations that we might not have been able to manage.

On July 15 the next day, after preparing for the trials at the Londonio head office, we departed in a 4WD-car. Members of the trial were Johnny Londonio, the owner of the Londonio Company, Mario Alango, engineer in charge of technical development and very fond of ultra light planes, and the two of us. Thus, four persons in all, went for the trial.

We intended to perform the trial in the Magdalena River. We performed the test at the bank of Antiokia in an area that the AIA Company owned. The AIA Company had a close relationship with the Londonio Company (Fig - 11). Just before the wide river entered a choppy, shallow, and narrow stream, we found a current speed of 1.5-m/s. We let the paddle wheel generator operate (Fig - 12) here. The water was calm, depth was 2 to 3 m, and wave height was between 2 and 3 cm. From a height of 3.6 m in a tree on the bank, we let flow the paddle wheel tied to a 60-m rope fastened to it and the tree. The carry-out system worked well, and the generator continued to rotate in the stream nearly 20-m away from the bank. At first, we had some trouble finding a suitable location, but in the end, the generator worked well and generated electric power of about 20 watts at 12 volts.

We stayed that night at a Spanish hotel in an old town said to have been built by the Spanish in olden times. The next day we went

Fig - 11 Test site in the Magdalena River (left) and members (right)
From left: Johnny Londonio/ Mario Alango/ Isao Toyama, photograph taken by Horiuchi

to the test site again and we increased the length of the rope to 95 m. Its carry-out distance became 35 m, and because speed of the current increased, we got 35 watts of electric power. The carry-out system and the planing plate, both of which were made in a hurry just before dispatch, worked very well on site enabling stable measurements, and I was very satisfied. Masato's efforts at the last moment proved to be fruitful.

As we found the paddle wheel generator could transmit electricity stably, we moved the same day looking for faster currents. We found a suitable place close to Boronbolo town where the current speed was 2.5 m/s or greater. We could get about 80 watts at about 20 volts by

Fig -12 Test in the Magdalena River

using a 60-m length of rope.

The tension in the rope however, reached 35 kg in a current of 2.5 m/s, and three persons would be required to hold the rope. Therefore, this current speed could be considered as the upper limit for operating the paddle wheel generator. We were very pleased with the tests and the stability at such speed. The paddle wheel generator gave a good performance and generated steady power, responding well to our expectations. After finishing the trial successfully, my concerns were how to decide the level of durability and suitable current speed.
On the 16th evening, we had a chance to meet Johnny and his father, and we asked them their frank impressions. I wished to get hints on how to decide the periodical overhaul cycle.

Johnny and his father were fully satisfied with its performance. They recommended an overhaul cycle every three months, which I thought was feasible. I had some concerns about the expenses of parts during overhaul, but this would depend on the efforts on the manufacturer's side and method of use by the user. I thought we should make efforts to attain the targets set at the start. The power generated by the prototype was too small for a current speed under 1.5 m/s, and it could not be started without manual aid at this speed. Therefore, the generator can be used in a current speed of 1.5 to 2.0 m/s. We even tested it in a current speed of 2.5 m/s, but this speed must be regarded as the maximum speed. Johnny wanted a new model for current speeds in the range of 1.0 to 1.5 m/s.

7) Second prototype (diameter of paddle wheel - 650 mm)

Masato was busy with the hydrofoil boat, but he gave careful thought to the paddle wheel generator, especially durability for a three-month overhaul cycle, performance below current speed of 1.5 m/s, and the cost considering the market conditions. Each task was difficult in its own way.

At the beginning of 1986, we compared many specifications by varying wheel size, gear ratio, generating capacity, cost, durability, and buoyancy, and finally we found one close to the ideal specification. It was a generator consisting of a wheel of diameter 650 mm, ferrite magnets, and 5:1 step-up gear using toothed belt (or cogged belt) (Fig - 13, 14). If a paddle wheel blade of 500-mm diameter was made, the generator could be used in current speeds of 1.5 to 2.0 m/s. Masato again designed it with these specifications and completed the second prototype with paddle wheel diameter of 650 mm. In July 1986, one year after the trial in

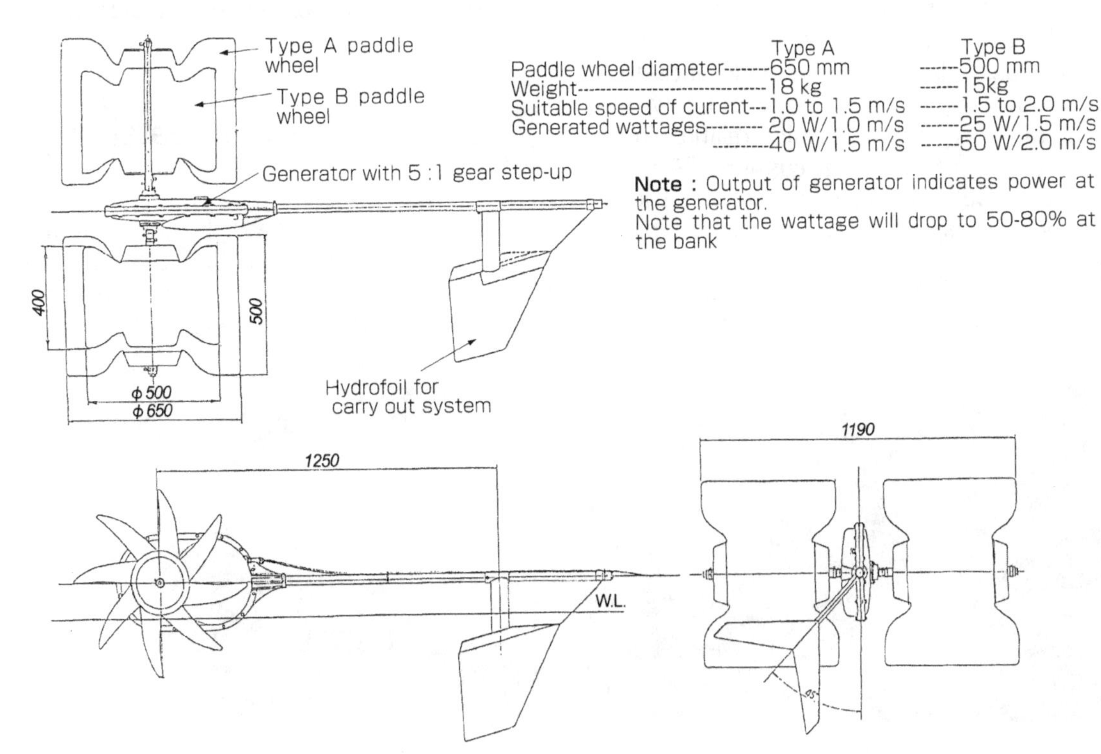

Fig - 13 General arrangement of second prototype

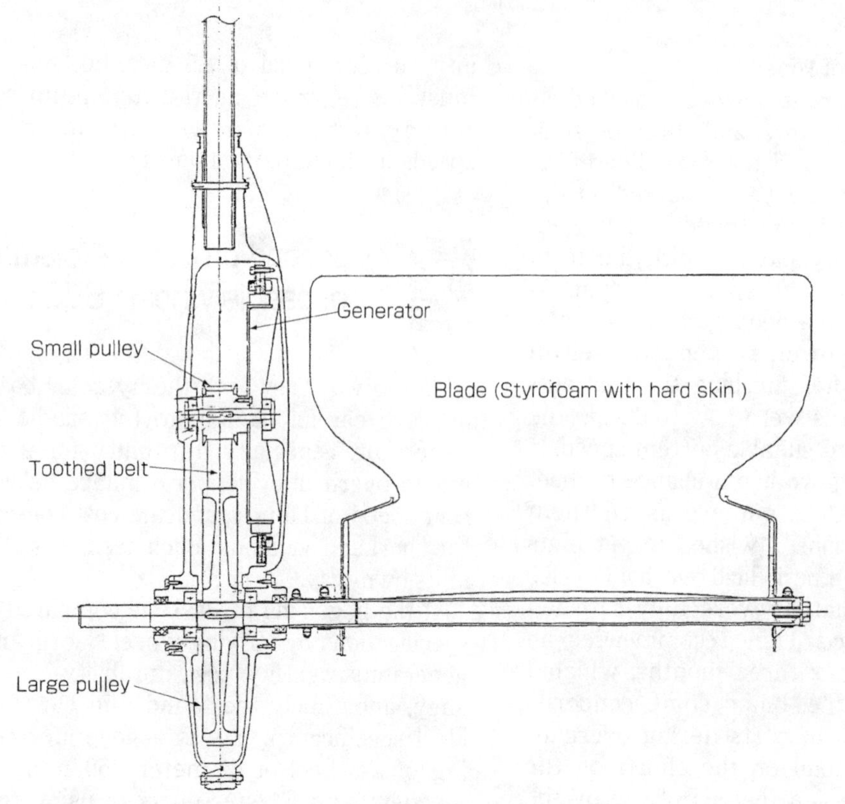

Fig - 14 Cross section of generator and gear for second prototype

Colombia, he started the endurance test at the Sasago camp area of Keta River (Fig - 15). There was a watch cabin at the camp area with a man stationed as watch. There was no electricity in the cabin, so after dark we supplied power from our generator to the cabin during the endurance test, which immensely pleased the man at the watch. During his usual job, he would also watch our endurance test, and should any problems occur he would inform us. This was a very nice relationship. I was not present at the discussions between him and us, but I am sure that Masato's friendly and sunny disposition brought about a pleasant relationship.

For one and half months, the paddle wheel generator worked smoothly. At the end of September, our endurance test ended because the watch cabin would close at the end of the summer vacation. I thought that the main parts of the structure had adequate durability after the endurance test. I also thought that it would be better to use a production model to confirm the three-month overhaul period.

8) Paddle wheel water pump

I would like to describe the water pump in this section. After completing the endurance test of paddle wheel generator, we had very little time to spare for the paddle wheel project.

Fig - 15 Second prototype
Floated well and rotated easily. Planing plate was not required.

But I still wanted to confirm the performance and feasibility of a paddle wheel water pump. I requested Masato to proceed with the manufacture of prototype by installing pumps in place of the generator. Two diaphragm[6] pumps used for bilge water discharge of boat were arranged in a V-shape (Fig - 16).

The two pumps were arranged in a V-shape to obtain uniform pump resistance and to lower the peak cogging torque. The pumps were operated through crank and rods fixed to the shaft of the paddle wheel. The paddle wheel water pump was completed at the end of 1986 and its test in the Tenryu River was successful. You can see Masato looking delighted with the success of the pump. (Fig - 17)

During the test, the speed of the current was 0.86 m/s, the pump head[7] was 1.3 m and its flow was 6.6 l/min. We could raise water up to three meters by pinching the mouth of the hose and increasing the pressure as shown in the photo. If the speed of the current is 1.5 m/s, its head increases to 2.3m, and the flow rate reaches 11 l/min. By pinching the mouth of the hose at the outlet, its head could be increased to over 5 m.

We arranged the piping of the two pumps in parallel. For piping in series, the flow would be half but the maximum head would be nearly 10 m.

At that time, Masato moved to the Design Department for motorcycle engines. He had worked for about three years at the Horiuchi Lab. In his place, Tsuide Yanagihara joined the Horiuchi Lab from another department in the company. Yanagihara's main job at the Horiuchi Lab was to develop a lean machine (see Chapter 12). Thereafter, he and I worked on the jobs related to Nepal that came occasionally.

9) In Nepall

At the end of 1987, Yanagihara was scheduled to make a business trip to Nepal to build the ferryboat there. During the meeting with Professor Kawakita before leaving Nepal, the professor requested Yanagihara to give a practical demonstration of both paddle wheel gen-

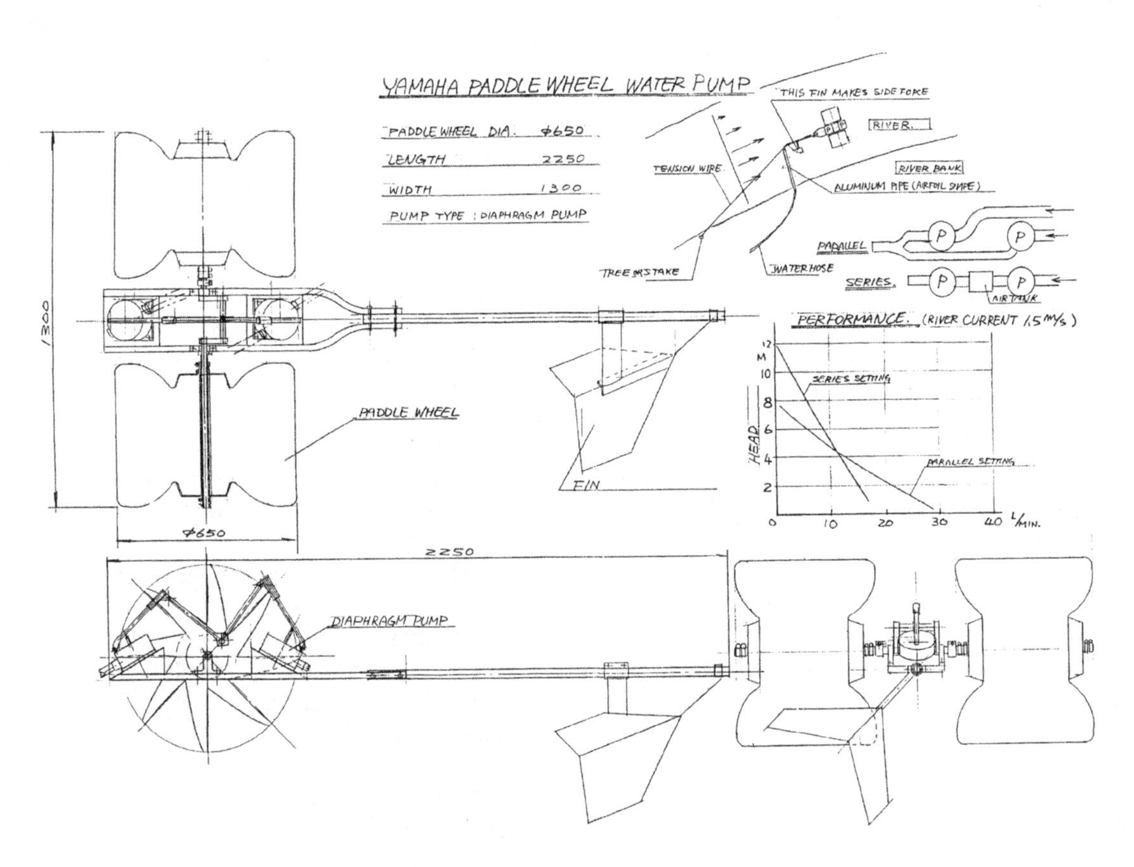

Fig - 16 General arrangement of diaphragm type paddle wheel water pump

6) Manually operated pump that wide rubber diaphragm is pumped by lever and get the big amount of flowing, normally used for bilge water discharge of sailboat.
7) Height that pump can raise water

Fig - 17 Masato testing the diaphragm water pump

erator and paddle wheel water pump together with the demonstration of the ferryboat. The applications that we considered are as shown in Fig - 18. We thought that the paddle wheel generator would not pose any problem, but the water pump would encounter problems in Nepal, with the need for raising water from deep valleys to houses located on top of cliffs. Therefore, the pump should be suitable for the increased head. The head required would be about 50 m, although the flow rate could be smaller.

The head for our diaphragm pump was too small. On the other hand, we could increase the head with a cylinder pump by reducing the diameter of the cylinder to reduce the flow. Yanagihara performed trial calculations and found that 50-m pump head could be obtained from a paddle wheel in a current speed of 1.0 to 1.5 m/s. In haste, he designed a high lift pump with two cylinders of 22-mm inside diameter arranged in V-shape and made the prototype. The paddle wheel of 650-mm diameter for generator was diverted and used for the pump. We were in successful at its trial in the Tenryu River on October 17, 1987, and this was just in time before Yanagihara's business trip to Nepal in the same month.

You can see its layout in Fig - 19. The photo of pump in operation is shown in Fig - 20. We performed trials of the pump under conditions that correspond to pump head of 50 m, and confirmed that it could pump up water at 1.34 liters per minute at a current speed of 1 m/s, that is, two tons of water per day. This amount was considered adequate for a village with 10 to 20 households, and we were satisfied with the results (Fig - 21).

Demonstrations of ferryboat, paddle wheel generator, and paddle wheel water pump were completed successfully in Nepal, and we entrusted further trials for practical use to the Royal Nepal Academy of Science & Technology (RONAST).

I heard that in January of the following year, the paddle wheel generator and the paddle wheel water pump were reported in the news on Nepal TV, attracting much attention. We received another report that the paddle wheel generator that we left behind in Nepal was

Fig - 19 High pressure pump (perspective drawing)

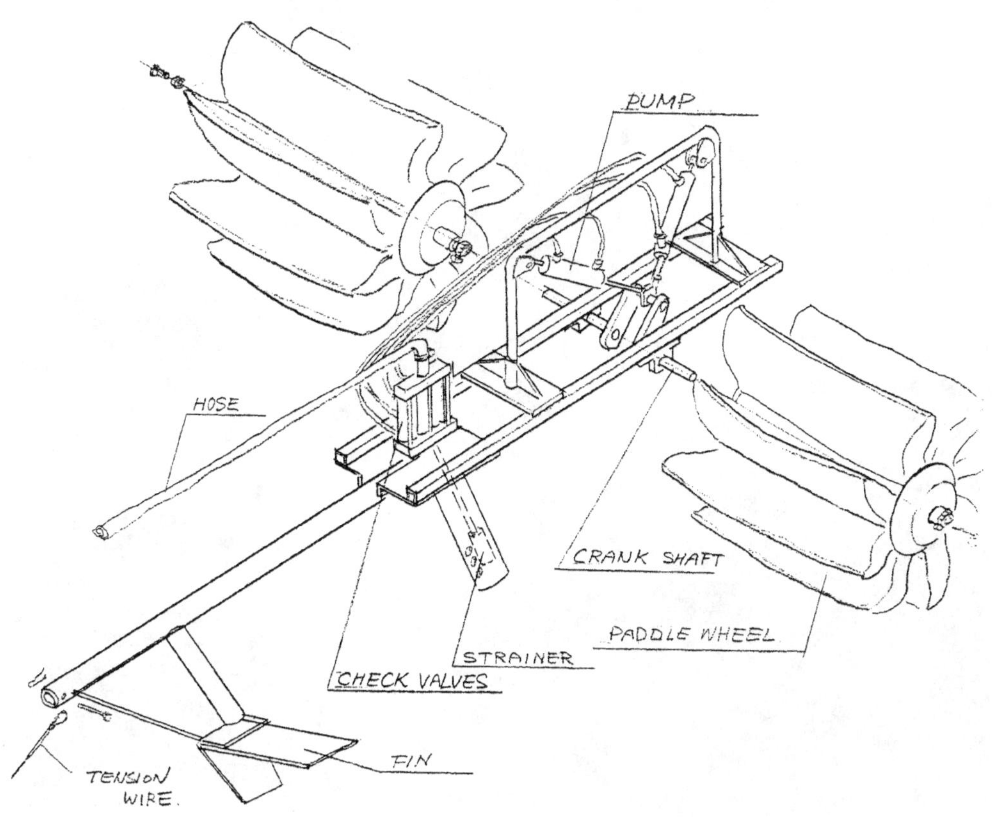

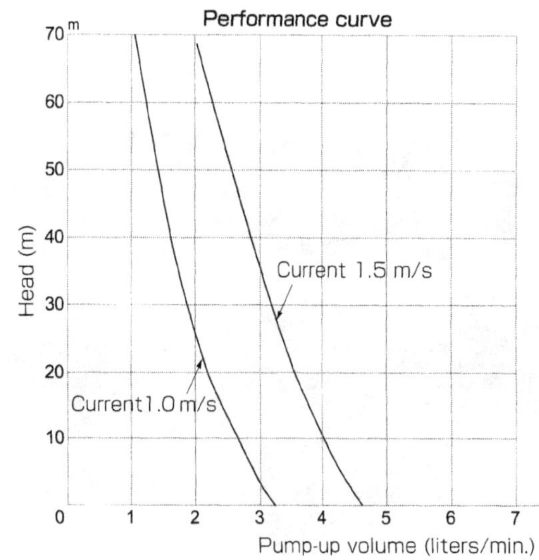

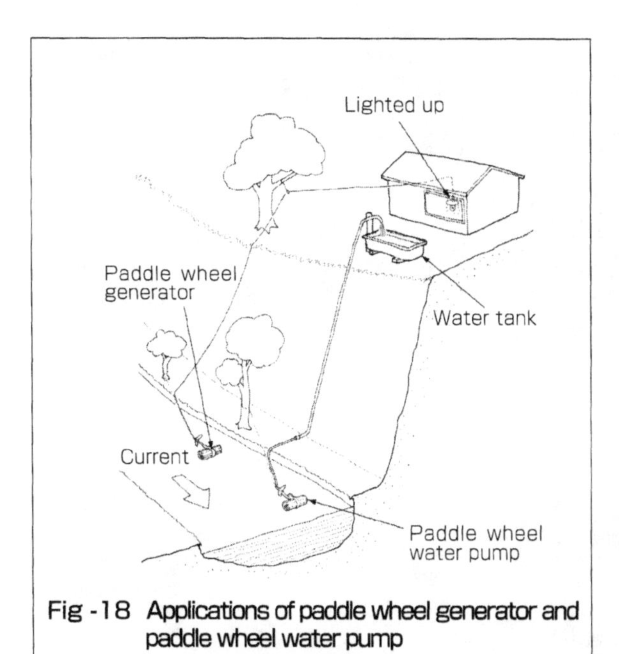

Fig -18 Applications of paddle wheel generator and paddle wheel water pump

Fig - 20 High pressure pump trial

Fig - 21 Overall view of high-pressure pump

used for lighting a roadside restaurant that had no electric light, and that the restaurant had prospered. Unfortunately, another report pointed out that after one month of prosperity, drifting wood hit the generator and damaged it. They could not repair it on site and it remained unused.

Our efforts were highly appreciated, and there were many expectations. However, we have not seen these products popularized subsequently. Our fully developed techniques were scheduled to be taken over by RONAST, which planned to do additional studies and popularize the items, but I have not heard of any progress.

Our efforts were to be introduced in the official report of RONAST, but to our regret, no local company came forward and took up our ideas to commercialize the prototypes. Under such circumstances, for the project to take root, a substantial amount of investment would be required to establish them so that there is no going back.

10) Conclusion

The paddle wheel world that began from Nepal's ferryboat project was attractive and very interesting, and we came to know the many good qualities of this clean, energy-saving, classical device .

Unfortunately, our contribution to Nepal did not proceed as we expected. There were some difficulties faced by related organizations in both Nepal and Japan, popularization was inadequate and, so we do not think that we contributed much to Nepal. On the other hand, we were constrained in promoting such activities by ourselves, since we were part of a profit-making company.

By chance, the dream meant for Nepal branched off to Colombia, thanks to Toyama's wide knowledge and guidance. This was mainly because of his desire to improve the standard of living in developing countries. On our part, we expected to popularize the products, as we in Japan generally look up on such matters from the commercial aspect. However, we were unable to bring it to the production stage although improvement of the machines progressed.

If the price of solar batteries decreased drastically, we predicted that the paddle wheel generator with many moving parts would be disadvantageous. This was one of the reasons why we were not very aggressive in bringing it to the production stage.

Since then, the performance of the solar battery has improved and its price has decreased. Nowadays, according to mail-order catalogs for boat accessories, the price of a 5-Watt solar battery is nearly 100 US$ and for 10-Watt battery, it is nearly 150 US$. Anybody can buy them. However, the solar battery does not work at night and is less effective in the morning and the evening. The average electrical power obtained becomes approximately one tenth of the face value. Therefore, a solar battery electrical system will only work at night provided it has large capacity and a charging device. This will be largely disadvantageous in Nepal and Colombia. Considering these points, the paddle wheel generator is still very competitive.

Also, an alternative to the paddle wheel water pump will not be available in places such as Nepal and I believe that if it became popular, it will be relied upon for long time in Nepal. Unfortunately, we have lost the chance for popularizing both paddle wheel generator and paddle wheel water pump, although we had successfully carried out trials there. We should probably make more efforts, and from now on, I think we should try to get in touch with appropriate organizations like the ODA[8], which is working to help developing countries.

The study and development of the paddle wheel generator and the paddle wheel water pump have been completed. It may be said that our efforts have not been of much use, but I am satisfied with our thorough studies on the paddle wheel. The studies on the carry-out system and effective planing plate were extremely interesting, since they were similar to designing aircraft. Masato, who joined the workforce for the first time, probably learnt a lot, and gained valuable experience that he might never have gained elsewhere.

I would like to give young people a chance in developments in which they can be engrossed. Regardless of the practical utility, it is important that young people try to do their best and aim for success through heir own efforts. I believe that providing such opportunities to young engineers is the best way to foster and cultivate talented engineers who can envision and dream up new technologies.

8) Overseas Development Assistance

16

HIGH-SPEED MOTOR SAILOR

With lightweight material for constructing the boat and a light outboard motor of over 200 HP, the general idea that a motor sailer has slow speed does not hold good. Actually, motor sailers with speeds of over 20 knots are selling well. My boat, Hateruma, was built under the same concept, and I considered using the same concept for 24-feet and 30-feet boats, which are popular sizes. I intend to install a new mast and an auto-sailing system particularly on the 30-feet boat. These boats will become popular in the future with people who do not wish to learn sailing, but want to enjoy the sea without working too much on a sailboat, and want to enjoy sailing in good water conditions within a limited time. Don't you think so?

1) Sail versus power

Mr. Akira Yokoyama, our senior in sailboat design, used to say that a motor sailer with a 50 : 50 ratio for sailing ability to powering ability, had never been successful. Recent motor sailers generally have normal sailing ability, similar to the usual sailboat, with 3 to 4 knots higher speed under power with a larger engine. The sailing ability to powering ability ratio may be 80 : 20.

Until now, it has been considered that in a motor sailer if the powering ability increases, the sailing ability decreases, so the total never exceeds 100%.

But around 1993, a motor sailer named MacGregor 26 with remarkably high powering ability appeared in the US market (Fig - 1). This boat had average sailing ability and a speed under power of over 20 knots, which stands out since conventional motor sailers a speed under power of about 10 knots only. I heard that this boat is selling so well all over

Main particulars

Length overall .... 7.874 m
Length waterline .... 7.010 m
Maximum breadth .... 2.388 m
Draught .... 1.767 m
Engine .... 5 to 50 PS outboard motor
Speed .... 20.73 knots
Ballast .... 636 kg
Displacement .... 1150 kg
Sail area .... 26 m²

Fig - 1 Sailing and powering of MacGregor 26 (from leaflet of MacGregor 26)

the world that MacGregor, the builder of the boat, has discontinued production of other models and is concentrating only on MacGregor 26 (recent annual production was 1000 units). The main point of attraction of this boat seems to be the superior powering ability. Also, its exceptionally economical price of under 3,000,000 yen has also helped the company to maintain good sales.

A long time ago, I heard from Mr. Yokoyama that there was no lightweight and powerful engine available in the market, and materials for constructing boats were limited. Therefore, boat weight could not be reduced adequately, so a total of 100% comprising sailing ability and powering ability was the only choice. Conditions now, however, are different. We can use powerful and lightweight outboard motors and lightweight structural materials for construction, so we should change our way of thinking. My assessment of the MacGregor 26 was that the boat had 85% sailing ability and 60% powering ability considering its size, so that total is over 100%. Giving weight to sailing does not hinder the powering performance, or installing a more powerful engine does not hinder sailing performance to a large extent.

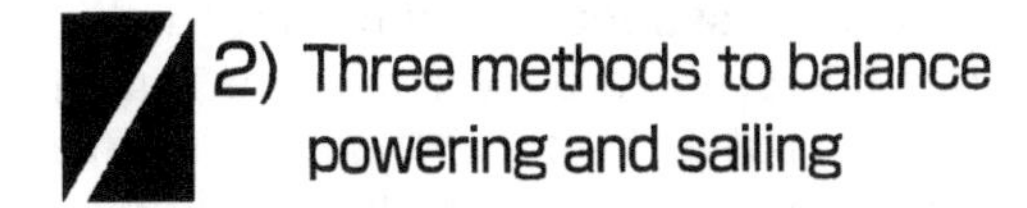

2) Three methods to balance powering and sailing

As sailing depends on wind and the speed with sails is only a few knots, a displacement type hull is suitable for sailboats. On the other hand, for high speeds, the boat must plane, and naturally, a planing hull shape is required. A good high-speed motor sailer is available only when a hull shape that adapts to these two modes is available.

As you are aware, the bottom line of a sailboat rises almost to the waterline at the stern. Therefore, when running, its transom face (vertical plane at the aft end of boat) does not touch the waterline. If it touches, eddy current occurs at that spot generating resistance, and the transom is called "dragging transom", which is not preferred. To identify the characteristics of a boat, a naval architect uses the area curve. It is a graph that gives the distribution of area of the cross section under the waterline from bow to stern in the form of a curve. For a sailboat, the sectional area is maximum around midship (middle of the boat length) and zero at both bow and stern (Fig - 2). When the curve shows some value at the stern, it indicates the existence of a "dragging transom."

For a planing boat, the sectional area increases as one approaches the stern. Since the shape indicates the "dragging transom", a planing boat has large resistance below 10 knots, but it has much less resistance when planing at high speed. The longitudinal position of center of buoyancy is the same as that of the center of gravity of area curve. As you may notice in Fig - 2, the center of buoyancy of the planing boat is positioned further aft compared to a displacement boat. As mentioned above, the longitudinal position of center of gravity coincides with that of the center of buoyancy, and the center of gravity of a planing boat is further aft than that of a sailboat.

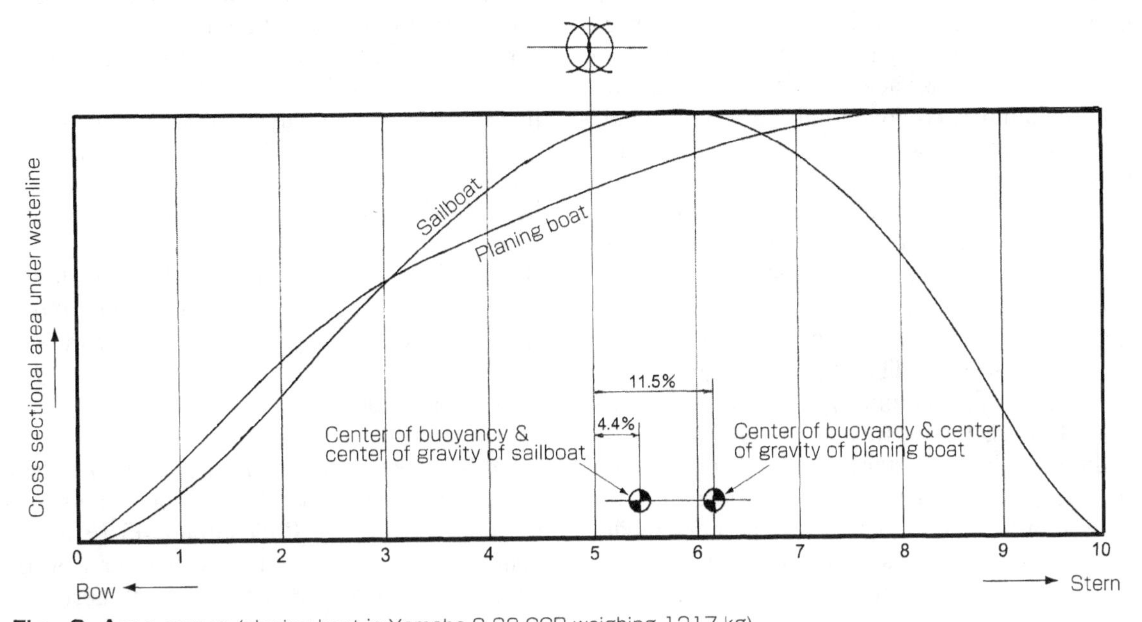

Fig - 2 Area curve (planing boat is Yamaha S-23-CCR weighing 1217 kg)

Fig - 3 Stern view of MS-21

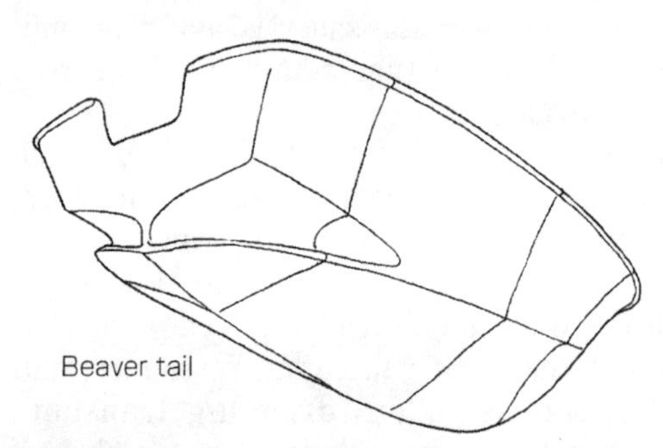

Main particulars
Length over all----------6.47 m
Length waterline------- 5.49 m
Maximum breadth------2.30 m
Draught-------------------- 0.3 m (1.73 m keel down)
Ballast---------------------- 150 kg
Sail area-------------------13.9 m²

Fig - 4 MS-21 on sailing

Considering these characteristics, basically there are differences between a displacement and planing type boat in the area curve and in the position of center of gravity. These differences have prevented the appearance of the high-speed motor sailer for a long time.

2-1 ; Flat bottom type boat

We can fill the gap between these two hull forms and make a hull that has moderate resistance at both high and low speeds. A long and lightweight boat has fewer gaps to fill. For instance, in the MacGregor 26, they have built a lighter boat compared to its length and moreover, eliminated the ballast keel to save weight. As a result, the boat can run at high speeds when planing, while its stability can be increased by filling water in the bottom of the keel during sailing. This is similar to a high-performance sailing dinghy, which has good performance as a displacement type boat under calm conditions, whereas under a strong wind, it has a good performance as a planing type boat. In case of a dinghy, instead of filling water, the crew move around shifting their weights. By hiking out, balancing, and trapezing, adequate transverse stability is ensured and at the same time, the longitudinal center of gravity is shifted to a position suitable for planing. If we use a hull like the MacGregor 26, we can call it a flat bottom type hull, as it has no special characteristics except a flat bottom. In addition to building lighter boats compared to their length to obtain improved performance at both high and slow speeds, several other modifications to hull shapes have been tried out.

2-2 ; Beaver tail type boat

A hull form designed around 1970 by Shigemitsu Aoki of Yamaha has the shape shown in Fig - 3. It has a part stretching under the bottom that looks like a tongue or a beaver's tail. This part extends the planing surface straight toward the aft. With this shape, the boat had good planing performance and its resistance did not differ much from that of the resistance of a sailboat at slow speed. We called this hull shape a beaver tail type hull. It looks strange, but as mentioned earlier, its area curve has a zero value at both bow and stern, and it has a smaller resistance. On the other hand, at high speeds, the boat rises on the water and the complex surface of the stern does not touch the water, so that it runs on a flat bottom at the aft end with the same water flow as that of planing type hull form.

Yamaha sold the MS-21 that used such a hull shape from 1974 (Fig - 4) onward. As the boat had a performance similar to that of a sailboat and gave a speed of 20 knots with a 55 HP outboard engine, it was well received in the market. More than a hundred boats of this type were produced. We can still see this model in the used-boat market today. This boat was probably the first of the 20 plus knot motor sailer that was mass-produced in the world.

2-3 ; Step type

I designed another hull form in 1987, which was used on my boat Hateruma and its production model, the Philosopher 45 (Fig - 5), about which I have written in Chapter 1 of this book. I would like to describe it again here.

This boat (Hateruma) has an ideal hull form and position of center of gravity of a planing boat, yet has a moderate underwater hull form and a position of center of gravity that prevents dragging of the tail for sailing, which were the targets for the boat. The powering ability to sailing ability of this boat is 100% to 30%. The characteristics of the hull form can be seen in Fig - 6. At slow speed, the water flows along the stern slope that rises toward the aft, therefore, its area curve is almost zero at the stern, and shape of area curve is similar to that of displacement type boat although the curve in the middle is not smooth. By making the hull shape so as not to drag the transom, I reduced its resistance at slow speed significantly. We decided to call this hull form a step type hull form.

When a step type boat is planning, the stern slope after the step separates from the water and the boat gets on the plane on only the planning surface forward of the step (Fig - 7). In this case, nothing spoils the shape of planning hull form. Additionally, as aft edge of planning surface is located considerably forward, the position of center of gravity comes a relatively suitable position aft of the planning surface. Thus the hull shape brings its planning ability into full play.

On the other hand, at slow speeds, the hull

Fig - 5 Hateruma

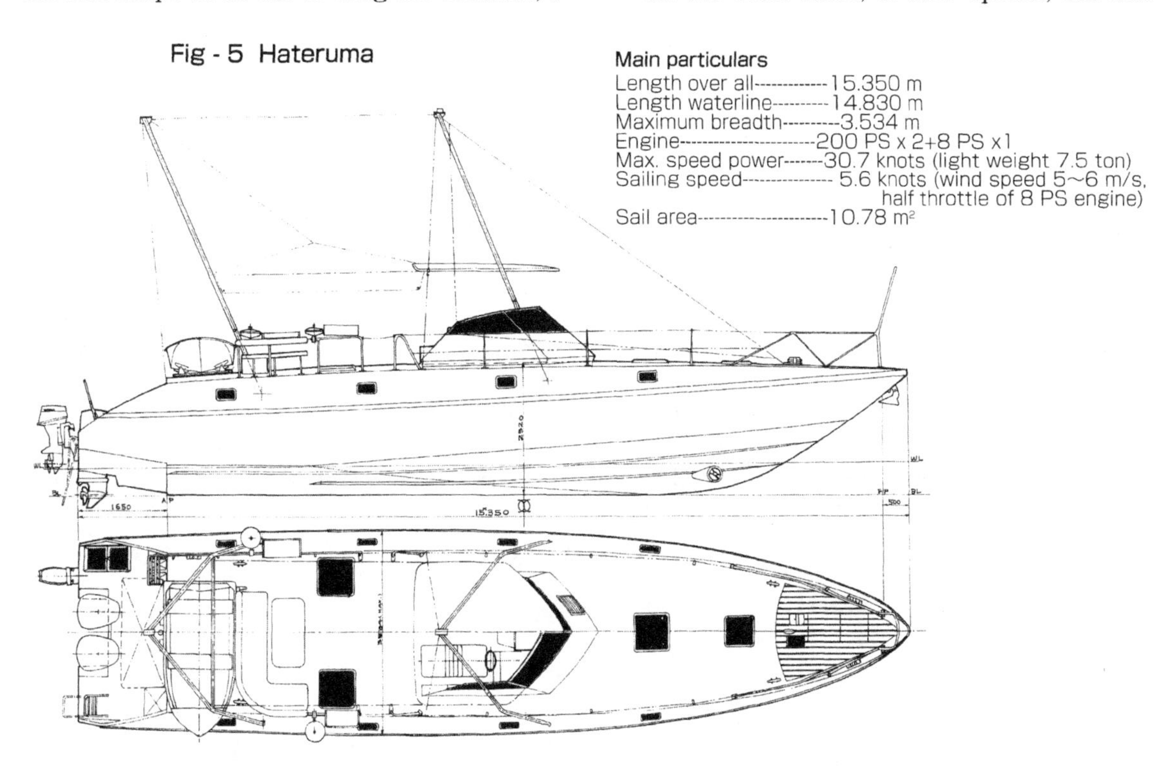

Fig - 6 Area curve of Hateruma vs. Sailboat

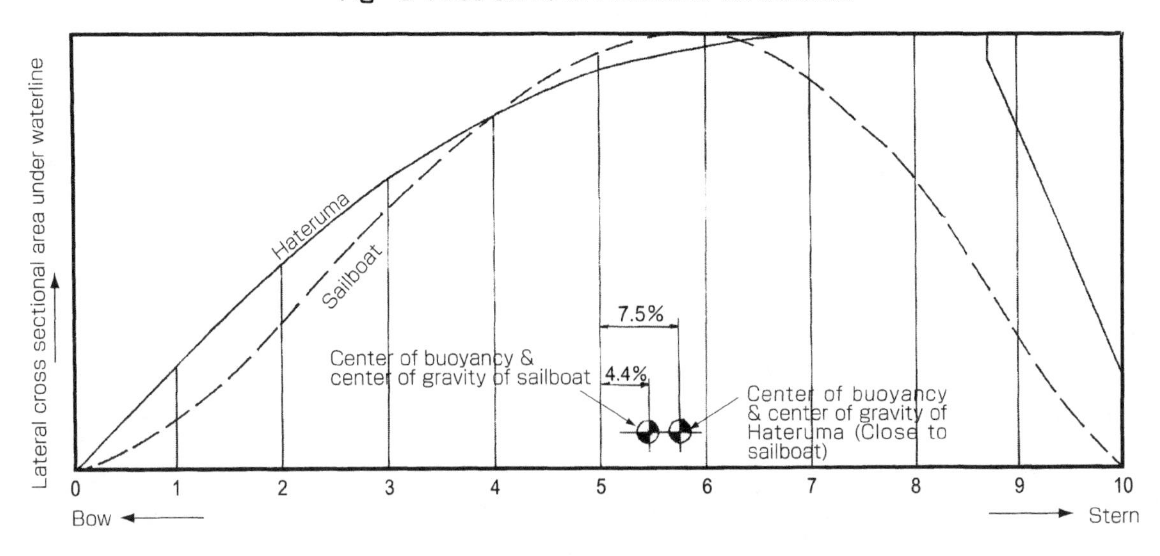

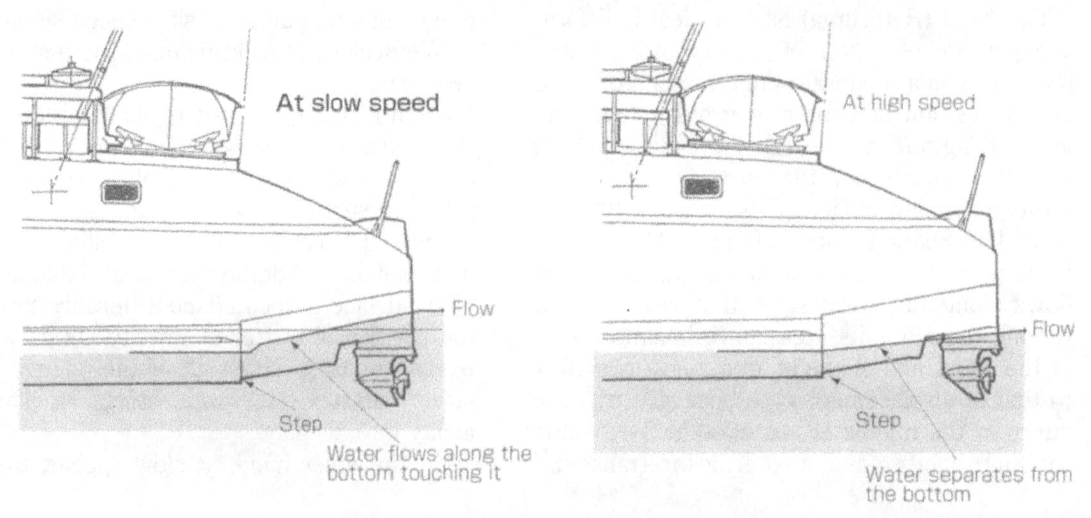

Fig - 7 Water flow at the stern of Hateruma

shape naturally has larger resistance compared to a pure sailboat, as the area curve of the boat is not smooth. You can see the difference in Fig - 8. The figure shows the comparison data of the Hateruma type hull form based on tank test results, and resistance data of a typical sailboat with the same displacement and waterline length as Hateruma. In Fig - 8, comparison of the resistance of bare hull without appendages such as keel, rudder, and propeller is shown. Additionally for comparison, I have shown the resistance data of the modified Hateruma type hull form, in which the aft part of the step has been cut off so that the transom became a dragging tail.

The resistance curve of sailboat is also shown on the graph③, and for comparison, the resistance curve of the sailboat with propeller is also shown④. In ④, the propeller is not a folding type but a fixed propeller. The propeller is locked to prevent rotation, and the resistance of the propeller is as much as 30% that of the bare hull. In Fig - 8, you can see the effect of the part aft of the step. At about 5 to 6 knots, the resistance becomes half, if the part aft of the step exists. On the other hand, the resis-

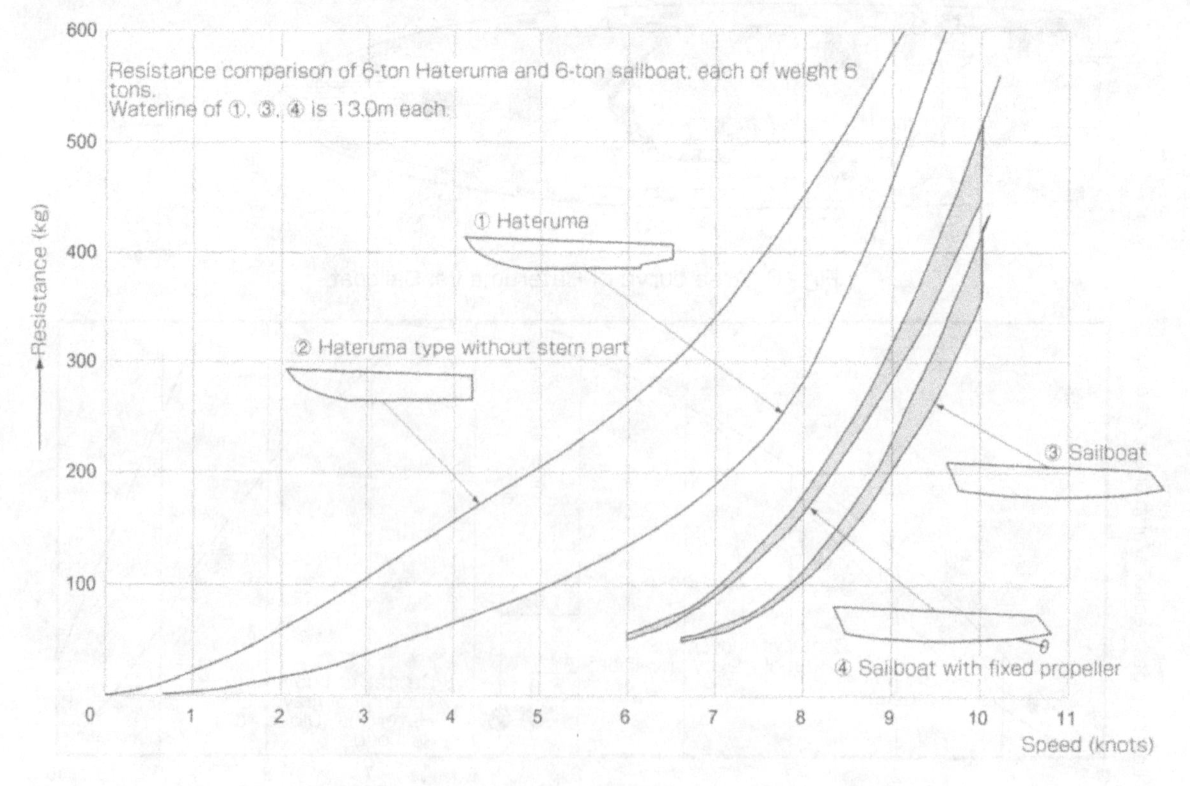

Fig - 8 Comparison of resistance of Hateruma and Sailboat

tance of sailboat with fixed propeller increases considerably. Consider a sailboat with twin fixed propellers, and its resistance will become almost same as that of Hateruma. The step type hull form is attractive because of its advantage that the hull bottom shape and position of center of gravity can be automatically adjusted to suit both displacement and planing type hulls.

3) MS-24 and MS-30

I launched Hateruma 13 years ago. The boat's performance was a good balance between the 30 knot powering ability and sailing ability. But with its length of 15.3 m, it needed two 200 HP outboard motors, and it was not a popular size. So, this time I wish to consider two boats with sizes that can be easily popularized. One is the MS-24, a bit smaller than MacGregor 26 with almost similar accommodation (Fig - 9). Since its length is less than 8 m, the charges for storing it in a marina, for example, in the Yokohama Bayside Marina, will be about 420,000 yen a year, which is economical in Japan. Considering marketing the boat overseas, I selected a trailerable size and sail rigs (Fig - 10).

The other boat MS-30 is of a larger size and is capable of both easy sailing and auto sailing, aiming to be an ideal motor sailer (Fig - 11). Each boat has 25 to 30 knots maximum speed at 100% power and about 20 knots cruising speed, yet the sailing performance is also considered to be good. Thus, the aim is powering ability to sailing ability ratio of 100 to 80.

Each boat has a step type hull shape and one large outboard motor. If a boat runs at high speed with the ballast keel down, it will heel outward when turning, which is dangerous; so the ballast keel must be retractable. In the MacGregor 26, a light centerboard is installed and water filled in its bottom for stability. But in our case, no water is filled in the keels of any of the boats, since each boat has a retractable ballast keel with a weight of 300 and 600 kg respectively. The weight is disadvantageous when planing because displacement increases, but this is covered by providing a larger horsepower engine. On the other hand, since the center of gravity of ballast keel is low when sailing, the ballast weight can be made lighter than the weight when water is filled in the bottom of the keel, so the sailing speed is greater. The ballast keel is moved up and down electro-hydraulically. When the ballast keel is raised to its upper position at high

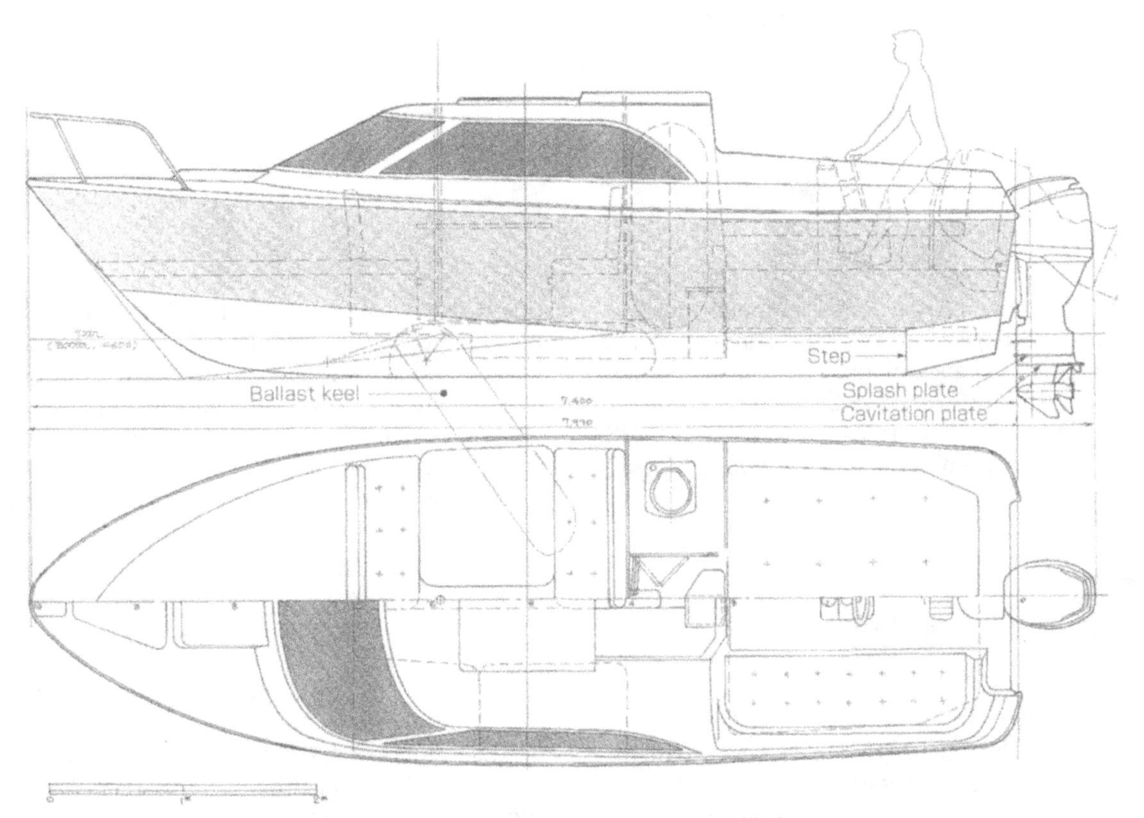

Fig - 9 General arrangement of MS-24

Fig - 10 Sail plan of MS-24

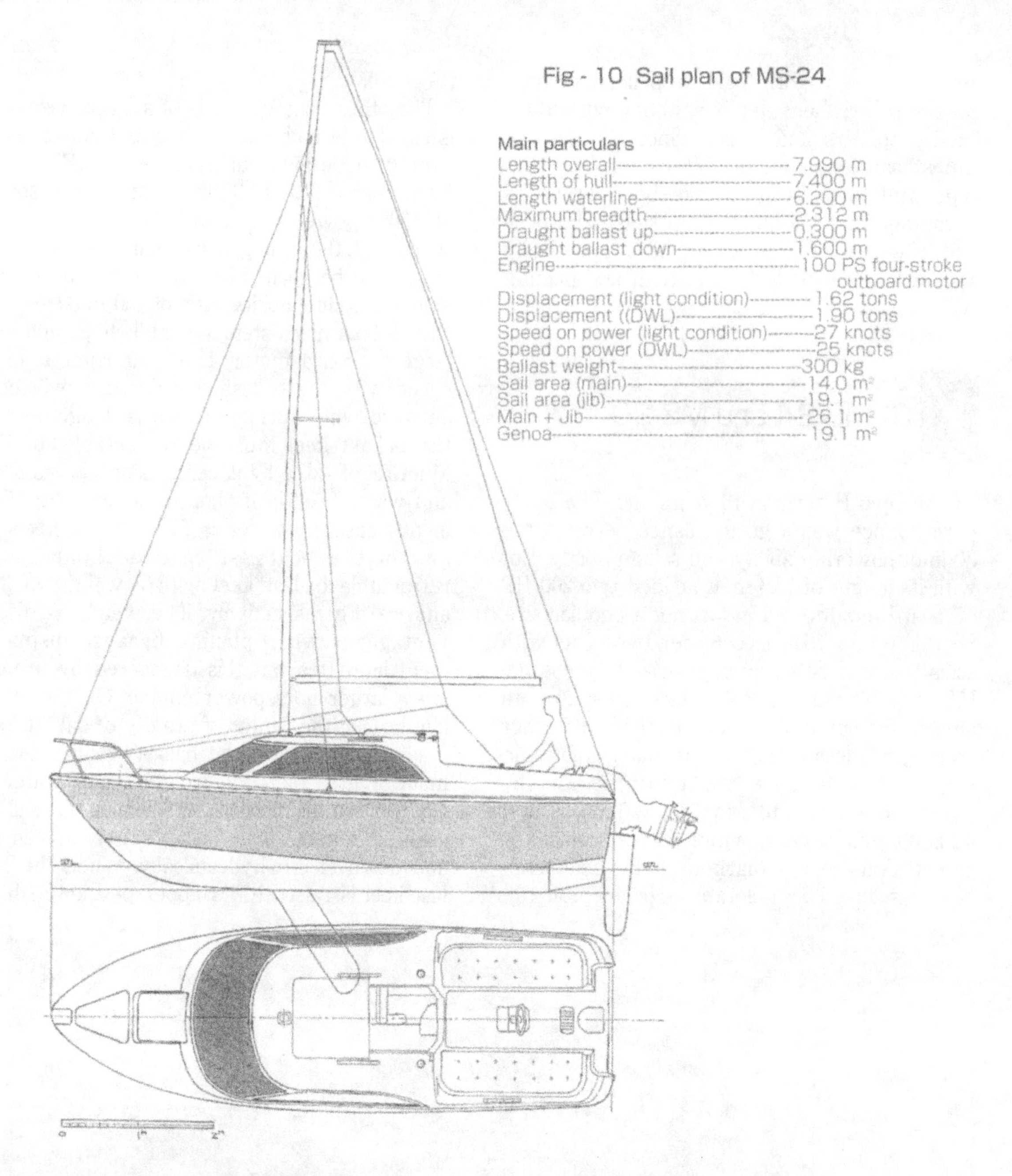

Main particulars

Length overall	7.990 m
Length of hull	7.400 m
Length waterline	6.200 m
Maximum breadth	2.312 m
Draught ballast up	0.300 m
Draught ballast down	1.600 m
Engine	100 PS four-stroke outboard motor
Displacement (light condition)	1.62 tons
Displacement ((DWL)	1.90 tons
Speed on power (light condition)	27 knots
Speed on power (DWL)	25 knots
Ballast weight	300 kg
Sail area (main)	14.0 m^2
Sail area (jib)	19.1 m^2
Main + Jib	26.1 m^2
Genoa	19.1 m^2

speed, a strong mechanical stopper must be provided to prevent the keel from dropping down due to wave impact. You can see the layout in the figures. MS-24 looks similar to MacGregor 26, and MS-30 is, as mentioned previously, looks similar to a conventional 30 ft sailboat.

4) Hull shape

According to the classification mentioned above, the hull shape of MS-24 belongs to the step type (Fig - 12) with the forward part of the hull having a so-called "double phase" shape that I recommend (Chapter 1, Fig - 6). When planing, the dihedral angle of the "double phase" underwater line is almost 20 degrees, maintaining a non-twisted ideal planing surface. The upper part of the waterline has a dihedral angle of about 40 degrees that produces soft riding. Accordingly, the hull shape is an ideal one, with a good balance between planing ability and soft riding ability. It has a sloping bottom aft of the step, as mentioned previously, that goes straight up to the waterline. As the distance between the aft end of the planing surface and the lower unit (underwater part) of the outboard motor is large, the boat has good longitudinal stability when planing and its bow rises to enable the boat to run at high speed. The outboard motor can be set higher,

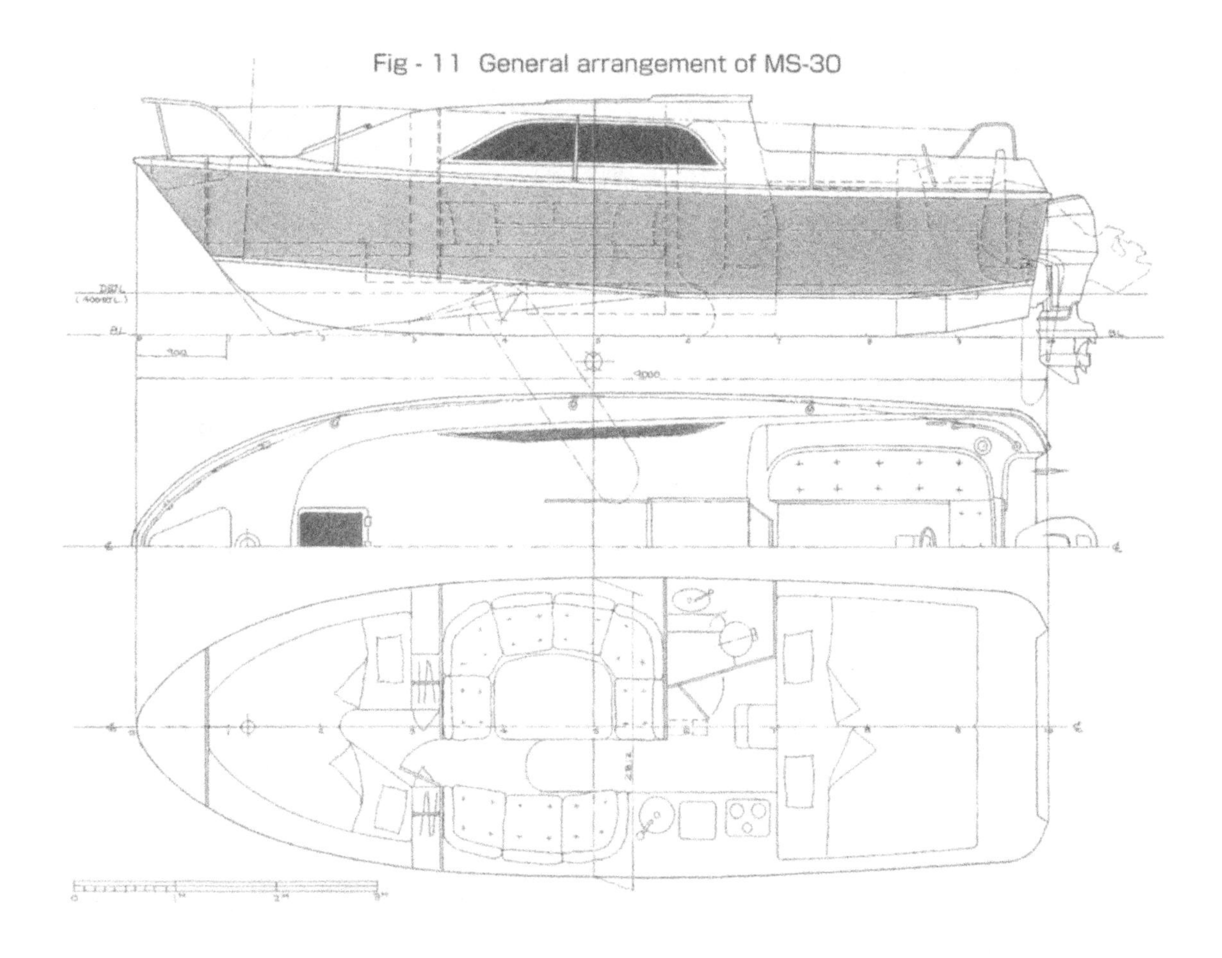

Fig - 11 General arrangement of MS-30

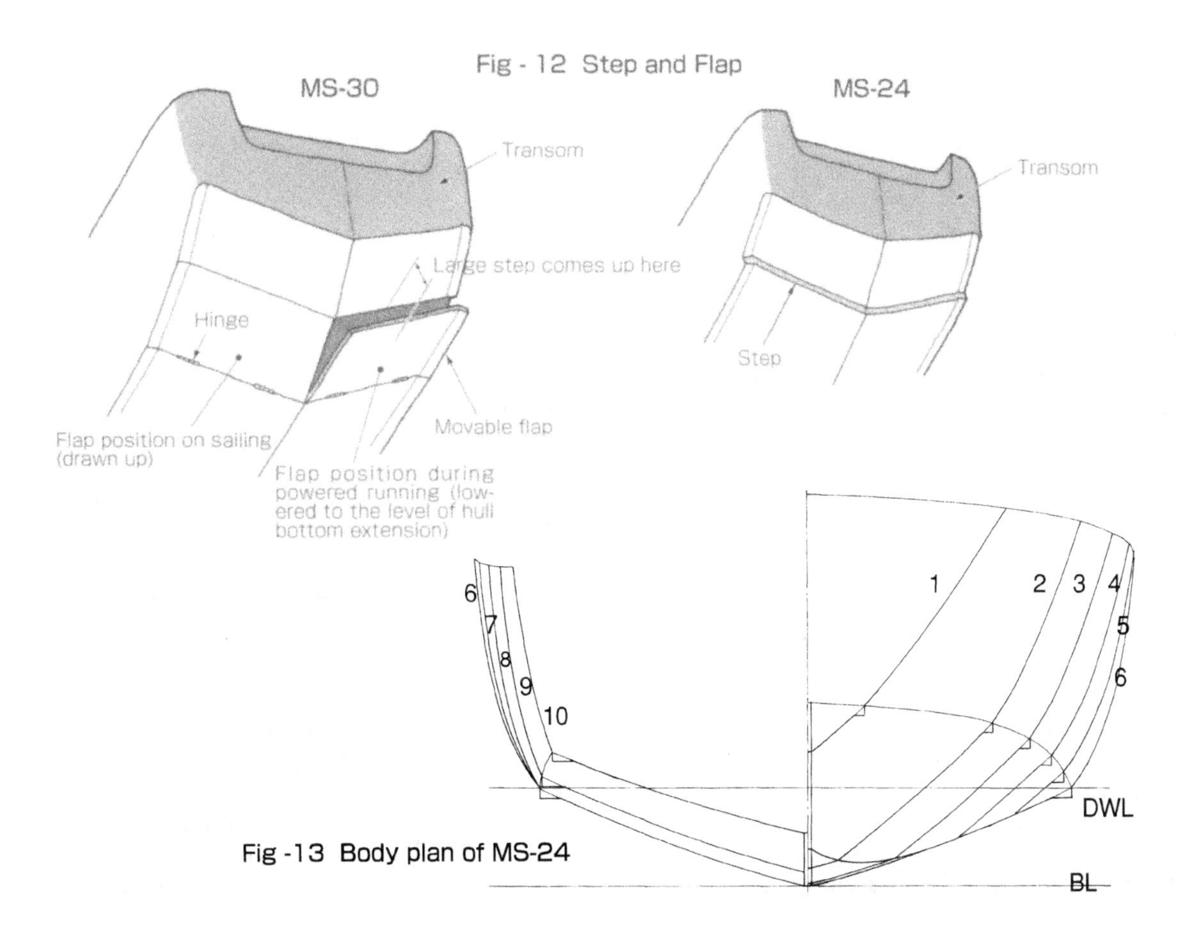

Fig - 12 Step and Flap

Fig -13 Body plan of MS-24

Fig -14 Sail plan of MS-30

so that while mooring the boat, the lower unit can be taken completely out of water by tilting it up. Thus, no special care is needed to keep the motor clean.

However, at intermediate speeds of 15 to 20 knots, water will hit the splash plate (Fig - 9, splash preventing plate positioned above the cavitation plate) of outboard the motor and raise splash, for which measures are necessary. The hull side above chine has an outward bulge (Fig - 13) that looks like the bulge in a sailboat. The bulge gives more stability at large angles of heel, and prevents the tail (transom) from dragging during heel and increases the internal volume.

Additional measures to reduce resistance when sailing have been adopted for MS-30. Power cruisers and sports boats are generally installed with movable flaps on both sides at the bottom of the transom; these flaps are called trim tabs. The quality of the boat's running performance, including the ease with which it gets on the plane, soft riding in head seas and correcting the heel caused by side winds improves considerably by varying the flap angles, so flaps were considered a necessity for large boats.

The same trim tab system is used in MS-30 to improve the running performance and to reduce resistance further at low speeds. You can see from Fig - 11, and Fig - 12 that there is no step on the aft bottom during slow speed running. The bottom shell has two knuckles, with a hinge line at the forward knuckle, positioned more forward than in the MS-24. The planes between these two lines form the bottom surfaces of the movable flaps. The forward knuckle become a hinge and when the trim tabs are lowered to the extended position of the bottom surface, the aft edges of the tabs form a large step and a faired bottom surface of a planing hull. When the flaps are closed, the hull can be used for sailing so that the bottom shell gradually rises to reach the transom.

Therefore, the hull shape is improved in three ways: firstly, there is no step, secondly, there are two knuckles lines and each angle at the knuckle reduces by half, and thirdly the position of the front knuckle is located more forward than the step in MS-24. Through these improvements, the area curve becomes smoother and the resistance during sailing reduces significantly.

As the hull form has not yet been model-tested, I cannot say up to what level the resistance will decrease. But I look forward to continue to improve the hull shape and spend adequate time on doing so. Its resistance will not be less than that of sailboat with a folding propeller, but since the outboard motor of MS-30 is tilted up while sailing (Fig - 14), propeller resistance is not taken into account. Therefore, I expect that its total resistance can be reduced to the same level as that of a sailboat with fixed propeller.

5) Engine and performance when running on power

I want both MS-24 and MS-30 to have good performance when running on power. I believe that their maximum speed, riding comfort and seaworthiness could be equal to or better than those of a pure powerboat. For a motor sailer to plane, the boat needs to be lightweight and powered by large horse-power engine. Fortunately, due to the recent severe exhaust emission regulations in the USA, large four-stroke or direct injection[1] two-stroke outboard motors with clean exhaust gas and largely improved economic fuel consumption, are appearing in worldwide markets.

I think we should use these newly developed products. I am considering the installation of one four-stroke outboard motor ranging from 100 to 120 HP for the MS-24, and the installation of the latest outboard motor ranging from 200 to 250 HP for the MS-30. You may feel the horsepower to be a bit excessive, but considering a car, the horsepower selected will be in the normal horsepower range. With these engines, both boats will attain a maximum speed of nearly 30 knots at light load, and 25 knots at full load. Since the cruising speed will be 20 knots, there will be adequate margin in the horsepower to ride out waves during a long cruise.

The cruising range will be a bit small because the fuel consumption of a large outboard motor is not very economical. However, the fuel consumption of recent outboard motors has improved significantly, so the cruising range of 400 to 500 km will be attained with 300 liters of fuel for the MS-30. In addition to the large outboard motor, I intend to install an 8 ps outboard motor[2] that delivers a large thrust at slow speeds. By using such outboard motors, the advantages of the hull form with small resistance at slow speed can be fully utilized. The boat may attain a speed of about 7 knots with the 8 hp outboard motor, and at a speed of 5-6 knots, its cruising distance will be over 2000 km with 300 liters of fuel. This speed is suitable for trawling, and the arrangement compensates for the short cruising distance of the large outboard motor, and may also be used in many different ways.

However, a large outboard motor, a small outboard motor, and twin rudders are to be installed on the transom, each of which must be of tilted up and steered; such work on a three-dimensional arrangement of the control system is not easy. The small outboard motor may be not be required for steering system, but it may be used together with the rudders. Such an arrangement will be successfully only after many tests based on trial and error. So I would like to keep the idea as a dream for the future.

6) Mast, rigging, and sails

Rigging (wires supporting mast) of MS-24 will be the same as in a conventional sloop[3] and structure and performance will be similar to those of the MacGregor-26. Considering the use of a trailer to transport the boat after folding down the mast, fixing it on deck by one person, we have to use a lightweight mast and lightweight accessories.

The MacGregor-26 was a well-designed boat that gave me many ideas. For the MS-30, I decided to install a futuristic mast considering various aspects from easy sailing to automatic

1) The method that injects fuel directly in the cylinder to have effective ignition and clean exhaust gas.
2) We can select YAMAHA FT8D outboard motor that has 108 kg of static thrust and electric tilt system.
3) Standard sail rigging consists of a mast, a main sail, and a jib sail.

sailing.

There are two obstacles for high speed cruising of a motor sailer: one is the ballast keel that must be completely retracted within the hull, and the other is mast and rigging. It is impractical to fold down the mast each time the powering is changed from sail to engine. However, cruising on power with mast and rigging as they are, is very inconvenient.

Based on my earlier cruising experience of MS-21, the rigging created excessive wind noise at speeds over 20 knots. Also, in head seas, the air drag acting on the top of the mast tends to raise the boat's bow making it difficult to get on the plane against a head wind. With 20 knots boat speed and 8 m head wind, the relative wind speed becomes 18 m/s, which hinders planing of the boat. However, for a comfortable cruise on power, we need to attack the problem and find a solution.

I used a self-standing mast without riggings for the MS-30. Similar types have been used earlier, such as the Freedom series in the USA, in which two thick masts with wishbone booms[4] are used. These boats sold well because of their good performance and because fewer persons were required to sail them. A new mast is the circular rotating mast of Loup De Mer (Fig - 15), said to be a collaboration between Mr. Uriu and Group Finot. I referred to this mast when designing our mast.

The mast has a circular section, is thick, and has no rigging. We can both reef and close a sail by furling (winding) it around the mast. This will be a typical system in which we can control sails by only mainsheets and furling ropes[5]. Mr. Tadami Takahashi, well known for his article "Tadami's favorite boats" in the Kazi magazine, has amply praised this system in his article (Kazi, 1999 May issue), mentioning that it was an ideal boat even for fishing because it had no rigging.

The Loup De Mer has a maximum speed of 12 knots, so the problem of air drag can be ignored, but if a thick circular section mast on a boat runs at 20 knots on engine power, its air drag will be large. Therefore, for our mast, I decided to use a streamlined section. It is not an easy task to make a rotating one-piece carbon fiber mast with a streamlined section. Since I had no experience of molding such a mast in one-piece, I designed it in two pieces, one of which was located under deck with a circular section and the other above deck with a streamlined section.

I also studied how to reduce air drag of oar shafts used for rowing (see Chapter 3). This data is shown in Fig - 16. You can see the change in drag cocfficicnt with speed. The drag at a thickness ratio of 50% decreases considerably at high speeds, but at low speeds, it does not decrease much because of the large influence of viscosity. We have to use a thickness ratio of 25% in this case to derive ade-

Fig - 15 Loup De Mer sailing (right) and running on power (left)
Source : Japanese magazine KAZI, December 1997

4) Looped frames used for wind surfing, which a rider grips.
5) Control rope for main sail and furling rope for sail

Fig - 16 Thickness ratio of oars and shafts

Converted to oar shaft dimension from the measured results of drag for rods that support mast, based on the results of wind tunnel tests for a Japanese boat for the America's Cup.

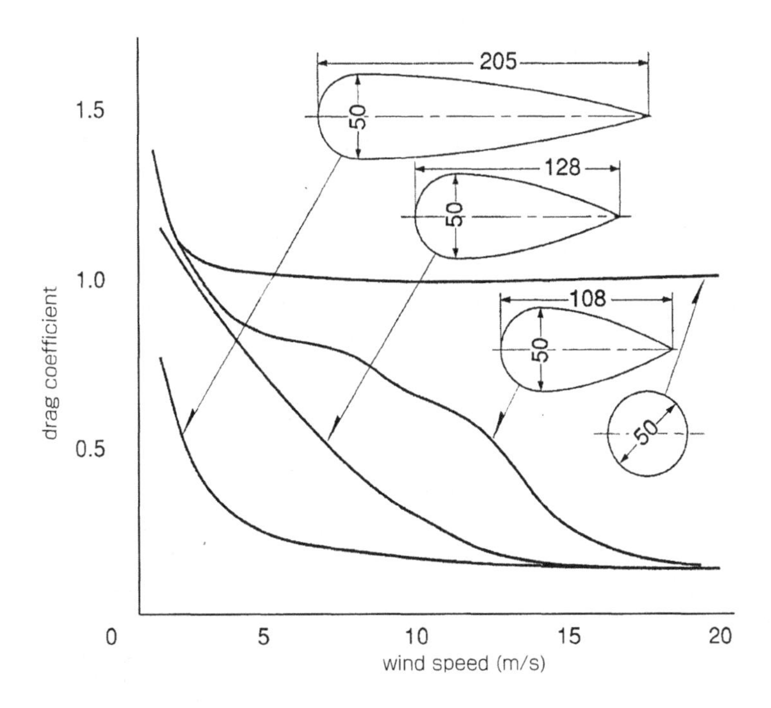

Fig - 17 Effect of streamlined cover on the oar shaft

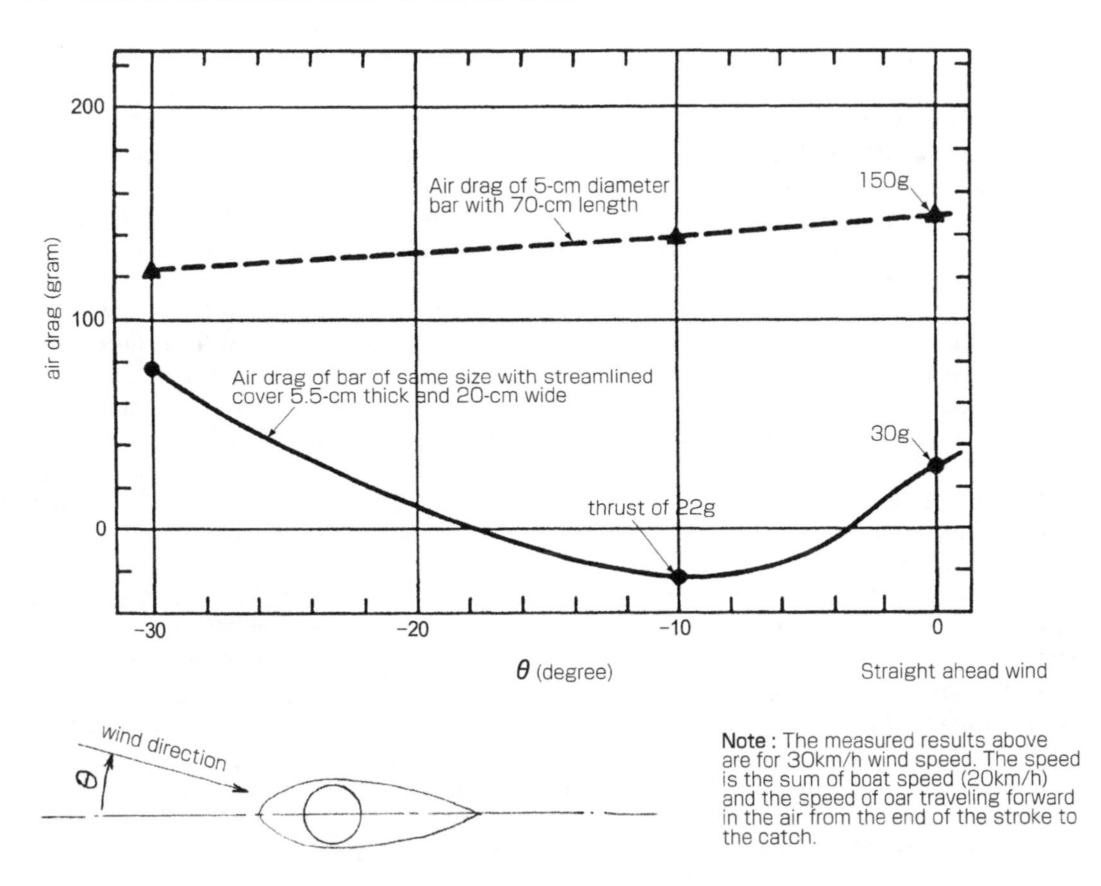

Note : The measured results above are for 30km/h wind speed. The speed is the sum of boat speed (20km/h) and the speed of oar traveling forward in the air from the end of the stroke to the catch.

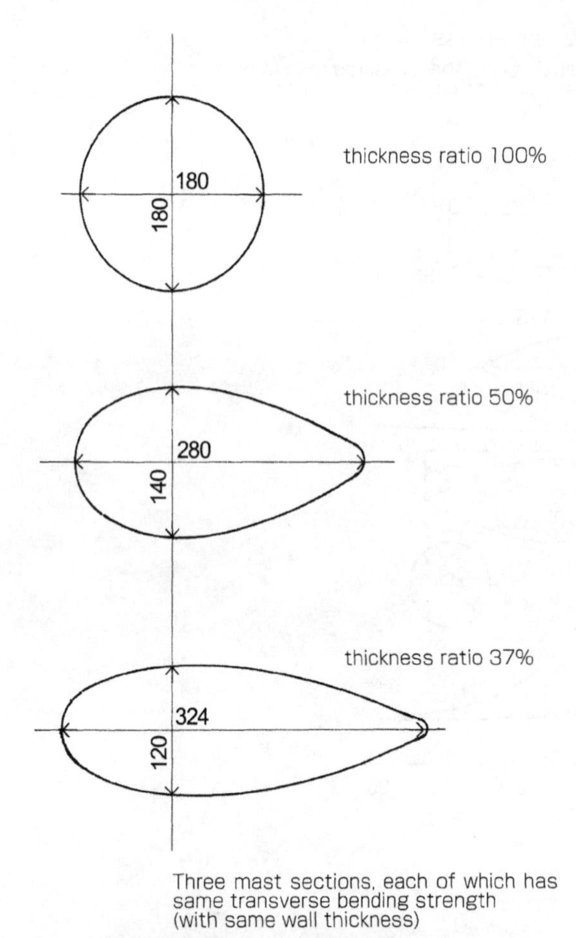

Fig - 18 Circular section and streamlined mast sections

quate effect.

According to wind tunnel test (Fig - 17), a round bar of 70 cm length catches 150 g of air drag in a 8.3 m/s(30 km/h) wind while a streamlined covered oar that has a thickness ratio of 25% catches only a 30 g air drag, which is one-fifth of 150 g. If wind flow direction changes 10 degrees from dead ahead, that is, if some amount of side wind is received, its air drag disappears and it produces a thrust of 22 g. This data is for 8.3 m/s wind speed (16knots). In our case, as the boat speed is 10 - 15 m/s (20 - 30 knots), and additional front wind increases the speed, if the length of the section is made shorter (high thickness ratio), its drag will not increase.

If we use a streamlined section, we can reduce the thickness compared to a circular section at the same strength. We need a diameter of 180 mm at the root for a circular section, instead of which we can use a streamlined section of 140×280 mm (thickness ratio 50%), or 120×324 mm (thickness ratio 37%). I have still not decided which one to use. (Fig - 18).

I calculated the drag of the mast. When the relative wind speed is 30 knots, drag of the streamlined section with 50% thickness ratio is 17% that of a circular section; if the thickness ratio become 37%, the drag decreases to 6%. Moreover, if there is a side wind, the drag will decrease further and some thrust will also be generated. As we have no rigging at all, I think we can cancel out the reverse effect of the mast during powered running by adopting this arrangement. However, I have some concern about mast drag while anchoring. If a side wind strikes a mast having a width of more than 30 cm, the heeling moment cannot be ignored. Especially if the ballast keel is in the pulled-up position, the roll due to side wind will be large. I have also thought of a wind vane on the mast so that it is kept in the heading wind direction always, but without performing trials, I am not sure whether it would work. I proceeded with a streamlined section having 50% thickness ratio, which was acceptable. If the space inside the thick mast is filled with urethane form, it gives a buoyancy of 200 kg, which gives some safety also. The buoyancy of the mast increases boat stability significantly at heel angles of 120 degrees to 140 degrees, and the range of stability increases up to 160 degrees. The normal range of stability is said to be good if it is above 135 degrees; so 160 degrees indicates good safety, and the boat cannot remain upside down even for a short time.

I hope to turn the mast using electric power because I want to turn it to both left and right sides freely with fine adjustments. With this method, we need no rope at the helmsman's position. In case of a power failure, the crew can go to the bow deck to operate. Also, mainsheets can be operated by electric winch, controlled preferably by pedal at the helmsman's position. More power will be needed, but the generator of a large outboard motor has a capacity of 500 kW and the battery capacity is also large. In addition, I would like to use a solar battery system. This system charges battery when the boat is not in use, and maintains the battery in good condition, so we also need not worry about battery charging.

In my boat, a small bow battery for the bow thruster that requires 200 amperes is charged by only a 6 W solar battery. If a battery drives the bow thruster, the weight of the battery cables will be as much as my weight because the boat's length is large. Since I did not want to do this, I selected the present system, because this system has worked without problems in the past so many years.

Fig - 19 Lazy sailing (mast facing forward at all times)

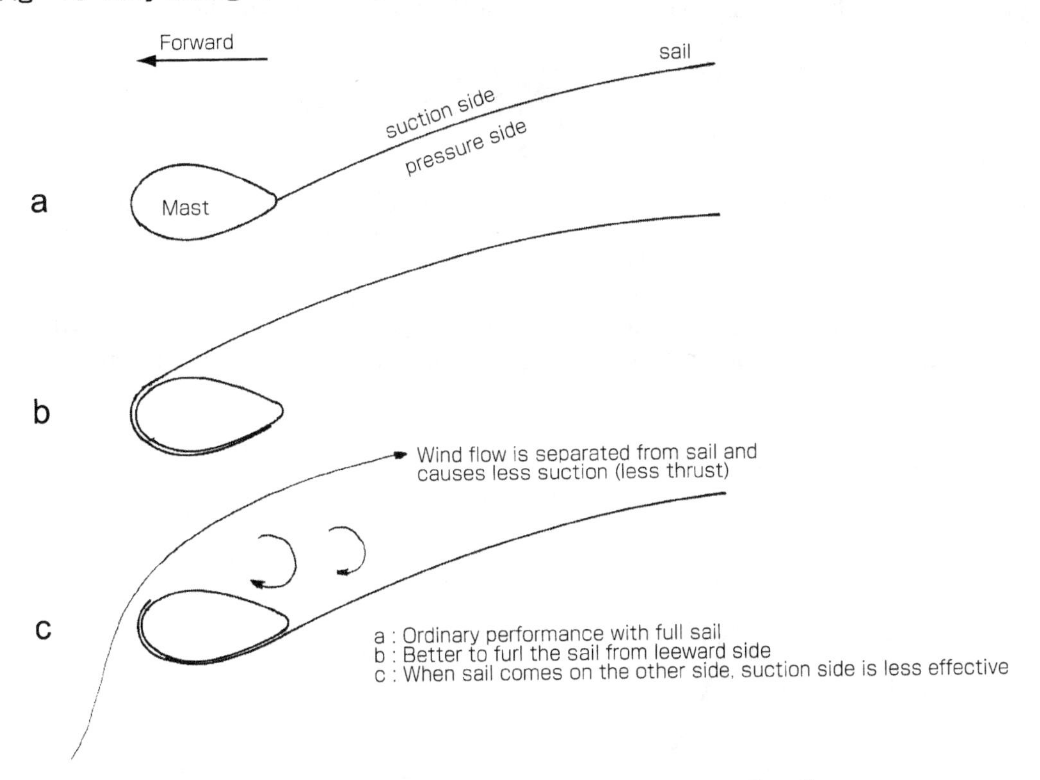

Fig - 20 When trimming mast direction

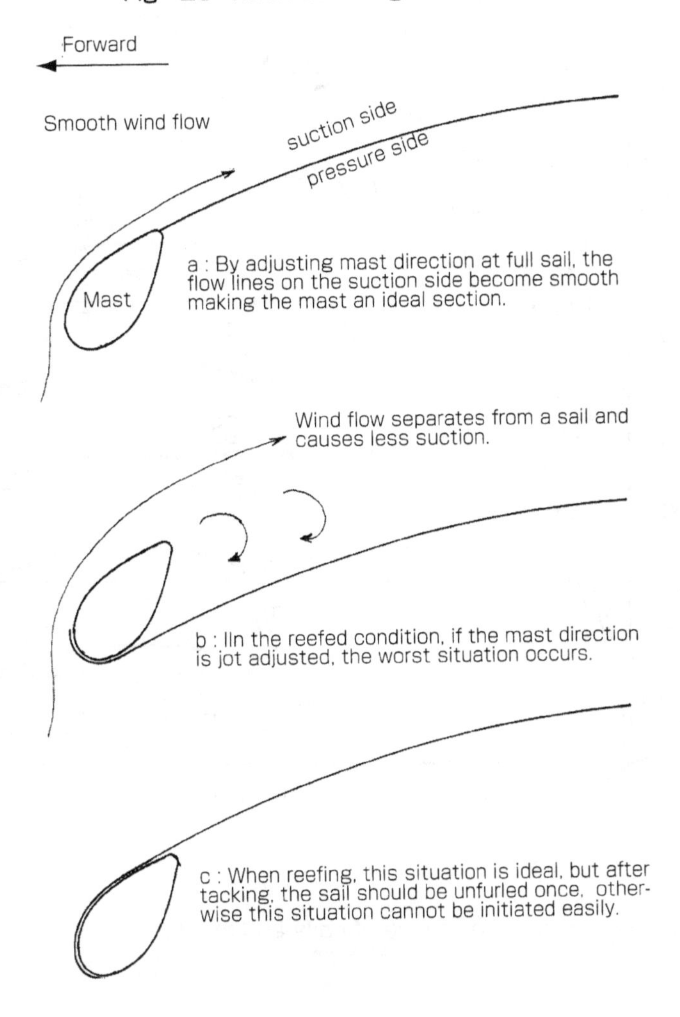

Let us consider mast control when sailing. Lazy sailing is to sail while keeping the mast in the same position without turning as shown in Fig - 19. When the mast is on the leeward side of sail after leafing, airflow around the mast is not satisfactory (Fig - 19c), but it will still work. On the other hand, when you trim the mast according to the sail direction (Fig - 20), its performance may be excellent. From this figure you can understand how the mast spoils the sailing performance. I think that the rigging in MS 30 can attain a performance close to "a" line in Fig - 21 by trimming the mast direction and the performance will adequately compensate for the disadvantages of the hull form.

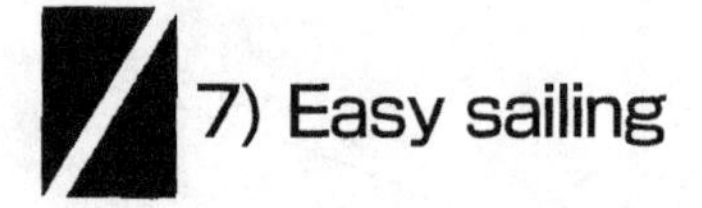

7) Easy sailing

My concept of sailing in a motor sailer is neither hard racing nor tough cruising in rough seas, but to enjoy it in a leisurely manner, such as dining or fishing in good weather. The majority of users will likely be interested in using a motor sailer in this way. Unfortunately,

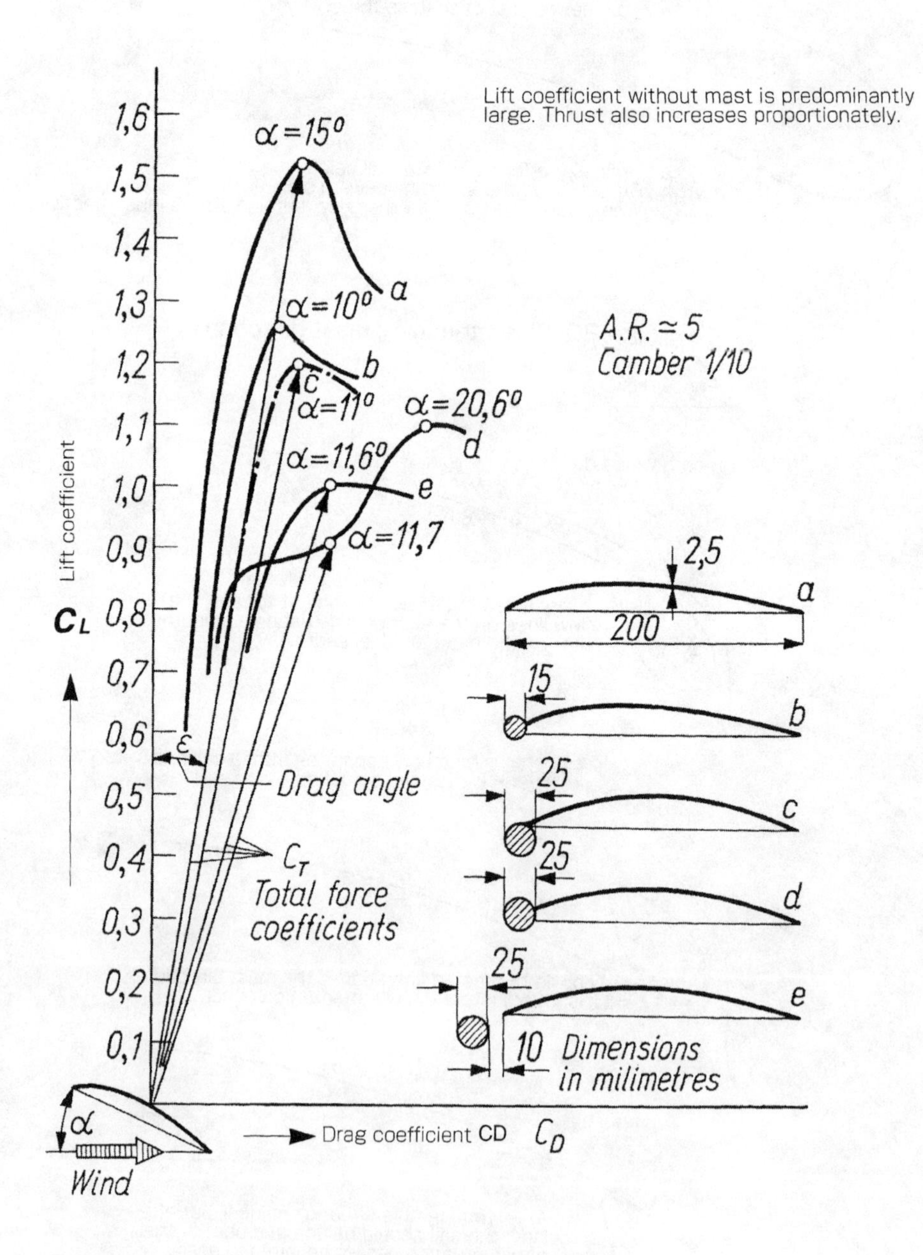

Fig - 21 Relationship between performance of sails and masts
Source : Sailing Theory and Practice, C.A. Marchaj

such leisure activities cannot be enjoyed in a motor boat. Many years ago, every time I got off the Yamaha 30-C, I used to take off its sails. Every time I had to fold the jib and main sail. This work bothered me as it took me considerable time. Later, the furling jib was fitted, and the main sail could be folded down and stowed above the boom, so the situation improved. I installed an ON/OFF device for autopilot and manual steering at the root of the tiller. When the tiller end was lifted to suspend it on the backstay, the manual steering was disconnected and the autopilot began to work automatically. Conversely, when the tiller end was replaced at its normal position, the autopilot was disconnected and the manual steering started working. When the auto-pilot was activated, we could use the wide cockpit because the tiller was tilted up. I also felt very happy that busy steering changes were not necessary when bringing the boat alongside the pier.

In 1991, we installed a system that could control sailing by merely pressing a button on the Philosopher and exhibited it at the boat show as an Auto-sailing System. The Philosopher does not have a centerboard or ballast keel, so the 10 HP outboard motor for sailing is run at half speed, and the boat run at about 3.5 knots. In this condition, when the boat is turned to an angle greater than 45 degrees to windward, and the main switch is turned ON, the folded sail is drawn out automatically and adjusted to offer suitable sail area and trim according to the wind direction and wind speed.

The skipper only needs to turn on the switch to start sailing in a following wind. Subsequently, he could go sailing in any direction except 45 degrees on two sides from dead ahead. He could continue sailing by only steering. I had already worked on and completed this design. Unfortunately, because of the poor performance of the rope winding apparatus, the range of sailing wind speeds was limited, and the automatic system did not go to the production stage. However, this is a problem related to a mechanism that can be solved with the passage of time. Eventually, such a system will appear on the market.

There are many elderly persons who want to enjoy sailing but do not want to learn how to sail. Such persons may like to own a boat like the MS-30. Also, this kind of boat is likely to become a dreamboat in the future for a powerboat user. There may be concerns that such a boat may result in a loss of interest in sailing, but people who wish to use and enjoy sailing techniques can always use conventional sailboats. Even such people will want to use a larger boat without much effort when they become older. In bad weather conditions, you have the advantage of using the engine. Aiming for such a boat, I think the present targets should be a dependable 100% power performance, and nearly 80% sailing performance, with a fully developed auto-sailing system enabling easy sailing.

8) Conclusion

When I retired from Yamaha Motor in 1996 at the age of 70, I had sketched several boats as dreamboats of the future. The high-speed motor sailer was one of them, and I have continued to study it considering it as my own personal boat.

The MS-24 is an economical boat. On the other hand, the MS-30 is a boat that satisfies many dreams so that the owner can enjoy it during his lifetime. Which should I select? I have designed two types, and I will continue to study them until I make the final choice.

I built Philosopher after studying the design for ten years. Now, I intend to enjoy the design work in a leisurely manner. Therefore, whether I will actually build the motor sailer, and which one I'll build, are issues to be resolved henceforth. Good ideas evolve and new components are developed after long-term considerations. Engine techniques will be become more advanced and materials will change. It will be a real pleasure to incorporate such advanced techniques and materials in the boats and to bring the boats to a higher level of completion henceforth.

17

SOLAR BOAT OR55

I have heard that electric boats are being used in harbors and rivers in the USA. At a slow speed of 5 or 6 knots, the electric boat is very quiet and it has an elegant arrangement. On the other hand, the cost of solar batteries is decreasing and its life is said to be 20 years. If solar batteries are installed in the electric boat, the boat can be used for 20 years without refueling. We had an inquiry for a boat intended to be used for daily short-period patrols around a dam. The Horiuchi Laboratory accepted building the prototype of the boat, which became the concept boat for the next Tokyo boat show.

1) Catalyst

In June 1991, I received an inquiry from Masanori Harada, at the Tokyo branch office of Yamaha Motor. This was an inquiry for developing a patrol boat for the Ministry of Construction, intended for use at the Arakawa control dam. They requested specially that the boat be an environmentally friendly boat, since it was to be used for controlling city water close to the metropolitan area.

Murakoshi, the General Manager, Planning Department, and Sato, in charge of environmental affairs, studied this inquiry, and decided on the development of this boat. At that time, I was the Director of Marine Operations, but since the boat was not an ordinary one, the Horiuchi Laboratory accepted the development work.

Since pollution was a major issue in those days, the Sales Department predicted that requests for such a boat would spread to dams and ponds under the control of the Ministry of Construction, and to dams and lakes under the control of the Water Resource Development Corporation, and was therefore, willing to take the order. The required main specifications were as follows:

Number of crew: 5 to 6
Intended Operation: Limited to surveillance and patrol
Speed: About 5 knots since patrol time was 1 hour for a distance of nearly 8 km
Style: Appropriate for the location

The first draft had to be presented in three months. The budget for the boat was to be compiled in 1992 and in 1993, two years later, when the dam was to be filled with water, and the boat was to be launched. I thought the project was well timed and was an appropriate topic for the laboratory. Yanagihara of the Horiuchi laboratory and Harada started the planning for this development.

2) Examples of electric boats

At that time, electric boats were already in use in English canals. You can see the examples, as in Fig - 1, which shows Salad Days with a teak deck of fine workmanship and a house on the FRP hull, looking like a design of the 1900's, a period when rich people probably enjoyed cruising the beautiful English canals with their families and friends on such a boat with a skipper. Now, such boats have been reproduced. The hull was made of FRP to

Fig - 1 <Salad Days>

Source : RIVER BOAT, motorboat monthly publication, August 1991

Fig - 2 <Amadeus 2>

Source : RIVER BOAT, a motorboat monthly publication, August 1991

reduce maintenance and the power was changed to electric motor for quiet operation. Its outside looks remain unchanged from the olden days, but the interior has been modified completely, and is a remarkable combination of plastic, teak, and electric motor.

There were many canals in England since the olden days, and were much used for transportation of goods and people, but now they have become important routes for pleasure boating. When I looked at the canal map once, I was astonished to see canal lines spreading all over the England like a net and I thought that one could go to any place in the country through the canals. The map appeared to have broken the country into many pieces. As the canals are connected to the Thames River, the water level is different in different areas and lock gates have been provided at key points. These gates are opened or closed by the onboard crew themselves, so that the boats can go to either higher or lower water levels.

Especially, there is a kind of boat called the narrow boat that is very slender and meant exclusively for canal use, with its proportions similar to those of an electric train. People rent such boats inexpensively and enjoy cruising canals for several days or several weeks with family. This is a normal marine leisure activity in their lives. At the English boat show, we can see many agencies chartering these boats and pamphlets of these boats. I have seen only some of these, but the scene along the canal is

Fig - 3 Daffy18

wonderful with nature at its best and in abundance. I hear that there are many associations that love the canals and devote much of their time to maintain and restore nature near the canals. Some other associations are trying to connect all isolated canals and create a network of canals, thereby expanding the cruise routes as well as adding to enjoyment. Quiet and clean electric boats are perfect for these canals.

Beside Salad Days, there is a little bit more friendly and intimate electric boat (Fig - 2). This is the 21-feet electric charter launch called Amadeus 2. The figure shows two couples enjoying the canal cruise, tasting caviar canape and champagne. The boat's style is the same as during the reign of Edward 7. The classical style is suitable for an electric boat. If the boat is provided with a full-length awning and side curtain, it can be used even in winter.

We have an example of an electric boat in the USA too. This boat imported by Japan was Daffy18 (Fig - 3), had a length of 18 feet and an eight-person capacity. This boat has several points similar to the Amadeus 2. Both have a displacement type hull form and approximately the same speed. Considering that their usage is approximately the same, a similar arrangement would be natural for a well-designed boat. Likewise, our boat had the same style.

Our boat was to have a power source using a combination of the solar battery and the normal battery; this is the point where it differs from the boats mentioned above. The awnings of Amadeus 2 and Daffy-18 would be convenient for carrying the solar battery, so it would be retained in our boat too.

I have seen boats like Daffy-18 being used for cruising around marinas in the USA. The layout is arranged so as to shut out the direct sunlight and to enable passengers to feel the cool wind on the water, and it is very comfortable. As these kinds of boats are propelled by electric motors, the noise is low. The only noise heard is the one caused by the bow wave. The hull, of course, is a pure displacement type form. The boat has a classical style suitable for its slow speed so that it looks elegant and make the people feel happy to be on board. The photo showing two pairs of a gentleman and a lady on board enjoying dinner on a quiet cruise with mansions on the shore, is picturesque.

The electric boat has also been used in Japan, but for a completely different application. It was used in an amusement park as a rental boat in a small area of the water. The boat was a small one-person or two-person capacity. When used on public waters, it is also regulated like a powerboat, requires boat inspection and a driver's license, which means it would not last as a rental boat. It also meant that the cruising would be limited to a small private pond. What worries people about the electric boat is the trouble they may face for charging the batteries. To charge many of the electric boats and bring them one by one to the pier must surely be troublesome job, especially in strong winds. If the battery discharges excessively, it breaks down soon. Once it breaks down, the physical labor required to carry a large heavy battery from or to the small boat is tremendous and the expenses also cannot be ignored. That means if the load for battery charging is not minimized, its practical value will decrease considerably. In western countries, rich people keep their favorite boat at private piers located in front of their house and they connect the electric cable to the electric socket when mooring the boat. In this case, there will be no problem, and charging the battery is not a burden. But there are not many such rich persons having private piers with a socket for charging boat batteries, and ordinary people who are not favored with such good fortune, may find it difficult to use electric boats, similar to Japanese rental boats. Battery charging may prevent the spread of the electric boat to some extent.

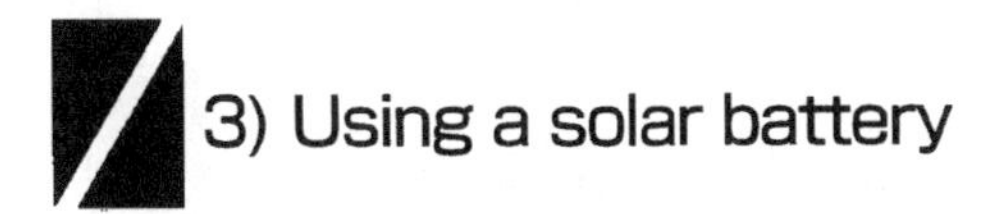

3) Using a solar battery

My boat Hateruma was launched three years before the inquiry of the patrol boat for the control dam was received. I used small solar batteries on the Hateruma for the upkeep of its main battery, and they were very useful and effective. There was no need to move the battery or to worry about the charging level of the battery, so much so that I forgot it completely. Ten years later, when I planned to install a bow thruster[1] I thought of acquiring an electric power source, and finally installed a battery of 80 ampere-hour capacity in the bow. This battery was charged by only a six-watt solar battery without any problem, for which I have been thankful.

I wanted to use the solar battery for charging our electric boat. However, for a boat like the rental boat that has long working hours each day, a solar battery for charging cannot be used since its charging capacity is inadequate. The patrol boat intended for use at the control dam was perfectly suitable for providing a solar battery to charge the main battery, since the boat was not intended for continuous use.

Firstly, Yanagihara investigated the statistics of daylight in the Tokyo district and how much electricity the solar battery on board would generate. Fig - 4 shows the results of the investigation. If 10 sheets of 55-Watt solar batteries were placed horizontally, the electricity generated would be 2443 W/h in May at the maximum, and 989 W/h in December at the minimum.

At that time, the life of the solar battery was said to be 20 years. Although the normal battery will not last as long, a boat that continues running without re-fueling for 20 years can be

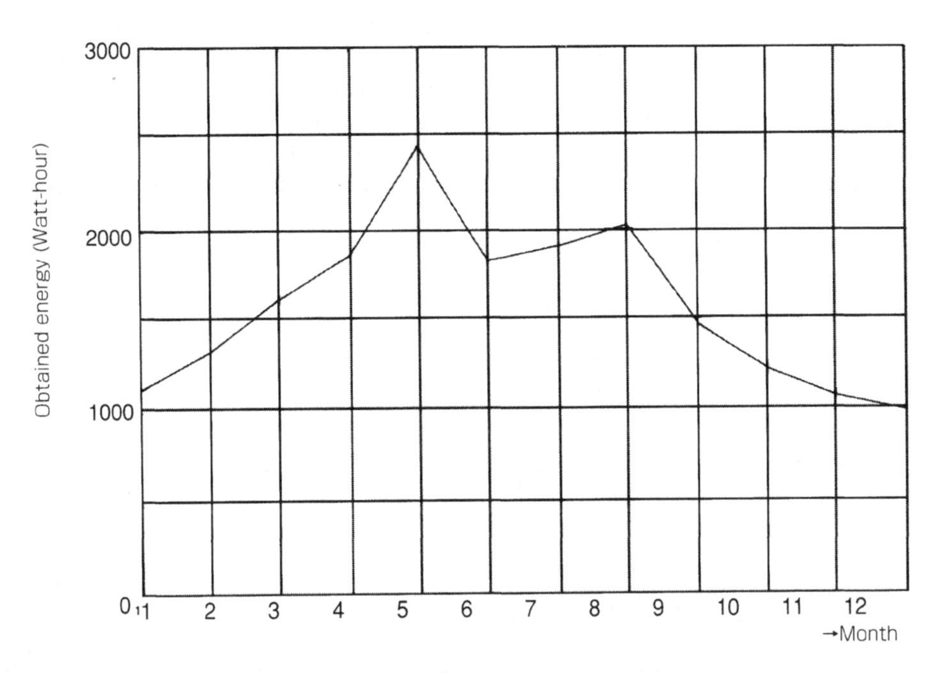

Fig - 4 Electricity generated by the solar battery in a day
10 sheets of 55 watt solar batteries, placed horizontally in Tokyo

1) Electric thruster to swing the boat's bow or stern when approaching or leaving a pier

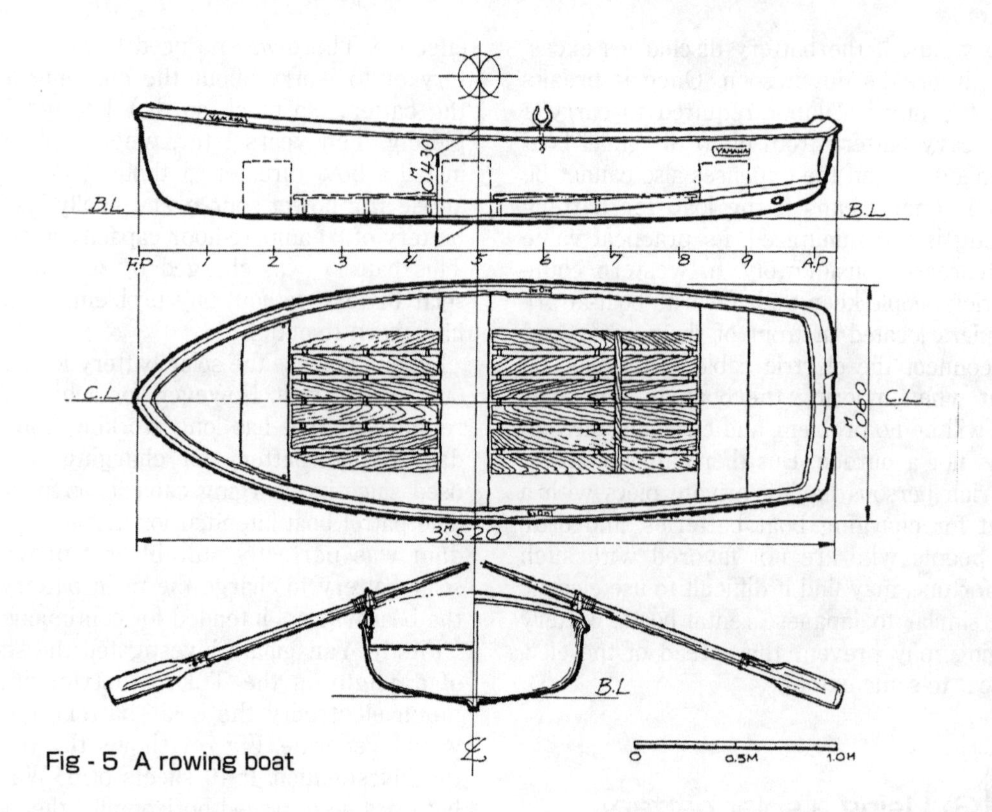

Fig - 5 A rowing boat

realized. Such a boat will not pollute the water, will run quietly, and will look elegant and match the surroundings. That would be splendid.

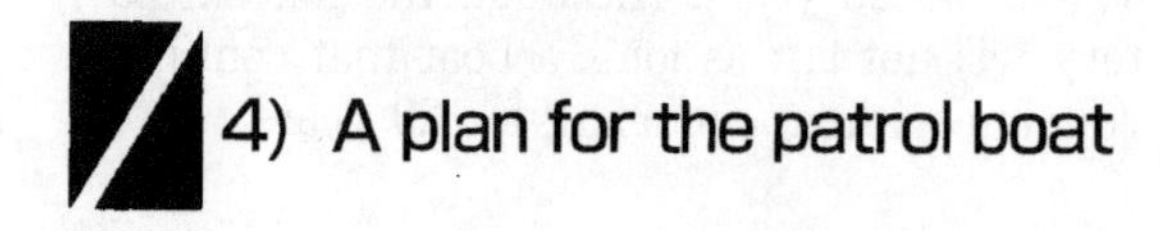

4) A plan for the patrol boat

The development number of the boat was decided as OR55.
A boat with a low-horsepower engine should naturally have a displacement type hull form. At the same time, the boat should have good stability because of its small size. If these conditions are fulfilled, the boat will take the shape of the rental rowing boat. Around 1960, I had a difficult time designing a rental rowing boat with a weight of 50 to 60 kg and with good stability. At that time, a planing dinghy type hull form was finally decided.

The transom width of the planing dinghy is wide and positioned above the water, but once the boat heels, the bilge at the transom touches the water and stability of the wide hull shape increases. Yamaha built 60 to 70 thousand rental rowing boats, and you may see some of them at any recreational water area in Japan.

We designed the complete hull (Fig - 5) start-

Fig - 7 The hull lines

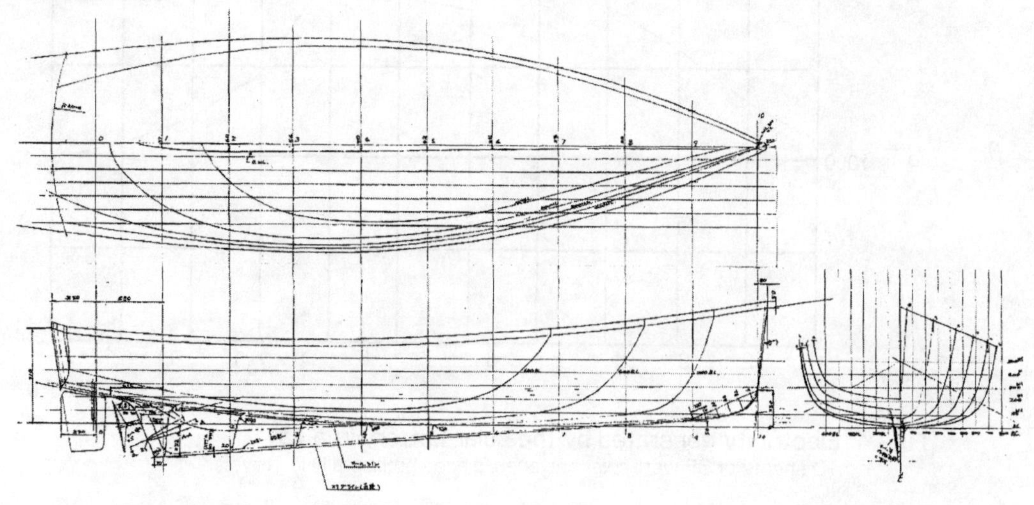

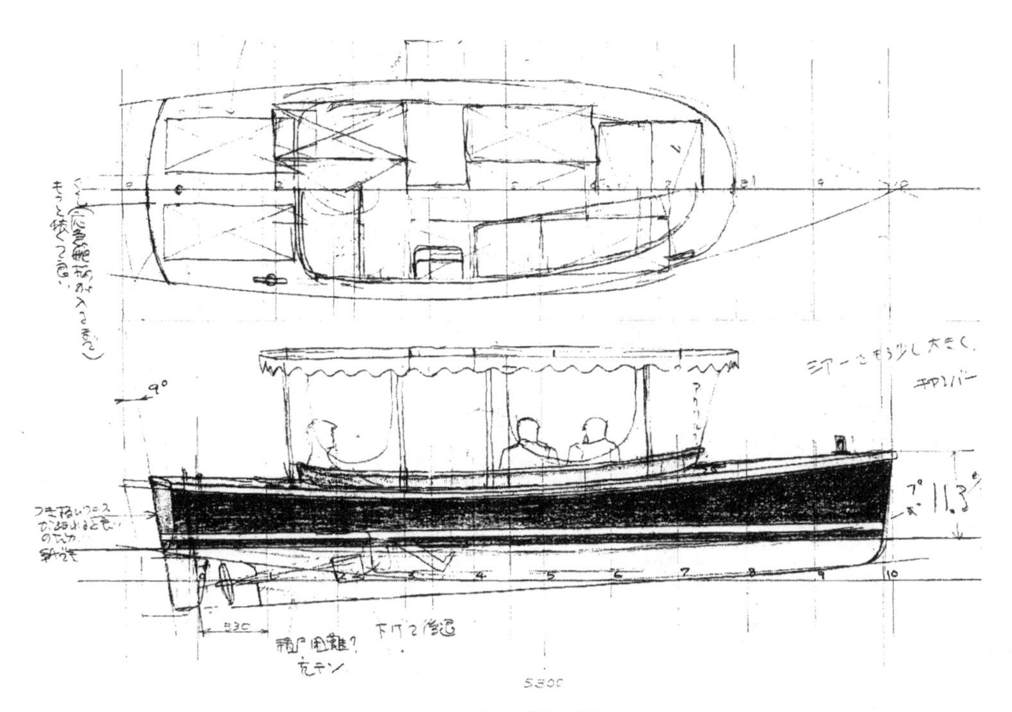

Fig - 6A Plan

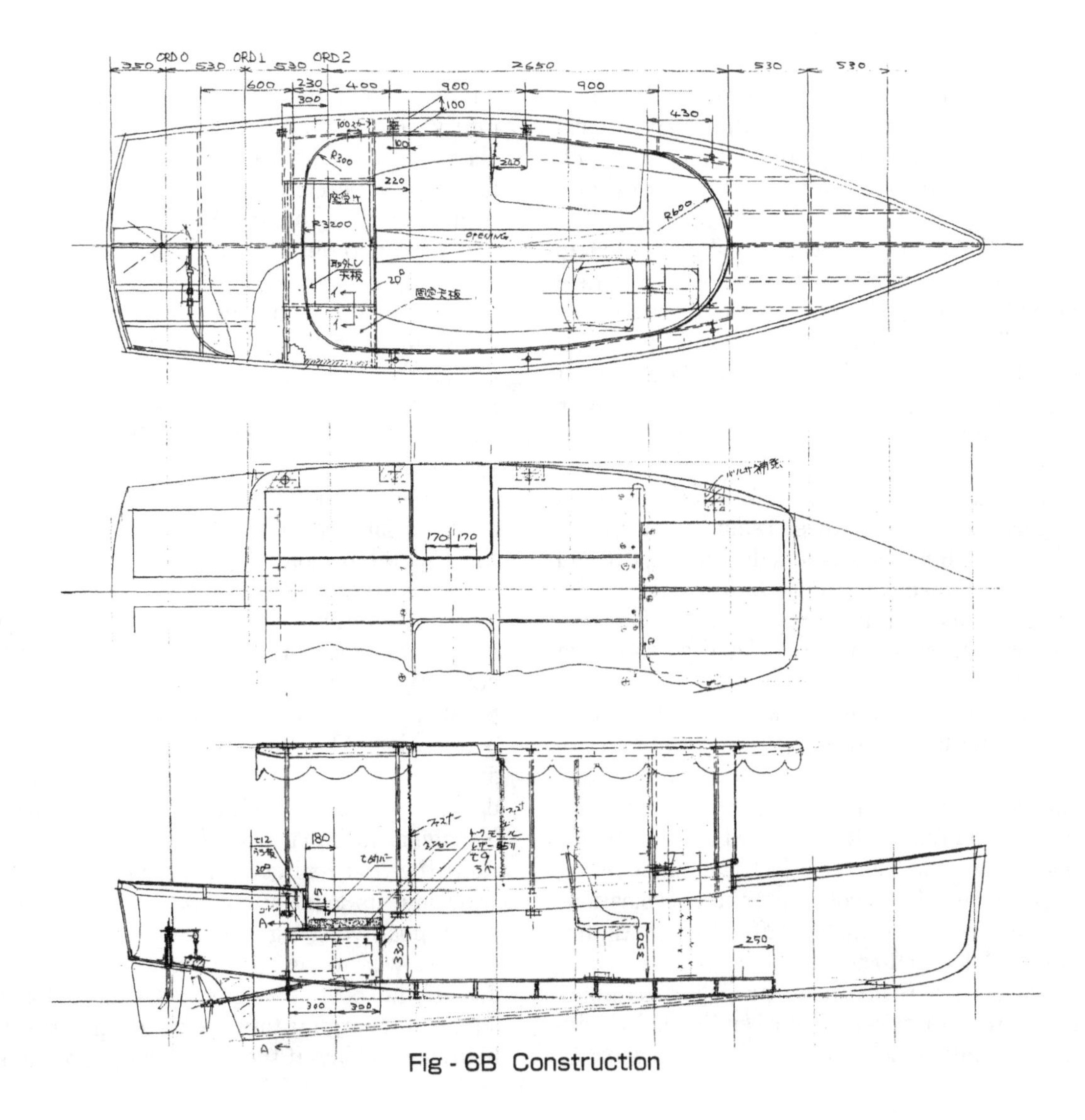

Fig - 6B Construction

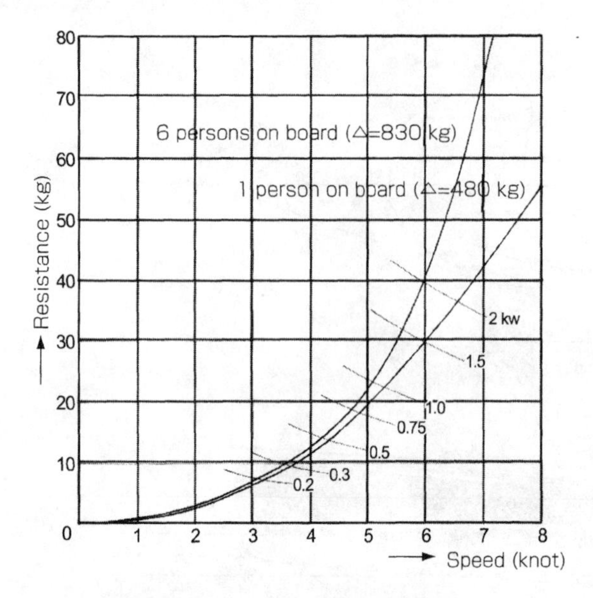

Fig - 8 Drag and the required shaft horsepower

Graph shows the required shaft horsepower of the motor for different speeds with 6 persons onboard and one person onboard. The propeller efficiency is assumed to be 60%.

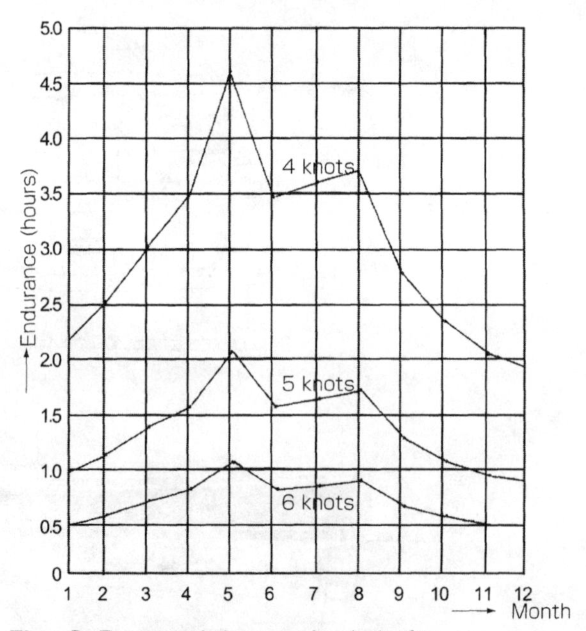

Fig - 9 Drag and the required shaft horsepower

Usable hours per day against the speed and the month

ing from the rowing boat hull form. It was a calm and composed shape with a conventional sheer (when looking at the profile, the deck line at side curves upward at the bow and stern) (Fig - 6A, 6B). Other important points were: the horsepower was small compared to the boat size, the hull had a deep bar keel extending to the skeg so as to increase maneuverability and to facilitate approach or departure from a pier in the wind. The keel area at the stern was made larger to derive adequate course stability and to reduce drift due to lateral wind (Fig - 7).

According to results of tank tests performed in parallel, the horsepower required for speeds of 4, 5, and 6 knots was found to be 0.4, 0.9, and 1.75 kW respectively (Fig - 8). From these figures, the distance that the boat could run in a day at each speed was calculated as shown in Fig - 9.

First, we planned to use a 2 kW-motor made by GE and used in the Yamaha golf cart in the USA, but a model that satisfied our intended specifications was not available. Therefore, we gave up on that motor and used another motor being developed in the Electrical Engineering Department of our company, the specifications of which were : Voltage : 120 volts ; Maximum power : 2 kw at 1000 rpm.

We used a ready-made propeller manufactured by the Kamome Propeller Co., and ready-made modules of NT744 made by Sharp Co., for the solar battery. The price of a 950 x 420-mm sheet for 55 Watts was ¥130,000 and 10 sheets cost ¥1,300,000, which accounted for a fairly large percentage of the total cost of the boat. Therefore, beside the solar boat, an electric boat without the solar battery might be necessary as a product.

To ensure that the solar boat can be left on the water without any maintenance or care, an all weather type boat cover should cover the cockpit and the solar batteries should be placed on the top of the roof. With these considerations, the layout will become more or less similar to the European electric boat. However, as our solar boat has solar batteries on top of the roof, the performance and usage must have been completely different.

Just before the boat show, the trials were performed. We measured the speed, the electricity consumption, the maneuverability, and other parameters with 3 people onboard at a strong 8-m/s wind. The maximum speed was 6 knots and the electricity consumption at 5 knots and 6 knots was almost the same as in Yanagihara's initial plan. Because of the large skeg, the boat had good maneuverability in the wind and its level of completion as a displacement boat was high (Fig - 10). The client's request was to run the boat for 30 minutes a day. If the boat is run at a speed of 4 knots, adequate charging time exists, even if run two hours every day. If the boat is run at 5 knots, it

Fig - 10 Solar boat OR55

Main particulars

Length overall	5.73 m
Maximum beam	1.60 m
Depth	0.65 m
Persons	6
Light condition displacement	410 kg
Maximum displacement:	830 kg
Maximum speed	About.6 knots
Cruising speed	About.5 knots
Main engine	DC motor 36 V
Motor output	2.2 KW/2450 RPM
Propeller	380 260P, 2-blade, 23 % EAR
Propeller RPM	700 RPM
Solar battery output	550 W
Reduction ratio	1/2.5(belt)

can be used for one hour a day. Of course, there will be cloudy or rainy days, but Yanagihara confirmed that on an average, charging from shore would not be necessary during the entire year.

However, only in December, as the performance drops a little, the boat's speed must be reduced proportionally. In any case, the demands of the Ministry of Construction were fully satisfied.

5) Tokyo Boat Show 1992

Like Salad Days, mentioned earlier, external appearance is important even for the OR55. The boat must attract people and make people wish to go onboard through excellent craftsmanship, even though the boat has a slow speed. After trials, the teak deck and the bright red FRP hull of the boat were polished until they shone. Skilled craftsmen, who had not been satisfied with building normal FRP molds all the time, worked on the boat assiduously as if it were a wooden boat.

Yamaha exhibited a highly advanced product at the boat show every year. The feature of that year's boat show was the solar boat OR55. The presentation at the exhibition and the finish of the boat were wonderful and the finish suited the boat. I was also satisfied (Fig - 11). Many sporty boats had been exhibited at the Yamaha booth until then, but only in that year, an elegant and slow solar boat was exhibited in the center booth. Fortunately, the solar boat was well received for both its shape and its concept, and the response to the boat was slow but steady. Other than the inquiry from the Ministry of Construction, there was an inquiry from Itako town, which asked us to for a

Fig - 11 Solar boat OR55 that attracted interest at the boat show

demonstration cruise during the Iris festival of Itako town as they wanted to use such a boat as their canal sightseeing boat, an inquiry from a golf course as they wanted to use the boat instead of constructing a bridge, and an inquiry from the Osaka Flower Exposition to use the boat for cleaning the pond. We were surprised at the range of ideas of the inquiries, but we never received an inquiry for personal use. This is probably a situation specific to Japan.

6) Noise and others

The exhibition of the boat at the boat show was splendid, but a serious problem was pointed out during the trials. The solar boat, which should be practically noiseless, made a clattering sound. The noise emanated from the area around the propeller shaft or from the propeller itself, and it marred the charm of the boat and its selling point. After the boat show, we began to investigate the source of the noise. There were indications that the two-bladed propeller was unsatisfactory and that the tip clearance (the clearance between the hull and propeller) was inadequate. Yanagihara finally solved the problem by changing the shaft diameter from 16 mm to 22 mm and by making many other minor changes.

First, we used the motor and controller that were developed by the Electrical Engineering Department in Yamaha, but due to concerns of electric shocks because of the high voltage and the uncertain cost, we changed these to ones made by GE, normally used in electric boats and golf carts. Their specifications were 2.2 kw/2,450 rpm at 36 volts. The propeller was driven through a belt drive at the reduction ratio of 1/2.5. All defects in the awning were removed, and the OR55 was completed in all respects.

7) Later

The policy of the youth association in the Chamber of Commerce at Itako town included taking the lead in the country by introducing an environmentally friendly new-concept boat instead of the traditional rowboat. We brought the OR55 to the Itako Iris festival held in July 1992, and actually operated it over a 2-km distance to study its suitability as a solar boat.

Business inquiries, which had been frequent after the boat show, became less frequent after the summer and talks did not proceed as smoothly as we anticipated. Itako town made a proposal in which they wanted a boat in the traditional rowboat style, with a larger 12-passenger capacity. In the meantime, business inquiries related to OR55 disappeared. With regard to the environmentally friendly patrol boat for the Ministry of Construction, the budget was not sanctioned and this project also went up in smoke although they were interested. Since then, perhaps due to the collapse of the economic bubble, other business inquiries also slowed down. Nevertheless, we tried to develop a series of solar boats ourselves and to expand its sales network so as to popularize the solar boat as one of our future products, but our efforts finally ended and the commercialization of the solar boat became a distant dream. The only one boat in operation currently is a cleaning boat sold to the Flower Exposition held in Osaka. Their request was for a small showy boat for a small pond. Instead of the OR55, we supplied them the 5-m catamaran boat OR10 (Fig - 12) with solar batteries on the roof of the boat. Later in 1995, Yanagihara and Shimizu of the Boat Department modified the OR55 again. Since then, every year a solar boat has been exhibited at the festival held in the Chubu Electric Power Company and also at the Solar and Human Powered Boat Race, for trial rides and also as a press boat. On such occasions, many people have praised the boat and I am pleased with their acceptance of the boat's charm, but still feel that it is unfortunate that these kinds of boats do not become popular in Japan easily, and do not result in further business.

8) Future dream

I believe that this attractive boat must be more used. As mentioned earlier, the boat can be used for 20 years without fuel and without much maintenance. The batteries may be changed every few years; the boat is really a boat that can be used with the little maintenance.

Let's consider the ways of enjoyment with a solar boat. People who have a private pier with electric sockets for the charging batteries by just connecting cables may be satisfied with a pure electric boat. But such the people are few and the remaining majority who are interested in the electric boat will became happy by using solar batteries in their boats. The dam observation solar boat was made maintenance free by limiting the working hours in a day and likewise the pleasure use solar boat can be made maintenance free with a long cycle by charging during week days and discharging in the weekend when the boat is used.

For example, even batteries that are almost discharged every weekend can be fully charged by the next weekend. The OR55 can be used for 5 hours at 5 knots and 11 hours at 4 knots before the batteries discharge which means the cruising hour limit will be almost zero.

At present, as there is no instrument that indicates the electricity consumption of the batteries and this is inconvenient. A voltmeter helps to see the remaining electricity. In the near future, the remaining cruising hours and the charging condition of the solar batteries in the solar boat will be seen more accurately than the fuel gage of the car or the boat.

If such a reliable instrument is fitted the condition of the charging and discharging will be understood quickly and the customer will use the boat without worry. According to the usage way of each user, the capacity of the solar battery and the number of batteries can be adjusted according to the budget or performance. Thus various arrangements or adjustments will expand the uses and minimize the expenses of the boat user.

How about using the solar boat for fishing? There are many fishing boats moored side by side at the quay in Lake Hamana. To board the boat from the quay, which has a different height bow is difficult; in addition to get onboard carrying the battery and fuel will be more difficult. In a solar boat, this does not exist problem because of it is maintenance free.

Young people may not be satisfied with the solar boat's speed, but for aged people fishing in the lake, the solar boat will be adequate. The point will be its price, but aged people nowadays are rich and they will not hesitate to invest in an ideal fishing boat. If that is correct, the solar boat may be a boat that may sell well in the market year after year, which I strongly hope.

At one time, I was excited about the solar boat I have hopes that a large number of solar boats will be developed in the future.

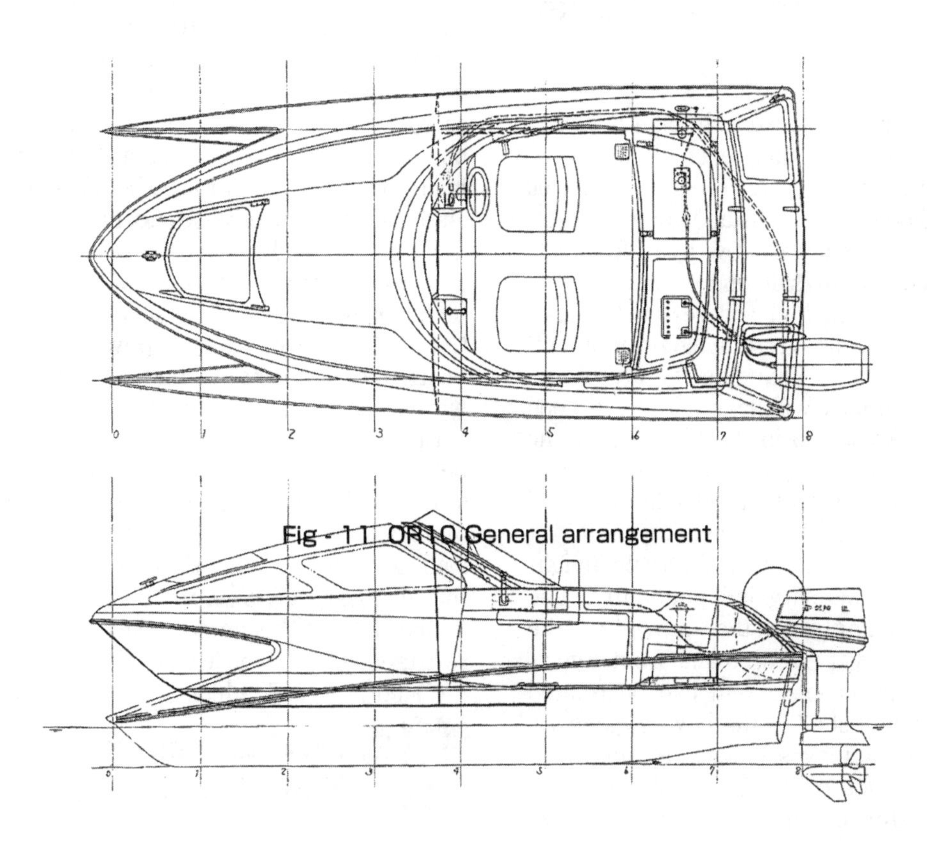

Fig - 11 OR10 General arrangement

18

RUNS OF TWIN DUCKS

A graduate student, who read the concept of the Twin Ducks, a fast hydrofoil dinghy in Chapter 2 of this book, chose it as the topic of his master's thesis. He ran a radio-controlled model of the hydrofoil sailboat at Lake Hamana in strong winds. The next student, who continued with the project, built a full-size boat and got the boat foilborne (supported on hydrofoils) in less than 4-m/s wind. Each hull (sponson) of the catamaran boat had independent longitudinal stability, which contributed to the successful layout of the torsion-free catamaran. These two students after many trials and errors, finally succeeded in operating the boat and presented their theses. As the topic was difficult, they were delighted with their success, and I also was very relieved.

1) Cocept of the Twin Ducks

In recent years, the speed of human powered boats with hydrofoils has reached nearly 20 knots. The running resistance of the hydrofoil boat is a fraction of the resistance of a displacement boat. If two such narrow hydrofoil boats are connected side by side and each is allowed to pitch freely, each hull will resist the large heeling moment (transverse inclining moment) caused by the sail by changing the attitude of the hull and the boat will have an overall good performance. As shown in the Fig - 1a, 1b, the rear beam was joined by pins at both ends and had a telescopic structure to enable each hull to pitch freely around the connecting shaft.

Each hull can maintain its foilborne height and longitudinal stability independently. The sensor is a planing plate that moves on the water surface. As the sensor and front hydrofoil are made in one-piece and can be rotated around the lateral shaft, if the bow rises excessively, the sensor moves downward relatively, the angle of attack of the front foil decreases, and its lift decreases to lower the bow. Likewise, if the bow sinks excessively, the angle of attack of the front foil increases and its lift increases to raise the bow.

The working of the sensor keeps the height from the surface of the water to the bow constant at speeds above 10 km/h. On the other hand, as the main hydrofoils are fixed on the hull, the angle of attack of the main foil increases with the rise in the bow to raise the entire boat. If the boat's speed increases, the lift of the main foil increases to raise the aft hull, and the angle of attack decreases to maintain balance at a specific height.

Thus, we started this project of building a high performance sailing boat by introducing the technique of human powered hydrofoil boats, the performance and stability of which were already guaranteed.

Effective, fully submerged hydrofoils are to be used in the boat to be developed. Since the skipper (crew) can balance the heeling moment with his weight and distribute the load on both left and right hydrofoils evenly, the necessary foil area for the boat to take off (foilborne) can be made smaller. The smaller resistance of the smaller hydrofoil gives a higher foilborne speed, which is an advantage.

2) Graduate student Yashuhiko Inukai

In March 1998, Mr. Takeshi Kinoshita, a professor at the Institute of Industrial Science, University of Tokyo (IIS), called and told me

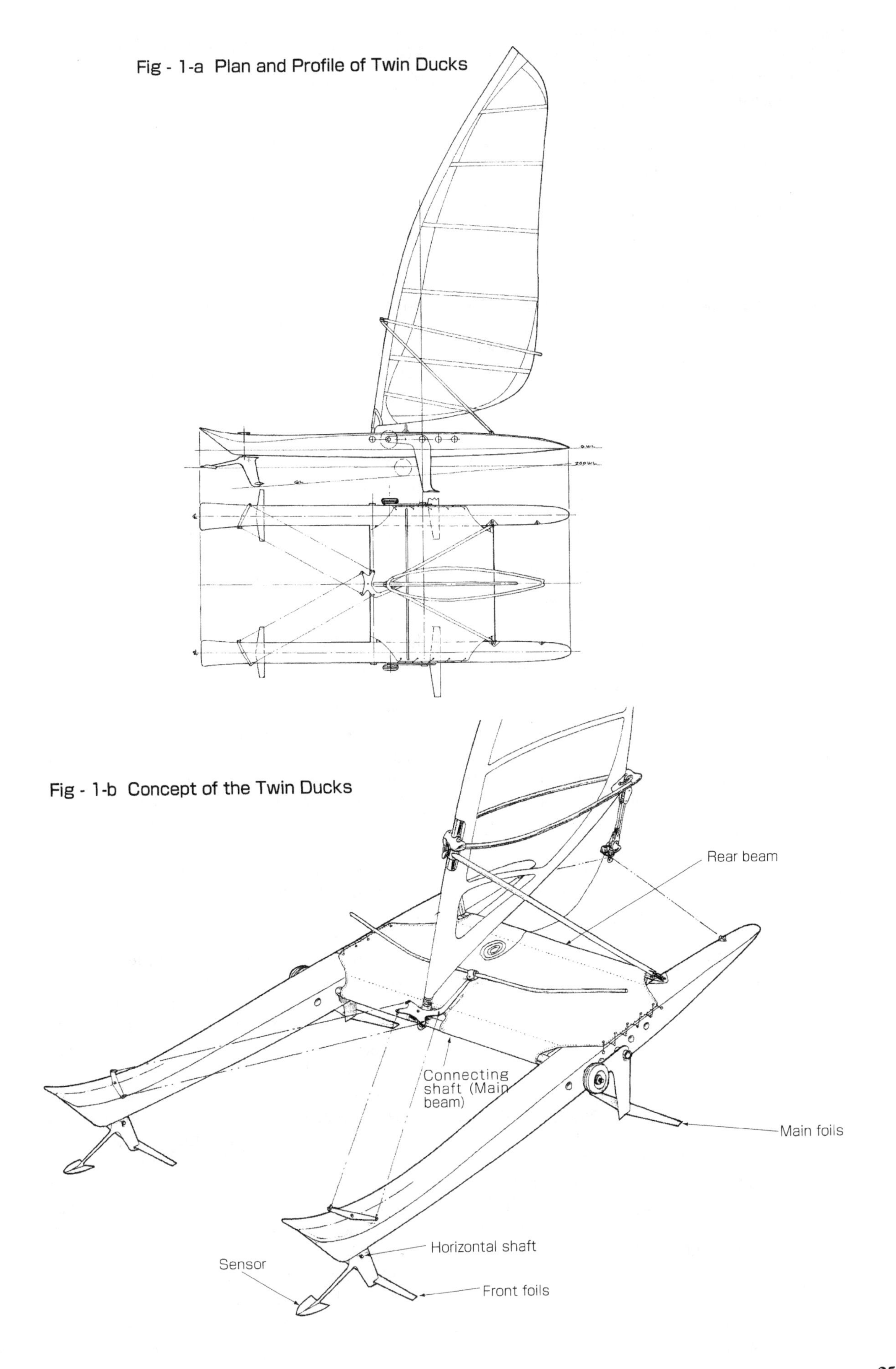

Fig - 1-a Plan and Profile of Twin Ducks

Fig - 1-b Concept of the Twin Ducks

that a graduate student wished to build a boat as the topic of his thesis leading to a master's degree, and requested my assistance.

I was pleased to hear about the student interested in building a boat, so I soon met him. On April 2, the graduate student, Yashuhiko Inukai, visited me at my house in Kamakura. His dream was to build a folding sailboat that could plane like the Sea Hopper[1], with which he wanted to sail all over the world. He already had sailing experience.

During my childhood, I have sailed on a Klepper[2] faltboat (folding boat), so before discussing the thesis, I related my experience and suggest that he talk to Ichiro Yokoyama, who was the designer of a folding boat and a canoe-type sailboat, Aquamuse. Inukai seemed disappointed because the boat he had visualized already existed.

In July, I called him since I had not heard from him. As I thought, he had lost interest in the folding sailboat and appeared to have lost his direction. I myself thought about other projects for him, but could not come up with a good idea. I just sent him a copy of the Twin Ducks, the concept of fast hydrofoil dinghy, suggesting that if he was interested in it, he could consider taking it up as his thesis.

Three days later, he called me and said that he was inspired after reading about the Twin Ducks and eagerly wished to take it up as his thesis. Later, Professor Kinoshita agreed and the development of the Twin Ducks began. After the decision was taken, I began to think about a procedure for Inukai that would enable him to succeed in his project, especially since he had never built a boat.

For a hydrofoil boat to become foilborne, delicate adjustment of the angle of attack of the foil is necessary. It is especially difficult to maintain transient balance while taking off on the water, because the boat's attitude changes every second. Looking at the lifework put in by people overseas on hydrofoil sailboats, I thought it appropriate that Inukai's thesis should terminate at the point where the scale model became foilborne.

For this project, in addition to wing theory, the phenomena of the boat taking off on the water, stall, ventilation (loss of the lift due to hydrofoil air draw), which are normally not familiar to us, should be understood well. Therefore, mere book learning is not sufficient to achieve a stable foilborne performance after resolving the problems caused by these phenomena.

3) Weight-propelled towing tank

To thoroughly study the phenomena mentioned above quickly, various water tank tests need to be carried out successively in a fast cycle before making improvements to the model. Fortunately, since the Twin Ducks has a layout that enables its skipper to balance the heeling moment (force that heels the sailboat) of the sail, the boat will run in most conditions with the same loading on the left and right hydrofoils during foilborne operation. Therefore, if the symmetrical model test is successful, many of the sailing conditions can be estimated.

To perform model tests at an early stage and to confirm that our concepts were generally correct, I was convinced that the most effective method was to test a small model boat in the small towing tank. I had previous experience with the small towing tank, and had worked on tank tests (Fig - 2a) while employed in the Yokohama Yacht Works.

The testing tank in Yokohama Yacht Works was built during the war for tests on flying boats. Although the test data measured using the tank was not very accurate, the tank was suitable for testing a small model boat in the high-speed range. Its towing distance was 25 m, and at both ends of the tank, two 200-mm diameter pulleys wound with 0.3-mm diameter endless wire were located. There was a string attached to the endless wire that towed the scale model. A small 40-mm diameter pulley was attached to one of the two 200-mm diameter pulleys. Another string reeled in the 40-mm diameter pulley was pulled upward by weights through another pulley on the ceiling. Using this mechanism, the scale model could be accelerated with one-fifth the force of the weights.

When a model is accelerated and reaches its maximum speed, the resistance of the model becomes equal to one-fifth of the weight. Therefore, by measuring the resistance at each speed for each weight, you can draw a graph that shows the relationship between speed and resistance of the model (Fig - 3). If the boat's intended speed is too high, the scale model cannot attain the top speed within the short course of the test tank. In such a case, adding more weight at the start of the acceleration can reduce the distance required for the acceleration.

When I moved to Yamaha Motor, I built an improved water tank of this type, which

1) The product name of YAMAHA 14 feet dinghy
2) World's first faltboote manufacturer in Germany

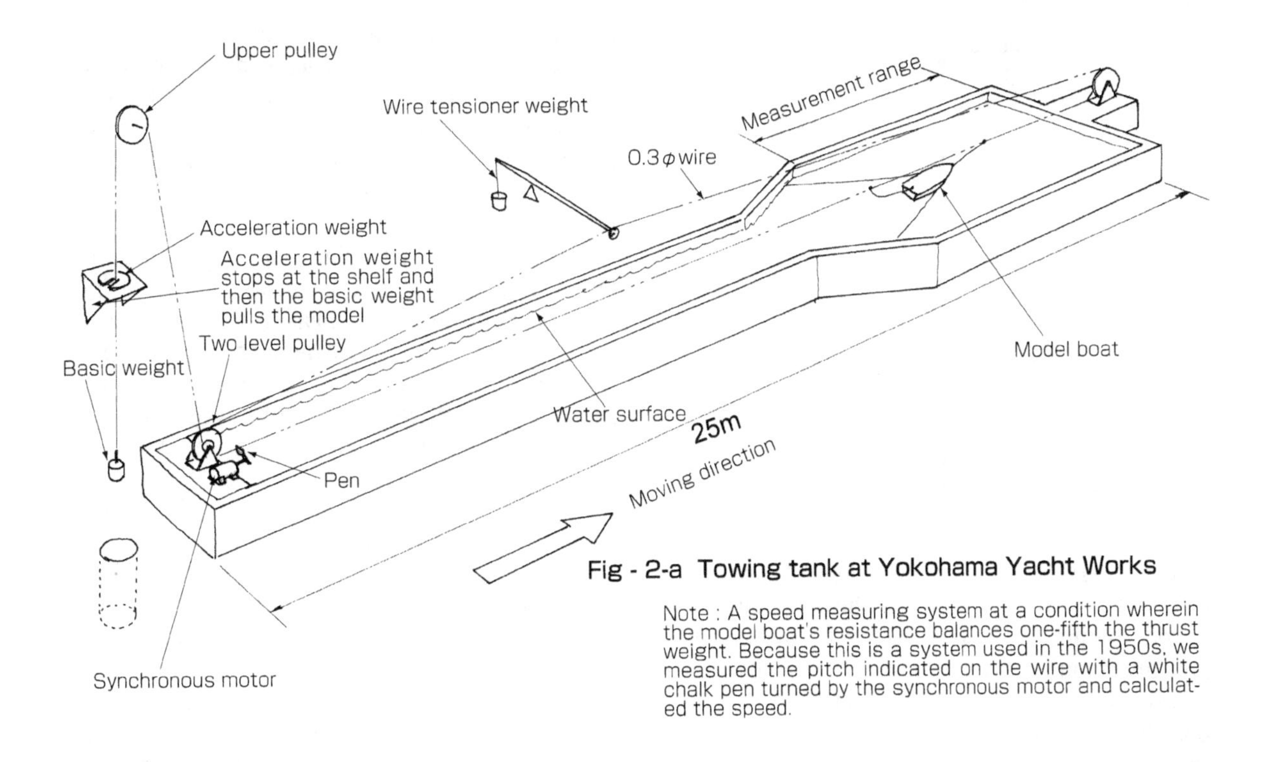

Fig - 2-a Towing tank at Yokohama Yacht Works

Note : A speed measuring system at a condition wherein the model boat's resistance balances one-fifth the thrust weight. Because this is a system used in the 1950s, we measured the pitch indicated on the wire with a white chalk pen turned by the synchronous motor and calculated the speed.

Fig - 2-b 1/5 model test system

Note : Test system for 1/3 model also has a similar arrangement. In this case, the pulleys at both ends are independently positioned and the towing line is a coil-up type instead of an endless type used in previous towing tanks. The direction of advance in this sketch is opposite to that of Fig-2-a.

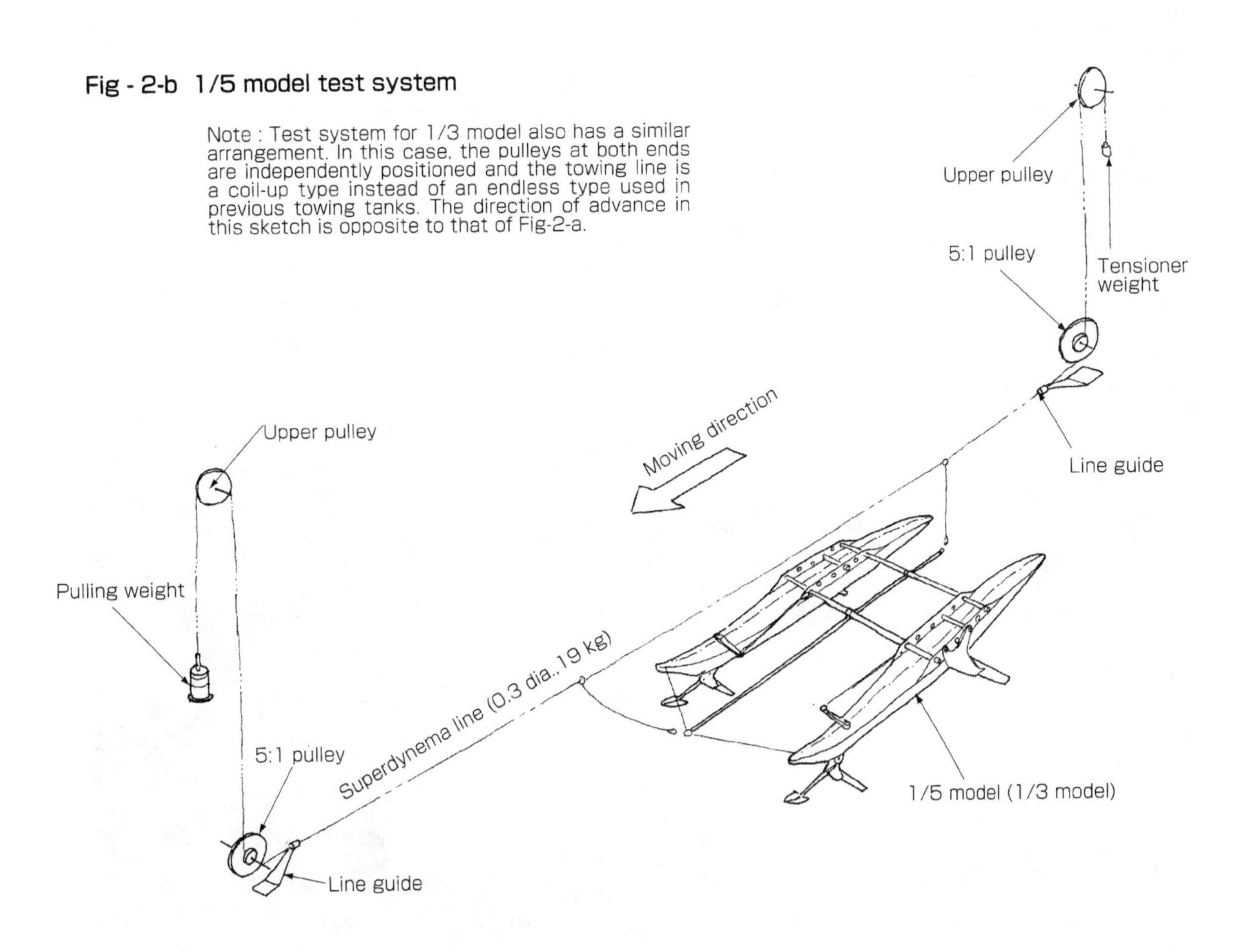

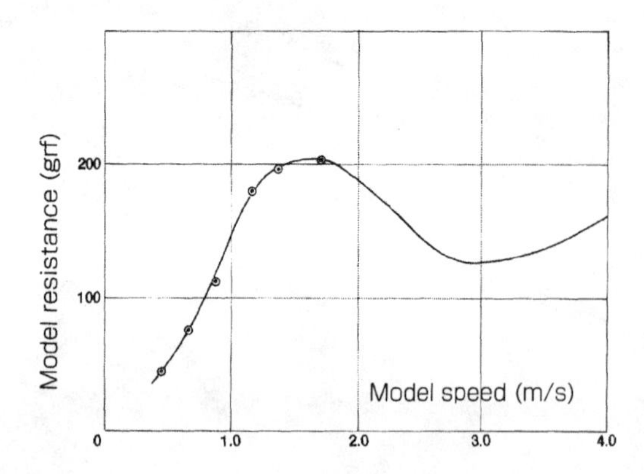

Fig - 3: Resistance curve of 1/5 model

(displacement : 1.13kg, the curve over 2.0m/s is estimated.)

worked very well for development of new hull forms. This time, it was the third weight-propelled towing tank. We looked for a suitable pool and found a fire fighting water pool with a length of nearly 10 m. We decided to use the pool for the boat take-off test after setting up the testing equipment (Fig - 2b).

The main purpose of this test was to improve the boat's take-off ability, so we did not need accuracy in the results related to speed and resistance. On the other hand, it was very important to observe every process during the acceleration so we decided to eliminate the additional weight system for acceleration and decided on a model scale of 1/5. In this way, the process from standstill to foilborne condition could be observed during the 7-m running distance. We used a low-stretch fishing line that called Superdynema instead of a wire. In this case, the towing line, which was a kind of fishing line, was not an endless wire but a coil-up type, by which we could decrease the tension in the line. This was helpful for the replaceable test system.

If we had not used a light and flexible fishing line, we probably could not have completed the system. For the speed measurement, we used a bicycle speedometer attaching it to the pulley. I made the drawings and the test equipment was produced at the prototype shop in the IIS. Inukai himself made the model hull and we began the tests. It took Inukai two months to make the catamaran hull on one side, and I was a bit worried; however, he took only four hours to complete the hull on the other side. I was surprised at his working speed.

The towing tanks at both Yokohama Yacht Works and Yamaha were under a permanent roof (a permanent indoor tank), but the towing tank this time was built outdoors, and it was the fall season. Also, we had to disassemble the test equipment every day. If a leaf got stuck in the strut or foil of the model boat, it would never get foilborne. Therefore, we collected leaves in the morning for 30 minutes every day before setting up the test equipment. Thus, it was hard work for Inukai.

The test was completed without many problems. The towing line often came off the pulley, but eventually we understood the main points for getting the model foilborne. The distance

Fig - 4 Foilborne operation of 1/5 model

during the foilborne operation was only 2 to 3 m, but the model became foilborne cleanly (Fig - 4). The main foil had a very small 2-cm chord, and its lift coefficient was half that of the actual boat. Also, since the resistance was large, the performance was not very good. The take-off resistance was nearly 20% of the total displacement. We had to admit that the performance of the foil was inadequate, but since the model was very small, the desired foil performance could not be obtained. Nevertheless, we were satisfied with the test results.

4) Sailing model

Inukai's aim was to operate the model boat by radio control. When we began planning the sailing model, we found that the weights of the servomotor of the radio control and battery were heavier than expected. In addition, we had to carry heavy movable ballast corresponding to the skipper's weight. With all these loads on the model, we found that it was impossible to make the total weight correspond to the displacement of the actual boat with the 1/5 model.

From the beginning, we intended that each of the left and right hydrofoils should support same load by the movement of skipper's weight for high performance. Therefore, the movable ballast was essential. In June 1999, Inukai studied various possibilities of weights for several scale models, and found that a 1/3 model was the best. This time Inukai made a drawing and began to make the model.

To reduce the total weight, the hull should be made lighter. We had used solid balsa for the first 1/5 model and this time, we made the 1/3 model by the strip planking method. The model was longitudinally planked on frames with strips of balsa of 2-mm thickness and 10-mm width. An extremely light hull having 2-mm outer shell was obtained for the model to simulate the behavior of the actual boat. Inukai made the model this time in nearly one month, to my surprise

The radio control system consisted of three items, namely the rudder, mainsheet, and transverse movement of ballast. We also wanted longitudinal movement of the ballast considering the actual boat, but to move the 2-kg weight, which was nearly half the total weight, in both transverse and longitudinal directions was difficult, and we gave up that idea.

Fig - 5 1/3 model at the initial stage
(the hull structure does not twist)

At that time, Hiromasa Kanou, a graduate student one year younger than Inukai, joined us to assist with the work. Kanou came to the graduate school of Tokyo University from Tokyo College of Science, and he had a strong desire to build boats. He mainly worked for making the hydrofoil and also helped with the test. He was a strong right hand to Inukai.

The 1/3 model had a length of 1.5 m. We had given up the idea of testing this model at the fire fighting water pool mentioned above. As the length of the model was large, the actual running distance was only 5 m. In this distance, we needed length for braking and some margin, besides length for pre-foilborne operation. This was almost impossible. But by chance, we found that there was another pool in the IIS. We came to know this through conversations between professor Kinoshita and Inukai. Inukai had not known about the pool and Professor Kinoshita had known about it but had thought it to be unusable.

The pool was located indoors and had a size of 1 m × 20 m. The actual running distance was about 10 m. There were many obstacles around the pool, but it could be used. There were no fallen leaves and equipment for the towing test was permanently installed, which was very helpful. Inukai again made test equip-

Fig - 6 1/3 model tested at Lake Hamana

Each hull is made rotatable around the forward aluminum tube beam in the photo. The square solid metal on the aft rail is the 2-kg balancing weight made of brass corresponding to the crew weight, and can be moved laterally by remote control.

Fig - 7 Comparison between the new and the old strut

The lower figure shows the old arrangement. The upper new one has a large rudder area aft of the rudder shaft

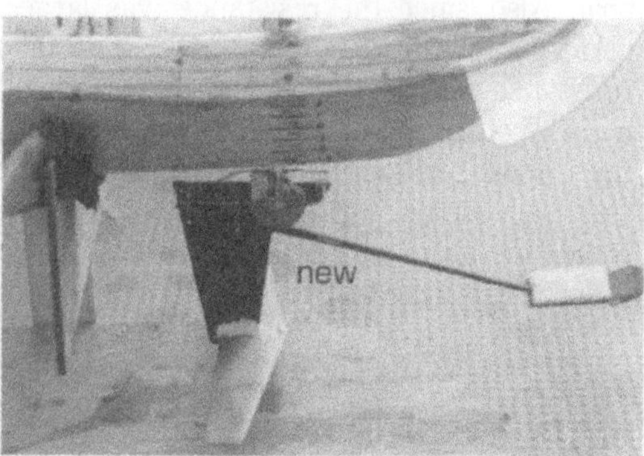

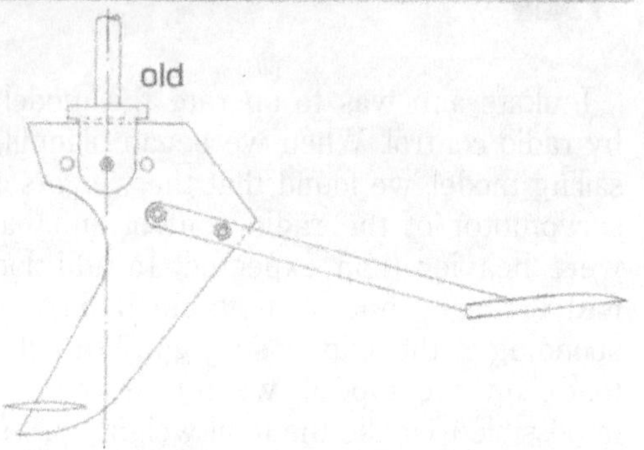

ment for this pool. After the test, we got good performance and precise data, since the scale of the model was larger. Through this test, we acquired reference data for foil angle and boat behavior, and we gained confidence in foilborne running.

Next, we performed the sailing test. We looked for several places such as the Shinobazu pond, the sea, and the Toda rowing race course as possible location for performing the test. It was not easy to find a suitable location if an escort boat was required during the test. In the meantime, with the help of Professor Kinoshita, a 50-m square pool that belonged to the Ship Research Institute, Ministry of Transport was used.

In early August, we often went to the pool to float the model (Fig - 5), but defects continually occurred. The model broke at several points, the rudder did not work properly, although we adjusted it many times assuming that the cause was wrong location of the center of wind pressure, but there was no effect. The movement of the rudder, mainsheet, and ballast did not work well.

Such failures were natural because Inukai was using a radio control device for the first time in his life. Kanou was no good at rowing. I could teach him the rowing technique, but all three of us were not familiar with radio control and we could not even judge whether we lacked in skill or whether the mechanism of the radio control was defective.

As there was no sign of the model becoming foilborne, we measured and found the maximum wind speed was only 2-3 m/s. The model boat needed 4 m/s wind speed to become foilborne. That was why the boat would not get

Fig - 8 Towing test

foilborne. Finally, we found some buildings surrounding the 50-m square water tank, therefore, it was impossible to expect a 4-m/s wind. Even if we could get wind of that speed by chance, it would not blow in the same direction and would not be suitable for the foilborne test. Finally, we decided to go to Lake Hamana for the test. Lake Hamana was a large lake, and naturally the wind speed would be adequate. I thought that if we select the Yamaha Marina, where I kept my private boat, as a test base, an escort boat would be available.

On September 19th, the three of us visited Lake Hamana for the first test, but the results were inadequate (Fig - 6). The rudder did not work as before, some parts were broken, and the model boat pitch-poled and capsized. When the wind blew, one side of the hull leapt out of the water but it was only for a moment. Nevertheless, there was a good result also. Thanks to the floatation attached to the mast, the heel angle of the model after capsizing was as high as 110 degrees, at which attitude, the center area of the deck, where electric accessories were fitted, did not become wet. This was not a major result, yet we were pleased.

Around this time, I found the reason why the rudder had not worked. It was due to the overbalance of the front strut (Fig - 7). The rudder was turned to port or starboard because the forward profile area of the rudder shaft was large. Therefore, a large load acted on the steering rod, which buckled and could not be used for moving the rudder. The fault in the 1/5 model, which I designed without paying excessive care, did not appear in the tank test, but appeared suddenly during the sailing test of the 1/3 model. I was ashamed to have made such a fundamental mistake.

Later, Inukai re-designed the front strut and the rudder worked well. We also corrected other mistakes and finally each part began to work well. On October 25, we again visited Lake Hamana for the test. The test boat this time had a better level of completion than before. We had a wind of 2-4 m/s that day and occasionally had some blows that raised the windward hull. Although the windward hull rose, the entire boat did not become foilborne. The reason could have been unintentional movement of ballast, or Inukai's unsatisfactory control technique, such as turning the boat suddenly while it was about to take off. Since we did not encounter a suitable wind most of the time, the result that the boat was on the point of take-off but did not take off. We had to give up the test because a storm was predicted the next day, and we departed Lake Hamana.

We again made full preparations and we scheduled the test on November 10/11. The time remaining for Inukai's graduate thesis was running out, so I thought this would be the last chance for the test and I wished to capture a video of the boat's foilborne operation. However, upon arrival at Lake Hamana, we found the wind was inadequate. The wind speed was only 2 m/s, not adequate for the boat to get foilborne. Reluctantly, we turned our efforts on getting adequate foilborne stability by the towing test (Fig - 8). Finally, the test model would get foilborne in a stable manner at high speed, but in the end, both left and right hydrofoils came off from the struts (vertical legs that support hydrofoils). The test was thus completed.

The time for Inukai to finish the graduate thesis was limited. He had to derive the mathematical formula for the behavior of the boat to analyze its stability in November, and had to measure the performance of the hydrofoil in the circulating water tank in December. So our schedule was that, after the return from Lake Hamana, Inukai would do his best to prepare his graduate thesis and if conditions improved, we would try to do the final test at Lake Hamana. The presentation of his graduate thesis was on the 14th of February. We scheduled the final test just before the presentation day, and after the day of the submission of the graduate thesis, that is, on February 11 and 12.

I could not attend the test, as I caught a cold. Inukai went for the test with his assistant, Kanou. I requested the staff at the marina to drive an escort boat, and I waited anxiously for the test results in bed at home. On the afternoon of the 12th, Inukai phoned me excitedly

Fig - 9 Successful sailing of the 1/3 model

that the test boat showed very stable foilborne operation in 5 m/s wind and in rough conditions (Fig - 9). This was splendid. I was convinced that with this result, he would graduate.

According to his report, the test boat finally became uncontrollable because the battery of the radio control was used up and the model broke when it collided with the escort boat. However, as they captured the motion of the model, which was their objective, it did not matter that the model broke. I really felt relieved that we tasted success at last.

Inukai informed me that at the graduate thesis presentation on February 14th, professors had taken keen interest in his project, and the presentation had been very pleasant. Anyhow, the video would have been very convincing.

Since we were all so busy, we did not find time to record numerical data. We found that the boat could go windward somewhat and could jibe[3] when foilborne, but it did not tack very well.

5) Inukai's graduate thesis

For the graduate studies, Inukai measured the performance of the hydrofoil that he used for the sailing test at the circulating water tank in the Chiba Campus of Tokyo University. He derived the equilibrium equation by setting up simultaneous equations of the ninth degree. Each of the left and right hulls of Twin Ducks was an independent pitch-free structure and the load on each hull was supported by the change in attitude of each hull independently. As the hull was designed to pitch so as to cause a negative lift on the windward main hydrofoil, the boat could withstand heeling moments even in a strong wind.

The calculations showed that the boat could run under strong wind speeds of more than 15 m/s. On the other hand, other calculations based on a rigid model showed that the boat would could not be used for foilborne operation at a 7-m/s wind speed because the height sensor came out of the water surface in such conditions. Therefore, it was evident that for the boat to become foilborne over a wide range of wind speeds, it should have a twistable structure.

Inukai graduated and got a job with the IHI Company. I believe his experience until he tasted success with the sailing of the Twin Ducks model must have been valuable.

The 1/3 model that had been designed by him referring to the 1/5 model, often broke due to lack of strength, but he had made the calculations for the strength and performance in earnest to solve the problems. Practical experiences such as these are valuable. Later, his pleasant experience with the Twin Ducks continued and he often attended tests of the actual boat and contributed to the success of the boat later, for which I was grateful.

6) Building the full-scale boat

After Inukai's graduation, Kanou took over the work of building the full-scale boat. Generally, it was too difficult a project for an amateur to build a manned boat, to run a complicated hydrofoil boat, and to write a graduate thesis based on the results.

It took Inukai two years to make a model boat, and Kanou was going to build an actual boat in only one year. Kanou acknowledged that Inukai had been pushed for time to complete his work, so Kanou wanted to spurt right from the beginning. Just after Inukai's graduate presentation, Kanou roused himself to start work immediately.

From the beginning, we intended to build a prototype with an ultra light structure using ACM (Advanced Composite Materials), because we thought an ultra light structure was very important to confirm the potential of this boat. Selection of appropriate materials that result in a good balance between the performance and cost, when the boat goes into production, would be very important.

I knew well that it would be difficult for Kanou, with no experience of building a boat, to build an ACM boat. However, he had some experience assisting Inukai in building a model boat for half a year. He must have studied the problems that could occur in an operating boat and how to solve the problems. This experience and knowledge would be valuable and helpful for the actual boat building work.

The first step to decide the structure of the boat was how to make the hull planking. For attaining light structure, we adopted an ultra light ACM method that we had used in the structure of a Yamaha single scull.

Takashi Motoyama of the Horiuchi Laboratory developed this method around 1990. It consists of two sheets of uni-directional carbon fiber prepreg[4] arranged on both sides of a three-mm thick acrylic form sheet with

3) To change the sail to the opposite side in the follow wind
4) A material that fibers dipped in the resin and semi-hardened so as to have easy workability. It has flexibility and the materials put on a mold through the vacuum bag method,are hardened and high quality structure can be achieved. This is a typical construction method of ACM.

the weight of the completed laminate per m² extremely light at only 700 gr. Prepregs were normally procured by special order for 0.05-mm thickness and the direction of fibers arranged (unidirectionally) so as to derive optimum strength and rigidity of the structure. This took a lot of work and needed a curing oven, which was the method originally used for the construction of aircraft. To build a construction facility would need more time, which we did not have, so we gave up on this method.

Another practical method was one that Mr. Kanai in Shichirigahama used for building a high performance sea kayak. Strips of 10 to 15 mm made from 3-mm marine grade plywood are planked one after another by glue on in-line temporary frames, and the necessary areas such as crew space were stiffened by laminating FRP from inside. This is the lightest version of the strip planking method. By this method, a weight per m² of 1.6 kg can be obtained. This is very light, and not even Motoyama's sandwich construction can achieve a lighter weight than this. If we take conventional 6-mm plywood in his method, its weight will be 4.2 kg per m², which means the plywood Mr. Kanai procures is much lighter and effective for lightweight structures.

Nevertheless, as the difference between 1.6 kg and 0.7 kg was not very small, we could not decide which to select, so we consulted GH Craft, which had in-depth experience with the development of racecar bodies, human powered boats, and solar cars. Until 2000, they were engaged in designing and building America's Cup boats, and presently (2001), they are working on the development of the body of a Japanese spacecraft. Thus, GH Craft was one of the companies in Japan with the most advanced CFRP techniques.

On 25th January, Kanou, Inukai, and Mr. Kanai who was interested in CFRP, Kobayashi of Kinoshita Laboratory, who developed new rowing devices for the Atlanta Olympics (1996) with us, and I, visited GH Craft to receive advice on the method of building a boat hull.

Manabu Kimura, the President, and Kiyoshi Uzawa, the technical executive, listened to the details from our side. President Kimura's final idea was as follows: Make the core with 4-mm balsa wood by strip planking method, laminate both sides with thin carbon fiber cloth; this was also a type of sandwich construction. With this building method, we did not need a curing oven or special prepreg materials. But we were not confident of the surface finish. After consultation, it was decided that we take the balsa-cored hull to GH Craft to get instructions on lamination. I was satisfied with this plan.

Fig - 10

Core made by 4-mm balsa strips on both sides of which carbon fiber is laminated to complete the hull.

This method would still require a lot of work and time. We had to transfer the boat to Gotenba, where GH Craft was located. To transport a weak balsa hull, we would need a special solid box.

I was especially concerned with the time delay. Whether to use Kimura's or Kanai's building method, depended on builder Kanou's intentions, and he had to decide. As a factor to help in the decision, we asked Kanai to make test pieces. Since Kanai was just considering which method to use for building his own boat, he accepted the request to manufacture the test pieces, which was helpful to us.

About the middle of March, the test pieces were completed. On the 17th of March, I visited Kanai with Kanou to confirm the test results. Kimura's structure was 1.17 kg per m²; this figure was just between the corresponding figures for Motoyama and Kanai's structures. After confirming the results, Kanou decided to use Kimura's method. I thought he would have a difficult time, but I did not tell him that. On the other hand, Kanai chose Kanai's method, considering it to be firm and stable.

Since then the progress was very slow. Kanou spent May, June and July job hunting. At last, he got a job in Mitsubishi Heavy Industries in July. In those days, he was very busy preparing for the examination, touring the factory, attending interviews, and therefore, he had no time to make plans for Twin Ducks.

Kanou had no experience in strip planking. It was difficult for him to fair the outside surface of the frames. As my initial explanations were inadequate, we had to communicate with each other by phone and talk about fairing work very frequently.

Finally, the strip planking with balsa was almost completed and the boat-shaped core was brought to GH Craft in July end (Fig - 10). Half a year had been taken for laminating the hull. There were many parts in the hydrofoil sailboat. The hydrofoil and strut were especially important parts located at four positions, namely at the port and starboard sides in the bow and the stern. Also, good accuracy in workmanship of these parts would be required, notwithstanding Kanou's inexperienced work. If the project progressed at this rate, it was obvious that the completion of the boat would be delayed considerably.

So I decided to request Yasuo Tugaya, who had organized a wooden mold shop, to manufacture the hydrofoils and struts. He had even made the male mold rudder for the America's Cup boat and was skilled in making foils. He accepted and I was delighted to send him drawings and explanations. After discussing some points by phone, I visited him on August 28, discussed the work with him in detail, and confirmed the plan to receive the parts at the end of September. Since the manufacture of the hydrofoil moved out of Kanou's territory, we could go ahead as planned, and I felt at ease. If we did not put too much of a load on Kanou, the boat could be on the water in the beginning of November.

I requested President Kikuchi of North Sail Japan to make the trampoline. Mr. Kikuchi was the Director of Operations for the Japanese team to the America's Cup and was extremely busy. I placed the order in the beginning of August, and he accepted the order willingly.

7) Names of parts

Fig - 11 shows the names of the each part for your reference. In the figure, "main beam" and "connecting shaft" may be used interchangeably.

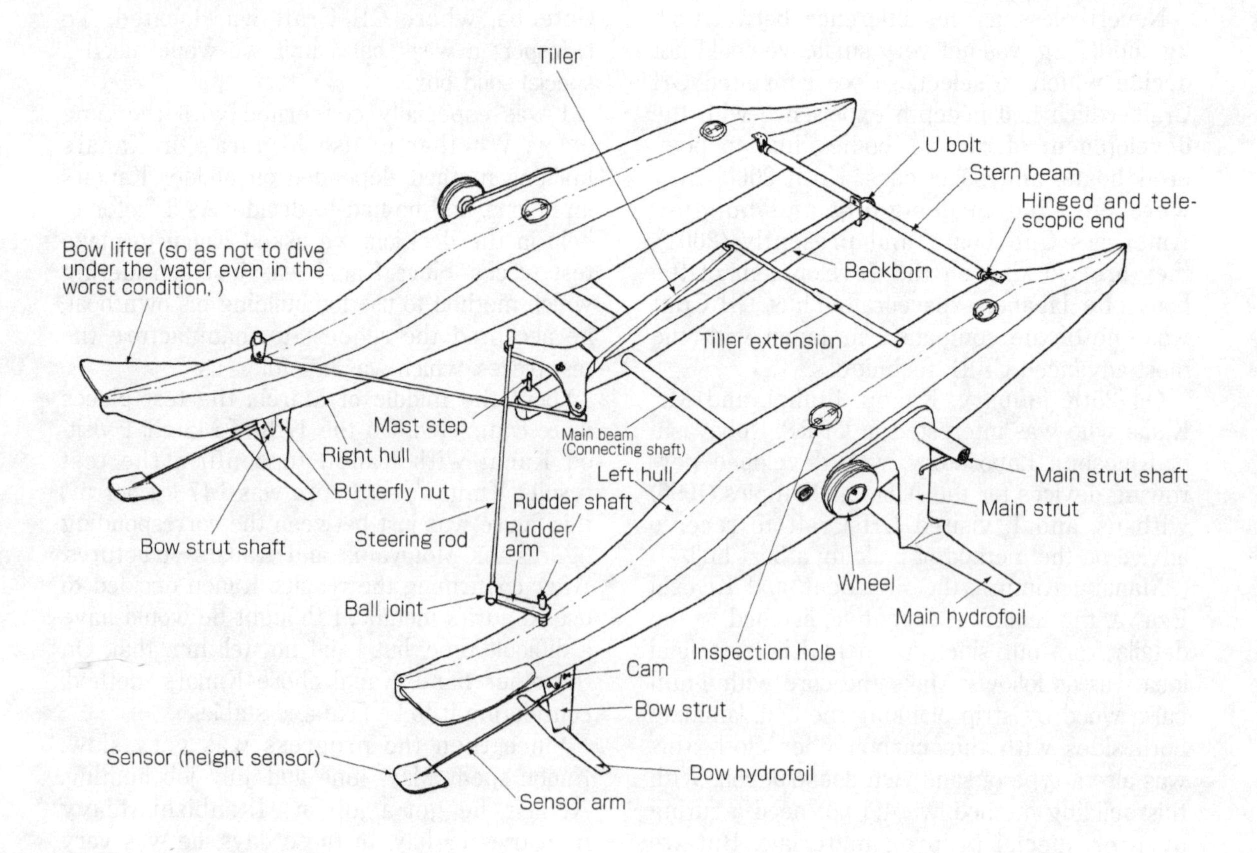

Fig - 11 Names of parts

8) Preparations for launching

After deciding the important parts, we had to prepare for the launching. The launch site was at Zaimokuza in Kamakura. We requested president Mithuharu Niijima, who ran a windsurfing school there named Seven Seas to launch the boat.

High performance at high speed was required from the sail of Twin Ducks, so we chose a ready-made windsurfing sail. In the early stages, we bought sails of area 6.4 m^2 and 10.6m^2 from Mr. Niijima's shop, studied the assembly process, and discussed how to erect the mast for our boat. We had a good relationship with him. The distance from Seven Seas to the beach was nearly 150 m, with some slopes and very uneven ground. Twin Ducks was designed such that the wheels came down when the main foil sprang up toward the aft, so that it could be transported by one person (Fig - 12). The wheel was pneumatic with 250-mm diameter and plastic hub. This kind of tire was originally made for carts intended to carry canoes and dinghies. As it had no ball bearings at all, it could be fully submerged in water.

Although there is no wheel on the bow strut, the part that extends below the hydrofoil acts as a skeg and prevents the hydrofoil from touching the ground. (Fig - 13).

This method of transportation however, was based on the premise of a flat concrete surface and slope in the marina. The clearance between the hydrofoil and the ground was only 10 cm, and this would not account for the uneven ground between Seven Seas and the waterfront. After considering these points, we made another wheeled cradle for transportation.

Fig - 12 Spring-up main strut

When the main hydrofoil springs up toward the aft, the tires come down facilitating transportation. The rubber sandals attached to the hydrofoil prevent it from tripping over.

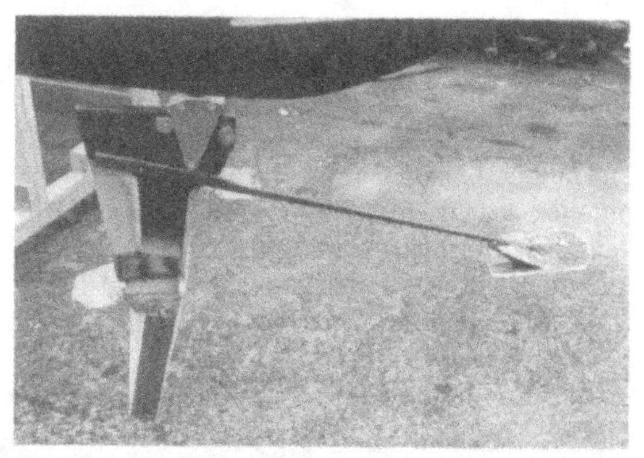

Fig - 13 Bow strut

The hydrofoil, strut, and sensor can rotate as a single unit around the bow strut shaft on the triangular metal plate. Cam stoppers are provided forward and aft of the triangular metal plate. By loosening the butterfly nuts positioned on the other side, the limit of rotation can be adjusted.

9) Strut stopper

Adjusting the foil angle (setting angle) of the hydrofoil was another problem that had always troubled me. We wanted to adjust the setting angle of the foil to an accuracy of less than 0.5 degrees, but measurement of the setting angle of the foil section was very difficult. It would be better to measure the foil angle at the front edge of the strut, but the connection between the foil and strut was not accurate in case of the model. The connection would break and be repaired sometimes. I was worried about such problems and unnecessary time was wasted.

If the hydrofoil broke off from the strut accidentally, our tests would stop and Kanou would not be able to graduate. Therefore, we made the connection stronger than necessary and the resistance of this area increased, but we thought that a strong connection was necessary.

How should we fix or adjust a strut that rotates? What mechanism should I use to stop the main foils after they spring up toward the aft? If the main foil hit a large obstacle and broke, how should we restrict damage to a minimum? I wanted the adjustment of the setting angle of the hydrofoil done with the boat afloat. Thus, there were many issues related to the foil fixing system.

I had to resolve these issues since August was approaching. The scheduled date of coupling the hull and deck was very near. The stiffening of this point in the hull would be inadequate after coupling the hull and deck.

After worrying about the problem for several days in bed, an idea suddenly came to me, which was to enable all adjustments at one time. I got up immediately and sketched my idea. Later, I made detailed drawings and submitted them on August 3.

The strut stopper was a simple mechanism weighing only 100 g (Fig - 14) with a length of 25 cm when folded in half. One end of the stopper was connected to the hull and other end connected to the end plate fitted to the strut arm. The end plate was made of a rectangular aluminum plate with 32 holes in three rows, and these holes were meant for making adjustments.

When the hydrofoil is lowered to the planing position of the boat, the strut stopper stops at its fully extended position at the upper end. A spring is provided at the knuckle of the half-foldable stopper, because of which the stopper motion stops slightly after its straight-line travel. The hydrofoil remains firmly fixed in position unless the stopper is pushed upward at its center to fold it. When the stopper if folded upward, and the hydrofoil springs up toward the aft, the lower end of the folded stopper again extends downward, and firmly fixes the position of the hydrofoil and wheel.

The end plate has three rows of holes. The clearance between the sprung-up hydrofoil and the hull can be adjusted by selecting one of the rows. Care is necessary to ensure that when transporting the boat, the fore and aft hydrofoils do not touch the ground. It is important to adjust such that the hydrofoils are retained as high as possible.

The main foil angle can be adjusted by selecting the appropriate hole in one row. The foil angle changes by one degree when you shift to an adjacent hole. Therefore, the foil angle and the clearance between the sprung-up hydrofoil and the hull on land could be measured accurately, and the foil angle and sprung-up height could be designated, for instance, as "A Row, #4 hole". Adjustments when afloat also became possible by pulling out and inserting two bolts on either side.

As another measure for collision of the hydrofoil, I made a 1.6-mm anti-corrosive aluminum plate stopper. When the hydrofoil hits an obsta-

Fig - 14 The strut stopper

The white rectangular plate with holes is the end plate and the inclining metal part connecting the end plate and the hull is the strut stopper at the middle of which is a round washer. If the strut stopper is pushed up by finger around the round washer, it folds to half and the hydrofoil comes down.

cle, the stopper is compressed, it buckles, and the hydrofoil springs up toward the aft. Thus, hull damage is prevented, and hydrofoil damage is minimized.

How should I select the thickness of the aluminum plate? If the plate is too strong, the hull-attaching point will be damaged; if it is weak, the hydrofoil will folded down at high speeds. After much thinking, I selected the thickness, about which I will write later. The selection was appropriate.

Five functions were fulfilled by this stopper: the fixing and adjustment of the foil angle, the fixing of the hydrofoil spring-up position, its adjustment and preventing its damage. I had forgotten the advantages of this stopper because I was so busy. When we assembled the complete boat at the Institute of Industrial Science, University of Tokyo (IIS), Kanou and Inukai were impressed with the stopper, and I felt happy at that time.

We used the wheels attached to the boat to take it to the beach. The cradle built later for this purpose, was not used at all. By taking care, we could go over uneven surface, and if we had many people, we lifted up the boat to clear the uneven surface (bumps). The swing down wheel arrangement that we provided was a huge success.

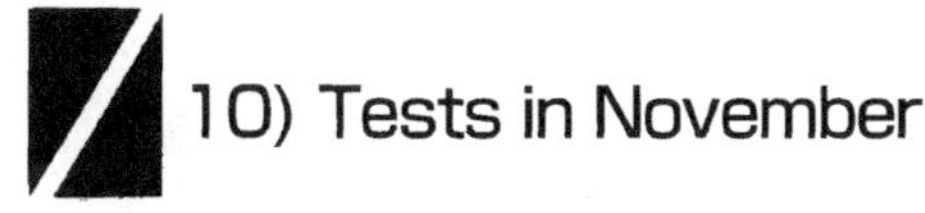

10) Tests in November

Since Seven Seas was busy during weekends, the test dates were decided as Tuesday, November 14 to Friday, November 17. Although Kanou had 10 spare days, many tasks had to be accomplished as the final date approached, and he slept very little before he transported the boat to Zaimokuza.

Kanou came to Seven Seas with his car Mazda Demio, with the two hulls and a sail assembly on the roof, and parts such as hydrofoils and tools (see Fig - 15) in the trunk of the car. Although many other tools and related materials were transported by another car, almost a complete boat could be carried in his small car. The initial concept of Twin Ducks was transportability to the marina in a Toyota Mark 2 and after assembly of the boat there, to be launched by a single person. This concept was almost realized. I am not sure however, if two sets of main hydrofoil and strut can be stored in other types of car with no rear door.

We were very busy for the tests of the boat

Fig - 15 Boat loaded on and in the car

The car is Mazda Demio (3.5-m length).

on November 14. Plenty of work had to be done and without Kanou, we did not know where the tools and materials were. As a result, the boat assembly continued until the 15th afternoon. At this stage, there was less wind and we tried to tow the boat without sails by a motorboat. Inukai who joined us on this day was onboard Twin Ducks, which was towed by a rope attached to a spring balance. Mr. Niijima pushed the throttle of the towing boat and Twin Ducks soon took off.

The model took off with great difficultly, but the full-scale boat took off very easily. The spring balance showed 13 to 15 kg at the hump (peak in the resistance curve before take off). After the hump speed, the resistance remained stable at nearly 8 kg, and then reached 15 kg with the increase in speed.

As the boat weight was 57.5kg and Inukai's weight was 60 kg, the ratio of the resistance to total weight was 12% at the hump and 7% after planing. The corresponding ratio for the 1/5 scale model was 20% at hump and that for the 1/3 model was 16%. Compared to these figures, the actual boat had a better performance. However, compared to an excellent human powered boat with a ratio of 5 to 6% up to 20 knots, there was still room for improvement.

We were very pleased with the success of the foilborne boat when towed. The foilborne sailing test of the 1/3 model was successful, after foilborne operation by towing was stabilized. We were sure that foilborne sailing (without towing) the next day would be successful. The next day, on the 16th, there was an ideal wind. Mr. Niijima volunteered to be the skipper. The test location was Mr. Niijima's home ground. He had extensive knowledge such as the sea depth, the locations of the fishing nets, and wind conditions. He was also a good windsurfer. Since Yamaha had been developing sailboards, I knew well about his skill as a windsurfer. With the ideal wind, the windward hull took off easily but the leeward hull did not rise (Fig - 16). The leeward hull kept its bow low. The boat went and returned several times from the east end of Zaimokuza to the west end of Yuigahama, but its attitude did not change.

Once, when Mr. Niijima tried to turn the bow to leeward, the boat suddenly began to take off. This time the foilborne condition did not continue maybe because he turned the tiller quickly. After that, there was no change in the attitude, although we changed the angle of attack of the foil, and made other adjustments. There was something wrong because the model test had been successful.

Soon, I found that the horizontal distance between the main beam (connecting shaft) that penetrated and connected the left and right hulls, and the point where the hydrofoil was located, was too large compared to that of the 1/3 model . If a large lift acts toward the aft of the main beam, the bow will naturally go down. We prepared three alternative positions where the hydrofoil could be installed. During the first trial, we used the intermediate of the three positions. Thus, we could shift the hydrofoil 15 cm forward, but this shift was inadequate. I was surprised to note that to maintain the same relation to the model, the main beam had to be moved toward the aft.

A large force would act at the points where the beam and the main hydrofoil were located in the boat, so these areas were made very carefully. If we move the beam and the main foil, we have to cut the hull over a large area. If we do that, perhaps some other areas may become weak; if we don't move these components, the bow-down condition will not improve.

It was my fault entirely since I did the layout drawing, but without any modification, there was no hope. Without bringing the boat back to the workshop and making major modifications, there was no prospect of improved performance. On the 17th we disassembled the boat. It was a miserable, rainy day and I felt woeful. After we sent the truck carrying the boat, and on my way home, I suddenly became aware that my assumption until then was that the trim (longitudinal inclination) in both left

Fig - 16 Leeward hull does not rise

With a trampoline stretched between the left and right hulls, the bow of windward hull rises more than expected but the bow of the leeward hull does not rise.

Fig - 17 Forces acting on the left and right hulls (for scale model)

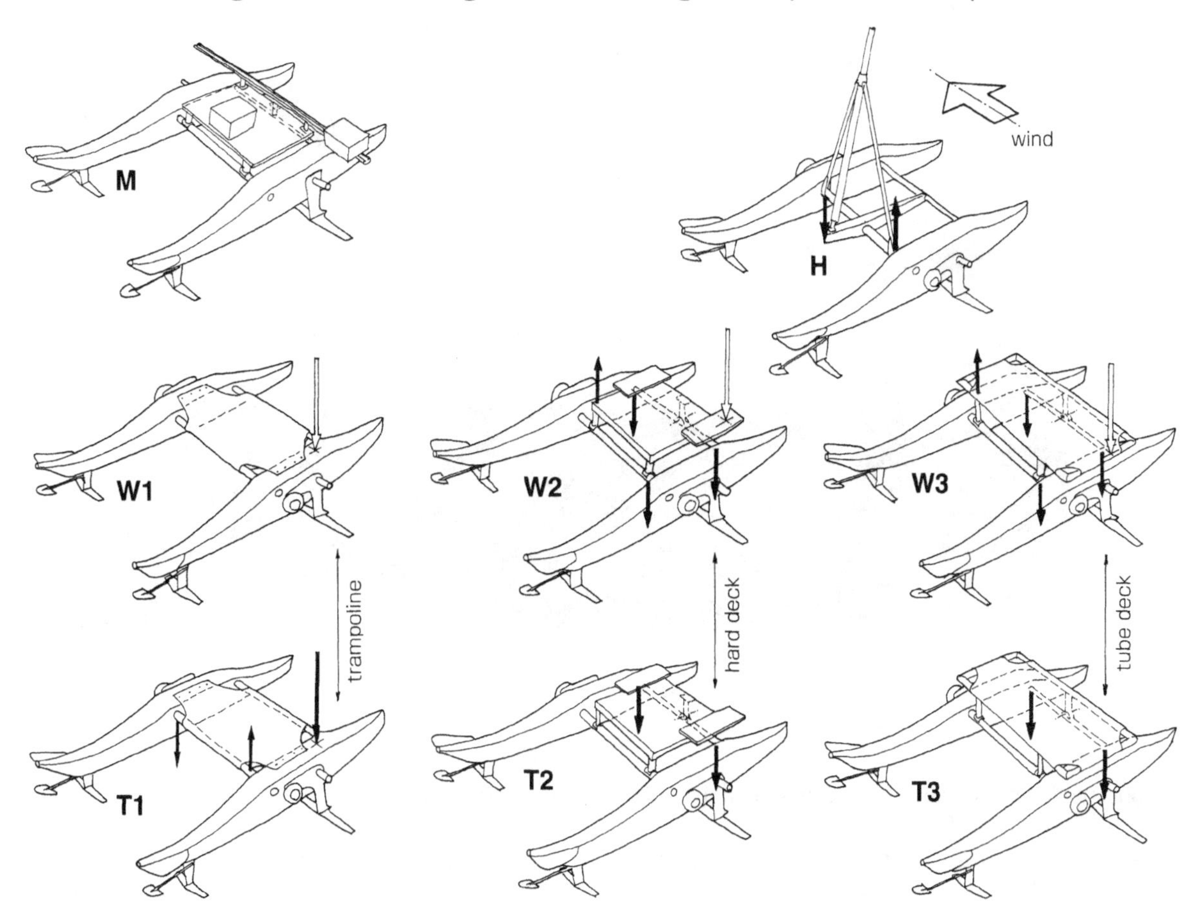

Forces acting on the left and right hulls are clarified on the premise that the heeling moment of the sail due to wind pressure just balances the moment due to weight of the skipper.
The upper right sketch (H) shows how the heeling moment due to the sail is conveyed to both ends of the main beam. On the other hand, each of three sketches W1, W2, W3 shows the forces due to the skipperls weight. In case of a trampoline, the skipperls weight acts on the hull directly (W1), and in case of both the hard and tube decks, the balancing moment is conveyed to the main beam through the deck. In addition, the middle point of the rear beam supports the aft end of the deck. Therefore, the skipperls weight is distributed equally to each hull (W2, W3).

Now the forces H acting on the hull due to the sail and the forces W1, W2, W3 due to the skipperls weight can be combined to T1, T2, T3. These are the forces finally acting on the hull. In the case of T1, total weight of the skipper acts on aft of the windward hull and causes the bow of the windward hull to rise and the bow of leeward hull to lower, as there is no force to push down the stern on this side. In case of T2 and T3, the heeling moment and the balancing moment cancel each other on the main beam. The forces acting on the left and right hulls are exactly same, which can be considered for our 1/3 model (M)

and right hull should be the same. Since the period of the model test was long, I might have got used to the condition of symmetry of left and right hulls.

However, the day before, only the bow of the leeward hull of the Twin Ducks was down. On the other hand, the bow of the windward hull was too much up. Hence, I thought that I should consider the balance around the main beam of the left and right hulls separately, and which I should have studied until then. I desperately calculated the balance between the left and right hulls to seek improvements. We could make the boat foilborne at least once, so if we could improve this situation without large modifications, such as to move the position of the main foil ahead by one point, to add a weight of about 5 kg at the aft end of the leeward hull, to change the position of the main sheet pulley from the mid point on the aft beam to the windward hull, and so on. By making these adjustments, we could expect considerable improvement. We should have tried to test these adjustments on the 17th, and I regretted having made hasty conclusions.

From the 19th, I performed calculations for the balancing moment around the main beam of the leeward and windward hulls for the 1/3 model. By comparing the results from the full-scale boat, I hoped to clarify the extent of the

Fig - 18 Hard deck
A box type deck with width equal to the width between the two hulls, thickness of 5 cm, was attached to the connecting shaft. Deck extensions provided to compensate the lack of width.

poor performance due to the installed position of the main hydrofoil.

The calculations helped me to attain a clearer understanding. The trampoline was not used in the 1/3 model. The scale model had a hard deck with high rigidity to resist twisting and it was connected to the main beam (Fig - 17). As the rail for the balance weight was on the deck, the balancing moment of the weight was conveyed to the main beam. The balancing moment was in direct equilibrium with the heeling moment due to the sail.

The aft end of the deck was supported at the center of the rear beam, so the load of the balance weight was distributed equally to the left and right hulls. Considering that the heeling moment due to the sail was just balanced by the righting moment due to the skipper's balancing weight, exactly the same force worked on each of the hulls, namely the left hull and the right hull. The actual boat however, did not have the same structure as that of the scale model. There was a trampoline fixed between the left and right hulls and the skipper took his position on the windward hull, so the skipper's weight did not push down the aft part of the leeward hull. Consequently, the tail of this side hull rose and its bow descended in the water due to the direct effect of the heeling moment due to the sail.

This was a fundamental concept, and I was surprised I had missed it. Since we had used the trampoline without much thinking, we had departed completely from the equilibrium of the scale model. Why had I not considered following the scale model completely? However, the layout was the same from the beginning.

The longitudinal position of the main hydrofoil and so on, were minor issues now. The issue was how to match the conditions of the 1/3 model. I immediately thought of a hard deck fitted on the main beam, which was similar to the scale model. As the hard deck needed rigidity for twisting, I considered its structure as a 50-mm thick box section made of 3-mm marine grade plywood, which could be supplied by Mr. Kanai (Fig - 18). The box weight would be in excess of 10 kg, nevertheless, we had to use this deck.

On the 20th, I asked both Kanou and Inukai to check the calculations, and they accepted my calculations. After setting up the new hard deck instead of the trampoline, the modifications would be complete. We would not need to cut off the hull and the test results would be guaranteed. Therefore, I wanted to perform the trials and to verify the result as soon as possible. Kanou set the next trial for the 11th to 15th of December.

11) Success

The twenty days for making the preparations passed quickly. The painting on the hard deck was finished on the night just before the trial, but the left and right extension decks and foot belts were not fitted until 11th of December. In such a condition, the boat arrived at the Seven Seas. We assembled the boat quickly but a lot of work still remained.

On 12th and 13th, Mr. Niijima was busy with another job and Takaichi, who belonged to the yacht club of Keio University, boarded the boat as a test skipper. On the 12th, we had no wind, so we performed towing tests with Takaichi onboard. The main foil was positioned forward a little, and the towing started, but something was wrong. The front hydrofoils rose excessively and the rudder did not have any effect if the skipper did not shift forward.

In the afternoon of the same day, we moved the position of the main hydrofoil to the rear end and tested the boat again, but it did not run in a stable manner. Just after becoming foilborne, the bow dropped. The lift force of the sensor seemed to be insufficient. The connection between the front foil and its strut had a flat frontal area, which probably contributed to increase in the resistance. Due to the large resistance at this point, the sensor would naturally receive a larger load and submerge, so I attached a small fairing piece in front of the sensor to check the effect (Fig - 19).

On the 13th, we towed the boat with the main hydrofoil in the middle, similar to the condition in November, which was the most stable condition. But the front hydrofoil tended to drop suddenly. The fairing in front of the front foil did not work. The boat's bow tended to drop when the rudder was steered mainly because of the wake encountered and the yaw of the bow. This phenomenon occurred only in the right hull, presumably because the strut drew air on its suction side and the air reached the upper side of the front foil because of over steering, resulting in a loss of lift, which is referred to as ventilation. I made a plate like a cavitation plate of an outboard motor by cutting off a 1-mm thick carbon fiber reinforced plastic plate, and attached it to the lower part of the strut, 50 mm above the front foil (Fig - 19). The aim was to stop the ventilation.

Fig - 19 Nose fairing (right) and cavitation plate (left horizontal plate)

In the afternoon, we performed the test with the alterations and found that the right hull no longer dropped but the left hull dropped instead. The cavitation plate did work, but the difference between the left and right hydrofoils was so small that the performance of the right hydrofoil improved and the left one deteriorated.

But I did not understand why the hulls dropped so easily. When we stowed the boat, Inukai suddenly remarked that when the boat turned, the rudder angle on the inside of the turn should be larger than that of the rudder angle on the outside of the turn, but this did not happen. How could this be? I thought that

Fig - 20 Crossing of the connecting rods for steering
(These rods do not cross in Fig-16). White parts of the rods are round bars of mops

I had checked everything thoroughly, and I was surprised when I saw the rudder motion. I asked Kanou to bring the drawings and found that the arrangement of the connecting rods was incorrect. The drawing showed the connecting rods between the rudder arm and tiller were crossed (Fig - 11) but they were not crossed on the boat (Fig - 16). How did the concerned persons miss this in spite of being with the boat for a total of 7 days in November and December? The sudden drop of the bow was evidently due to the error mentioned above which caused excessive steering on one side.

Soon we tried to correct the mistake but the length of the rod was 200 mm short, so we had to extend it. We found several pipes in the waste area of Seven Seas but the diameter did not match. In the meantime, I found a stick used in a mop that was just right. We were pleased with it and we connected parts of the stick to the rods to extend them (Fig - 20).

From the beginning, the tiller was designed so that the boat turned to leeward when the tiller was pulled and turned to windward when it was pushed, to make it similar to tiller in a sailboat (the arrangement in which was opposite to that in our boat). Wouldn't the skipper who tested the boat with the incorrect arrangement have had a difficult time controlling the boat? This repair would certainly resolve the problem of bow drop. On the 14th, Takaichi again boarded the boat. As it was less windy, we expected the boat to run perfectly when towed. However, when we launched the boat, we felt a good wind. The boat ran by itself without being towed and sometimes, the bow rose.

The wind speed was 3 m/s. When the wind speed increased, Twin Ducks suddenly began to get foilborne. Kanou shouted with joy (Fig - 21, Fig - 22). Takaichi, who was busy swinging his body until the leeward hull had taken off, relaxed and sat down after the take off. The boat took off at a speed of 10 km/h and soon its speed increased to 15 km/h to 20 km/h. After the wind stopped blowing, the foilborne operation continued for a while. For some time, the wind speed decreased and the boat dipped its tail in the water. Even during that time, the boat again took off when Takaichi swung the upper part of his body as required. This time the boat continued sailing for about 1 km for 3 minutes in a light wind. During this time, Kanou continued to capture the boat motion on video. This was proof of foilborne operation, and I was sure that Kanou would graduate.

According to Takaichi, after rudder control became normal, he felt the steering was working well. Looking back at the situation, the steering system had been terrible until then. Naturally, the phenomenon of bow drop stopped. Takaichi also said that the helm had changed. The boat tended have weather helm before take off (turn to windward) and tended to have lee helm after take off the boat. Increasing the rake of the mast would correct the lee helm during foilborne condition. How should I cope with the weather helm? I had no

Fig - 21 First foilborne operation (1)

Fig - 22 First foilborne operation (2)

chance to try that.

According to Takaichi, even if the wind speed was less than 4 m/s during take off, the boat could get foilborne, although the 470 class (sailing dinghy of 470-cm length overall) could not get on the plane at the same wind speed. It seemed that the boat could take off at a smaller wind speed than the wind speed at which a sailboard could plane with a large sail, so getting to the foilborne condition in a light wind, which was the main topic, was a success, and I was very happy.

On 15th, the next day, Mr. Niijima took the helm. His weight was 80 kg, which was 15 kg more than Takaichi's weight. If the wind were strong, there would not be any problem. But looking at the running boat, I became a bit worried. The hard deck of the boat twisted because of the additional weight of the skipper and tended to touch the windward hull. If this occurred in a strong wind, as in the case of the trampoline, the skipper's weight would push down the windward hull and the bow of leeward hull would plunge into the water.

In such conditions, it was difficult for the boat to take off although the wind speed was higher compared to the previous day. The boat sailed to Zushi offing looking for appropriate winds. During foilborne operation, the leeward bow occasionally plunged into the water. A heavier skipper and strong wind inevitably brought about same tendency in the boat as in the boat with the trampoline. I was sorry for Mr. Niijima because the boat did not get foilborne comfortably compared to Takaichi on the boat. In the meantime, Mr. Niijima stood up to control the boat. I suppose that sailboarders would also feel at ease standing up on the board so that they could move quickly. But he stumbled and fell overboard (Fig - 23). Looking at the photo, you can see that his leg has shifted from the hard deck to the hull, and the moment his weight shifted on to the hull, the leeward bow plunged into the water.

Fig - 23 Falling overboard

Mr. Niijima fell overboard and the boat overturned as it lost the skipper's weight. After the skipper fell off the boat and was some distance apart, and when he got on the strut of the main foil, that sail top was already submerged 2 m below the waterline and it could not be restored; the boat capsized fully. We ran a rope around the far side of the hull and pulled it slowly, and the boat recovered easily to its upright position. During the recovery, the root of tiller broke, so we returned to the harbor towing the boat.

When lifting up the boat, we found the strut stopper completely bent (Fig - 24). During the recovery, a strong force had been exerted on the strut by the rope. We had unintentionally carried out a capsizing test, thanks to the accident. When the main foil hits an obstacle during the foilborne operation, the stopper works as a safety device to protect the hull. My selection of plate thickness proved to be correct.

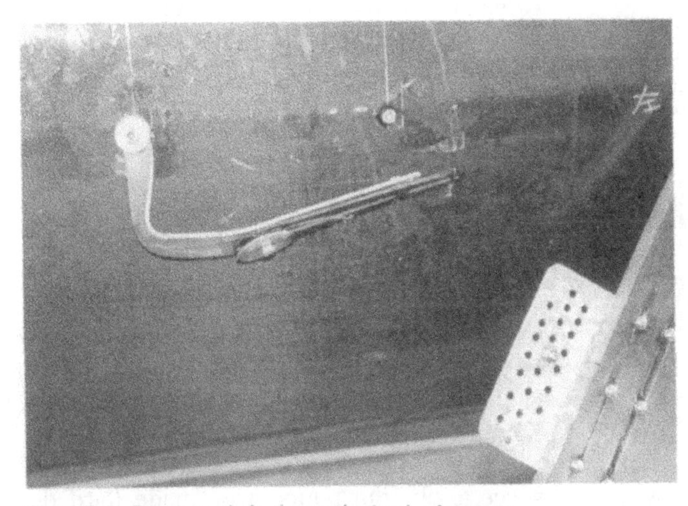

Fig - 24 Bent and damaged strut stopper

All the tests were finished. We had taken splendid video shots and photographs. The remaining important issue was Kanou's graduation thesis, which had top priority. For the time being, the boat remained as it was.

12) To Inukai and Kanou

I am happy that Yasuhiro Sudo, a new face in the Kinoshita Laboratory and a member of the sailing club of Sophia University, came to assist us in our test in November. If he succeeds with the project, we will have another chance to run the Twin Ducks.

For future tests, we will exchange the hard deck, that tended to twist easily and be relatively heavy, to a wider deck like that of the 49er (one of a dinghy class used in the Olympic Game), which will be rigid and light and made of aluminum tube frame with canvas. If there is adequate space between the frame and hull, there will be no problem like the one we faced on the 15th ; therefore, Mr. Niijima will be able to enjoy foilborne operation of the boat.

If the boat takes off easily at wind speeds of 3 m/s to 10 m/s, the cruising between Zaimokuza and Enoshima can be covered quickly, this distance will be covered in less than 10 minutes at a 40 km/hr speed if there is an 8 m/s wind.

The dream of the boat is overtaking other dinghies at a fairly good speed, which I had when I thought of the concept of Twin Ducks, seems to be in front of my eyes, and I am happy. I will be happier if both Inukai and Kanou graduate after attaining their dreams.

They had exceptional experiences when they developed a vehicle that had never existed before. They formulated the plan by themselves, designed, prepared the materials, built, asked many questions, received assistance, learned, and acknowledged the value of cooperation. They encountered many difficulties and cleared them to go forward. As a result, they gained valuable experience and they should bear in mind that with this confidence that they can take up bigger challenges and attain success however difficult the challenges are. Congratulations to both of them!

13) Press release and the Tokyo boat show

At the end of the year, I heard from Professor Kinoshita about the press release on the 17th January, 2001, where the results of the research at IIS would be announced. On the day of the announcement, about 20 press officials attended although 3 or 4 only were expected to attend; so, we could not give the press release to all. The mechanism that supported the heeling moment by twisting the Twin Duck hulls seemed to interest the press, who belonged to the science department of newspapers. Professors Kinoshita's exciting announcement and the impressive video of the 14th December stirred up considerable interest and the meeting room was filled with excitement.

In the morning the next day, I found a copy of the Twin Ducks article that had appeared in the newspaper, Asahi Shimbun in my facsimile machine. The fax was from Yoshiaki Murakoshi, ex-Yamaha Motor and working in the Japan Boating Industry Association at that time. He also requested me to exhibit Twin Ducks in the Tokyo boat show next year.

Soon after receiving the fax, maybe around 8:30 a.m., Murakoshi phoned me to my surprise, and requested me again to exhibit the Twin Ducks at the Tokyo Boat Show that would start 20 days later. Recently, due to economic recession, the sales of sailboats were dropping, and they were looking for measures. They thought the Twin Ducks was an exciting and eye-catching product. As the show was just before the presentation of Kanou's thesis, I hesitated; but Professor Kinoshita agreed, to my inquiry and reported that all members of his laboratory would support it. Thus, the plan to exhibit Twin Ducks at the boat show this year was decided the same day.

Since the location where Twin Ducks was exhibited at the boat show was very good and close to the entrance, many people gathered around it every day. There were many people who had heard of it through the news media and a large number came to the boat show to see only the Twin Ducks. I enjoyed having conversations with the people who were interested in the boat and I was pleased that so many people were interested in it. Although I designed the boat out of my own interest, I felt reassured that the boat would sell fairly well if in production.

14) Aluminum Tube Deck

When the prospects of Kanou's thesis became favorable, Professor Kinoshita told me that another sea trial should be done in March after Kanou graduated. I soon agreed.

Looking back, the test in December was a success, but the temporarily made hard deck lacked rigidity and was heavy. The boat also did not work as intended with Niijima's weight. Additionally, for foilborne operation against the wind, it was preferable to have a wider deck and increase the balancing moment. So, I decided to make a deck with good performance before the test in March (Fig - 25).

An American company named Aircraft Spruce & Specialty sells materials and parts for homemade aircraft by mail order. From them, we can purchase aluminum alloy tubing of small thickness and large diameter. Actually, the connecting tubes that connected the two hulls and the rear beam were supplied by this company.

To make a rigid and light deck, I designed the deck frame made of 3-inch thin tubes (76-mm diameter, 0.89-mm thickness, 6061T6). It looked sturdy because of the large diameter, but actually, its weight was only 7.5 kg. Even if we add 2-kg as the weight of the trampoline cloth stretched on the frame, the new structure is much lighter than the earlier hard deck of 15-kg weight.

But the small-thickness tubing was unbelievably expensive, the price of which worked out to $500 per boat. On the other hand, if I use a slightly thicker 2.5-inch tube (63.5-mm diameter, 1.25-mm thickness, 6061T6), the weight will be 1.2 kg heavier but the price will be only $150 per boat. During the prototype building, we can afford tubing of higher price but in production stage, the price will play an important role. Thus, I decided to use the 2.5-inch tubing. If we weld a heat-treated T6 material, the strength near the welded point reduces significantly. To prevent the strength reduction, the welding was limited to only 4 corners of the rectangular frame and the remaining connections were glued or clamped with hose bands. The trampoline to be stretched on the frame was to be provided by President Kikuchi of North Sail Japan. We were ready several days before the test in March (Fig - 26).

15) Speed polar curve

Inukai, who had carried out the model test of the Twin Ducks, verified the stability and ana-

Fig - 25 Change in the deck

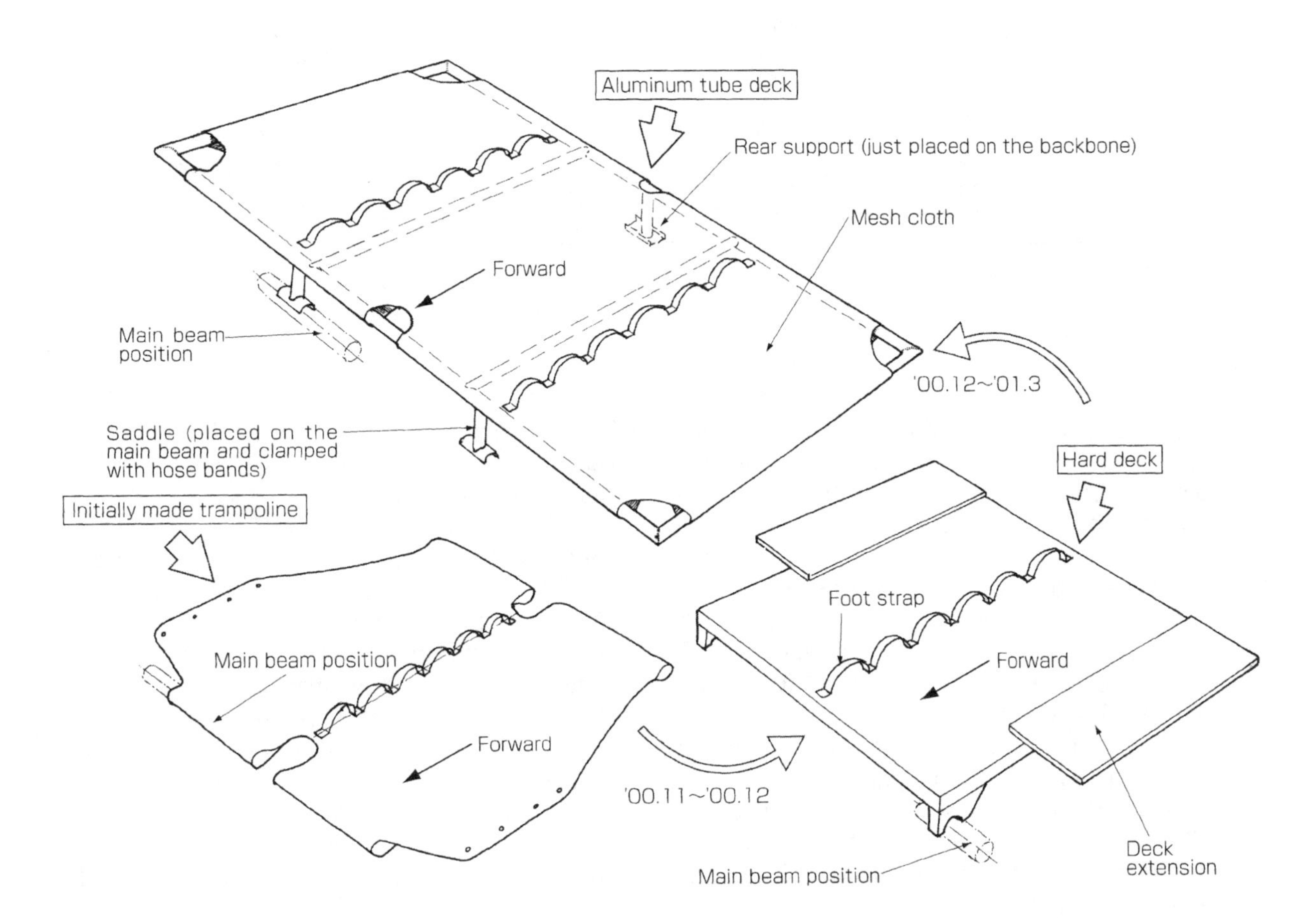

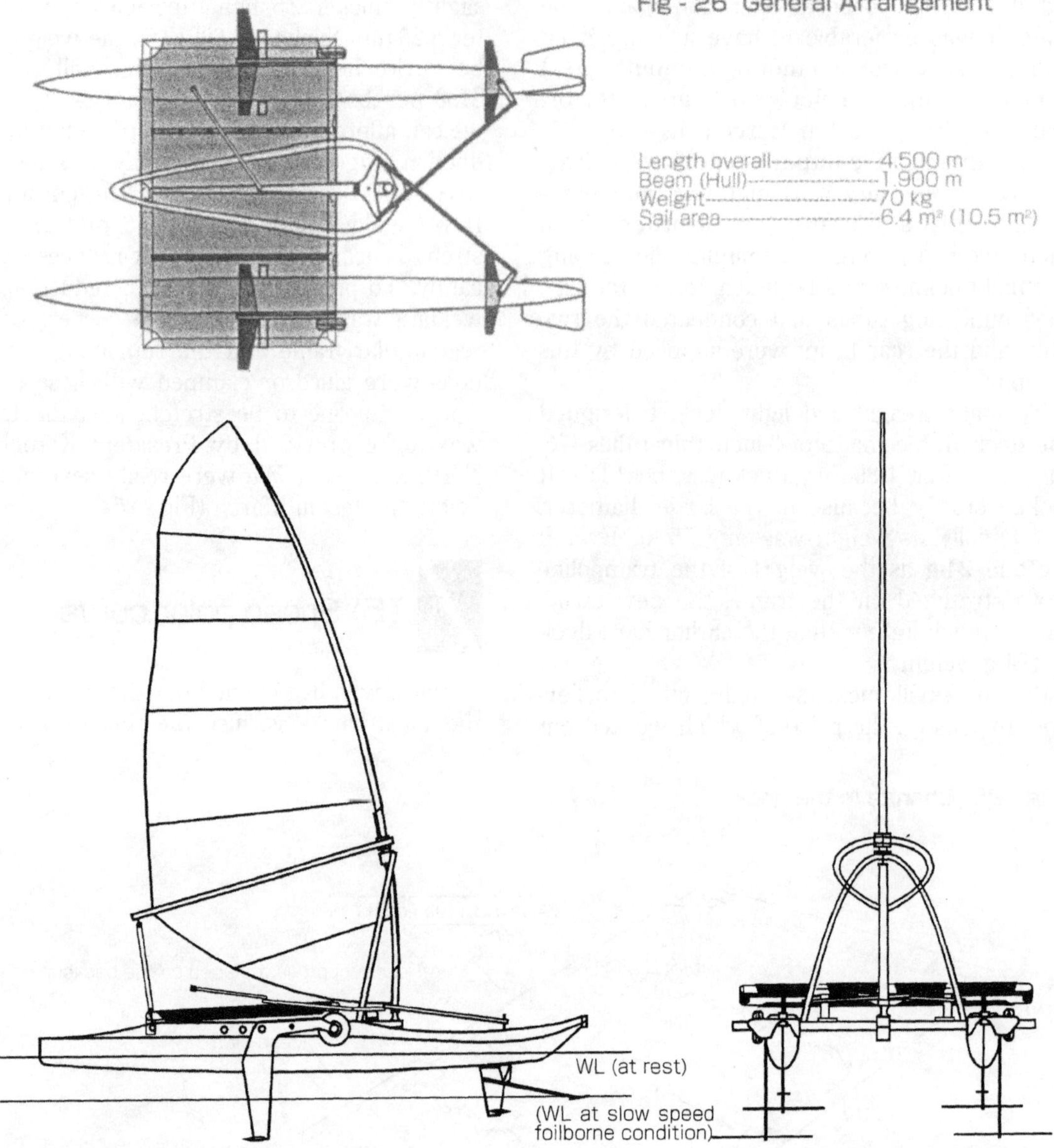

lyzed the performance of the boat in his thesis, and then graduated. Kanou, who continued with the project, took over the model and produced a polar curve by including the actual boat conditions. This was a part of his graduate thesis, and is shown below (Fig - 27).

The right hand side of the graph shows the case of the boat with 6.4-m² sail and the left hand side shows the case with a 10.5-m² sail. The wind is from the top.

The graph is read as follows: If the boat has a 6.4-m² sail in an 8-m/s wind, the boat speed at a heading of 106 degrees is 11 m/s (21.4 knots). The limiting close-hauled direction is about 55 degrees, in which case the boat speed decreases to 7 m/s (13.6 knots). Likewise, when heading 150 degrees in the downward direction, the boat speed decreases to 7 m/s. Boats naturally cannot overtake the wind.

And if the boat has a 10.5-m² sail in a 4-m/s wind, and the heading is 105 degrees, its speed reaches 6 m/s, but it can become foilborne only from 89 to 121 degrees. Generally, the boat speed abeam, that is, 1.4 - 1.5 times the wind speed seems inadequate. The limit of the close-hauled direction was also not satisfied probably because our focus was more on the boat becoming foilborne in a light wind. As we do not have the data of the actual boat, we cannot compare the differences. Anyhow, the objective of the boat has been attained, I believe.

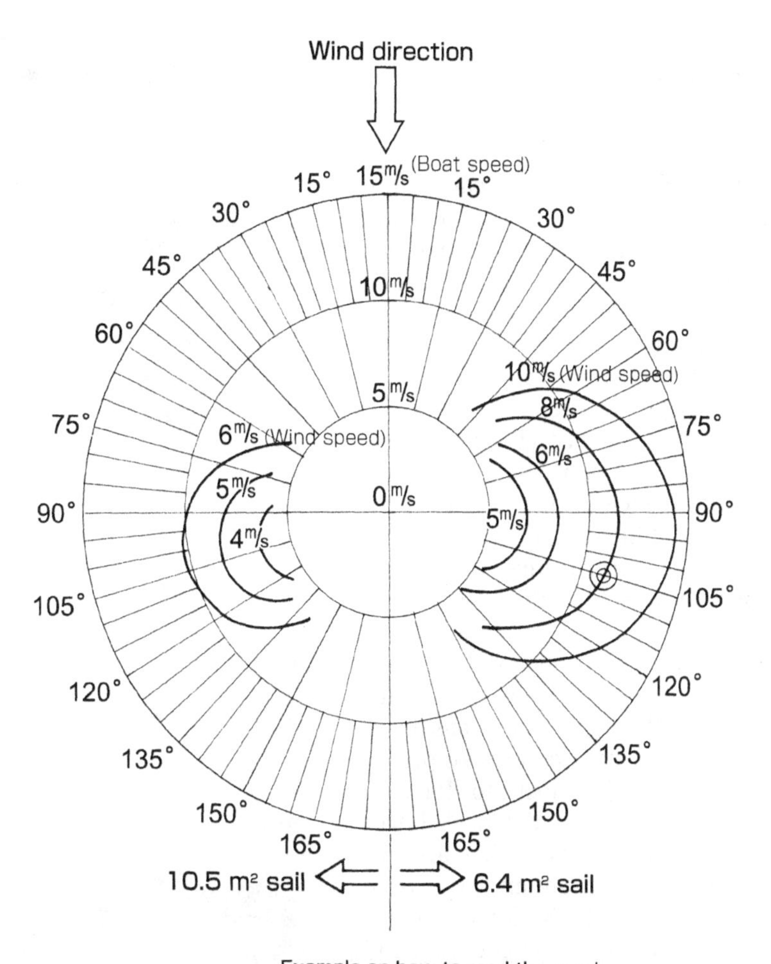

Fig - 27 Speed polar curve

Example on how to read the graph
If the boat carries a 6.4-m2 sail (right hand side), when the boat runs in the 105-degree direction in an 8 m/s wind (mark ◎), the boat speed reaches 11 m/s (21.4 kt). Note 1 m/s = 1.94 kt

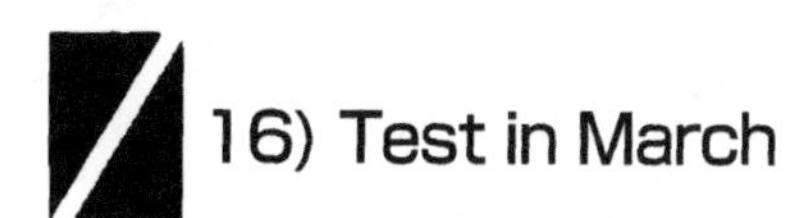

16) Test in March

On March 19, 2001, almost the same members as in the previous year carried the Twin Ducks to Seven Seas at Zaimokuza. I heard that NHK would come to film the boat running on 21st and to broadcast it in their program Good Morning Nippon on the early morning of 22nd, and they would also broadcast live from the boat on the beach.

Until then, the skippers of the boat were Mr. Niijima and Takaichi. But this time Yasuhiro Sudo, who would continue with the project, was going to be the skipper of the boat, so he had to become familiar with sailing the boat.

We had good winds of 5 to 7 m/s that day. Firstly, Niijima took a ride and Sudo was on the escort motorboat. The boat got caught in seaweed that day, and could not continue its foilborne operation. Later, Sudo took a ride and suddenly the boat ran for about 100 m in the foilborne condition before it was stuck because of seaweed. The strut stopper was bent and the angle of attack of the main foil deviated. Seaweeds are in abundance every spring, which is a cause for concern. However, the new aluminum tube deck worked very well (Fig - 28).

20th was Sudo's training day. Due to this training, NHK could shoot a nice film on the 21st. The live broadcast on the 22nd morning went well. Everything was finished (Fig - 29).

We tested the boat for three days the next week, but the wind was not so good. During this time, Sudo became accustomed to sailing the boat, which was a good thing. In the meantime, we extended the skeg of the front wing downward (Fig - 30), by which the rudder worked and it gave us a good feeling. This was an important step forward for the boat. However, the performance data measurements were postponed until the next test.

17) Conclusion

The activities of fiscal year 2000 came to an end as mentioned above. Next year, I would like to measure the performance at each attitude of the front and main foil units independently in the towing tank. After improving the performance of each unit, they will be assembled in the hull and we will be able to improve the performance of the boat to a fair extent.

Especially, the performance of struts in the lateral direction will be important, because the close-hauled performance is dependent on the ratio of lateral lift and the drag of the strut.

I want to see the performance of the boat after assembling it with the improved hydrofoils and struts. I also want to generate the polar curves for each combination of boat specifications, by which I will be able to extract the best performance of the boat. When considered from these aspects, the development of performance of Twin Ducks is just about to start.

For example, if we reduce the main foil area to half keeping the other parts as they are, calculations show that the maximum boat speed at 10 m/s wind will exceed 35 knots. In this case, the ratio of maximum speed to wind speed becomes 1.8, which is fairly high compared to the present figure of 1.4 to 1.5, but it may become difficult to get foilborne in a light wind at this reduced foil area.

I would like to maintain the advantage of this boat, which is the ability to become foilborne in a light wind. The point of the boat is not to aim for speed records but to enjoy the foilborne speed sailing over a wide range of wind speeds, especially in light winds. The key point of this boat is that it gives the skipper the opportunity to extract the maximum performance by balancing his weight against the heeling moment of the boat. On the other hand, such balancing by the skipper is also dangerous at speeds of over 30 knots (Fig - 31) since he can easily fall overboard.

We can get a speed of 10 to 25 knots at a wind speed of 3 to 8 m/s from the boat. What are the prospect for its performance henceforth? I'd really like to see the boat reach its upper limit of performance in light winds.

Fig - 28 Aluminum Tube Deck
We made a frame of 6061 T6 round aluminum tubing of 63.5-mm diameter and 1.25-mm thickness, highly rigid for twisting, and then stretched a trampoline on it (the person seen is Sudo)

Fig - 29 Foilborne operatin in March 2001
The rigidity of the pipe deck was adequate and looked dependable. The boat became foilborne in a wind speed of less than 4 m/s

Fig - 30 Large skeg
Large black skegs were attached

Fig - 31 Foilborne sailing in March 2001
NHK broadcast on the boat of a beautiful sunny day

19

HIGH SPEED SEA KAYAK K-60

High-speed displacement boats with slender hull form have small resistance. The ultimate in slender boats is a rowing shell that has a beam of only one by twenty six times its length. Mr. Kanai, who often participates in races with his self-built sea kayak, had the same idea and he built slender boats with beam equal to that of the human waist and with a side float. He got good results in races with this boat.

One day I wondered why the boat beam should be the size of the human waist and thought of designing a slender boat like a scull. Resistance calculations of this boat clearly indicated its advantages.

I showed Mr. Kanai the drawing and he wanted to build the boat; the result was the K-60. He installed a swiveling seat on the boat and it was successful. In the future, I am considering installing a sliding seat and wing sail to make the boat more enjoyable.

1) Passport 5

Since 1989, I have been associated with races of solar and human powered boats. Both these races are principally races for home-built boats. I was concerned that people who wanted to participate in the race might find it difficult to build their own boats because of lack of knowledge and techniques of boat building

It was decided to hold a joint race of solar and human powered boats at Lake Hamana in 1994. My responsibility was to edit the bulletin. In 1997, while wondering how to collect articles for the bulletin, I decided to design a boat that an amateur could build easily. I thought many people would be pleased to read an article related to boats built by amateurs.

Fig - 1 Model of Passport 5 made of paper

The Stitch and Glue method is one such method to easily build a boat and is described here. To my regret, I have never tried the method myself, but it is simple to build a knuckled boat with plywood.

The name Stitch and Glue comes from the construction method in which plywood is stitched with wires and fixed by glue. Small holes are drilled along the edges of plywood boards, which are cut out exactly according to contour patterns, and the adjoining cut plywood boards are connected to each other by copper wires to form the boat shape. The outside and inside of the connection lines are filled with epoxy putty and both surfaces are laminated with FRP to complete the boat. This is just the right method for home-built boats.

The contour of the cut out plywood decides the boat shape definitely, but it is not easy to decide the contour. This is because the contour of the plywood boards determines the actual three-dimensional boat shape, and unless the complete boat is built, one cannot say for sure whether the bow and the stern parts have appropriate contours. On the other hand, once

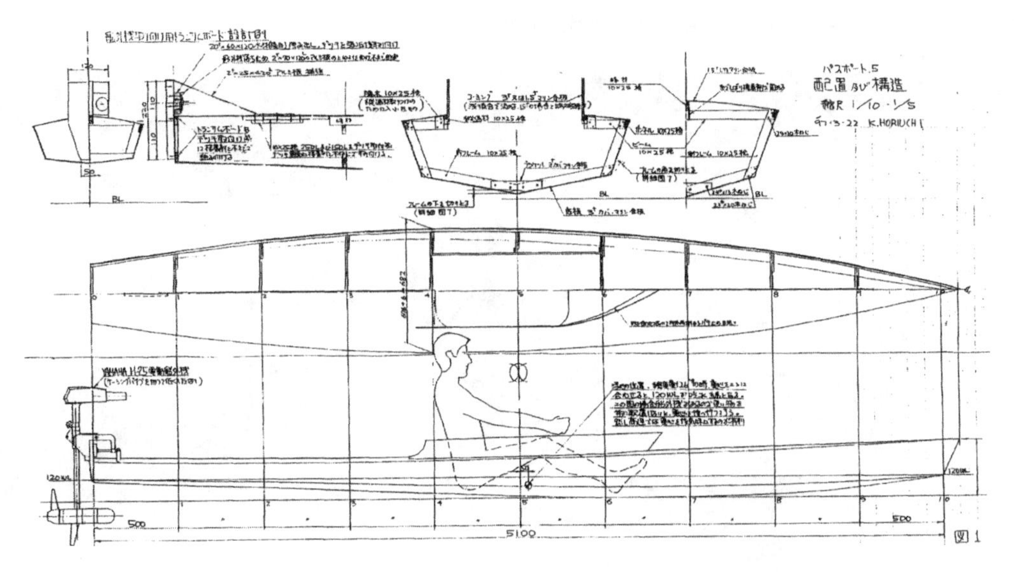

Fig - 2 Structural drawing of Passport 5

the contour shapes are decided, building the boat is easy.

I included the contour patterns in the bulletin. For confirmation, I made a small paper model (Fig - 1) out of the contour patterns. I confirmed that the boat could be correctly made as I had intended.

After building the boat, we can decide the propulsion means. If we need a solar boat, an electric outboard motor available in the market can be used. On the other hand, for a human powered boat, we do not have a ready-made propulsion system. The builder must design it from the beginning.

Thus, what I presented finally in the 12th bulletin of Solar and Human Powered Boats was Passport 5 (Fig - 2). It was a 5-m boat designed for the amateur builder. I asked Yamaha Motor to supply the electric outboard motor M25 at a discounted price to only those people who built the solar boat. The propulsion system of a human powered boat could be any means from a pedal to a propeller, and I requested Tsuide Yanagihara, a successful designer of very fast human powered boats, to design the mechanical system. Yanagihara's article describing the method of manufacturing a drive unit and propeller appeared in the 13th bulletin.

2) Kanai Design

After I published the article on Passport 5 in the bulletin, and while looking at a canoe magazine for my next design hint, I found an advertisement from a canoe factory named Kanai Design in Sichirigahama, Kamakura. The factory was close to my house, so I thought it would be interesting to visit it.

In the April of 1997, I visited Kanai Design after making an appointment by phone. Mr. Kanai, who had retired at the age of 55, began building canoes after retirement. He sold drawings, kits, and complete boats to be built by the Stitch and Glue method, which came from Chesapeake Light Craft Company in the USA. Beside these, he also sold boats of his own design.

When I visited Mr. Kanai, he was at work in his factory, which was a remodeled cottage separate from his main house. It had space for building a slender 8-m boat. It was not a large space, but felt warm and pleasant, and was a small factory where Mr. Kanai himself enjoyed building boats.

As I had not seen wooden boat for long time, I was very impressed after seeing the varnished grains of plywood and also impressed with a sea kayak with fair lines. I had often heard the phrase Stitch and Glue but the building method I saw for the first time looked really simple (Fig - 3). When the edges of several sheets of plywood are stitched to each other, we can soon discern the shape of the boat. It gave me a warm feeling to see the smoothly faired sheer and chine lines and the curved and faired surface of the hull. Fig - 3 shows details of the building process.

The plywood Mr. Kanai used was excellent. It was made of beautifully grained Okume (Gaboon), of dimensions 4 feet x 8 feet (1220 mm x 2440 mm) and was 3-mm thick of marine grade (appropriate for boat structure)

Fig - 3 Stitch and Glue Process (see in numeric order)

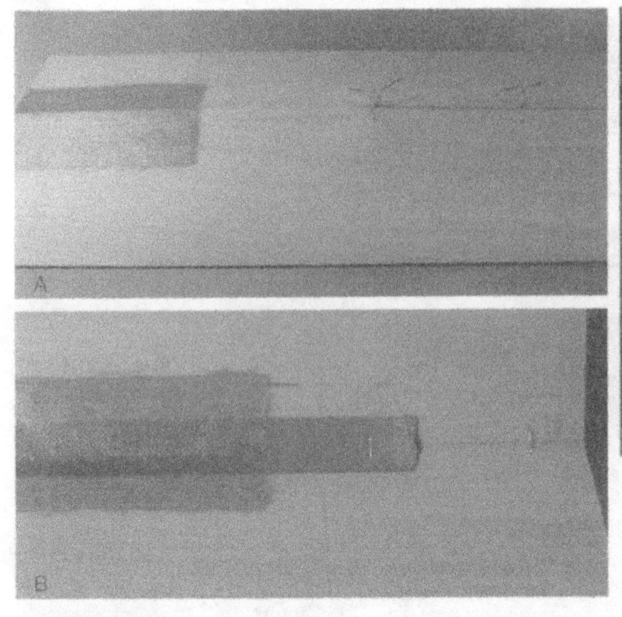

note : Circled number shows order of work.

A=Hull outside
① : Connect two adjoining plywood boards with copper wires (right).
⑤ : After cutting out copper wires, fill the connecting line with epoxy putty that includes wooden powder (middle).
⑥ : Laminate the lines with 75-mm wide fiberglass tape (left).

B=Hull inside
② : Crush and flatten copper wires inside connecting line (right).
③ : Fill epoxy putty in the connecting lines (middle)
④ : Laminate the lines with 75-mm wide fiberglass tape (left).

C=After connecting plywood boards with copper wires, the boat shape appears. Six sheets of plywood were used in the hull in this photo used, and in Fig-1, 4 sheets were used. Mr. Kanai is using a small boat as a side float made with only 2 sheets of plywood. Naturally, if number of sheets used increases, the boat shape becomes round.

plywood. The weight per sheet was 5 kg, and it was very light with specific weight of 0.55, same as that of solid wood of spruce. The combination of the building method and the plywood was so good that Kanai's beautiful boats and his building method are etched in my memory. I thought about the design of Passport 5, which indicated Stitch and Glue method but actually had frames and used heavy plywood made of birch. Since I had no experience in this building method, I was afraid to eliminate the frames.

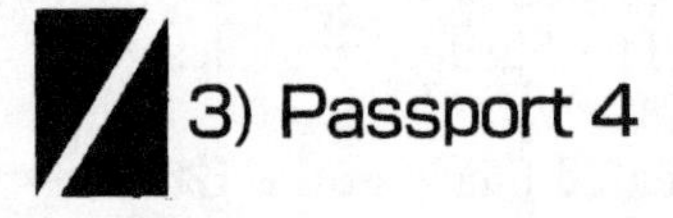

3) Passport 4

After about one month, the 12th bulletin was issued. I received a letter from Mr. Junichi Miyawaki, a teacher in the Omuta High School in Kyushu. He wanted to build the Passport 5 that had appeared in the bulletin and with the boat he wanted to participate in the solar boat race to take place at Yanagawa in August. However, Passport 5 was a 5-m boat and it did

Fig - 4a Daijayama

Fig - 4b Big Dragon

not comply with the Yanagawa boat race rules (less than 4 m). He asked me to redesign the boat so that the length was less than 4 m. There were only a few days before the race. I made the drawings of Passport 4 with its length shortened to 4 m in a hurry and sent them to the teacher on the 3rd of June. At the Omuta High School, they arranged two teams, each of which consisted of ten students who were interested in building the boat. They built two boats named Daijayama and Big Dragon (Fig - 4a, b) by the end of July and participated in the race held between August 1 and 3.

In the race, they had given a good fight for the 3rd place in the student class and the 3rd place in the free style class. The teacher Miyawaki reported that the race results were excellent considering it was their first attempt and said that his students had experienced great pleasure by building the boat together.

4) Canoe factory in Naguri village

In August 1997, the Solar and Human Powered Boat Race took place at Lake Hamana where one of boats was a catamaran type human powered boat named Naguri Tyuke (Fig - 5) from the team Water Gospel.

It was a catamaran built at a canoe factory in Iruma-gun, Saitama Prefecture. The team had been participating in the human powered boat race since 1995. The builders spend most of their time in building FRP hulls, and had very little time left to make the propulsion systems, so they were struggling with the propulsion systems year after year. But this year, it was different.

The team members went to Naguri village where they built a wooden boat in the canoe factory under the instructions of the factory manager, Mr. Naoyuki Yamada. The boat was named Naguri Tyuke. They proudly said that they could build a hull in three days.

At the award ceremony after the race, the Naguri method for building a light boat easily was acknowledged as a splendid boat building method and received a special award. Mr. Yamada showed us how to lift and support the slender hull vertically at one end with only one hand demonstrating the lightness of the hull, which was well received.

Some time later, Mr. Yamada did invite me to see his canoe factory in Naguri village, and I visited it with Mr. Takaaki Toda, a race com-

Fig - 5 Naguri Tyuke

Fig - 6 Canoe factory in Naguri village

Fig - 7 Method of building the Naguri Village canoe

mittee member. The 660m^2 nice factory included an exhibition hall and office where about 20 people were enthusiastically building canoes (Fig - 6).

These people paid approximately one hundred thousand yen ($800) as instruction fees, which included the material cost of the canoe; they were building the hand-made canoes every weekend under instruction. I saw some couples were working earnestly. Perhaps they

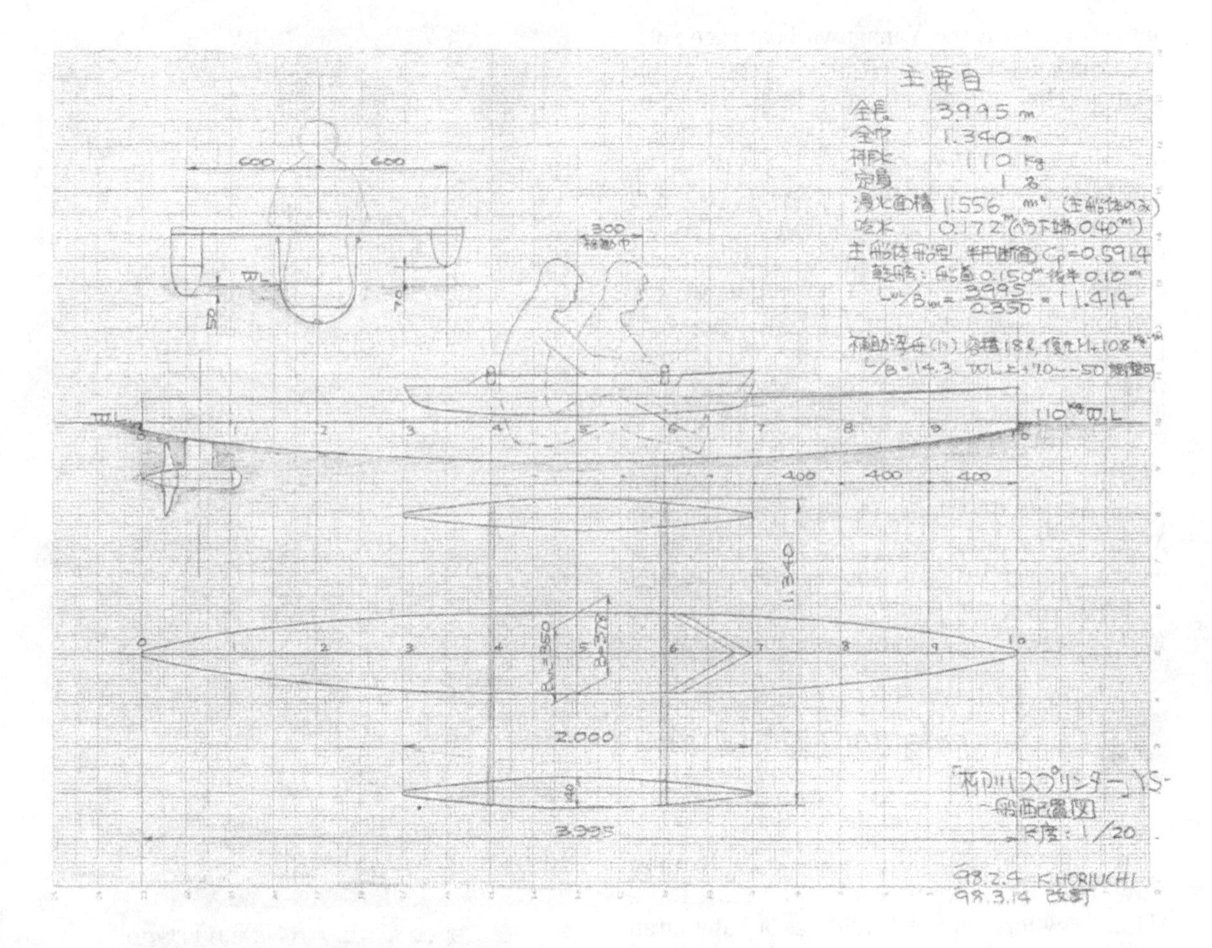

Fig - 8 Basic layout of Yanagawa Sprinter

were looking forward to complete the canoes and take them back.

They used strips 5-mm in thickness and 15-mm in width from straight-grained Japanese cedar, which was a special product in that region. They fixed the strips on the lined frames with thumbtacks and tackers, and then glued them using acetic vinyl resin (Fig - 7). This work looked easy even to a beginner and was also pleasant. After fixing the strips and finishing the surface with sandpaper, a fine wood grained Canadian canoe appears. This is the strip planking method.

The boats become strong after they are laminated with FRP on both sides of the wooden hull, but it did not appear to be easy to smoothen the FRP surface. Nevertheless, if the boat is to be used for races and treated carefully, the FRP laminating work can be eliminated, the boat becomes lighter and a lot of work is saved.

You can build a round bilge boat by this building method, and the process becomes simpler if FRP lamination is eliminated. Since a lighter, beautifully-grained round-bilge boat can be built, this was the most appropriate method and easy for building a solar boat or a human powered boat for races. I thought this method should be recommended for building a slender round-bilged hull. Mr. Kanai's Stitch and Glue method and the Strip Planking method are the two best methods.

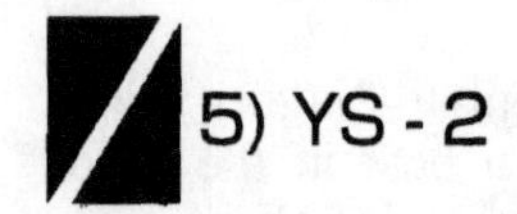

5) YS - 2

Probably in January 1998, the next year, I received a telephone call from Miyawaki, the schoolteacher of Omuta High School, who asked me to design a smart and faster speed-boat for that year.

The boats for the Yanagawa solar boat race are limited to a length of less than 4m, and hydrofoils are not allowed. The boat must be a displacement type boat with approximately 400-W propulsive power.

I studied the question carefully and made several drawings (Fig - 8). To minimize the wetted surface area, the ideal under water hull section is a semi circle (Fig - 9). Assuming a

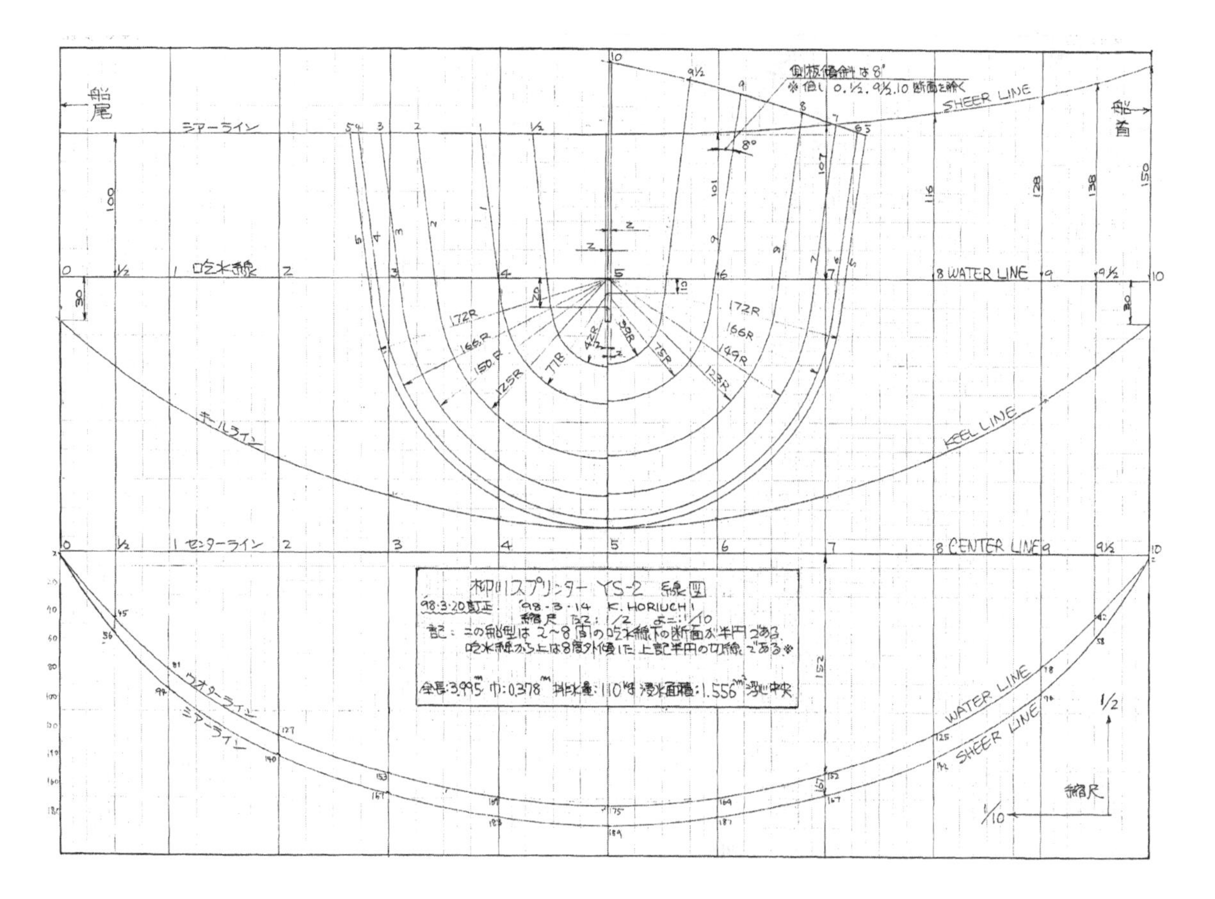

Fig - 9 Lines drawing of YS - 2

displacement of 110 kg, I made the preliminary design to find the boat's beam as nearly 34 cm, which was just the size of a person's waist. I was pleased with this result. Therefore, I named the project YS-1, short for "Yanagawa Sprinter."

The boat was so slender that the single hull would capsized without doubt. So, I considered attaching side floats to both sides of the main hull. The boat had a beam of 34 cm and if a person sat down so as to touch the hull bottom, the side floats could be set above the waterline, and the boat could run on a straight course without the side floats immersed in the water. Although, the single scull for rowing has a beam of nearly 30 cm and its seat level is 10 cm above the water, its skilled crew can keep the boat in a stable condition holding the oar blades above the water. Compared to this situation, YS-1 had better transverse stability.

The only item of concern was that the waist size of the actual skipper in the 34-cm beam of the boat. The skipper had not yet been decided. If the beam was too narrow for the skipper and its seat had to be set higher, the boat's center of gravity would rise and the stability would deteriorate; then the design of the boat would not hold good. Although the beam can be reduced according to the skipper's waist size, a margin in the beam that would increase its resistance was not recommended.

I could not decide what to do, but I designed another hull changing the longitudinal distribution of the hull section and calculated its resistance (Fig - 10). The first YS-1 had 33.7cm as the maximum beam, while YS-2 had 34.4 cm as the maximum beam. As there was little difference in resistance in these two boats, I decided to use the wider YS-2 (Fig - 11).

The schoolteacher Miyawaki visited Mr. Yamada's factory in Nagri village and was taught the Yamada method of construction. As they had built two boats in the previous year, the building work this time appeared to go smoothly.

The new boats of the Omuta High School were named Daijayama 2 and Big Dragon 2 respectively. Their carpentry work was good. Since some materials remained, a group of teachers of the high school built another boat (Big Dragon). Finally, three boats from Omuta

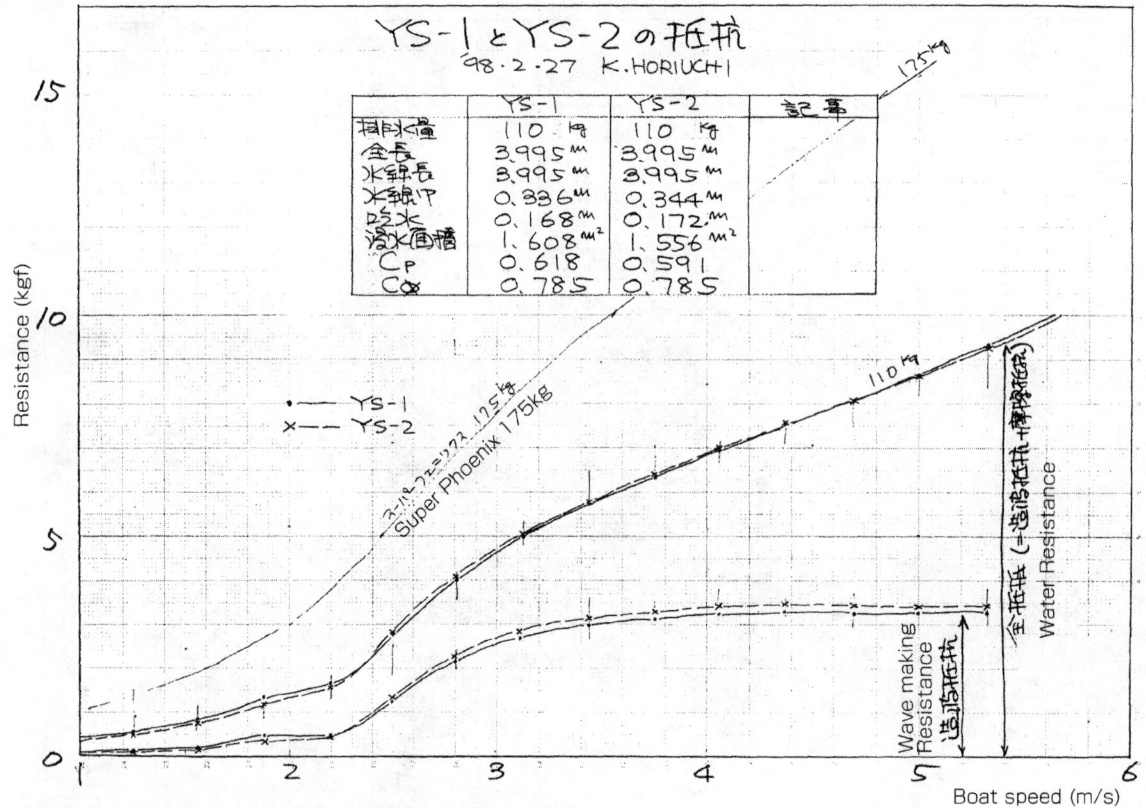

	YS-1	YS-2	記事
排水量	110 kg	110 kg	
全長	3.995 m	3.995 m	
水線長	3.995 m	3.995 m	
水線巾	0.336 m	0.344 m	
吃水	0.168 m	0.172 m	
浸水面積	1.608 m²	1.556 m²	
Cp	0.618	0.591	
C⊗	0.785	0.785	

Fig - 10 Resistance curves of YS - 1 and YS - 2

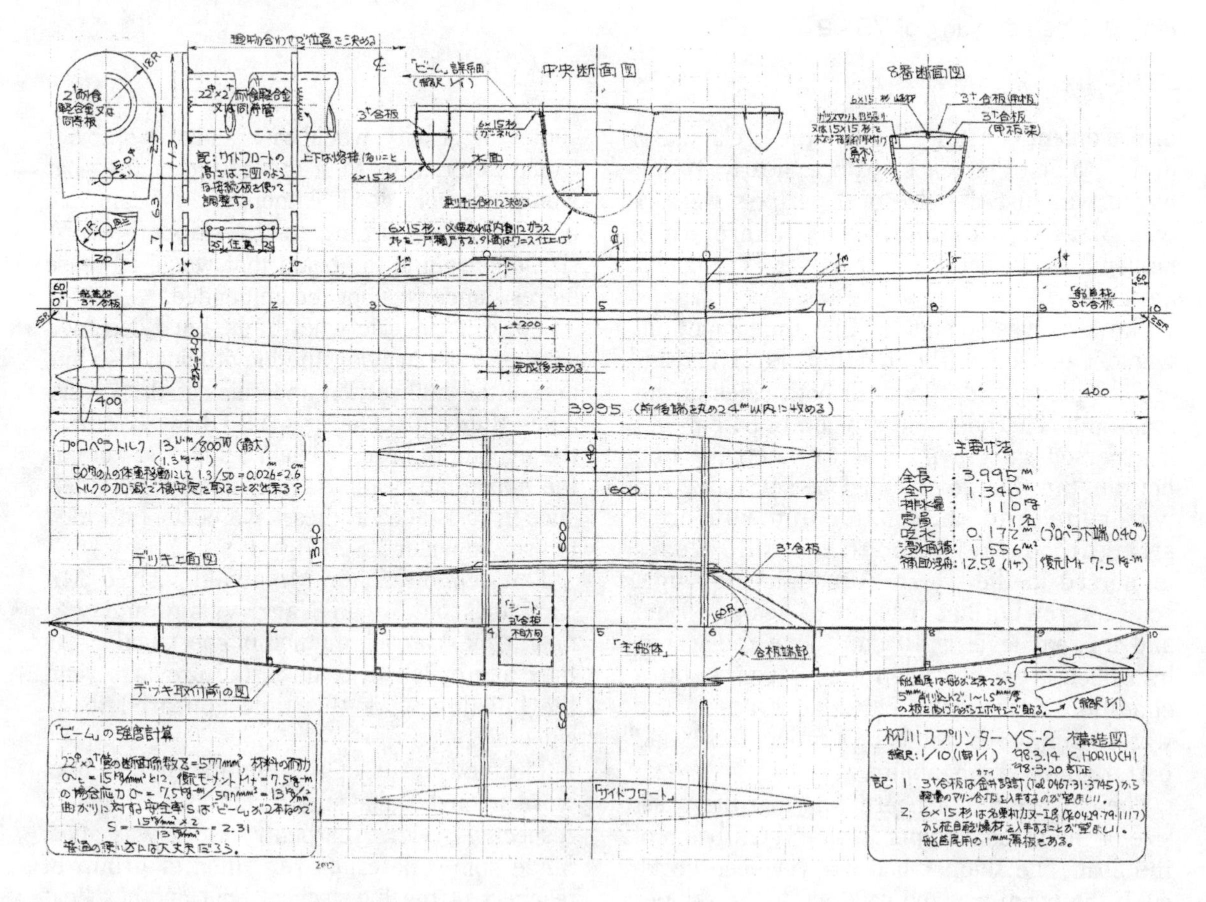

Fig - 11 Construction Plan of YS - 2

Fig - 12 Big Dragon 2

High School participated in the Yanagawa Solar Boat Race in 1998.

From the 60 boats that participated in the race, all the three boats of Omuta High School reached the final round. Big Dragon 2 came first (8th overall) and Daijayama 2 was third (11th overall) in the high school division, and then Big Dragon came 17th. The school principal came to cheer the school on all the days and all the school members looked very happy.

Big Dragon 2 took 67 minutes for completing the race of 3 rounds in the 3-km canal, which was considerably more compared to the 40 minutes in the initial plan. The reason was that they laminated FRP on the wooden parts of the hull so the boat became heavier, the boat surface was not very smooth, the resistance of support pipe with no fairing for the outboard motor was too high, and the skipper could not run the boat keeping the side floats above the water. However, this boat still had potential. It made a good run at the Hamanako Solar Boat Race that summer (Fig - 12).

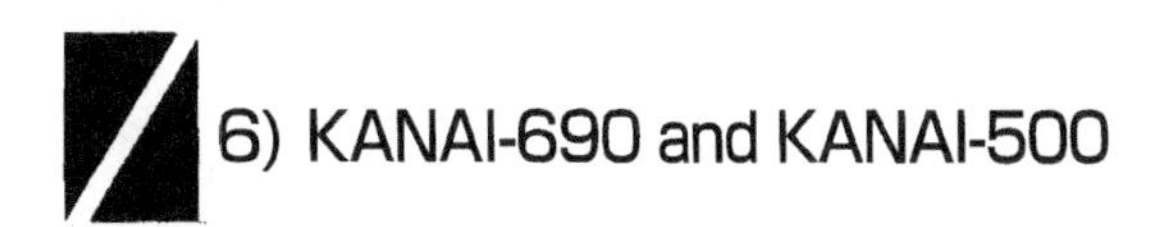

6) KANAI-690 and KANAI-500

I became friendly with Mr. Kanai and enjoyed discussing canoes with him. Mr. Kanai used to join me at the Amami Sea Kayak Marathon at Kakeroma Island of Amami in Kagoshima Prefecture, where he used his own designed boat and got good results. I was very interested in listening to his stories.

In 1998, his boat, the KANAI-690 (Fig - 13), a slender hull of length 6.9 m and beam 0.36 m, with one side float only on the port side of the main hull participated in the race. It came 25th out of 114, a vast improvement in the record of the previous year. However, KANAI-690 had a

Fig - 13 Kanai - 690

The boat took the 25th place out of 114 boats at the Amami Sea Kayak Marathon in 1998. The side float is only on the port side, which is unique.

Fig - 14 Kanai - 500

The boat's length was made shorter than that of Kanai-690 to reduce its wetted surface area. Its float position was moved aft and the boat began to surf easily. The boat also did well, taking the 27th place out of 125 at the same race in 1998, and took the 20th place out of 129 in 2000.

larger length compared to its beam and a larger wetted surface area.

In 1999, Kanai-500 (Fig - 14) was newly built. The boat length was shortened to 5.0 m and its beam was 0.35 m, just the size of a person's waist. Its section was nearly semi-circular and the wetted surface was reduced, and its concept became the same as that of YS-2. Kanai-500 had a side float on the port side, and it came 27th out of 125 boats with a time of 4 hours 11 minutes 29 seconds.

The race results were excellent considering Mr. Kanai's 53-kg weight and his age of over 60; the slender hull also contributed to the good performance at the race.

There is also no chance of capsizing of this boat except when rolled over by a huge wave. Even if the boat capsizes, it will be easy to get on board again. That means it must be a fast and safe boat. I have always admired the concept of this boat. The only point of concern is that the one-sided float adds a certain amount of resistance to the main slender hull.

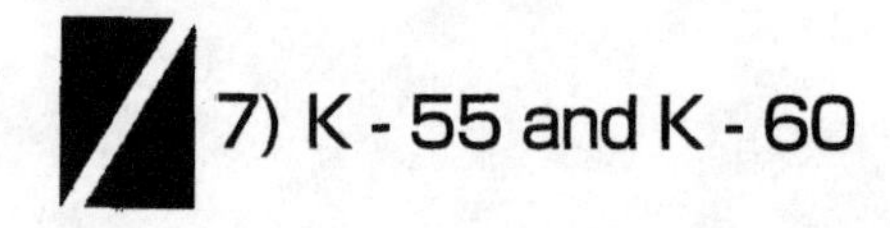

7) K - 55 and K - 60

The normal kayak until then was designed to maintain stability, with the crew's weight deep down in the hull. However, as mentioned previously, we can make a rowing scull slender by raising the seat level 10 cm above the waterline. If the crew gets used to it, he can maintain transverse stability, even if the boat beam is only 30 cm.

The slender sea kayak has a side float and the conditions are more relaxed, and there is no worry about capsizing. Then why not build a very slender hull without bothering about boat beam to be made the same as a person's waist size? On top of that, by reducing the resistance of its side float and its air drag, wouldn't it be possible to reduce the total resistance significantly? In case of a scull, it takes one second to return the blade after finishing the stroke to catch the water. The crew must keep the boat stable for about one second with the blade held above the water. On the other

Fig - 15 Studies on stabilizers

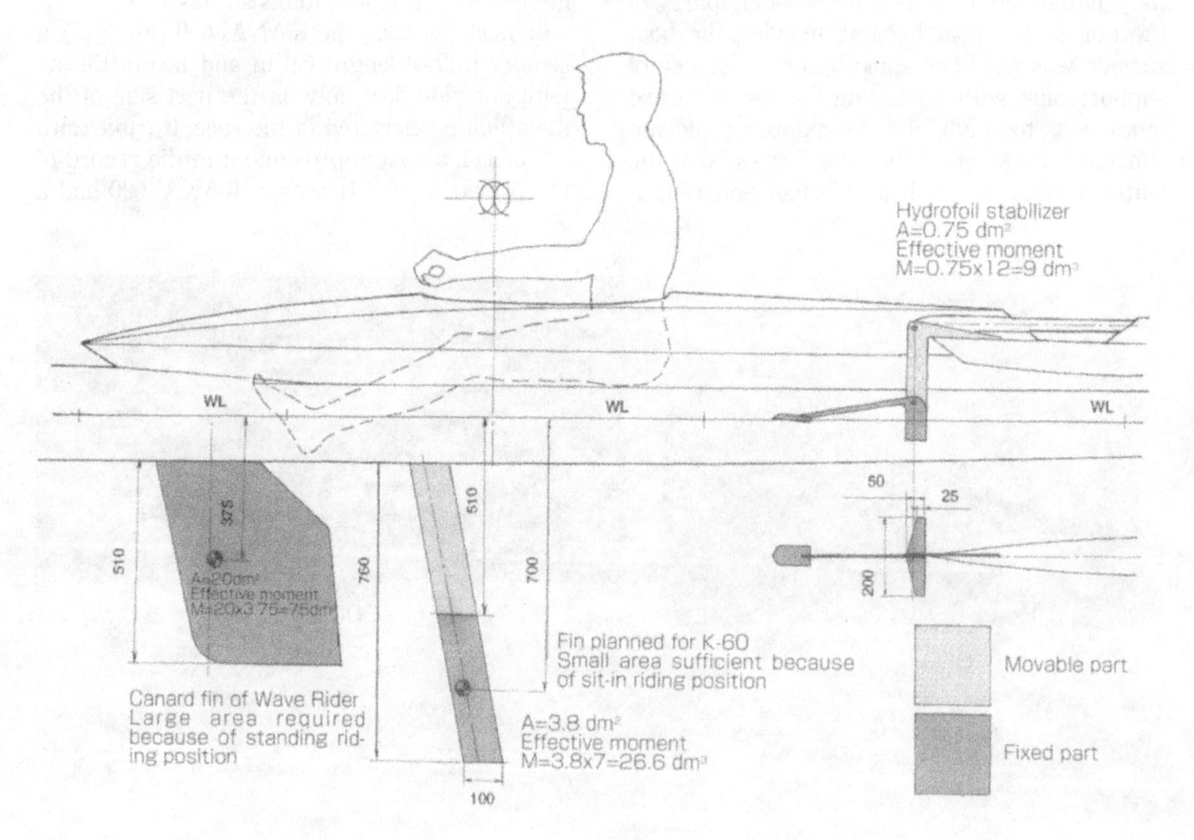

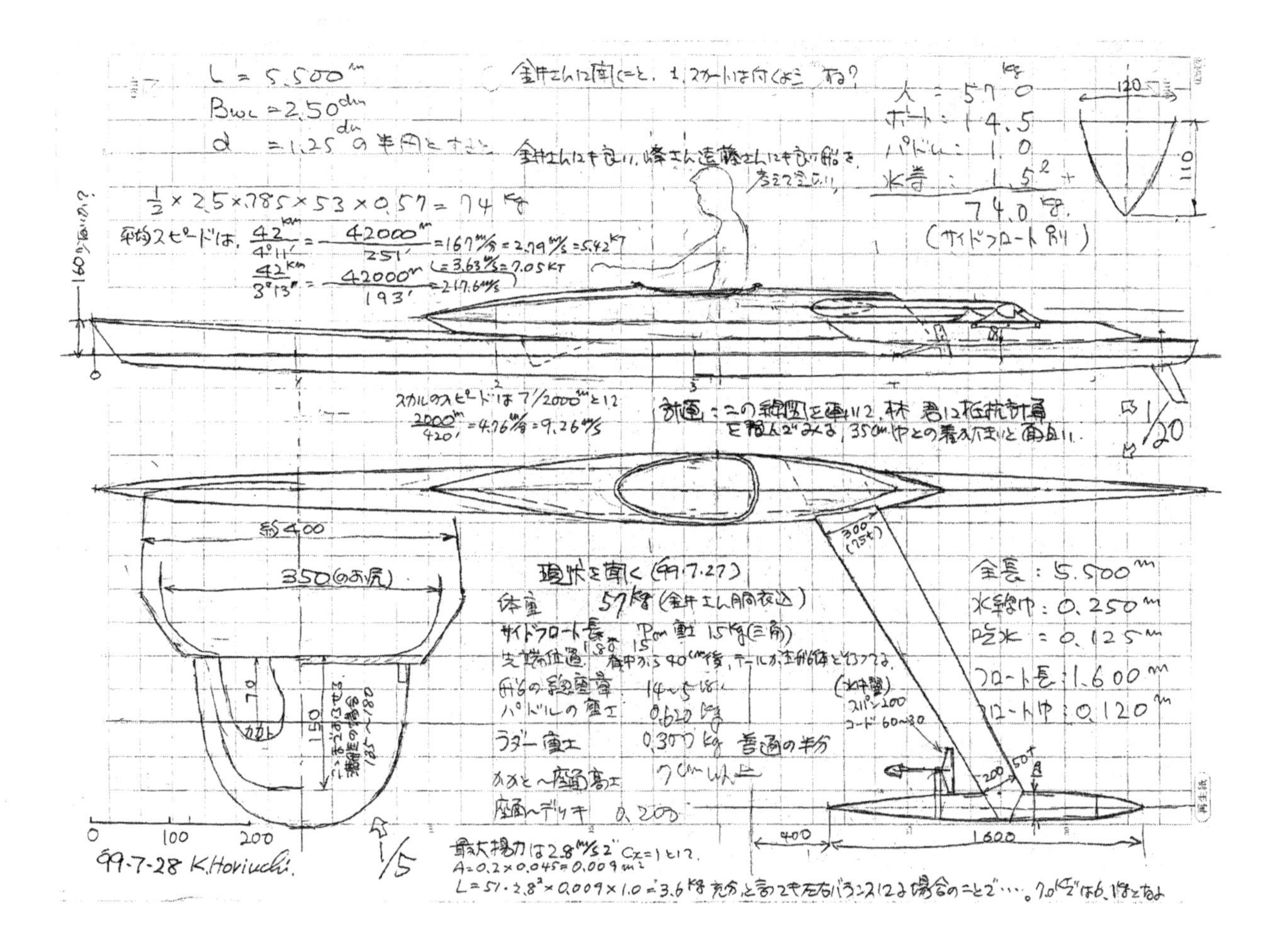

Fig - 16 Design sketch of K - 55

hand, in case of a kayak, the time that for the right blade coming out of the water and its left blade going into the water is very short, maybe about 0.1 or 0.2 seconds. This time is too short for the balance to be disturbed. That means, if the kayak is paddled smartly so as not to cause it to roll, there should not be any need to rely on the side float on a straight course.

If the side float can be kept above the water when running straight, it is an ideal situation. That may be possible in 500-m sprint race, but the side float cannot be kept above the water during a 4-hour race. The crew's attention may be lost, and we have to assume that the side float should be relied on. Nevertheless, I wanted to eliminate the increase in resistance when the side float immerses in the water. I considered running the kayak with the side float raised above the water with hydrofoil. But a large amount of lift is required to raise the side float fitted only on one side, and it will be accompanied by a large increase in resistance. To prevent this situation, it would be preferable to arrange side floats on both sides of the boat and maintain balance. As the left and right floats balance each other, the required lift for the hydrofoil is not appreciable. Beside the hydrofoil, I studied a centerboard type fin used in the Wave Rider made in the USA and also studied an advanced type in the Wave Rider (Fig - 15).

I probably made this design sketch when I visited Sapporo (Fig - 16). The length of the boat is 5.5 m and the beam is 0.242 m. The ratio of waterline length to waterline beam is 22.5, which is not as large as that of the scull but close to it. I arranged swept-back wings with streamlined section, on both sides of the main hull and at the end of each wing, I fitted a side float as small as possible.

I wanted to fit the wings to a cockpit[1] with some width and height. It was essential to adopt the swept-back wings so that they did not touch with the paddle when paddling the boat. I showed the drawing to Mr. Kanai and

1) The beam of main hull is too narrow for the skipper's waist, and then a wide area to accept the skipper's waist was prepared on the deck, which was named a cockpit. The cockpit prevents waves coming over the deck and has functions to fix the skipper's skirt and to fit the removable wings.

Fig - 17 Comparison of resistance of 4 different kayaks

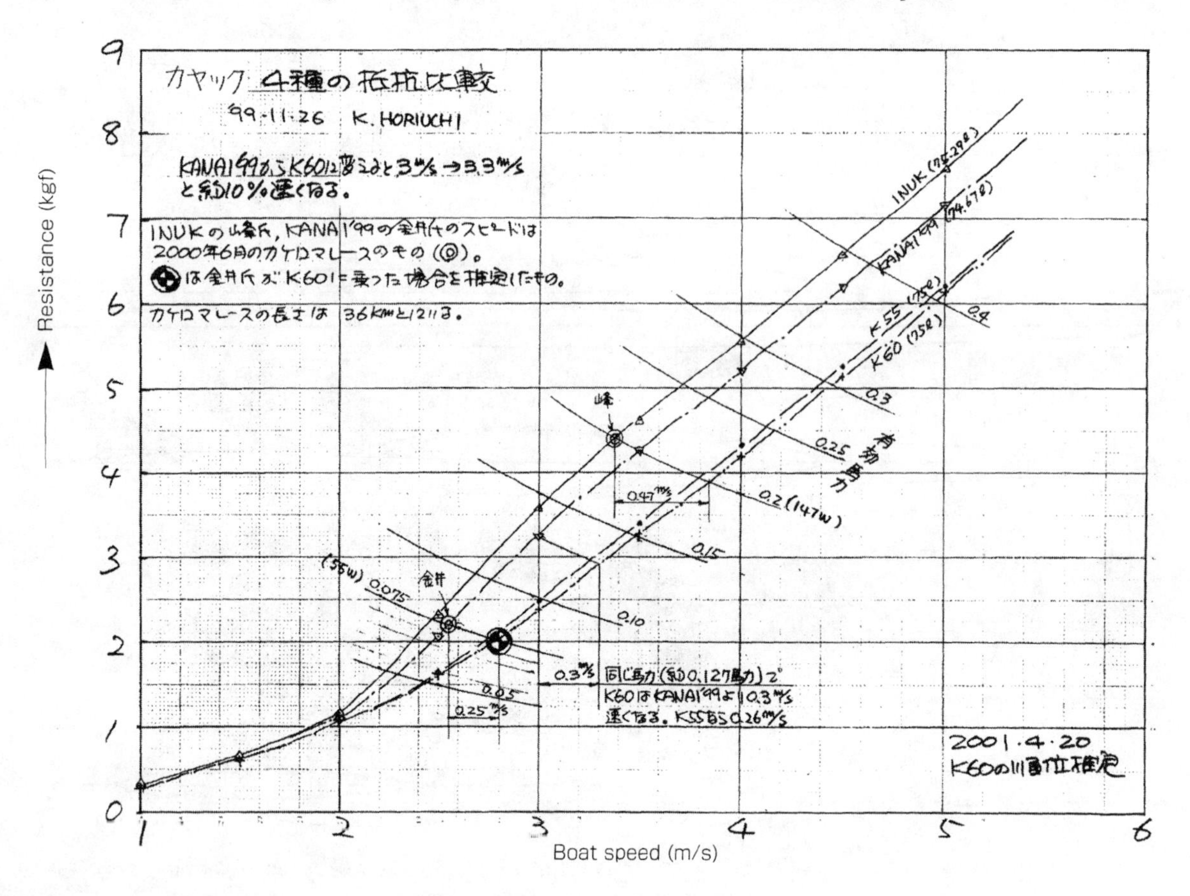

Fig - 18 General arrangement of K - 60

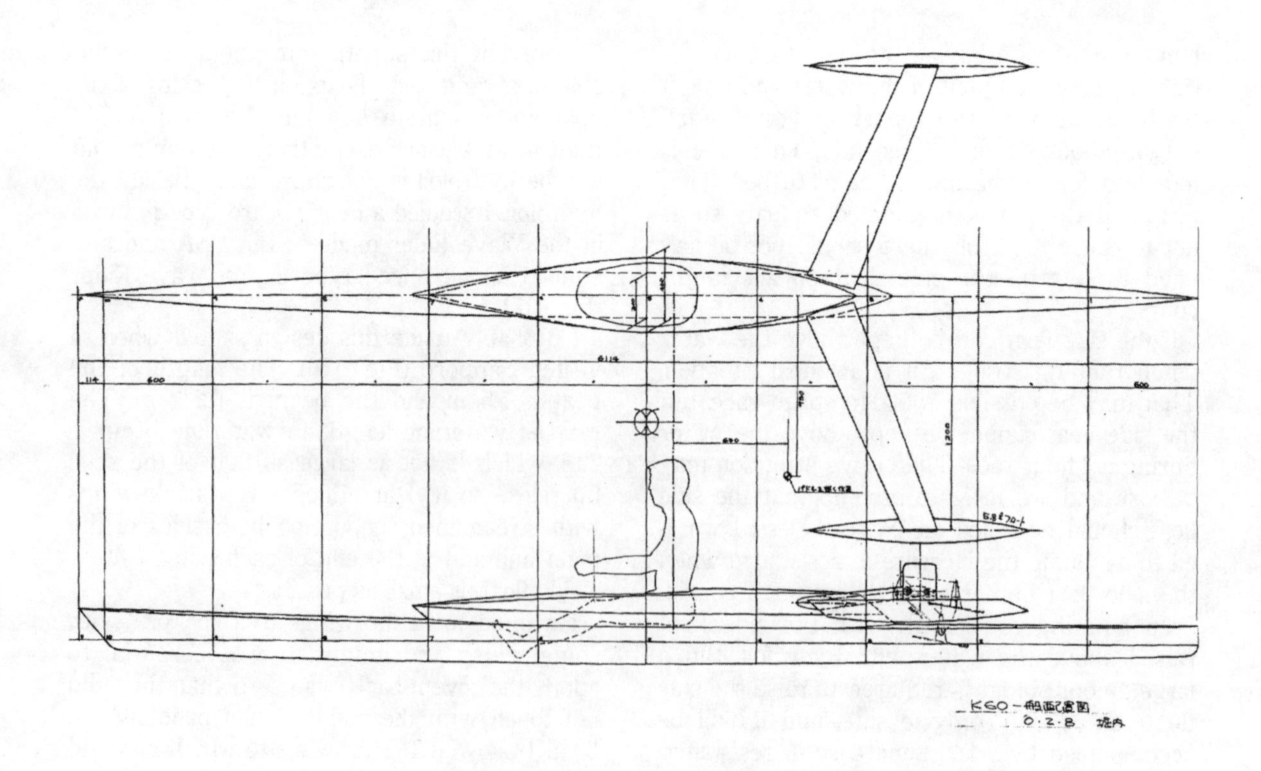

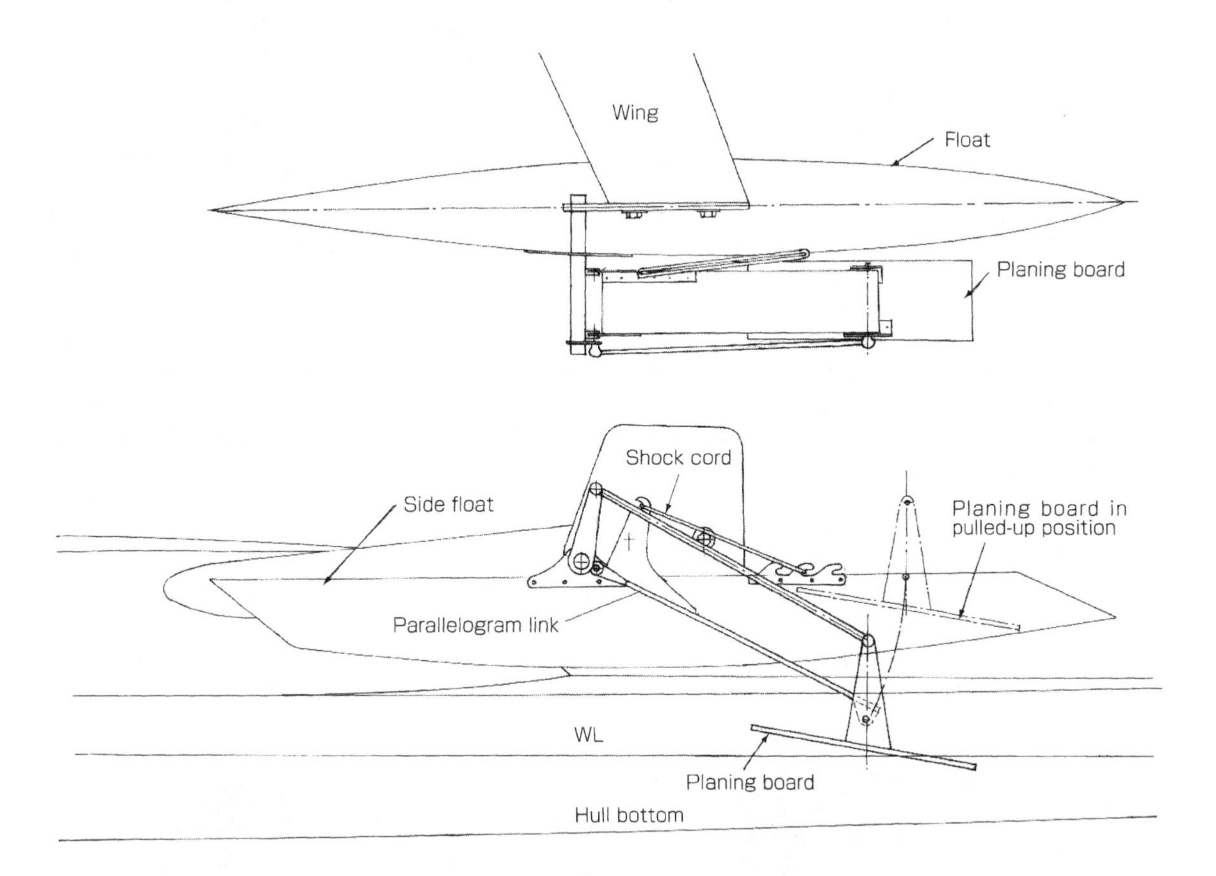

Fig - 19 Planing board type stabilizer

he said that it looked smart – like a jet plane – and that he wanted to build the kayak, and I was very pleased. To re-confirm, I checked the performance of a more slender hull form. If the length becomes 6.0 m, the beam becomes 23 cm at the waterline and the length to beam ratio becomes 26.09, which is about the same as that of a scull. I named the designs K-55 and K-60 respectively, and proceeded with resistance calculations. Fig - 17 shows the results.

Though the difference was very small, it is evident that the resistance of K-60 was smaller than the others over the entire speed range. As K-60 was a little longer, it was not a convenient size for drawing on a normal A2 or A3-size sheet. But Mr. Kanai looked at the design and immediately decided to build the K-60.

The drawings of K-60 were completed (Fig - 18) at the end of January 2000. One concern that remained was whether the hydrofoils would hold the side floats above the water. The worry was that the hydrofoils might become entangled in seaweeds and the skipper would have to remove them or the hydrofoils would be damaged. I thought that I had to devise a system that pulled up the hydrofoils with the end of the paddle, to be fitted in the left and right floats. If both the left and right systems were damaged during a long run, what to do? I was worried about this.

As an alternative plan, I considered preparing planing boards. If the planing board is fixed, its reaction in the wave is stiff; when submerged by a strong force, its resistance is large. I used a system by which the planing boards are gently pushed down by spring, so that transverse stability is restored and the boards move upward in case of a strong reaction (Fig - 19). In this system, the planing boards maintain initial transverse stability and also help to recover sudden unbalance with the assistance of the buoyancy of the side floats. When the planing boards do not touch the water, their resistance is zero; on this point, this system is superior to the hydrofoil system. There was no need to hurry to decide which plan to use – hydrofoil or planing board. Any system can be fitted later. Ideally, such a system should not be fitted; the right choice should be made if such a system is absolutely essential, and the system fitted after careful considerations. In early February in 2000, I submitted the drawings and calculations to Mr. Kanai.

8) Building the K - 60

Earlier, I have mentioned that the boats at the Kanai factory were built mainly by the Stitch and Glue method.
Mr. Kanai was also building other boats by the strip planking method, in which 20-mm x 5-mm wooden strips were planked side by side.

Especially, Mr. Kanai's race boat had a light structure. The previously mentioned 3-mm marine grade plywood was cut into strips of 10 to 15-mm width. The strips were placed and glued one by one on the mould like the canoes in Naguri Village. As the thickness was 3 mm, the boat was lighter than the canoes of Naguri Village without the FRP lamination. After painting and finishing, the weight per m^2 was nearly 1.6 kg. As the surface area of K-60 was 5.2 m^2 including deck and cockpit, the hull weight would be 10 kg. With this weight, a round-bilge boat can be built, and this is what I preferred.

But Mr. Kanai was very interested in CFRP (Carbon Fiber Reinforced Plastic) structures. This was just before building the Twin Ducks and we were considering whether to use CFRP or 3-mm plywood for the round bilge Kanai method for the outer shell. We visited GH Craft together to hear their opinions, and after the meeting, Mr. Kanai made the test specimens. In the middle of March, we studied the results of the test pieces and decided to build Twin Ducks of CFRP sandwich construction and K-60 to be built by the plywood strip planking method, with which Mr. Kanai was familiar.

Mr. Kanai started building the boat, and after the outer shell of the hull and side floats were planked (Fig - 20), he was forced to stop work on this boat because he had orders for many other boats that were to be completed first.

Fig - 20 Narrow hull of K - 60

Hull sides were planked with wide plywood strips and bottom semicircular sectional areas were planked with narrow plywood strips.

Fig - 21 Rotating seat

This photo shows a rotating seat turned 45 degrees in the clockwise direction. The string at starboard is for closing the rear hatch and the string at port is for lowering the tilted-up rudder.

Finally, he gave up participation in the Kakeroma race with K-60 that year; but he did participate in the race with Kanai-500, the same boat as the previous year.

Mr. Kanai's results in the race improved tremendously, placing 20th out of 129 entries, and rising by 7 places compared to the previous year. Mr. Kanai mentioned that he could extract the full performance of the boat in the race. His time was 3 hours 54 minutes 44 seconds just below 4 hours, and this result could be attributed to his familiarity with handling the boat and because the pace at which he paddled suited the boat.

9) Rotating and sliding seats

While watching the canoe race in the 2000 Sydney Olympics on TV, I was astonished to see the skipper's left and right knees move up and down as he pulled on the paddle. I had never seen such a scene before. Perhaps the skipper's seat turned around the vertical axis. I consulted with Mr. Naoya Mathushiro, who was a specialist in racing kayaks.

As I guessed, his answer was that either the seat surface was made wet so that the waist turned easily or the person might actually be using a rotating seat. Normally, the kayak skipper paddles by turning the upper body, but if the skipper's waist also turns, the force exerted by the feet helps in adding more power to the stroke.

I showed Mr. Kanai the videotape of the Olympic canoe race. Immediately Mr. Kanai brought a TV turntable with a thrust ball bearing and installed it in the Kanai-500 to participate in the race in October 2000. He felt he could paddle easily with the seat. As I guessed, it was natural to rotate the waist. Later Mr. Kanai installed a combination of rowing boat seat and turntable in the K-60 (Fig - 21). Since the turntable could be stowed in the narrow hull, the seat height would never be higher even if a turntable were introduced.

Basically, the seat of K-60 was at a high position because the hull was too narrow to accommodate the skipper's waist. Therefore, skipper's foot was at a relatively low position facilitating the posture for paddling. The vertical distance between heel and seat was 16 cm, close to that of rowing shells. This arrangement was comfortable for the crew.

I thought that using a sliding seat (as in a rowing shell) in addition would help get a better performance. In the Olympic canoe race, the paddling pitch is so high that there is no time to move the sliding seat. However, in a 3 to 4 hour long course, if the paddling strokes are distributed not only to the upper part of the body but also to rotation of the waist, swing of the body, and stroke of the feet, and if the speed of each of these factors is reduced, then the tiredness in the muscles can be reduced and dispersed. Fortunately, the rails and wheels for this purpose can be stowed in the hull, and even with such devices, the seat height did not rise further. As such devices can be fitted later, we can improve them during the tests. With a speedometer and a pulse meter on board, and maintaining the boat speed at a constant level, I should be able to see how the turning and sliding of the seat affect the pulse and to know the approximate result.

The sliding stroke may probably be almost zero when the paddling pitch is high. When the skipper is tired during the latter part of a long course, the longer sliding stroke may be used. If the skipper feels that his arms and shoulders are tired, he may paddle using only his legs and by the body stroke. Interesting variations are possible.

10) New facility

Mr. Kanai always looked busy. Although he had many orders, he wanted to build his own boat, and in addition, he had orders from his family members. He had to delay building his own boat, and because of this, he could not participate in the race with K-60 in 2000. As his factory had limited space, he found a new facility 200 m away from his house and he moved his factory there. He wanted to restore the previous factory to residential spaces and rent it, so that he could pay it for the rent of the new facility. To improve the new and old facilities, he need considerable manpower.

I think that those days were difficult for Mr. Kanai. The new facility was splendid. There were several tens of sea kayaks and accessories exhibited in an elegant room used as clubhouse, having an area of nearly 150 m^2, and there were also a conversation lounge and a kitchen with counter, where a small party could be held. There was also a small guest bedroom.

Under these rooms, the entire space was devoted to a workshop where several kayaks could be built at the same time. Since he had moved to this facility in July 2000, every time I visited it, I saw people using the space and building their own kayaks. In the future, many people will build kayaks by themselves here and become kayak skippers.

Mr. Kanai could build the K-60 in this facility since it had adequate space. When we exhibited Twin Ducks at the Tokyo Boat Show of February 2001, Mr. Kanai assisted us by giving explanations on Twin Ducks throughout the 5 days to visitors. During this period, we decided to exhibit the K-60 at the 2002 Boat Show.

As Mr. Kanai built the K-60 carefully for the exhibition, the finish of K-60 was splendid. The beautiful grain and varnished finish was a good example of his fine handiwork. The contrast between black shiny wings and grain looked dignified and emphasized its shape; Mr. Kanai said it looked like a jet plane.

Mr. Kanai also thought about how to exhibit the boat at the Boat Show. He said that he wished to keep the boat vertical, so that the boat's plan view, especially the narrow hull and swept-back wing shape would be visible and the boat's attractive points could be conveyed. This would also require little space and the boat would stand out. I think Mr. Kanai got this display idea from the displays of boats in his new facility.

11) Launching

On 26 April 2001, the K-60 was launched. The location was at Niihama in Misaki, which was a small bay with a beautiful beach, adjoining the Aburatsubo Bay on the west side. Mr. and Mrs. Kanai, and I gathered there, and the K-60 was launched (Fig - 22, 23). Its floating attitude was fine, and the trim and waterline were as planned. Mr. Kanai said it was easy to go on board, probably because of the boat's adequate transverse stability and the high seat position.

Until that time, we had not decided whether to use floats, hydrofoils or planing boards to maintain the transverse stability of the boat in the running condition. What kind of rolling would the skipper's motion cause to the kayak? I could not decide the system before understanding the motion of the boat. As I was new to kayaks, I had never experienced such an issue before.

The last target was to reduce the increase in resistance to practically zero and maintain the transverse stability. For this purpose, the height of the side floats was made variable over a range of 15 cm. The floats on both sides were adjustable from just touching the water to a high level above the water and supported by planing boards or other means. The ideal situation was to balance the boat balancing by paddling it such that the floats, planing boards, and hydrofoils were not immersed in the water. As mentioned earlier, after estimating the time the paddle was out of the water and the stability of the boat, this situation was achievable. The floats were set at a higher level when the boat was launched. When paddling started, the float on the side of the paddle in water dropped on the water with a splash, and this was repeated on the other side also. The rolling angle of the boat was about 10 degrees, and as the floats plunged into the water, 70 to 80% of the floats were submerged in the water. The submerged time was short but its resistance might be appreciable.

Gradually the float height was lowered so that the floats were very close to the water. At this position, the roll angle reduced to 5 to 6 degrees and the skipper could paddle the boat comfortably. Mr. Kanai paddled the boat pleasantly and went to the offing far from the shore to test the boat in waves. He said that the narrow hull cut through waves smoothly and there was very little impact.

The boat should have 30% less resistance compared to the usual sea kayak if we neglect the resistance of the side floats. To Mr. Kanai the boat appeared to have smaller resistance after he started paddling it, but since the seat was higher and the blades did not fully catch the water, we did not know how what speed the boat had actually attained.

We saw a speed of 3.2 m/s on the speedometer compared to the target average speed of 2.8 m/s, but we were not sure, as we did not have measured data of other boats for comparison. However, I did believe the statement made by Mr. Kanai that it felt very easy when he paddled at a speed of 2.8 m/s. But it did surprise me that after each paddle stroke, the paddling side suddenly lowered. If this happens all the time, it would not be possible to run the boat by balancing it so as to eliminate the resistance of the side floats, which was the ideal method. I did not think of asking Mr. Kanai to change his paddling method, because I was afraid his paddling might become weak; moreover, I was still new to kayaks. I decided to review the hydrofoil system.

Fig - 22 Test paddling 1 Starboard float is high above the water

Fig - 23 Test paddling 2

Fig - 24 Damper

Fig - 25 Stabilizer

12) Stabilizer and Damper

There are two kinds of hydrofoil systems. One is the system in which a hydrofoil with zero fixed angle of attack, is just immersed in water, that is, the hydrofoil merely reduces the rolling of the boat, or acts like a roll damper. I have abbreviated this name to Damper (Fig - 24)

By fitting this system in front of the floats the lift of the hydrofoil resists the sudden rolling with a force proportional to the angular velocity of roll. And it is interesting to note that similar to the theory of oar fairing (Chapter 3-5), a thrust sometimes occurs in the hydrofoil if the rolling is sudden. When rolling is moderated by the damper, the skipper can move his weight and adjust the heel to port and starboard so that the boat runs with the floats above the water. However, it should be noted that this effect is proportional to the square of the speed, it decreases when the skipper is tired and the boat speed reduces.

The other system is one that keeps the float at a constant height above the water by a water surface sensor that adjusts the angle of attack of the hydrofoil. The system may be called a hydrofoil stabilizer with sensor, and I shortened it to Stabilizer (Fig - 25)

This system eliminates roll control by the skipper, but at slow speed, the effect decreases similar to the damper system.

Both systems are affected by the seaweed problem. When seaweeds attach to the system, they must be removed using the paddle. Moreover, if the system hits or gets entangled in seaweeds, it may break or the angle of attack may deviate considerably, so it must be brought above the water to eliminate the resistance. Systems must be provided on both sides, and with spares for both sides. Since seaweeds are in abundance, I had to give up on the hydrofoil system. I had thought of an idea to protect the planing board system from seaweeds, but when I saw the launched boat and its hard rolling, I immediately felt that the planing board system, which pushed gently on the water surface with the help of a spring, would not be suitable. As soon as I returned home, I began to design the Damper and the Stabilizer systems. These systems were made of CFRP plates by hand and delivered to Mr. Kanai on the 5th of May.

Both Damper and Stabilizer were made so that they were fitted on the front end of the side floats with gummed tape. The test took place on the 9th of May at Kuruwa harbor in Hayama. I joined Mr. and Mrs. Kanai as they launched the boat from the beach.

It was a windy day, and although it was difficult to clearly identify the effect of the damper system, I was certain that the damper decreased the rolling of the boat (Fig - 26). However, it was not possible to operate the boat keeping the floats above the water. Had the foil area been a little larger, the situation

Fig - 26 Trial run with Damper

Fig - 27 Trial run with Stabilizer

might have been different. But the rolling moment was greater than I had estimated. Unfortunately, we could not test the Stabilizer because the support rod for the sensor was weak and it was bent laterally (Fig - 27).

13) Hayama Race

Several days later, there was approximately a 15-km round trip race from Hayama to the lighthouse in the Sajima offing. Mr. Kanai participated in the race with the K-60. His position was good at 3m/s speed on the forward course but on the return course, he was tired, the speed was below 2.6 m/s speed, so that finally he did not attain a good result. Because of fine weather that day, he did not take adequate clothing and became wet. On the return leg he was shaking with cold and could not paddle with all his strength. The damper system was fitted initially, but finally it was taken off before the start of the race start because of the abundance of seaweeds at the site and because the system did not work. He lost some time to take off the system and this might the reason he could not make preparations to take proper clothing.

Another point was that the seat height was 20 cm higher compared to the usual sea kayak. The skipper was probably not used to paddling from this height and might have been tired. Mr. Kanai expected that and used a longer paddle, but it might not have been adequate. He had to paddle from an unusual position, and it might have been necessary for him

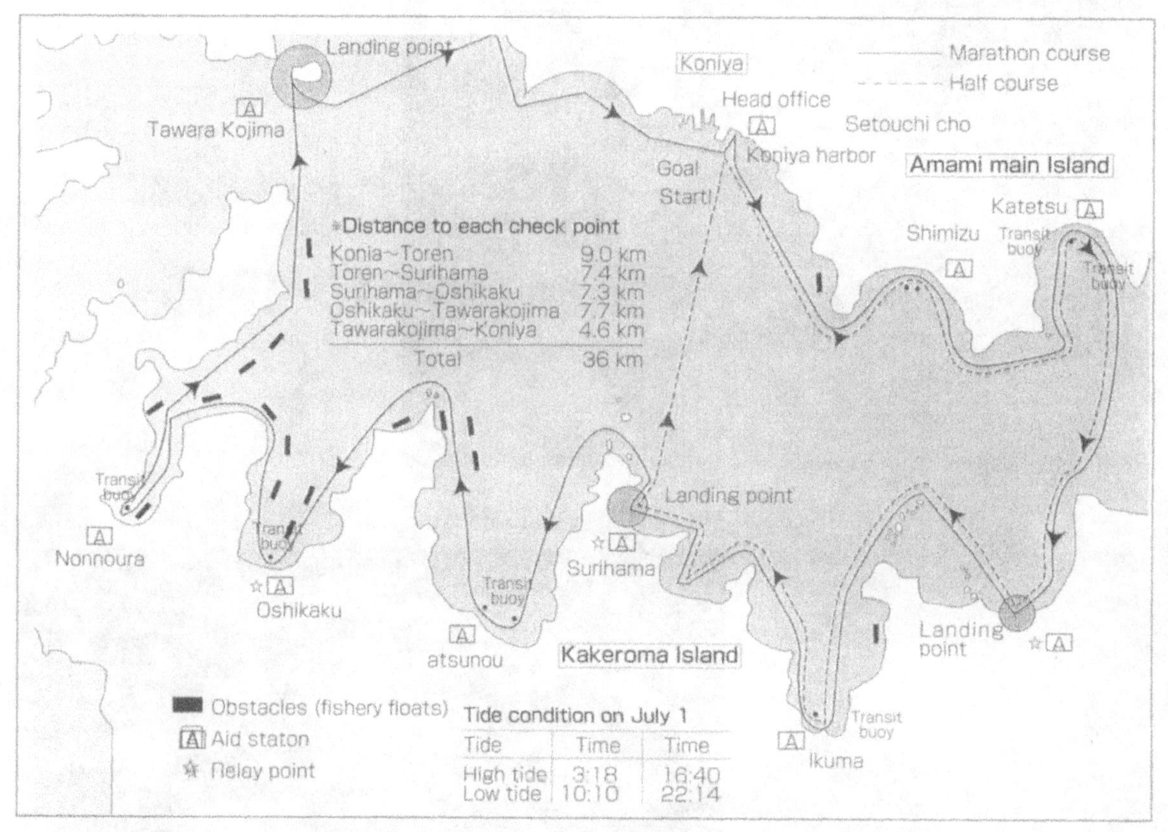

Fig - 28 Amami Sea Kayak Marathon course

to become accustomed to the geometry. I had realized early on that the position of the skipper in relation to the water level was higher. The paddling force becomes smaller as the distance to the water level increases. My thinking was that as distance to the far water level increases, the skipper uses a longer stroke, the speed at his hand reduces, and thus overall the arrangement would not be a disadvantage. This idea was probably a bit too optimistic.

14) Amami Sea Kayak Marathon

Mr. Kanai targeted this race with the K-60. For him, participating in the race was similar to taking the summer holidays, and at this period, he enjoyed visiting Amami with his family. He does not take holidays at all during the rest of the year. A chartered ferry for this race makes the trip both ways between Nagoya and Amamioshima, so many people enter the competition just for enjoyment. The 2001 race was on July 1 with 300 boats and 562 persons participating in it. There were 138 boats in the single sea kayak class (including the K-60) and the marathon race course was for a distance of 36 km round trip between the Amami main island and the Kakeroma Island (Fig - 28).

About 2 or 3 days after the race, Mr. Kanai phoned me from Amami to inform me the race results. He was not feeling well and was down with diarrhea, but he was very happy with the race result since as he had reduced his time by about 9 minutes compared to 2000. His time in 2000 was 3 hours 54 minutes 44 seconds, while his time in 2001 was 3 hours 46 minutes 11 seconds, a reduction of 4 %.

Compared to the time of the Kanai-500 in 1999, which was 4 hours 11 minutes 29 seconds, the time in 2000 just reduced by 10%, as we had estimated. On the other hand, his position fell to 25th, although he was aiming to be within the first ten. He said that this was because there were ten more people who could cover the race within 3 hours 30 minutes compared to the previous year, so it was inevitable that his position would be 25th.

The Amami Sea Kayak Marathon is very popular and the level of competitors is becoming high so Mr. Kanai's race position will not climb higher. Comparing the results of 2000 with those of 1999, he cut down his time by about 17 minutes with the same boat to which he was accustomed to, so I was keen to know his time in 2002 with the K-60. He has been shortening the time every year since 1998, with the time shortened to about 27 minutes total up to this year. I am happy to note that this is due to the progress made in the boat as well as to his handling skills in spite of becoming older.

15) Conclusion

The project is idle for the time after the Amami Sea Kayak Marathon. Tests on the damper, stabilizer, and planing board will take place henceforth. It is more interesting for me to see Mr. Kanai become familiar with the new boat at Amami; it may be possible for him to balance the boat and ride it without hydrofoil or planing board, as he himself said once. Recently, I received a photo from Mr. Kanai, which shows him paddling with both floats above the water. I am expecting this boat to be exhibited at the Tokyo Boat Show in 2002, and also expect to develop and fit a sliding seat in it.

FRP CAR OU68

Fiberglass Reinforced Plastic (FRP) materials are used in boats. Advanced FRP composites are used in airplanes. Structures using advanced FRP composites are remarkably strong, rigid, and light. It will be wonderful if we can develop an ultra lightweight car using FRP techniques. As FRP does not decay, the car can be used life long. The advantages of the car are light in weight, compact, long life, and low pollution; these will be the most desired characteristics in the future as automobiles spread to all Asian countries also.

It was not easy to adopt advanced laminating techniques, currently used in the aircraft industry, for mass production of cars while ensuring good productivity, and was a very huge challenge.

1) In the beginning

Honda and Suzuki were originally motorcycle manufacturers, but now they also manufacture cars, with their main sales dependent on cars. On the other hand, Yamaha did not enter the car-manufacturing field, but we did try to develop cars several times. Since I participated in one of the developments, I am writing about it here.

In the 1960s, Yamaha Motor developed the forerunner to Sylvia in cooperation with Nissan Motor and also developed the 2000 GT with Toyota Motor in 1966. Being engaged in boat development at that time, I was not directly associated with the development of these cars. However, boat production engineers were called up because FRP was used in some parts of the car body, and I listened to some of the difficulties they faced.

Since then, Yamaha Motor has been supplying Toyota Motor with sporty engines and has even participated in the F1 races. In addition, during the bubble economy, Yamaha developed and announced a luxurious super sports car OX99, which was a single seater with an F1-like engine. Thus, we were also associated with four-wheeled vehicles.

This article is the story of the period when Yamaha Motor was defeated after the HY (Honda-Yamaha) battle for marketing and developing the motorcycle business. At that time, Yamaha' business was at its lowest point.

Yamaha Motor considered entering the small car business with a view to recovery. I resigned together with some colleagues from the Board of Directors during the downsizing of the company at that time, became assistant to the Executive Director, and was given the charge for development of the project.

That was a long time ago, my memories are a bit dim, and documents and records in my hand are limited, however, I only write what I know to be correct here. Therefore, in this article, I would like to present only my ideas, and the photos that remain with me, and explain the progress and the processes of the project within the limits of the existing documents.

At that time, everyone in the company was being urged to work hard for its recovery; this project would also help in the recovery. Since the company's business was at the lowest point, it did not have the capacity to enter the car business in the normal way. We wanted to develop a new product even if its scale was small in those hard times, to encourage people within and outside the company and to show

Fig - 1 One-fourth scale model of OU68

them that the company intended to make a speedy recovery (Fig - 1).

2) Charm of the FRP body

The FRP body is a good choice considering small investment and production. Compared to the investment for press-formed steel plates, the investment for FRP molds is very small. We have some successful examples of sports car bodies made of FRP, such as Lotus Seven and GM Corvette. Moreover, we already had the experience of building 20,000 FRP boats a year.

Another point that adds to the charm is the strength of FRP. Normal FRP, which you see as the outer shell of a boat, has a bending strength of nearly 20 kg/mm². Its specific gravity is 1.6 and its specific strength, that is, the strength divided by the specific gravity is 12.5. On the other hand, steel plate has a bending strength of 35 kg/mm² and its specific gravity is 7.8; which means that its specific strength is 4.5, which is only one third that of FRP.

Perhaps you are aware that the strength of FRP plate glued to the surface of a ski is greater than 100 kg/mm². With the same glass fibers, the strength becomes as much as 5 times by arranging the fibers in the same direction and thoroughly squeezing out the resin. Its specific strength can be as high as 60.

This is because of the synergistic effect brought about when the amount of glass fibers in the load direction is increased, the glass fibers are non-woven, and the resin content is extremely small due to the laying glass fibers in a dense arrangement. By proper orientation of glass fibers, the strength of FRP can be increased significantly.

Since a car has many openings and its surface is uneven at many locations, stress concentration occurs in some areas. If this strong and light material is intelligently used in such areas, we can obtain a very strong and lightweight structure with long service life.

In addition, the number of parts can be reduced considerably because of the excellent moldability of FRP (Fig - 2).

However, the productivity of hand lay-up is not good. On the other hand, while the SMC method that uses metal male and female dies is suitable for mass production, the molded FRP is made by short glass fibers and its direction cannot be arranged as intended. As a result, the strength of FRP is less than the strength of FRP obtained by hand lay up.

I visualized development of a new molding system that enabled glass fibers to be arranged in the intended direction, yet had

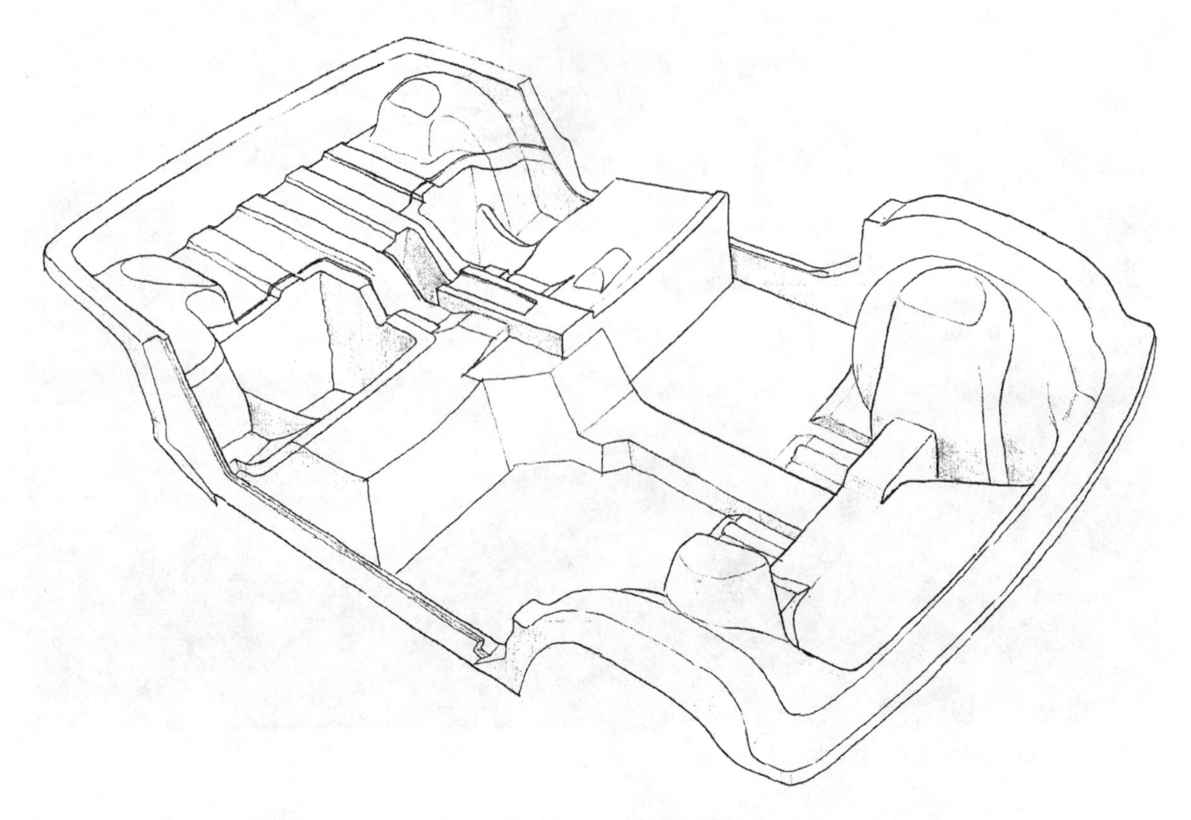

Fig - 2 The number of parts can be drastically considerably by using a one-piece frame.

good productivity. However, this was too high a target to aim for since it was an unknown area, and it was inadvisable to advance it in parallel with the development of the car.

3) Durability of FRP

Another attractive feature of FRP is its durability. Twenty years had already elapsed since Yamaha Motor began building FRP boats. At that time, by chance, we investigated the strength of the outer shells of several boats that had been left in the open, and we found no decrease in strength. Moreover, various data show that the strength of many of such boats become stronger than the strength of boats just after shipment.

Scratches on the surface must be repaired with filler, but in case of chalking, when the outer shell becomes whitish, by merely buffing the gel coat, the surface finish becomes almost similar to that of a new boat. I have even heard that our sales people were embarrassed because they could not distinguish used fishing boats from new ones when selling them.

The surface of FRP is coated with 0.4 to 0.7-mm thick gel coat, which is a colored layer also meant for UV protection. Thus, the durability is excellent and there is no thickness reduction even if the gel coat is removed by buffing, if about 1/100 mm is removed. In addition, there is no rusting as in steel plates, therefore the service life of an FRP body would be several times longer than that of a steel body.

4) Preliminary prototype of the OU68

We, at Yamaha Motor, were going forward with several projects related to new vehicles lying between two-wheelers and four-wheelers. The OU68 was one such project, and I was in charge of this project.

This car was lighter and smaller than normal mini-cars, and was to be made of FRP. The production quantity was small, so there was a limit to reducing the cost. Obviously, the car will not be attractive if we developed it in the normal manner. If the car does not stand out

Fig - 3 Package layout

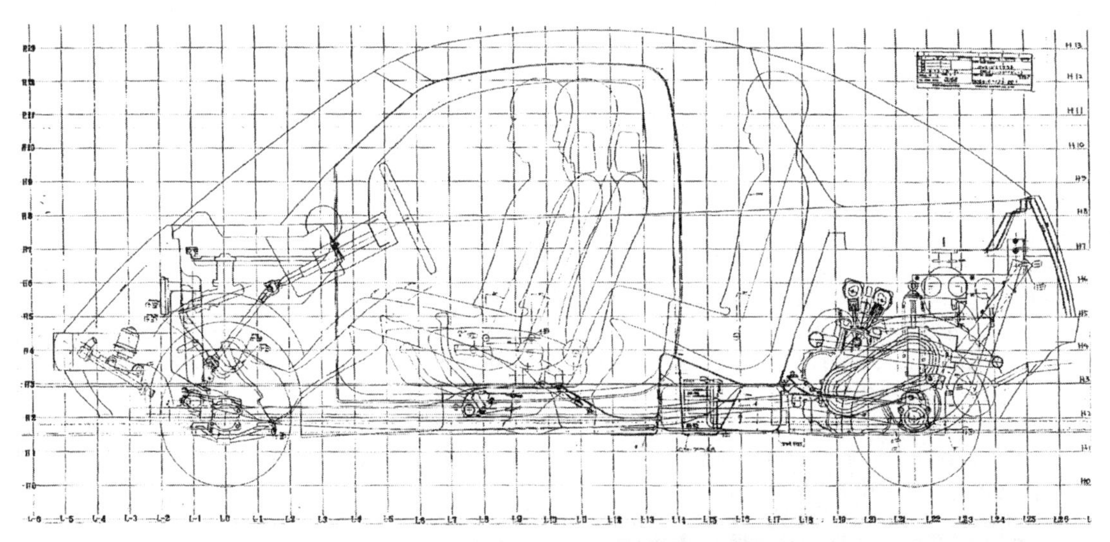

Fig - 4 Wind tunnel test

	OU68 (PT280)	Compared to Suzuki ALT MS-QG
Total length (mm)	3190	3195
Total breadth (mm)	1390	1395
Total height (mm)	1360	1335
Wheel base (mm)	2150	2150
Tread (F) (mm)	1220	1215
Tread (R) (mm)	1200	1170
Min.road clearance (mm)	165	175
Car weight (kg)	300	555
Persons (no.)	4	4
Climbing (tan)	0.27	0.26
Min.turn radius (m)	4.3	4.4
Fuel consumption at 60km/h(km/L)	45	24.1
Engine	Water cooled 4 stroke one cylinder	Water cooled 4 stroke 3 cylinders
	DOHC 4 valve	SOHC
Total engine displacement (cc)	276	543
Steering	Rack&pinion	Rack&pinion
Suspension (F)	Strut	Strut
Suspension (R)	Semi trailing	Rigid axle reef spring
Brake(F)	Two leading	Two leading
Brake(R)	Leading trailing	Leading trailing
Tyre(STD.)	160-8(16x6.5-8)	5.00-10-4PR ULT
Clutch	Dry multi-plate	Torqe converter
Transmission	V-belt automatic variable speed	2speed full automatic

Fig - 5 Main specifications

from the ordinary mini-cars and does not have its own special attractive features, however small, it will not be successful. It was difficult to design such a car. All the persons in the design team had the same doubts -- if we design the car in the normal way, it will not sell well. What should we do?

I looked for answers in the FRP body. A radically light car weight, aeronautically smart body shape, and a well curved and attractive external shape that was the opposite of the box-shaped normal mini-cars, could be designed. With these characteristics, I looked forward to develop the car that had a radically different design, superior drivability, and good running compared to normal mini-cars.

The project started in May 1983 and we planned to exhibit the car at the Tokyo Motor Show in October the same year, which gave us very little time for development. In addition to the initial members of the project including Katsumi Shimizu, Hiroshi Tanaka, and Hiroshi Nagai, three industrial designers namely Yorito Sagara, Kotaro Imai and Kenji Santoku also joined the team. General Manager Susumu Shimamoto was in charge of the project, and with the assistance of the Technical Affairs Department and Car Engine Operations, about 10 persons worked full time on this project.

In practice, however, on the 30th of September, it was decided not to exhibit the car at the Tokyo Motor Show, but the project would continue aiming for production (Fig - 3).

The development went ahead almost as planned. We completed the preliminary prototype and we presented it to the relevant people in the company. Later, we performed the rigidity test of the body, the wind tunnel test (Fig - 4) and running tests.

Although the prototype had many problems, I thought that it was successful on the whole (Fig - 5). It had the image of a small cute animal as intended by the designers. The rigidity of the body was also as planned. Its drag coefficient as observed from wind tunnel test was 0.34, a value with which I was not satisfied, and which I intended to reduce to 0.3 by fairing the under body.

The car weight was over 300 kg and was heavier than planned, and there were many issues that were inevitable because of a prototype developed in such a short time. As the weight increased, its power performance reached the same level as that of normal mini-cars.

Fig - 6 Preliminary prototype

Fig - 7 Body of second prototype
Rear corner windows were added for rear visibility

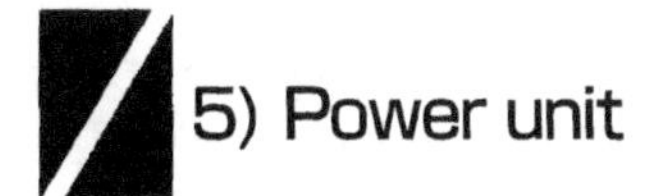

5) Power unit

We asked the Automobile Engine Division to develop a prototype of the power unit. Noboru Sakamoto took up this job under the direction of Manager Jiro Watase. There were only four months remaining for the project to be completed. Sakamoto planned to use half of the engine of the motorcycle XZ550 and a V-belt variable transmission used in Tracy, a 125-cc scooter. The idea was to reduce costs as well as to shorten the development period. He designed and completed the power unit that included a reverse and differential gear. The power unit was a compact, one-piece unit with a low center of gravity. It was a splendid job. This engine just focused on the exhibition at the motor show and the subsequent test run. For the production model, we were going to use a completely new design of the power unit to suit the planned performance of the car, after re-designing each part of the car to enhance its attractive features.

6) Presentation for people outside the company

Nearly a month before the motor show, we ordered a consultancy company to study the marketability of the OU68 (Fig - 6). The company invited total of 35 men and women who were interested in cars or were customers in general, and then showed them the photos of the car and a video film, then invited them do group discussions.

The results showed that the car was considered a unique next-generation design and it was cute, and its design generally scored high points. The survey showed that persons with a matured sense of values tended to have a high acclaim for the car.

On the other hand, some men and women in their early twenties were severe and their comments were that its style was too pushy, and it lacked a quality image. Many of these people also felt the interior to be poor. Some of young women were concerned about its safety during impact and its small-sized tires.

However, thankfully, there was no opposition to the FRP body. Overall, the style was acceptable and supported by the people who were knowledgeable about cars, and we were happy. At the same time, we adequately acknowledged that it had problems.

From the beginning, we were aiming to use an ideal monocoque structure for the ultra light body. Therefore, we made the outer shape of the car roundish, made the corner radius of openings large to eliminate stress concentration, and designed the external shape such that FRP molding became easy. That is to say, the shape was dictated by the car's functions. The final shape expressed a good harmony of the functional (aerodynamic and structural) shape and the cuteness of a small animal. It was natural that it became conspicuous by its look in the 1980s when box type cars were popular in the market.

7) Second prototype of OU68

After deciding not to exhibit the car at the motor show, we decided to restart the OU68 project. Tadao Suzuki, the chief of the Technical Administration Dept., personally handled this project from November 1983 and Takuji Kamo, who had a good knowledge on FRP and was also engaged in boat design work, was in charge of the design of the body. I was a consultant and offered my opinions for the project indirectly. As the main point of second prototype was to resolve the problems found in the preliminary prototype and to introduce items that arose during external presentations, my contributions became fewer and reduced.

Early next year in January 1984, the Horiuchi Laboratory started. I spent more time in the laboratory than before. On the other hand, in addition to the tests and measurements for noise, visibility, brake, acceleration, aerodynamic characteristics, power and running characteristics were being measured in the preliminary prototype. At the same time, the design and manufacture of four units of the second prototypes, in which necessary improvements were made, continued through the year (Fig - 7).

In the March 985, the improvements were completed and crash tests were performed.

8) End of the project

At the end of April 1985, the OU68 project finished after the crash tests of the second prototype. All the planned tasks were finished and the project was maintained in the list as one of new products, but it never restarted.

Unfortunately, I cannot remember the situation at that time. I suppose that President Eguchi and Executive Director Komiya might have felt some uncertainty in the marketability of the OU68 concept and the mass production of the FRP body. That might have been the reason the project did not move forward.

Actually, during that period, I did not find any time to involve myself in the development of the FRP manufacturing system. One year later, we started to develop a high speed, high-strength FRP manufacturing system called SMS in the Horiuchi Laboratory. Chemical engineer Takashi Motoyama continued to work hard on this system for several years. The system was once applied to the production of Marine Jets but there were some unsolved items in the system, and it was suspended before the issues were solved.

We were very successful in developing a device that could raise and lower the temperature of the mold within a minute, but were not successful in obtaining a very smooth surface for the molded parts. We did not have the resources to improve the resin. Since I was not very familiar with chemistry, I forced Motoyama to work hard, which I regret now. The decision taken by our president and executive director was probably correct.

I am grateful to the excellent engineers who worked hard on the design and prototype building, the designers, General Manager Shimamoto, Chief Suzuki and other persons who developed the engine, and many other concerned personnel of related departments. I would also like to apologize to them since the project did not materialize, more because of my inadequacies rather than because of their efforts.

I would like to extend my heartfelt thanks to Executive Director Takehiko Hasegawa (later President), Director (former) Shunji Tanaka and Mr. Norihiko Hayashi, who have always assisted and supported me.

BACKGROUND

1926 — Author Kotaro Horiuchi, born in Tokyo.

1947 — Graduated from the Department of Applied Mathematics, Tokyo University. Hoped to study Aeronautics but after the war, the Department of Aeronautics was changed to the Department of Applied Mathematics. (Aeronautics was prohibited by the US army)

1950 — Was employed by the Yokohama Yacht Co., and worked on the design of patrol boats, sightseeing boats, hydrofoil boats, air propulsion boats, cargo sleighs (for South Pole exploration) .

Duaring this period, he was temporarily transferred to Okamura Mfg. Co., for one and a half years. He was part of a design team of the first postwar airplane and a two seater sailplane (soarer).

1960 — Joined Yamaha and worked to set up boat business as a naval architect. He worked for 36 years in the company, and was a director for 17 years in Yamaha Motor Co. He retired as Senior General Manager of Marine Operations in June 1993.

1960 & 1964 — At the Olympics in Rome and Tokyo, he worked as the head coach of the Japanese rowing team. He won many races with eight oars shell as an oarsman and coached the Tohoku University crew for 12 years.

1961 & 1962 — He won the 1000-km Motorboat Marathon from Tokyo to Osaka.

1984-1993 — Director of Horiuchi Laboratory, the R&D team in Yamaha. Here, he developed many new vehicles. This laboratory fostered young designers.

Persons belonging to the Horiuchi Laboratory flourished in Yamaha. They are developing human powered and solar-powered boats, and human powered airplanes with excellent performance in their spare time.

1993 — Appointed as a regular adviser to Yamaha and worked to enhance devices for rowing shells for the Japanese team in the 1996 Atlanta Olympics.

1996 — Retired, lives in Kamakura (50 km south of Tokyo). He is still active in promoting boats and races as the President of the Japan Solar & Human Powered Boat Association. Privately. He has developed hydrofoil dinghies, sea kayaks, kite sails, and ultra light powerboats. Now, he is developing a hydrofoil sailboard in his yard.

Books "A Locus of a Boat Designer","A Locus of a Boat Designer 2" (both in Japanese)

CPSIA information can be obtained
at www.ICGtesting.com
Printed in the USA
BVHW01s1632240818
525517BV00014B/185/P

9 780971 264649